MW01231936

Springer Geography

Advisory Editors

Mitja Brilly, Faculty of Civil and Geodetic Engineering, University of Ljubljana, Ljubljana, Slovenia

Richard A. Davis, Department of Geology, School of Geosciences, University of South Florida, Tampa, FL, USA

Nancy Hoalst-Pullen, Department of Geography and Anthropology, Kennesaw State University, Kennesaw, GA, USA

Michael Leitner, Department of Geography and Anthropology, Louisiana State University, Baton Rouge, LA, USA

Mark W. Patterson, Department of Geography and Anthropology, Kennesaw State University, Kennesaw, GA, USA

Márton Veress, Department of Physical Geography, University of West Hungary, Szombathely, Hungary

The Springer Geography series seeks to publish a broad portfolio of scientific books, aiming at researchers, students, and everyone interested in geographical research.

The series includes peer-reviewed monographs, edited volumes, textbooks, and conference proceedings. It covers the major topics in geography and geographical sciences including, but not limited to; Economic Geography, Landscape and Urban Planning, Urban Geography, Physical Geography and Environmental Geography.

Springer Geography—now indexed in Scopus

More information about this series at http://www.springer.com/series/10180

Radomir Bolgov · Vadim Atnashev ·
Yury Gladkiy · Art Leete ·
Alexey Tsyb · Sergey Pogodin ·
Andrei Znamenski
Editors

Proceedings of Topical Issues in International Political Geography

Editors
Radomir Bolgov
Saint Petersburg State University
St. Petersburg, Russia

Vadim Atnashev
St. Petersburg State University
St. Petersburg, Russia

Yury Gladkiy
Herzen State Pedagogical University
of Russia
St. Petersburg, Russia

Art Leete
Department of Ethnology
University of Tartu
Tartu, Estonia

Alexey Tsyb
Peter the Great St. Petersburg
Polytechnic University
St. Petersburg, Russia

Sergey Pogodin
Peter the Great St. Petersburg
Polytechnic University
St. Petersburg, Russia

Andrei Znamenski
University of Memphis
Memphis, TN, USA

ISSN 2194-315X ISSN 2194-3168 (electronic)
Springer Geography
ISBN 978-3-030-78689-2 ISBN 978-3-030-78690-8 (eBook)
https://doi.org/10.1007/978-3-030-78690-8

© The Editor(s) (if applicable) and The Author(s), under exclusive license
to Springer Nature Switzerland AG 2021
This work is subject to copyright. All rights are solely and exclusively licensed by the Publisher, whether the whole or part of the material is concerned, specifically the rights of translation, reprinting, reuse of illustrations, recitation, broadcasting, reproduction on microfilms or in any other physical way, and transmission or information storage and retrieval, electronic adaptation, computer software, or by similar or dissimilar methodology now known or hereafter developed.
The use of general descriptive names, registered names, trademarks, service marks, etc. in this publication does not imply, even in the absence of a specific statement, that such names are exempt from the relevant protective laws and regulations and therefore free for general use.
The publisher, the authors and the editors are safe to assume that the advice and information in this book are believed to be true and accurate at the date of publication. Neither the publisher nor the authors or the editors give a warranty, expressed or implied, with respect to the material contained herein or for any errors or omissions that may have been made. The publisher remains neutral with regard to jurisdictional claims in published maps and institutional affiliations.

This Springer imprint is published by the registered company Springer Nature Switzerland AG
The registered company address is: Gewerbestrasse 11, 6330 Cham, Switzerland

Preface

This volume contains the papers submitted (and accepted for publication) to the International Conference "Topical Issues of International Political Geography" (TIPG-2020). The volume focused on specific aspects of contemporary political geography and international relations. The volume provided a platform for discussion and collaboration of academicians and experts in the fields of political geography, human geography, geopolitics, urban studies, demography and population studies, migration politics, natural sources politics, international organizations and integration, conflicts and security, international law and other related areas of studies. TIPG is a continuation of the International Conference "Topical Issues of International Relations in the Current Geopolitical Context", which has been held in St. Petersburg annually since 2016.

The Program Committee comprising of the recognized researchers from 15 countries had conducted a rigorous peer review.

The volume consists of seven parts. The titles of parts 1, 2, 5 and 6 were represented in the last TIPG conference of 2019. Discussion of 2019 has generated new works collected in the proceedings under this umbrella. At the same time, the volume discovered new directions for the studies. The titles of parts 3, 4 and 7 are new topics of TIPG. These parts consist of the papers presented on new sections of TIPG.

The chapters of the Part 1 "Ideologies of Regionalism and Globalization in Historical Context. Philosophy of politics" focus on the spatial aspects of two parallel processes—globalization and regionalism. The authors discuss the effects of globalization on the ideology, identity and symbolics of the nations and communities. The authors discuss such trends as Asianization and shift of the power from West to East. The context of the section is designed by historical framework of geographical issues with the use of historical geographical approach to studying the politics. Following Élisée Reclus, "geography is history in space while history is geography in time."

The Part 2 "National Policies & International Politics" moves from cases at the local level to national, bilateral, multilateral and global ones. The chapters cover such issues of domestic and international politics as anticorruption policies, digital strategies, elections, etc. The main idea of the section is to discuss effects of national policies on the international politics and effects of international politics on the national policies. The geographic space is a background for this discussion. The section contains a set of examples of comparative analysis focused on national cases (for instance, a comparative study of eHealth strategies in EAEU countries). Some studies represent the international platforms as a context of politico-geographical processes (for instance, Eurasian Economic Union and Belt and Road Initiative as platforms for Russia–China cooperation).

The publication of Part 3 "Geopolitics & Security" became possible due to cooperation with the Research Committee on Geopolitics and Security (Russian Association of Political Science). The chapters of the section discuss the international and domestic security issues related to the power distribution depending on the geographic location. The contemporary background is the COVID pandemic. The chapters study Eastern countries cases such as China, Kazakhstan, as well as non-traditional threats to national and international security, including food security.

The chapters of the Part 4 "Sustainable Development & Environment Protection" present the studies of international organizations and countries' activities in the field of sustainable development as well as theoretical issues (for instance, the concept of Geo-Eco-Ideology). The authors analyzed the cases of the United Nations, European Union, BRICS and Arctic Council as well as national cases (for instance, Germany).

The chapters of the Part 5 "Migration and Socio-demographic Processes" deal with the migration and socio-demographic issues. They focus on cases of European countries, China and Russia. The authors make conclusions about challenges of migration to education and tourism, as well as the multiculturalism issues. In the context of COVID pandemic, online education became more and more topical. It changes our views on the place and time and effects on the migration flows.

The chapters of the Part 6 "Cultural Dimension of International Relations" represent the culture as a factor of geography (for instance, international cultural exchanges, cultural policy, etc.). Can Korean pop culture affect Chinese politics? What is the place of the concept of cultural diversity in the cultural policy of the European Union? Is the cultural expansion of the Gulf states an instrument of their religious and economic influence in Europe? The section is intended to answer these and other questions. The studies focus on national, regional and local cases in the EU, Asian and post-Soviet countries.

Finally, the Part 7 "Discourses of Political Geography" starts with the chapters which discuss theoretical and historical issues on the crossroad of the politics, geography and discourse. Then, the authors study the cases of the USA, China and Caucasian countries with demonstration of discourse analysis application to political studies of geographic space.

We would like to thank those who made this event possible and successful. We especially express our gratitude to the Program Committee members for their contribution to the event. We thank the authors for submitting their papers. We are proud to attract a great team of scholars from different countries and disciplines. We will work further to sustain and expand the TIPG community through joint research and collaboration.

We will keep monitoring the evolution of COVID-19. We hope that TIPG-2021 will take place in St. Petersburg on time and everyone can safely make it through this global issue.

Radomir Bolgov
Vadim Atnashev
Yuri Gladkiy
Art Leete
Sergey Pogodin
Alexey Tsyb
Andrei Znamenski

Contents

Ideologies of Regionalism and Globalization in Historical Context. Philosophy of Politics

The "Non-west" Methodological Concept for Building a System of International Relations in the Eurasian Space

Natalia Vasilieva[1], Zeinab Bakhturidse[2], Nikita Ivannikov[2], and Alexey Tsyb[2,3](✉)

[1] St. Petersburg State University, Saint-Petersburg, Russia
[2] Peter the Great St. Petersburg Polytechnic University, St. Petersburg, Russia
[3] Sociological Institute of FCTAS RAS, St. Petersburg, Russia

Abstract. The article deals with the activities of the project where the authors explore the political concept of "non-West". The task of the research is to methodologically substantiate the use of non-Western approaches for the on the Eurasian continent. It is necessary to pay attention to the current world political situation in this region in the first half of the XXI century. The purpose of this study is to define the main trends that condition the formation and influence on international relations in the Eurasian space in the early XXI century. It includes the development of methodological principles of conceptual content of the terms "Eurasian space", "Eurasian segment of world international relations"; revealing the peculiarities of formation of political relations of large and small states on the Eurasian space under the conditions of global regionalization; structuring the main Russian theoretical developments on the problems of Eurasian regionalism; determining the degree of influence of digitalization on strengthening the role of civil society in international relations of the Big Eurasian region; determining the role of civil society in international relations of the Big Eurasian region. Such a study has both theoretical relevance and practical application, as it provides an opportunity to predict and analyze trends in modern history.

Keywords: International relations · Geopolitical strategy of development of Russia · Greater Eurasia · Westernization and Eurasian space · Eurasian segment of global international relations · Digitalization and role of civil society

1 Introduction

The fundamental objective of the research is to convey methodological substantiation for applying non-Western approaches in order to define the main trends that condition the formation and influence on international relations in the Eurasian space in the early XXI century:

1. The transition from uni- or bipolarity to multipolarity [1];
2. The transition from globalization (westernization) to regionalization (centric state) (the policy of D. Trump and Covid-19 have put an end to globalization);

© The Author(s), under exclusive license to Springer Nature Switzerland AG 2021
R. Bolgov et al. (Eds.): *Proceedings of Topical Issues in International Political Geography*, SPRINGERGEOGR, pp. 3–10, 2021.
https://doi.org/10.1007/978-3-030-78690-8_1

3. International relations in the Eurasian space, which are in the stage of formation due to significant geopolitical changes – from the collapse of the USSR to the contemporary rise of Asian countries.

Internationally, the emergence of the Greater Eurasia concept was caused by objective factors. First, it was necessary to define a new modality of interaction between a large group of friendly states in the conditions of growing global uncertainty. Secondly, the historical process proved the impossibility of integrating the continent "from above" based on the model and methods that were created in Europe during the Cold War and found there a form of European integration close to perfect. Third, the objective need to adapt China's growing power to the peculiarities of regional international relations, characterized by the presence of other superpowers and a large number of medium-sized and small states with a relative free choice.

There is a demand to define actual tasks required to implement the concept. Difficulties preventing the formalization of such tasks were experienced by those who support the concept and tried to implement which is a fact worthy of mentioning. The most important obstacle is, of course, the diversity of foreign policy objectives of the main participants and potential partners.

These factors require scientific work in the Russian expert community, since the Eurasian space is the basic one for forming the geopolitical strategy of Russia's development in the XXI century. Thus, it is necessary to change the direction in the development of modern domestic political scientific thought. This is due both to the end of the period of yielding adoption of Western political thought, and with the awareness and acceptance of the Eurasian identity. Feature of the latter is the adoption, revaluation, addition and development of existing experience of Eastern civilization. In modern times, the development of Western political science leads to convergence with the theoretical concepts proposed by scientists from the East (mainly India, China, South Korea), as well as other researchers representing the countries of the non-Western world. The success of Asian countries in the last decades in socio-economic, technological and scientific fields is the obvious reason for the present changes. This success has integrated the common desire of the Eastern scientific elite to actively participate in contemporary and relevant science, and to influence the global development of the modern world. In practical terms, this may mean that the concept of Great Eurasia is capable of becoming a theoretical element that would bring a possibility for the intraregional states to conduct a proper identification in the context of a single, common space in which their interests can not be separated from those of their neighbors.

2 Contemporary Development of the Issue

In recent years, new conceptual approaches to understanding the nature of transformation of the world political system from Western-centric to polycentric type of development have appeared. In this connection, conceptual discussion problems about the need for integrative development of the theory of international relations in the Eurasian space, where the field of scientific reflection includes both Western and non-Western approaches to understanding the multifactoriality and multivalue of international relations in the XXI century were raised.

At the turn of the century, the creators of the theory of social constructivism enter into discussion with representatives of the main paradigms of Western schools of international relations [29], who to a large extent predetermined bringing to the fore the problems analysis of modern international relations using non-Western and Western approaches. According to the opinion of constructivists, analysis should be conducted using such categories as: identity, values, political views and institutions, that is, those created by actors themselves. T.A. Alekseeva notes the "remoteness" of the system of international relations from humanity as something that does not exist by itself. The system of the IR emerges only due to the human interest and, consequently, consists of ideas, and not being formed by material forces or actions [4, 5]. If this idea is to be developed, it may lead us to a fundamental conclusion that the nature of the international relation system should be understood as the result of human thought, theoretical concept that is produced by the system of certain ideas and norms applied to the political environment at certain times and places.

M.M. Lebedeva believes that the attempts to find solutions for shaping contemporary state-system model using non-Western civilization are caused by the crisis of the existing Western model [16]. B. Buzan and A. Acharya unanimously agree on the growing role of non-Western political thought, noting both the approach of Asian leaders to the interpretation of the world order as well as a significant impact of traditional political views of the Eastern people [2, 3].

If we consider the paradigm of postmodernism from the point of view of methodology, it can be largely determined by the principles of "dialogical imagination", which dictate the rejection of Euro- and ethnocentrism, replacing them with acentrism. The latter should be understood as a fundamental rejection of any privileged pole of superiority (e.g. the supremacy of Western civilization in world politics). According to J. Deleuze and F. Guattari, the "rhizome" type of thinking implies recognition of diversity and discursiveness both of human history and actual social reality, as well as the methods and ways of cognition. According to this, it can be concluded that there is a significant number of individual actors whose activities are independent of any pole of power or possible existing hierarchy. This picture emerges from an understanding of the impulsiveness, discreteness and polycentricity of historical development. It is these statements of the post-modernist paradigm that make it possible to declare fundamental changes in the theory of international relations.

Another idea that may be regarded as truly relevant, belongs to A.Panarin, who characterized the Western political science as "unconscious regionalism correlated with the development of European civilization but not quite adequate to other civilization worlds" [17]. Following this idea, we can consider as fair the statement of A.I. Kozinets and A.M. Kuznetsov about the need to promote the formation and development of alternative, non-Western concepts. It is the thing that may actually result in creation of the new theoretical approach in the theory of international relations [23].

The assessment of the position of contemporary Russian scholars on this issue is indeed necessary to ensure an objective analysis. According to A.D. Bogaturov, the search for development has led to an appeal to the West, and as a result, since 1991 scientific research may be considered as prevalence of simple translation and description of existing Western concepts rather than analysis [8, 9]. This formed the informal task of modern Russian political science – to catch up with the years of imitation and copying, and, as a result, to return the West "intellectual duty". This is possible through the introduction of original ideas and concepts into the theory of international relations by Russian scientists. They, in turn, will allow us to move away from the use of familiar Western theoretical constructs, which critical flaw is strong belief that the influence of the cultural, scientific and historical development of Russia is a deviation. In reality, the analysis of all theories, both Western and other, is necessary to understand one's identity and to understand one's position. The result of this analysis may become a scientific and methodological basis for foreign policy of Russian Federation and its geopolitical strategy.

3 Objective and Goals of the Research

The purpose of this study is to determine the main trends that may have a significant impact on international relations, particularly in the contemporary Eurasian space, as well as to develop methodology regarding theoretical research of "Eurasian space" concept. The other objectives are to conduct identification of the peculiarities of the international state-system within the Eurasian space under the conditions of global regionalization; to structurize the main Russian theoretical developments on the problems of Eurasian regionalism; to determine the degree of influence of digitalization on strengthening the role of civil society in international relations of Greater Eurasia.

4 Materials and Methods

Globalization as a civilizational development vital feature is increasingly losing its meaning, both in theoretical and practical terms. It is important to distinguish between the concept of globalization, where the principle of unification prevails, i.e. universality is understood as westernization, and the concept of global regionalization, where the mosaic principle prevails, i.e. universality is understood as a way to create unity in diversity. This is not changed by nowadays global agenda. Space of "Greater Eurasia" still consist of states upholding different and sometimes opposing trends, e.g. supra-nationalism and disintegration. Thus, the modern development of Greater Eurasia has no permanent direction, it is impulsive, discrete and polycentric, which implies the coexistence of a large number of international actors.

Within research a complex approach was used due to its capability of provision a set of different research methods and tools. Their synergetic effect allows to combine different descriptive characteristics of the research object.

The research strategies used in the project include author's approaches and general scientific methods:

1. System approach, which allows to consider a complex structural and functional object;
2. Use of comparison, modeling, situational analysis, methods of qualitative and quantitative analysis;
3. Creation of a model of possible ways of evolution of modern international relations.

The certain system expressing object in the characteristics important for studying and capable to replace object was understood as models. That allowed studying of model to receive new data and to create exhaustive representation on subject of the research. As a method of applied research, modeling promotes scientific result and opportunity to complete future studies. That may be of much use while adapting obtained results to practical matters. Such model may be created in a process, which should include both descriptive and analytical data, which is why functional and behavioral approaches should be applied. While conducting modelling process all empirically obtained characteristics (both quantitative and qualitative) were used. The data was collected via usage of the following methods:

1. Desk-research. Finding and analyzing the results of existing research and studies, as well as data, which may be obtained via Internet- and other open and accessible sources of information;
2. Structural analysis of the content presented on the official websites of both objects and subjects of research.

Advantages of the selected methods: ergonomics, efficiency, versatility of research.

5 Expected Results and Their Scientific and Applied Relevance

The expected results include: a rationale for the theoretical significance of non-Western approaches to the studying international relations in the region under consideration. Identification of fundamental trends in the development of the system of international relations in the contemporary Eurasia, development of methodological principles of conceptual content of the term "Eurasian space", theoretical substantiation of the idea that the synergy of cooperative action of the West and East of Greater Eurasia will make appropriate adjustments in the system of international relations of the XXI century, the peculiarities of formation of political relations of large and small states on the Eurasian space in the conditions of global regionalization; structuring of the main Russian theoretical developments on the problems of Eurasian regionalism; determination of the degree of influence of digitalization on strengthening the role of civil society in international relations of Greater Eurasia.

6 Conclusions

In previous years, the team members carried out analytical studies of the concept of "Greater Eurasia" in the modern scientific and political lexicon, its project function as an ideological and theoretical construct implemented in program documents and agreements of the coalitions of the continent's states on an intercivilizational basis, analysis of the "Greater Eurasian Partnership" project as a logical continuation of the "Greater Eurasia" project together with the idea of "Greater Europe" proposed for several decades [11–13]. It also seems possible to analyze global security problems in a broader sense, understanding them not only as international, but also in the ontological sense of global security, which can be characterized by general models of development, structure and functioning, interaction with the biosphere, human community and technosphere. Definitions of theoretical components of the new emerging integration paradigm have been developed by the example of EEU. In the framework of the research performed different concepts were proposed such as the concepts of global regionalization, multipolar world, "middle space" [15, 23]. It also revealed the emergence of new spatial and temporal forms of the post-Soviet space, caused by the impact of global transnational processes. Theoretical analysis leads to the conclusion that the result of this influence is the emergence of a new global neo-Eurasian region. And global regionalization is regarded by contemporary political science as an essential condition for the successful establishment and functioning of the Eurasian Union. Such an association will be larger than the USSR, as it will develop in new directions under the influence of transnational processes. Now we can consider integration as a way to create a qualitatively new space, not only as a form of interaction in the post-Soviet space. Such integration allows new participants to appear in the global neo-Eurasian region, and also has a positive impact on the intraregional ties between states, which may be regarded as restoration of these connections, development and usage of which were hindered by the dissolution of USSR [22, 24, 25]. Global processes, their impact on modern international relations, as well as the differences in their assessment are considered one of the key elements of modern world politics. Therefore, studying their theoretical aspects may have an important impact on modern scientific thought in the field of international relations [6]. A comparative approach was used, as well as an analysis of sources and statistical data on PRC policy issues in Central and Eastern Europe (CEE) on the "16 + 1" cooperation platform with 11 member countries and 5 EU candidate countries based on both EU and PRC documents, as well as on 16 + 1 summit declarations and expert assessments [20].

Different conceptual approaches to understanding the nature of transformation of the world political system from the Western-centric to the polycentric type of development have been studied, in this connection the discussion problems of conceptual nature about the need for integrative development of the theory of international relations, where the field of scientific reflection includes both Western and non-Western approaches to understanding the multifactorial and multivalued international relations in the XXI century, have been analyzed [21]. Regional and sub-regional security issues were the focus of more targeted research, in particular on the South Caucasus. In these studies, the region appears to be an extremely important strategic object in the context of transformation of the existing system. It is also the subject of growing interest and confrontation of major geopolitical authors in view of the special current situation – the crisis situation in certain areas, unresolved problems of self-determination, separatism, exposure to outside influence, as well as attempts to expand its influence using soft power tools [10] were researched.

The roles of the both Russian and Chinese efforts affecting the current stance of the state-system and world order were also studied. Both countries largely support the United Nations as a unique actor addressing global challenges of our time, in particular, the security of the international community. Both leading states show solidarity in addressing the most important security issues, in particular, such items of the modern agenda as the nuclear status of DPRK and Iran, the Iraqi question [18, 19].

References

1. Acharya, A.: The end of the American world order. The Russian Truth. http://ruspravda.info/Konets-amerikanskogo-mirovogo-poryadka-5258.html. Accessed 20 July 2020 (in Russian)
2. Acharya, A., Buzan, B. (eds.): Non-Western International Relations Theory: Perspectives on and beyond Asia, 1st edn. Routledge, New York (2009)
3. Acharya, A., Buzan, B.: Why is there no non-Western international relations theory? An introduction. Int. Relations Asia-Pacific **7**(3), 287–312 (2007). https://doi.org/10.1093/irap/lcm012
4. Alexeyeva, T.A.: Think like constructivist: discovering a polyphonic world. Comp. Polit. **5**(1(14)), 4–21 (2015). https://doi.org/10.18611/2221-3279-2014-5-1(14)-4-21. (in Russian)
5. Alekseeva, TA., Lebedeva, MM.: What is happening to the theory of international relations. Polis. Polit. Stud. **1**, 29–43 (2016). https://doi.org/10.17976/jpps/2016.01.03. (in Russian)
6. Bakhturidze, Z.Z.: International relations in the epoch of “post”: possibilities of evolution. Cent. Global. **3**(23), 113–120 (2017). (in Russian)
7. Bakhturidze, Z.Z., Vasilieva, N.A.: Space of international relations: concepts and judgments. Int. Relations Dialog. Cult. **6**, 7–17 (2018). https://doi.org/10.1870/HUM/2304-9480.6.01. (in Russian)
8. Bogaturov, A.D.: Ten years of development paradigm. Pro Contra **5**(1), 195–202 (2000). (in Russian)
9. Bogaturov, A.D., Kosolapov, N.A., Krustalev, M.A.: Essays on the theory and methodology of political analysis of international relations. Scientific and Educational Forum on International Relations, Moscow (2002)
10. Ivannikov, N.S.: EU food policy humanitarian projects in the states of the South Caucasus. Eurasian Integr.: Econ. Law Polit. **2**, 79–86 (2020). https://doi.org/10.22394/2073-2929-2020-2-79-86. (in Russian)
11. Kefeli, I.F.: Big Eurasia: civilization space and design of the future. Eurasian Integr.: Econ. Law Polit. **3**(25), 60–68 (2018). (in Russian)
12. Kefeli, I.F.: Geopolitical chances and risks in the process of formation of the “Greater Eurasian Partnership.” Geopolit. Secur. **2**(38), 38–46 (2017). (in Russian)
13. Kefeli, I.F., Kuznetsov, D.I.: Eurasian Vector of Global Geopolitics. Moscow, Yurite. https://sziu-lib.ranepa.ru/new_book/Html_01.19/evrvektor/evrvektor.html (2018). Accessed 18 July 2020 (in Russian)
14. Kuznetsov, A.M., Kozinets, A.I.: Non-Western international relations theories – from marginality to recognition. Oikumena **4**, 8–23 (2016). (in Russian)
15. Lagutina, M.L., Vasilyeva, N.A.: Concept “Eurasian Economic Union” as a new integration paradigm. Admin. Consult. **10**, 78–90 (2013). (in Russian)
16. Lebedeva, M.M.: Non-Western theories of international relations: myth or reality? Vestnik RUDN. Int. Relat. **17**(2), 246–256 (2017). https://doi.org/10.22363/231306602017172246256. (in Russian)
17. Panarin, A.S.: Global Political Forecasting. Algorithm, Moscow (2000).(in Russian)
18. Pogodin, S.N., Tarakanova, T.S.: Problems of cooperation of the People’s Republic of China, the Russian Federation and the United States of America in the Central Asian region. Eurasian Law Journal **7**(98), 317–319 (2016). (in Russian)
19. Pogodin, S.N., Tarakanova, T.S.: Strategic partnership between Russia and China at the SCO site. Geopolit. Secur. **4**(40), 67–71 (2017). (in Russian)
20. Pogodin, S.N., Bakhturidze, Z.Z., Vasilyeva, N.A.: “16+1” cooperation platform role in strategic relations between EU and China. Izvestia Ural Fed. Univ. J. Ser. 3: Soc. Sci. **14**(2), 108–117 (2019)

21. Vasilieva, N.A., Bakhturidze, Z.Z.: Space of international relations: concepts and judgments. Int. Relat. Dialog. Cult. **6**, 7–17 (2018). https://doi.org/10.1870/HUM/2304-9480.6.01
22. Vasilieva, N.A., Lagutina, M.L.: Philosophical Questions of World Politics Science. Lambert Academic Publishing, Saarbrucken (2011).(in Russian)
23. Vasilyeva, N., Lagutina, M.: The Eurasian idea from the modern political perspective. Wschodnioznawstwo **7**, 257–268 (2013)
24. Vasilyeva, N.A., Lagutina, M.L.: Global Eurasian Region: Experience in Theoretical Reflection on Social and Political Integration. Polytechnic University Press, St. Petersburg (2012).(in Russian)
25. Vasilyeva, N.A., Lagutina, M.L.: Formation of the Eurasian Union in the context of the global regionalization. Eurasian Econ. Integr. **3**(16), 19–29 (2012). (in Russian)
26. Wendt, A.: Social Theory of International Politics. Cambridge University Press, Cambridge (1999)

Asianization as a Main Trend in World Politics

Zeinab Bahturidze[1(✉)], Natalia Vasilieva[2], and Ziad Shahoud[2]

[1] Peter the Great St. Petersburg Polytechnic University, Saint-Petersburg, Russia
[2] Saint Petersburg State University, Saint Petersburg, Russia

Abstract. The last 500 years was marked by the political and economic dominance of European civilization, which is reflected in the concept of Westernization of the world. In the first half of the twenty-first century, there is a change in the geopolitical map of the world, where the complexly structured world of Eastern civilizations, which geographically belongs to the Asian region, comes to the fore. In the international political science discourse, the theme of 'Asianization', 'Asiancentrism' in the 21st century came to the fore, which is methodologically developed in the concept of polycentrism, global regionalism and a non-Western world political picture of the world. As a result, on one hand, Eastern civilizations are returning the political and economic positions lost in recent centuries, and on the other hand, being technologically westernized, they introduce non-Western mental and value attitudes into world politics.

Keywords: Asianization · Asiancentrism · Westernization · World politics · Asian region · Mentality · Civilization

1 Introduction

In the 21st century, the main indicators of global economic growth are determined by the growing giants of Asia: China, Japan, India, Indonesia, Vietnam, South Korea, and a number of other eastern countries, which indicates the departure of European countries, the USA, and Canada to the second positions. Back in 2012, the US National Intelligence Council asserted that "in a tectonic shift by 2030, Asia will collectively surpass North America and Europe in terms of global power based on GDP, population, military spending, and technological investments" [1, p. 15]. In modern conditions, we can talk about the renaissance of Eastern civilizations (Chinese, Japanese, Hindu, Islamic), whose millennial historical and cultural background turned out to be a very important addition to the technological modernization that we borrowed from Western civilization. It is important to emphasize that Asian countries have a significant strategic resource in the context of the rapid development of a digital society - and that is their human capital. The top ten most populous countries in the world include China, India, Russia, Indonesia, Pakistan, Bangladesh, and Japan, which makes up more than sixty percent of the world's population. It is natural that, because of the socio-economic successes, the leaders of Asian countries strive to make their own significant adjustments to the global political order. Thus, the 'Asianization' of world politics most likely presupposes

© The Author(s), under exclusive license to Springer Nature Switzerland AG 2021
R. Bolgov et al. (Eds.): *Proceedings of Topical Issues in International Political Geography*, SPRINGERGEOGR, pp. 11–21, 2021.
https://doi.org/10.1007/978-3-030-78690-8_2

the reform of those structures of world politics and economy that are clearly dominated by Western countries today. In particular, these are such institutions as the UN Security Council, endowed with great powers, or the International Monetary Fund which has serious power, and controls financial flows throughout the world.

2 Objectives of the Research

Against of the rapid economic and social progress of Asian peoples, the crisis of the Westernization model of the political organization of the world is clearly visible, which makes scientists look for other solutions.

As noted by M.M. Lebedeva, it was the Eastern civilizations that gave a lot to mankind in medicine, mathematics, literature and other fields of knowledge [2]. According to the well-known theorists of international relations A. Acharya and B. Buzan [3], it is time for scientists and researchers to turn their attention to the origins of the Eastern political and philosophical tradition, which gives an intellectual impetus to find answers to the world political problems of our time. There is a need for new political constructs based on the many principles of theoretical approaches, where the political science developments of scientists from China, India, Vietnam and other countries of the non-Western world occupy a worthy place. According to T.A. Alekseeva [4, p. 9], “the international system is a product made by people consisting of a set of ideas and a system of norms created at a specific time and place”, therefore it is logical that the successes of Asian countries gave rise to the desire of the scientific elite of this macro-region to contribute to the theoretical understanding of the transformation of the international system, based on their mental and cultural traditions. It is difficult to disagree with the view that if events keep unfolding in the same way, “the two-century domination of Europe, and then of its giant North American offspring, will come to an end. Japan was the herald of the Asian future. she turned out to be too small, too introverted to change the world, but those who follow her, and above all China, are free of these shortcomings” [5, p. 17].

In fact, we can talk about a new philosophical worldview where the idea of ‘dialogical imagination’, and the rejection of the principle of euro-and-ethno-centrism come to the fore. Postmodern philosophical terminology (a-centrism, as the denial of any privileged domination “centers”, particularly of the West, in world politics) has proved to be in demand within the framework of the political discourse on the equivalence of western and eastern components of the contemporary political map. The methods of cognition and the forms of culture are diverse, and historical development has no constant direction, it is impulsive and discrete.

Therefore, the paradigm of Westernization, which has absorbed the worldview postulates of Western society, loses its theoretical universality in modern conditions, and what remains, according to A. Panarin, is an unconscious regionalism, correlated with European civilizational development, but not quite adequate to other civilizational worlds. It seems that this is also why it is so important to promote and contribute to the creation of non-Western concepts. After all, this is precisely what can help to emerge a new integral theoretical understanding of the system of international relations [6].

3 Methodology of Research

When considering the problems of the Asianization of world politics, it is important to find appropriate methodological approaches to the analysis of this political and economic phenomenon. It seems that the most adequate methods can be considered the historical, constructivist, and civilizational approaches.

From a historical point of view, natural and climatic conditions influenced the nature of political and economic relations in Eastern societies. In particular, the system of irrigated agriculture, prevailing in Asia, could not but have an impact. The so-called "rice-growing civilization" required the observance of civil, political and social discipline, which led to the formation of rigid administrative structures with a large number of officials. Such qualities of social behavior as collectivism and subordination to the state power came to the fore. There is an opinion according to which the essential difference that distinguishes the positions of power in the west, from positions in the east is indicated. According to L. Vasiliev, "if in antique-bourgeois Europe, power depends on the balance of contradictory tendencies in society, then in the East the authority of power never depends on anything of that kind. Ultimately, everything is decided only by the force of the power itself, the effectiveness of a well-established administration and a regular inflow of a guaranteed rate of income into the treasury" [7, p. 710].

The constructivist approach assumes to operate with such concepts as identity, cultural and religious values, political views, political institutions, which are constructed by the actors themselves It should be noted that in Asia, all worldview systems and all social institutions formed a tendency to strengthen the central government. As the researchers note, "the states of the Eastern type are based on the principle of sacred justice, that is, statehood in them is messianic, freedom is not individual, but collective, this is the freedom of the people, which is constructed collectively, for the people, respectively, there is a collective destiny, a collective vocation (one cannot be saved, only all can be saved together), statehood is sacralized, and it is reproduced in local communities as the highest value" [7, p. 32].

The civilizational approach assumes, according to M.A. Barga, the application of the principle of space-time deployment of each civilization. This means that any civilization is immanently aspiring to its historical and logical "limit", which is predetermined by its potential for material and spiritual growth. In this regard, it is important to understand whether the Western civilization is lagging behind the Eastern civilizations, based on the assumption that there is an inconsistency crisis with the demands of modernity. It seems that at the moment, Eastern civilizational characteristics associated with the principles of collectivism, solidarity, and not individualism and liberalism are in demand, which was especially evident in the era of the "coronavirus", when the Asian paternalistic state systems proved to be the most successful in the fight against new "challenges" of the 21st century. According to K.S. Hajiyev, in the Asian civilization code, along with paternalistic tradition, and other archaisms, such components as loyalty to traditions, philanthropy, a sense of duty, respect for elders, observance of the rules of social, intra-family, and group relations, etc. are of considerable importance…. [8].

4 Results of Research

Consideration of such a multifaceted phenomenon as the Asianization of world politics requires its analysis from different points. It seems logical to investigate the factor of Asianization in the world economy, politics, and culture.

As for the forms of economic modernization of Asian countries, the expert community is unanimous in the opinion that in Asia the effectiveness of various forms of economic development has been proven in practice. A number of new forms of accelerated entry of Asian countries into the corps of modernized states can be named (Deng Xiaoping's concept of modernization, Lee Kuan Yew's paternalistic practice, etc.), which did not correspond to the Western textbook methods of modernization. In fact, a model of socio-economic development that is authentic to Eastern civilizational foundations has been formed: Asian identity and scientific and technical Western innovations have successfully complemented each other.

According to Deutsche Bank in the early 2010s. World's highest economic growth rates, except for those recovering of complete destruction like Libya and Iraq, were demonstrated by such a country as Mongolia - 12–17% per year. It also set a world record for industrial production growth - 37.4% in 2012 - primarily due to the launch of the Oyu-Tolgoi copper and gold mine. The explored coal deposit Tavan-Tolgoi4 also opens up great opportunities [9, p.11]. According to experts, among the ten fastest growing economies are such countries as Macau, Papua New Guinea, Laos, Cambodia, Turkmenistan, Bhutan, and a number of African states [10, p. 93]. According to N. Chan, the consumption demand in Asia "will grow from $ 4.5 trillion in 2009 to $ 44 trillion in 2030, which will be half of global demand, and 1.6 times higher than this indicator for all OECD countries. But even then, out of 4.2 billion people who will live in the developing part of Asia outside of China, 780 million will have an income of less than $ 2 per day, and another 1.7 billion - $ 2–10 per day." [11, p. 34–35].

Probably, it is especially significant that the Asian economy is being transformed from an object of globalization into its subject, largely determining its direction and results. A good historical example can be seen in the early 70s. You can give a historical example of the early 70s. XX century: a number of oil-producing states of the Middle East, and with them, as V. Nikonov notes, Libya and Algeria announced a reduction in oil production, and imposed an embargo on oil supplies to the United States. Such a reaction to American support for Israel in the war against Egypt and Syria led to the disappearance of cheap oil. A new trend began to appear - the development of economic ties between Asian countries and states along the "South-South" line. These are not only African countries, but also the Central Asian region with its vast natural resources. China and India are actively starting to get involved in investment projects in the energy, mining and agricultural sectors.

An increase in the Asian share in the world economy requires the establishment of regional good relations between neighbors, therefore, for example, despite significant historical differences, the Chinese and Japanese business circles are in favor of improving relations, since Japan's industrial future is increasingly dependent on the huge Chinese market, and technological and investment proposals are extremely important for the Chinese digital economy.

The most significant integration successes are observed in Southeast Asia. The active development of the Association of Southeast Asian Nations (ASEAN), ASEAN + 1 (plus China), ASEAN + 3 (plus Japan, China, South Korea), as well as APEC (Asia-Pacific Economic Cooperation) creates an opportunity for economic interaction, both large and small Asian states.

Asian countries, having acquired a decisive economic weight on a global scale, are striving to actively shape the agenda of world politics. All the prerequisites for this are available, given the current international status of China and India. Along with Russia, they became the creators of a new, but already very influential informal structure in the non-Western world - BRICS. It is important to emphasize that the political weight of the Asian countries is enhanced by the presence of three nuclear powers. At the same time, the Asian countries have a common idea of how the world order will be transformed. First, it is assumed that in the field of foreign policy no state can dominate in Asia. Secondly, they all have the right to an independent course and resist pressure from outside. And finally, thirdly, cooperation between states is encouraged.

If we talk about the role of leaders in advancing Asia to the heights of world politics, then, undoubtedly, China plays the main role. Implementing a strategy to increase its influence in the world, China focuses on creating a "zone of good-neighborliness and zones of mutual prosperity" along the perimeter of its borders. It is interesting that in this direction of their activities, the Chinese authorities focus on those territories where there are serious reserves of raw materials, energy resources and high technologies.

According to S. Huntington, China can count on attracting the diaspora around the world to solve the tasks set by the Chinese leadership. "Come to the mirror and look at yourself" is the prompt of the Beijing-oriented Chinese to those compatriots who are trying to assimilate in foreign countries [12].

Migration processes have largely affected the Indian society, especially after gaining independence. The Indian diaspora is very influential in Western countries, especially in Britain and the United States.

The researchers note the fact that Indians in the United States have an average income 25% higher than the American average. It is also noted that "about 65% of Indians living in the United States have higher education, they own 15% of new firms in Silicon Valley, make up 10–12% of all US doctors, control about 40% of the American hotel business" [13].

Active promotion of specialists from Asian countries in the Western economic system is largely due to the quality of education. In recent decades, the governments of leading Asian countries have made quality education available to large numbers of young people. A number of Asian universities are in the first lines of the world rankings, which indicates the desire of Asian states to create their own research and technological structure of the world standard. Against this background, the emergence of a new migration trend is noteworthy. As noted by K. Brutents, we are talking about a "recurrent brain drain", about the widespread return of graduates of foreign, primarily American, universities to their original countries [14].

As noted by Western experts (for example, D. Moisi [15, 66–67]), in comparison with China, India does not restore its historically lost world positions, it forms the image of a modern democratic power that is looking into the future. There is an opinion that,

"Indians will not consider foreign policy as a crusade, just as they do not see the spread of democracy as their main national task. Indian way of thinking says, 'live and let live'. Indians abhor obligations. India does not feel very comfortable being called America's 'main ally' in Asia or an integral part of a new 'special relationship'. This discomfort in the precise and clear definition of friends and enemies is an Asian trait in particular" [16, p. 172–173]. It is important to emphasize that the political elite holds the balances quite delicately in the foreign policy and economic spheres, based on their national interests. So, on the one hand, close and diverse contacts with Western countries are maintained, but at the same time, relations with Russia are developing steadily, in particular, in the military-technical sphere. India's successes in the economy and world politics give reason to believe that the Indian potential (demographic, cultural, civilizational, etc.) will enable the country to take an important place among the major actors in modern world politics.

Conceptually, India's foreign policy is based on the principle of "three strategic rings", but the tasks of foreign policy influence relate to South Asia, where India itself has the largest territory, population and GDP. The Indian leadership emphasizes that the solution of regional problems should remain within the competence of India itself, without the intervention of any external forces.

As for world politics, in the 21st century India has taken a number of decisive steps towards a non-Western version of building the world system. Over the past decades, India has been putting into practice the conceptual ideas of mutual assistance along the South-South line. Back in 2003, the Indian leadership initiated the signing of the Brasilia Declaration, which marked the starting point of the institutionalization of the IBSA format (India, Brazil, South Africa). According to the expert, the officially declared goal of IBSA was to stimulate dialogue and cooperation along the South-South line in such areas as poverty alleviation, development, global climate change, dialogue of cultures (peace as "unity in diversity"), healthcare, education, promising areas of energy, scientific and technological progress, and investment [17, 122–123].

When we look at the Asianization process of World politics; we see that the main two countries are India and China. Nevertheless, there are political, economic and territorial problems hindering the relations of these two giants. These two countries competed for centuries over the control of Central Asia, Burma, Tibet, and the states of the South and Southeast Asia. In modern times, the two giants have a competition over raw materials and markets not only in Asia but in many parts of the world, like Africa, and Latin America.

Another issue between India and China is the naval presence of China in the India Ocean. This is a painful topic for India. A. Volodin points out that many in India perceive China's activities in the Indian Ocean as part of the Chinese "Pearl String Strategy". According to this doctrine, China seeks to surround India with naval bases. In India, they suspect that China is willing to take their country in a "geopolitical grip", from the coast of Myanmar to Pakistan (from east to west). India defends its leading role in the Indian Ocean, from Suez to Singapore, and this does not please China [17, pp.282–283].

Returning to the topic of the civilizational eastern "offensive" against the position of western modernity, it should be noted that an important component of this process is tradition in its broadest sense. As F. Braudel pointed out, "in contrast to the West, which sharply divides the human, and the divine, the Far East does not know this difference. Religion permeates all forms of human life: the state is religion, philosophy is religion, morality is religion, social relations are religion. All these forms are related to the sacred. Hence their tendency towards immutability, towards eternity" [18, p.183]. Power is traditionally sanctified by divine powers. For example, in modern Japan, the person of the emperor is sacred, he is directly connected with higher powers. The same can be observed in modern China, where the Confucian ethics of submission to the power of ruler-father determines the behavioral stereotypes of the inhabitants of the Celestial Empire. As the Russian orientalist, L. Vasiliev notes, traditional ideas are very tenacious in the modern political life of Asian countries: those in power must fulfill the sacred duty of protecting their subjects. And the people must follow the universal - established by heaven - laws. Therefore, ideas about the hierarchy of society, relations of domination and subordination are natural [7].

For example, the role of Islam can hardly be overestimated in countries such as Indonesia, Pakistan and Bangladesh. More and more grounds are found for the assertion that Islam and Confucianism have common values on which the achievements in the economies of Asian countries are based. In particular, Lee Kwang Yew, constantly contrasted the ethical norms of Confucian culture (order, discipline, family responsibility, hard work, collectivism, abstinence) to Western attitudes towards individual success (self-confidence, individualism, disrespect for power), which led to the "decline" of Western civilization.

As known, all world religions represented in Asian society speak of peace and human solidarity, however, in the practice of international relations in the Asian region there are still many unsolved problems, one of which is the creation of a collective security system. Undoubtedly, certain successes in this direction were achieved in the second half of the 20th century when the countries of Southeast Asia announced the creation of a nuclear-free zone, which was a complete surprise for the USSR and the United States. In modern conditions, much depends on the active peacekeeping position of India, China, Vietnam, Indonesia and other states to increase the level and scale of Asian security cooperation. The South China Sea islands have never had a permanent population. The question of their jurisdiction is related to the possibility of extracting energy resources from the continental shelf.

Another problem that concerns the global security system is the threat of terrorism. Many Asian states faced with it, in which the activities of extremists and separatists are carried out, including the training of new militants of these organizations (we are talking about countries such as China, India, Pakistan, the Philippines, Indonesia, Thailand, Myanmar). This problem is even more acute for the states of Arab Asia. As T. Friedman noted, "the term 'Arab spring' should be abandoned. There is nothing in it that looks like a spring... The 'Arab awakening' is also no longer appropriate, given what exactly awakened" [19].

5 Discussion Questions

The term 'Asian civilization' is a very broad and therefore imprecise definition of the features that are observed in the economic, political and cultural life of Asian peoples.

At one time, S. Huntington wrote, "in East Asia alone, there are countries belonging to six civilizations - Japanese, Chinese, Orthodox, Buddhist, Muslim and Western, and taking into account South Asia, Indian is added to them...... The result is an extremely complicated sample of international relations" [12, p. 344]. However, for all the diversity of nations, languages and beliefs, one can find some common matrix characteristics that unite Asian peoples into a collective Asian civilization. And, probably, the most important characteristic can be considered mental attitudes based on traditions and beliefs. Moreover, religious values are often opposed to Western ideas and values, primarily liberal, and sometimes even social democratic. In this regard, the Western expert community discusses a lot about the democratic foundations of Asian society. According to J. Kurlanczyk, the spread of democracy principles in Asia has significantly slowed down if not reversed in recent years, therefore the prospects for establishing civilian control in the young democracies of South and Southeast Asia are very uncertain. Particular disappointment is associated with the failure of the 'Arab Spring' in the Middle East, because as a result, the power positions of the military and Islamists only strengthened, and the liberal intelligentsia was a losing party. Everywhere in Asia, there is "growing" disillusionment with democracy" and "authoritarian nostalgia" [20, p. 12–13].

There are many reasons for the rejection of Western democratic values, which, in particular, is associated with the growth of Asian self-awareness, primarily as a result of There are many reasons for the rejection of Western democratic values, which, in particular, is associated with the growth of Asian self-awareness, primarily as a result of the modernization in the economy. This modernization was conducted based on its own economic and political models. This enabled the "Asian Tigers" (Hong King, Singapore, South Korea, Taiwan) to transfer themselves from an agrarian economy to advanced industrial states within a generation. Then the two giants of Asia, China and India, followed them.

According to the famous Russian researcher A.D. Voskresensky, "Asian democracies can be quite different from the European and American models, but still remain democracies. They are not "better" or "worse", they are simply "different" and, perhaps, are better adapted to solve their own political problems" [21, 20–25].

The ideas of "historical revenge", the restoration of historical justice, trampled by the colonial might of the Western powers, are gaining great popularity. As S. Huntington notes, Asian society is firmly convinced that economic success is due to Asian culture that surpasses the culture of the West, which is in a postmodern crisis. Such sentiments are gradually but steadily strengthening the belief of Asian political elites that the role of the United States, and the West, as a whole, as "Big Brother" in the political scenario of the 21st century no longer exists. In this regard, the viewpoint of Japanese political scientists H. Sato and T. Inoguchi is interesting to consider. They came up with it at the end of the 20th century, and it says that over the next 50 years, the "PAX AMERICANA" model is unlikely, as there is an "accumulation of fatigue" from a confrontational foreign policy. Therefore, the United States will gradually lose motivation for its policy of sole intervention in the world and regional processes. The transformation of the United States into the world's largest debtor leads to the fact that the American administration is no longer seeking to cover the costs of global goals alone. This is clearly seen from the decline in US interest in providing assistance to foreign states and in financing the

UN. This forecast is very much in line with the current political course of D. Trump. Therefore, Japanese political scientists considered promising the era of "global-post-American world order" [22, p. 102], whose contours are seen by modern Asian politicians in the century of Asia concept, which testifies to the overcoming of such idea as the superiority of the West.

It is important to emphasize that, for example, in modern China, an active state campaign is being conducted to promote the values inherent in Asian culture among young people through cinema and the internet, such as the superiority of group interests of the group over individual interests, which contributes to the general group potential that is necessary to raise the economy and culture of the country. Thus, the state seeks to "overcome" Western cultural and moral-ethical values, which are incompatible with some Asian traditions.

Thus, we can talk about the offensive of Asian countries in all areas of civilizational development. The emergence of the Asian region on the first roles in world politics and economics gives unprecedented importance to interaction with it. In this regard, Russia has a special place, which in Asia has 80% of the territory and 70% of the borders, Eastern Muslim and Buddhist traditions are widespread.

After the collapse of the Soviet Union, Russia's geostrategic and geopolitical center is located in the Krasnoyarsk region. Russian prospects are inextricably linked with the Far East and Siberia. The impetus for stepping up interaction with Asian countries was the form of APEC 2012 in Vladivostok. Even a new term has appeared - Russia - a Euro-Pacific power, which is quite logical, given the importance of the Asian and Far Eastern regions for the future Russian state. As D. Trenin wrote back in 2003, "the present - and even more so the future - Russia is no longer" Eurasia "and even less: Euro-Asia". Most accurately it can be described as a Euro-Pacific country open to the outside world..." [23]. Thus, the rise of Asia is directly related to the national interests of Russia.

For the Russian Federation, the Asianization trend is reinforced by the tendency in the West to form the image of Russia as an aggressor prone to conflicts and far from the tolerance, humanism, democratic values inherent in Europeans. Besides an active sanctions policy led to ousting the Russian Federation from the western space of the world economy and international relations. A comparative analysis of the Russian Federation foreign policy concepts demonstrates the transformation of theses, and aspirations: from a clearly expressed inclination to be included in a single world space (the 1993 Concept), and familiarization with European values, to a clearly recognized serious crisis "in relations between Russia and the Western states", and the need for an "independent foreign policy dictated by the national interests of the country" (Concept of 2016). It seems quite logical and justified to turn to the east in order to find partners and allies to solve the problems of ensuring regional and global security (the threat of world terrorism, economic, and environmental problems) [24], to establish a new system of international relations based on a holistic picture of the world of civilization where Asia with its centuries-old traditions and priorities will find a worthy place.

6 Conclusions

Summing up, it should be noted that the rise of Asia does not mean the return to the unipolar world model of the late 20th century. The conclusion of the American scientist of Indian origin A. Acharya about the formation of a multipolar world seems logical since "A multifaceted world presupposes the presence of many world and regional powers, united in complex forms of interdependence. It's like a multiplex cinema, where different films are shown under one roof with their own directors and actors" [25].

Philosophical understanding of the tendency to move away from the polarity syndrome towards the principle of polycentrism in international relations is reflected in the words of S.V. Kortunov, who quite correctly noted that "an attempt on the versatility of the world is an encroachment on the laws of the universe. The desire of one center to achieve, at the expense of others, the predominance of that meaning, the keeper of which it is, dominance over other centers creates a threat of degeneration of the entire world system of centers" [26, p. 543]. According to the researcher, the universe, in fact, is based on the diversity of civilizations, where each of the centers "embodies one of its ideological manifestations, and this creates and maintains the world civilization as a whole" [26].

It is difficult to disagree with this point of view, indeed, there are different civilizations, a long-standing cultural and confessional conflict between the West and the East, the economic abyss between the North and the South, but hostility is not an immanent feature of the interaction between them. It is worth finally recognizing the fact that they are sources of other versions of the explanation and understanding of the world order, but in fact this is a certain stability of the world, and "any fusion of meanings will destroy their combination in diversity, and there will be no peace" [26].

In the modern world, peoples are trying to preserve their originality and uniqueness, because this is what unites them, but at the same time, it seems that in conditions of unity in diversity, they do not at all seek to isolate and isolate themselves, because such behavior is based on fear and creates only the illusion of security.

References

1. Global Trends 2030: Alternative Worlds. December, 2012–01, National Intelligence Council, N.Y. (2012).
2. Lebedeva, M.M.: Non-Western theories in international relations: Myth or a reality? Vestnik RUDN series: International relations **17**(2), 246–256 (2017)
3. Acharya, A., Buzan, B. (eds.): Non-Western International Relations Theory. Routledge, Perspectives on and Beyond Asia (2009)
4. Alekseeva, T.A.: Thinking constructivist: discovering a polyphonic world. Comp. Polit. **5**(1/14), 4–21 (2014)
5. Ferguson, N.: A world without power. Free Thoughts **21**(1), 21–30 (2005)
6. Kuznetsov, A.M., Kuznets, A.I.: Non-Western theories of international relations – from marginalization to recognition. Oikymena **4**, 8–23 (2016)
7. Vasiliev, L.S.: History of the East, Vol. 1, 6th edn. Urait, Moscow (2013)
8. Voskresnovo, A.D. (ed.): East and Politics: Political Systems, Political Cultures, Political Processes. Aspect, Moscow (2011)
9. Hajiyev, K.: Westernization or a special path of modernization. Policy **4**, 148–162 (2008)
10. Dokachev, D.S.: Secrets of the steppe dragon. Echo Planeta **8**, 11–13 (2013)
11. The World in 2015. The Economist, p. 93.
12. Chun, N.: Opportunities and Challenges from the Rise of Asia's Middle Class. Poverty in Focus 26, October, pp. 34–35 International Policy Center for Inclusive Growth (2013)

13. Huntington, S.: Clash of the civilization. AST, Moscow, pp. 158–161 (2003)
14. Nikonov, B.A.: The Modern World and Its Origins. Moscow University Press (2015)
15. Brutents, K.N.: The continent of the future. International Life **3**, 36–56 (2013). https://interaffairs.ru/jauthor/material/829. Accessed 1 July 2020
16. Moisi, D.: How Cultures of Fear, Humiliation, and Hope Are Reshaping the World Translated from English. Doublday, A Division of Random House, Inc., New York (2009), Moscow School of Political Research, pp. 66–67 (2010)
17. Zakaria, F.: Post-American World of the Future. Europe, Moscow, pp. 172–173 (2009)
18. Volodin, A.: India in BRICS. Strategy of Russia in BRICS: goals and tools. Sat. articles. Nikonova, V.A., Toloraya, G.D. (eds.), pp. 119–124. RUDN, Moscow (2013)
19. Braudel, F.: Grammar of Civilizations. Translated from French. Ves Mir, Moscow (2008)
20. Friedman, Th. L.: The Arab Quarter Century. International Herald Tribune, April 11, 2013
21. Kurlantzick, J.: Democracy in Retreat: The Revolt of the Middle Class and the Worldwide Decline of Representative Government. New Haven, London, pp. 12–13 (2013)
22. Voskresensky, A.D.: Comparative analysis of political systems. In: Voskresensky, A.D. (ed.) Political Systems, Political Cultures, Political Processes. Aspect Press, Moscow, pp. 20–25; 28–29 (2011)
23. Chugrov, S.V.: The world "after the cold war" in the concepts of Japanese political scientists. World Econ. Int. Relat. **9**, 98–107 (1999)
24. Trenin, D.: Euro-Pacific Power. Russia in Global Affairs. No. January/March (2003). https://globalaffairs.ru/articles/evro-tihookeanskaya-derzhava/. Accessed 2 July 2020
25. Khudaykulova, A.: Explaining the security of the global South: Western and Non-Western approaches Vestnik of Saint Petersburg University. Int. Relat. **13**(3), 394-417 (2020). https://doi.org/10.21638/spbu06.2020.307
26. Acharya, A.: The end of the American world order. https://newsland.com/community/5134/content/konets-amerikanskogo-mirovogo-poriadka/2779448. Accessed 2 July 2020
27. Kortunov, S.V.: National Identity and Comprehension of Meaning. Aspect Press, Moscow (2009)

“Digital Panopticon” as an Objective Reality of the Global World: The Dilemma Between Social Control and Civic Engagement

Nikolai Baranov(✉)

North-West Institute of Management, RANEPA, Saint-Petersburg, Russia

Abstract. The spread of the new technologies has a great influence on modern political processes. On the one hand, they expand the capacities of socially active citizens, but on the other hand a threat of an interference with privacy increases. New communication technologies influence the civil activity of the modern society, involving in their orbit mostly the youth of large cities and megacities. The new forms of a social control called “Digital Panopticon” facilitate the youth’s political mobilization. The people respond to these forms of control with online mobilization and other technologies, suggested by the opinion. The implementation of the digital technologies and the artificial intelligence programs to control over human behavior and actions called digital totalitarianism facilitate change in perception of policy agenda by some social and democratic groups, in particular, by the youth. The constructivist approach underpins this article. According to this approach the cognitive activity is a construction of the ideas about the world of politics. These ideas are realizing through the establishment of a behavior on the basis of the activating political subject. The civic engagement in response to the tightening of the social control by the public institutions is switching from offline space to online. The objective process of the technological changes in all spheres of the social life dramatically influences the establishment of a relationship between the government and the population.

Keywords: Civic engagement · Social control · Digitalization · Digital rights · Political online mobilization · Digital totalitarianism

1 Introduction

The world is changing so quickly that humanity is often forced to just belatedly reflect on the changes taking place, without going deep into the essence of what is happening. Technological innovations, according to the founder and President of the World Economic Forum in Davos, Klaus Schwab, “are on the brink of fueling momentous change throughout the world – inevitably so” [1, p. 14]. The pace of innovation development is unprecedented, but it is impossible to ignore the negative consequences of such a high dynamic of changes. According to St. Petersburg neuro-linguist Tatyana Chernigovskaya, “digital reality is already a sign of selection in society. If you imagine a country

© The Author(s), under exclusive license to Springer Nature Switzerland AG 2021
R. Bolgov et al. (Eds.): *Proceedings of Topical Issues in International Political Geography*, SPRINGERGEOGR, pp. 22–30, 2021.
https://doi.org/10.1007/978-3-030-78690-8_3

that cannot afford to enter the digital world, you may consider that it doesn't exist at all" [2]. In fact, the compliance with the modern innovational requirements is dictated by the time, it is not only a fascination with technologies.

Modern technologies are demanded both by the person and by the state, which has a great resource capacity. It can impose the society its own understanding of security and suppress online freedoms using the administrative resources. The new technologies increase human capacities, as they are used for their self-development and the comfort of life. People use the technologies to control the authority's ant to solve their own problems. According to A.A. Kosorukov "the digital worlds of the government and the population are still separated: the government may need to catch up to the modern trends in the field of the digital technologies development, and the society may underestimate the threats and challenges to the digital sovereignty of the state" [3, p. 142–143].

The Fourth Industrial Revolution, described by K. Schwab, may help bring people and state closer together. He mentions three factors to justify the coming of the revolution. 1. Velocity: Contrary to the previous industrial revolutions, this one is evolving at an exponential rather than linear pace. 2. Breadth and depth: It builds on the digital revolution and combines multiple technologies that are leading to unprecedented paradigm shifts in the economy, business, society, and individually. 3. Systems Impact: It involves the transformation of entire systems, across (and within) countries, companies, industries and society as a whole [1, p. 8–9].

Adaptation of the latest information and communication technologies to the needs of the individual, society and the state is becoming of a great importance. The potential scope of innovation may be used to meet both the interests of an individual and the state, which are not always identical. If a person's technological innovations expand opportunities for development, improvement, implementation of their plans in economic, social or humanitarian areas, then for the state there will be new opportunities for control over society, over individuals, and unauthorized access to the private sphere. Finding the optimal relations between the state and the citizens is the most important goal of the social sciences, including the political studies.

2 Purpose and Methodology of the Study

This article is based on the constructivist approach, according to which cognitive work is the construction of ideas about the social world. According to Peter Berger and Thomas Luckmann, "Man is biologically predestined to construct and to inhabit a world with others. This world becomes for him the dominant and definite reality. Its limits are set by nature, but, once constructed, this world acts back upon nature. In the dialectic between nature and the socially constructed world the human organism itself is transformed. In this same dialectic man produces reality and thereby produces himself" [4, p. 204].

One of the major representatives of the constructivist methodological direction, the American scientist Nicholas Onuf connects social relations with the interaction of active political actors who influence each other and thereby change each other. In the book "World of our making", he highlights hierarchical principles in politics based on social relationships that are founded on power [5]. And the American political scientist Alexander Wendt proclaims the basic principle of constructivism: people both consciously and unconsciously build political reality, and do not find it as it is [6].

According to the authors of the theory of social constructivism, social order is a product of human activity. As political practice shows, in the modern digital world, civic engagement is being reformatted from offline to online in response to the tightening of social control by public institutions. The objective process of technological changes in all spheres of society has a fundamental impact on building relationships between people and the state. Public institutions, most often, form new rules and norms in the new digital reality, and citizens adapt their behavior to them, showing support or initiating changes in the emerging social structure.

Therefore, the purpose of this article is to identify current trends in increasing control by the state and society caused by the development of digital technologies that restrict civic engagement and the exercise of human rights and freedoms.

3 Pros and Cons of the Modern Technologies

The Internet occurred to be the first technology that penetrated into the lives of citizens faster than into public structures. The public activity started following different technological innovations. For example, in 2000s the "open data" movement appeared, within the framework of which the principles of open data, making the basics of the digital control – accessible, complete, primary, timely, machine processable, non-discriminatory, non-proprietary, and license-free – were developed [7]. Such movements stimulate power structures to create an atmosphere of informational openness and transparency of decisions and contribute to the democratization of the political process.

In 2014 in Estonia the system of electronic citizenship, supporting the integration of the government and the numerous data warehousing servers was launched. On the web-site E-Estonia it is said that "'the most advanced digital society in the world' was created in Estonia" [8] with an efficient, secure and transparent ecosystem that helps citizens to solve their problems.

The role of citizens in producing public services increases: they are now treated not only as the recipients of the services, but also as the co-producers, which means the higher responsibility of the citizens in the sphere of production and delivery of public services. The number of applications, which let the citizens to meet their needs in medical assistance, educational needs, while receiving social services has increased.

The "'Big data' technology became an integral part of the digital control and cooperation. "Big data" refers to datasets whose size is beyond the ability of typical database software tools to capture, store, manage, and analyze" [9].

It became possible to use the "Big data" technology due to the new technical and programme elements of distributed data processing systems. It is widely used by the public authorities, expert communities and think tanks, educational structures, civil society, business. But there is a threat of abuse of control over citizens, interference with privacy, using personal data for criminal purposes. The story of the British company Cambridge Analytica, which used the "Big data" analysis technologies for developing strategic communication on the Internet during the election campaign in the USA and during the referendum on the British exit from the European Union testifies to the broad possibilities of influencing voters and voting participants.

There is a strong lack of information credibility both to the information from private sources and from the public sources, including that from the independent sources. The post-truth and Fake News epoch so strongly discredited the information field, that people believe only in facts that prove their position and which are not in conformity with their opinion. Adjusting to people's opinion became a trend on the news agenda even when the news is neutral. But the people who are not able to move easily through such a flow of information prefer to believe nothing, or to distrust their semantic priorities. According to T. Chernigovskaya, "We must form the ability to live in a digital world and not to lose the humanity" [2].

It can be stated that on the one hand, "modern technologies, built in the 'figure', allow you to quickly respond to the daily problems of residents, respond to their initiatives, to their appeals, to respond accordingly, and therefore more effectively and quickly solve the problems that people face in everyday life" [10]. On the other hand, there is an ethical aspect to using big data. The state and large corporations get access to personal data without the explicit consent of citizens or bypassing the laws. Moreover, the applied technologies of big data analysis in the field of security can make mistakes, for example, in the process of identifying criminals or terrorists, which leads to additional checks of ordinary citizens or the use of violence against them by law enforcement agencies [3, p. 153].

4 The Threat of a Digital Totalitarianism

Increased securitization will rise interest among experts towards restrictions of human rights and functioning of democratic institutions. Digitalization, along with opportunities to expand democratic procedures, brings with it the risk of software and hardware vulnerabilities, increased conflict in cyberspace, and restrictions in ensuring the protection of digital rights and digital sovereignty of both the state and citizens. Access to the Internet is internationally recognized as a basic human right and is enshrined in international law, which has led to the creation of digital rights movements in some countries. Thus, the UN General Assembly "recognizes the global and open nature of the Internet and the rapid development of information and communication technologies as one of the driving forces for accelerating progress towards development in its various forms" [11]. At the same time, the same document confirms the right to privacy.

Threats emanating from cyberspace affect the interests of businesses and every individual who uses modern technologies. In Russia, as well as in many countries of the world, new technologies are often perceived through the prism of the new threats and challenges they generate, rather than the new opportunities they create. However, it depends only on people for what purposes information technologies will be used – for good or for evil [12, p. 31]. In the context of the COVID-19 pandemic, the problem of security is becoming more urgent, which draws the attention of the Deputy Chairman of the Security Council of the Russian Federation D. A. Medvedev. In his article "Security cooperation during the novel coronavirus infection pandemic", he writes: "Digital technologies will undoubtedly become the most important driver behind the socioeconomic and political development in the post-pandemic world. But it is extremely important to draw a clear line between the benefits of digitalization and the threat of a 'digital Big Brother', a restriction of fundamental human rights and freedoms. The economic efficiency that digitalization brings cannot be bought at the cost of 'digital totalitarianism'" [13].

The threats that scientists, experts, and politicians focus on are already in our lives. In social and political practice, the term "digital totalitarianism" has appeared. It means total digital control over human behavior and actions using video cameras, gadgets, digital applications, and artificial intelligence programs to further build its rating in society.

UN Secretary General Antonio Guterres, addressing the General Assembly in early 2020, identified four threats that faces humanity: global geostrategic tensions, the existential climate crisis, deep and growing global mistrust, and the "'dark side' of the digital world. Despite enormous benefits, new technologies are being abused to commit crimes, incite hate, fake information, oppress and exploit people and invade privacy" [14].

Speaking in Davos at the world economic forum in January 2020, Israeli futurist Yuval Noah Harari named three main threats to humanity in the 21st century: nuclear war, environmental crisis and technological disruption, especially highlighting the latter problem as still unknown. Technologies, in his opinion, carries a lot of risks, among which he considers the most important the appearance of a new "'useless class' – people who are useless due to the disappearing of the old jobs; unprecedented inequality not just between classes but also between countries; the rise of digital dictatorships, that will monitor everyone all the time; as humans will rely on AI to make more and more decisions for us, authority will shift from humans to algorithms (Facebook, Google, Netflix, Amazon, Alibaba); the destruction of humanity, which means technology might disrupt not just our economy, politics and philosophy – but also our biology" [15]. The same problem issues F. Lukyanov. He writes about the loss of "control over information technology giants that are able to interfere in people's lives and dictate how to behave" [16].

The report, prepared for the start of the Davos economic forum in 2019, focuses on technological instabilities and introduces the concept of "Digital Panopticon", which refers to new forms of social control: "facial recognition, gait analysis, digital assistants, affective computing, microchipping, digital lip reading, fingerprint sensors – as these and other technologies proliferate, we move into a world in which everything Digital Panopticon about us is captured, stored and subjected to artificial intelligence (AI) algorithms [17]." An Artificial intelligence (AI) is already being considered as the new reality of e-justice [18], which may lead to the digital dictatorship of AI in the future.

Domestic scientists who study politics also analyze the problems associated with digitalization. Thus, D. Trenin writes about the latest tool that modern states have mastered – "digital technologies that provide an unprecedented level of control over societies." In his opinion, "the opposition between democracy and authoritarianism is increasingly being replaced by differences in the quality and effectiveness of governance. Citizens ' rights are balanced by their responsibilities, which are ensured by an all-encompassing control system. The most successful countries are those where the level of social solidarity and mutual responsibility of the ruling elites and societies is higher" [19].

The most famous practice of digital totalitarianism is the social credit score system in China, which was introduced by the State Council of the People's Republic of China in 2014. China's social credit system is a database or set of databases that keep up-to-date information on the behavior of individuals, corporations, and governmental entities across China. It uses established by the public body rules and regulations which form some special parameters collected through mass surveillance tools and "big data" technologies [20, p. 88].

The project became fully operational in 2014 after the publication of the "Planning Outline for the Construction of a Social Credit score System (2014–2020)", according to which by 2020 all companies and every resident of mainland China will be monitored and evaluated by this system in real time. The main task of the Program is as follows: "those who justify trust receive every available benefit, but those who lose trust will not be able to take a single step" [21]. The Chinese social credit score system (SCS) is a social concept based on the differentiation of citizens in terms of the ability to receive social and economic services, depending on the individual "rating". The rating is determined by scores that can be either accrued or deducted depending on the actions of a citizen, his or her social status, circle of contacts, and various other factors. There are many publications about the Chinese SCS in Russian scientific periodicals [22–24, p. 114–121], which indicates the interest of the scientific community in new digital practices that are widespread in the world.

The analogs of the social credit system are used in different versions in other countries, for example, in the United States or in a number of European countries – Great Britain, Germany, and France. In the United States, each resident is assigned an individual Social Security Number (SSN), which is associated with both their personal data and some other information, such as the ability paying capacity.

Collection and processing of personal data in Europe is focused on calculating credit scores for individuals and organisations. This is achieved by cooperation between European Central Bank and economic authorities in member states. In some countries, such as Germany, private agencies collect and process credit information.

The Russian Federation is also in a similar digital trend. Federal law No. 123-FZ of April 24, 2020 begins an experiment in Moscow on July 1 on establishing special regulations in order to create the necessary conditions for the development and implementation of artificial intelligence technologies [25]. In accordance with the law, an experimental legal regime is being established in Moscow for a period of five years. The goals of this regime are to improve the quality of life of the population, improve the efficiency of state and municipal management, increase the efficiency of economic entities, and form a comprehensive system for regulating public relations in the course of implementing artificial intelligence technologies. Despite the fact that the result of the establishment of a pilot legal regime may not be the restriction of constitutional rights and freedoms of citizens, the legal control over the citizens is evidently tightened. This happened due to the imposition of additional duties, violating the unity of economic space in territory of the Russian Federation, this can be clearly seen from the fifth article of the law. Therefore, the future will show whether the experimental legal regime will benefit people.

5 Conclusion

Technological turning points in the modern global world, called by K. Swab deep shifts have a contradictory social impact, accompanied by both positive and negative effects. Among the negative aspects of drastic changes are such as privacy/potential surveillance;

decreased data security; privacy concerns; increased manipulation, violation of confidentiality, and even an existential threat to humanity [1, p. 110–157]. More than a half of deep shifts are subject to negative effects associated with interference with privacy by both public and corporate structures. The bureaucracy is always ready to prohibit rather than allow, but the pandemic has shown an unexpected trend: the consent of state structures to remove some regulatory barriers to digital technologies.

Nevertheless, the control over society is being strengthened. According to experts, none of our actions on the Internet goes unnoticed. Russian Internet entrepreneur, head of the international company Reputation House Dmitry Sidorin, reflecting on big data, writes: "The growing amount of data makes it possible to predict the course of events. Based on information from social networks, they build models of the consumer demand evolution, the stock markets behavior, models of the epidemics spread, forecasts of public activity on a particular occasion, including rallies, and much more. The ability to predict creates a desire to change the course of events and interfere in the Internet space. Therefore, data analysis is of interest to major global companies and public structures" [26].

The political and cultural context of modern countries is intertwined with the needs of society to create a secure digital environment in which both citizens and the state are interested. There is a conflict related to the search for a balance between the powers of public bodies to ensure the security of individuals, society and the state, on the one hand, and the inadmissibility of their intrusion into the private sphere, on the other hand. This contradiction is relevant for almost all modern states, regardless of the political regime and ideological priorities. The possible solution seems to be the combination of a secure information space, created to a decisive extent by the efforts of the state, and the maximum use of the digital technologies' opportunities for the benefit of people through the provision of wide information rights.

Thus, the modern person has a dichotomy of choice between security and freedom, and each of these components is of equal importance. The priority will be determined by citizens, based on their ideas about the expected desired future, superimposed on the political and cultural characteristics of the perception of existing problems.

Acknowledgments. The reported study was funded by RFBR and EISR, project number 20-011-31753 "Youth of Metropolis as a Social Basis for Public Protest: Prerequisites, Technologies, Forms, Risks and Effects of Political Online Mobilization".

References

1. Schwab, K.: The Fourth Industrial Revolution. World Economic Forum, Geneva, Switzerland (2016)
2. Chernigovskaya, T.: "Everything collapsed in the world at once". Tatyana Chernigovskaya on the civilization of idleness and distrust of information [in Russian]. In: Center for Strategic Assessments and Forecasts. http://csef.ru/ru/nauka-i-obshchestvo/445/v-mire-ruhnulo-srazu-vsyo-tatyana-chernigovskaya-o-czivilizaczii-prazdnosti-i-nedoverii-k-informaczii-9165. Accessed 04 July 2020

3. Kosorukov, A.A.: Digital Management Model: Open and Big Data [in Russian]. In: Solov'ev, A.I., Pushkareva, G.V. (eds.) Politics and State Administration: New Challenges and Development Vectors: Collection of Articles, pp. 142–159. Aspekt Press, Moscow (2019)
4. Berger, P.L., Luckmann, T.: The Social Construction of Reality. A Treatise on Sociology of Knowledge. Anchor Books, Garden City, NY (1966)
5. Onuf, N.G.: World of Our Making: Rules and Rule in Social Theory and International Relations. University of South Carolina Press, Columbia, S.C (1989)
6. Wendt, A.: Anarchy is what states make of it: the social construction of power politics. Int. Organ. **46**(2), 391–425 (1992)
7. Open Government Data Principles: https://public.resource.org/8_principles.html. Accessed 04 July 2020
8. E-Estonia: https://e-estonia.com. Accessed 04 July 2020
9. Big data: The Next Frontier for Innovation, Competition and Productivity. McKinsey Global Institute. Report. https://www.mckinsey.com/business-functions/mckinsey-digital/our-insights/big-data-the-next-frontier-for-innovation (2011). Accessed 04 July 2020
10. Local Government Development Council Meeting [in Russian]. http://www.kremlin.ru/events/president/news/62701. Accessed 04 July 2020
11. Resolution of the UN General Assembly of January 21, 2013 No. 68/167 "The right to privacy in the digital age". https://undocs.org/pdf?symbol=en/a/res/68/167. Accessed 04 July 2020
12. Baranov, N.A.: Openness vs security: priorities for the state and civil society in the context of digitalization [in Russian]. Manage. Consult. **10**, 28–36 (2019)
13. Medvedev, D.A.: Security Collaboration during the novel coronavirus pandemic [in Russian]. In: Russia in Global Affairs, vol. 4. https://eng.globalaffairs.ru/articles/security-cooperation-pandemic/ (2020). Accessed 01 July 2020.
14. Guterres, A.: Remarks to the General Assembly on the Secretary-General's Priorities for 2020. https://www.un.org/sg/en/content/sg/speeches/2020-01-22/remarks-general-assembly-priorities-for-2020 (2020). Accessed 04 July 2020
15. Yuval Harari's Blistering Warning to Davos: https://www.weforum.org/agenda/2020/01/yuval-hararis-warning-davos-speech-future-predications/. Accessed 04 July 2020
16. Freedom is slavery. Fedor Lukyanov – on the Results of the World Economic Forum in Davos [in Russian]. In: Kommersant. https://www.kommersant.ru/doc/4228069 (2020). Accessed 04 July 2020
17. The Global Risks Report 2019, 14th edn. World Economic Forum, Geneva. http://www3.weforum.org/docs/WEF_Global_Risks_Report_2019.pdf (2019). Accessed Accessed 04 July 2020
18. Numa, A.: Artificial Intelligence as the New Reality of E-justice. https://e-estonia.com/artificial-intelligence-as-the-new-reality-of-e-justice/ (2020). Accessed 04 July 2020
19. Trenin, D.: Virus and World Order [in Russian]. Kommersant. https://www.kommersant.ru/doc/4307968?utm_referrer=https%3A%2F%2Fpulse.mail.ru&utm_source=pulse_mail_ru (2020). Accessed 04 July 2020
20. Razumov, E.A.: Digital dictatorship: features of the social credit system in the People's Republic of China [in Russian]. Trudy IIAJe DVO RAN **24**(3), 86–97 (2019)
21. Notification of the State Council of the PRC on the Program for Creating a Social Credit System (2014–2020). http://www.gov.cn/zhengce/content/2014-06/27/content_8913.htm. Accessed 04 July 2020
22. Larina, E., Ovchinskij, V.: The Chinese Social Credit System: Traditions and Technologies [in Russian]. Zavtra.ru. http://zavtra.ru/blogs/kitajskaya_sistema_sotcial_nogo_kredita_traditcii_i_tehnologii (2019). Accessed 04 July 2020

23. Avseenko, I. Social'nyj kredit sformiruet neravenstvo ravnyh [Social credit will form equal inequality]. In: Information and Analytical Agency "East of Russia". https://www.eastrussia.ru/material/sotsialnyy-kredit-sformiruet-neravenstvo-ravnykh/ (2019). Accessed 04 July 2020
24. Galiullina, S.D., Bresler, M.G., Sulejmanov, A.R., Rabogoshvili, A.A., Bajramgulova, N.N.: China's social credit system as part of a digital future [in Russian]. Bull. USPTU Sci. Educ. Econ. Ser.: Econ. **4**(26), 114–121 (2018)
25. Federal Law of April 24, 2020 N 123-FZ On conducting an experiment to establish special regulation in order to create the necessary conditions for the development and implementation of artificial intelligence technologies in a constituent entity of the Russian Federation – a city of federal significance Moscow and amending Articles 6 and 10 of the Federal Law "On Personal Data" [in Russian]. In: Rossiyskaya Gazeta. https://rg.ru/2020/04/28/tehnologii-dok.html (2020). Accessed 01 July 2020
26. The crisis has broken the boundaries of privacy. Dmitry Sidorin – about how big data analysis from a sales tool has become a big policy tool [in Russian]. http://plus-one.vedomosti.ru/blog/krizis-slomal-granicy-privatnosti. Accessed 01 July 2020

Greek Myth as the Beginning of Political Geography: Heracles in Ancient Spain

Vadim Atnashev[1,2] and Alexey Tsyb[3,4](✉)

[1] Saint Petersburg State University, Saint Petersburg, Russia
[2] North West Institute of Management, RANEPA, Saint Petersburg, Russia
[3] Peter the Great St. Petersburg Polytechnic University, Saint Petersburg, Russia
[4] Sociological Institute of FCTAS RAS, Saint Petersburg, Russia

Abstract. The article examines the question of the beginning of political geography in Europe as a type of early geographical descriptions. Like any field of knowledge that once became a scientific discipline, the political geography develops from the mythology. And the main civilizing factor for Europe in this respect was the mythology of Ancient Greece. Here we find the origins of all aspects of European culture. This idea was based on the myth of the tenth labour of Hercules, which is directly related to the first political and geographical descriptions of Spain. The myth of Hercules' journey to the land of Erythia, the "red country" at sunset, where the sun, deified as the titan-Helios, descends daily into the gloomy reaches of Hades, has extensive connotations. At its early stage (Hesiod and Stesichor), it was a cosmological myth, telling about the limits of the oecumene known to the Greeks (the earth was flat, surrounded on all sides by the Ocean River), in which the Hercules path was a reconnaissance operation to survey the boundaries of space. But it became soon "mundane" geographic, colonial and political ideas culminating in the imperial era of Rome. The article also investigates the real history of the penetration of the Hercules cult into the Iberian Peninsula and the connection of the Greek myth about the abduction of Geryon's oxen by Hercules with possible historical myths. The questions posed are about whether the myth of the Hercules' tenth labour can explain the history of his cult in ancient Spain, as well as about its possible connection with local Tartessian-Celtic theologies and with the Celtic cults of the tricephalic deities in southern France. Or was the original myth of Hercules' journey to southern Spain to seek Geryon's wealth "anticipated" by the Greek mythological culture and imported into the ancient Spanish culture along with the syncretic cult of Hercules-Melqart?

Keywords: History of Spain · Greek mythology · Hercules cult · Geryon's herds

1 The Beginning of Political Geography

European political geography, like any other type of humanitarian knowledge, has its origins in myth. The first Greek land descriptions (Periegesis and maritime Periplus) were as fantastic stories as the Greek epic. For example, its Cycles included descriptions of

© The Author(s), under exclusive license to Springer Nature Switzerland AG 2021
R. Bolgov et al. (Eds.): *Proceedings of Topical Issues in International Political Geography*, SPRINGERGEOGR, pp. 31–43, 2021.
https://doi.org/10.1007/978-3-030-78690-8_4

countries and lands inhabited by magical creatures and peoples of a different nature, and that ideas, which form and conserve opposition "friend or foe" (precisely, "completely alien"), kept and developed the Greek civilization surrounded by more ancient and stable, but "barbaric" cultures of the ancient Mediterranean.

Greek *ethnocentrism* is as an idea of the existence of "*not just a people*" in some hostile or *opposing cultural environment* (or "a chosen people" among other less significant ones), but the only people in general living in a genetically alien environment (since they all come from different, less significant gods (Plato Alc. I 120e-124a). The idea manifested itself at the ideological level up to the entry of Hellas into the empire of Rome, as it also did in the "enlightened age" in the stories of Herodotus about the one-eyed Arimaspi and the vulture people, the Hyperboreans (Herod. III 116; IV 13) and Amazons (IV 110–116), living on the other side of the Euxinos Pontos, about Indians stealing gold from ants who were "almost the size of a dog, but smaller than a fox" (III 102–105), about their custom of eating corpses of sick tribesmen (III 95), and in the famous statement of Aristotle on «barbarians as natural slaves» (Arist. Polit. 1252b 8–10). From the point of view of this ethno-theo-centrism, the circumstances and characteristics of the culture, life and religion of other tribes and peoples were subjected to Greek rethinking. For example, it was in the Herodotus' story about the strangeness of king election among the Persians (Herod. III 80–88), reminiscent of a literary farce, or in the hellenised Phoenician theogony of Sanchuniathon-Philo of Byblos, in which alien gods take on names and functions of the Greek ones [25]. And thanks to the phenomenal rise of literary and cognitive culture in Greece, starting from the VIII century BC, this view of the world remains in the general Mediterranean environment as dominant, with significant losses in the oral and literary traditions of the surrounding cultures.

At the same time, myths about the strangeness of other gods and their religions, about the curious way of life of their followers remain completely organic parts of Periegesis and Periplus until the very end of the Ancient World, and the *political background of geography*, like no other field of knowledge, retains this genetic connection with myth. The last statement is true to this day: politics and geography, political and geographical myths go hand in hand. Our task is to trace the Spanish history of the Greek hero Heracles as one of the first mythologemes of Spanish political geography.

2 Political Tasks of Greek Heroism

Heracles is a character representing a separate generation in the hierarchy of Greek mythology – generation of heroes. By their origin and essence, they are half gods, half human beings, because they are born from the alliance of divine and human beings. In the absolute sense, this applies to representatives of the very first generation of heroes, and their descendants, living among humans and entering into the continuation of the genus with humans, gods, and among themselves, carry in themselves the diluted concentration of divine blood, portion of which defines a hero's social status [1]. Like many others, the notion of heroism in the Greek religion is not strict. Thus, from the point of view of "genetics", even the Olympic god Dionysus, son of Zeus and Semela, is a hero of a special level (Apoll. III. 4, 3).

The "first line" of heroes, which includes Perseus, Theseus, Bellerophon, Jason and Heracles, performs a special function in the mythological history of the Olympic space. It consists in purifying the world of all kinds of residual, rudimentary, ugly phenomena generated by the dominant gods of previous eras – Uranus, Gaia, the Barren Sea, Night (Hes. Theog. 116–239). According to the Greek myth logic, these undesirable phenomena – giants, some titans, Chimera, Medusa-Gorgo, Lernaean hydra, Nemean lion, Minotaur, and other monstrous entities can be wiped off the earth face only by efforts of the heroes, their "works", their exploits. For example, so says Apollodorus about gigantomachy (Apoll. I 6, 1).

In addition, the heroes commit other accompanying actions and completely human actions, just and unfair, often unseemly or criminal, thereby showing the so-called "ethical ambivalence", in the essence of which, according to A.I. Zaitsev, the meaning of "ethically neutral power" is in the first place [2, p. 462–466]. Unlike gods and humans, the heroes have a different ability of immortality. If the gods are immortal by definition, and the humans fall into the infernal Hades, then the heroes become stars in the firmament, take a place on Olympus, like Heracles (Hes. Theog. 951–955), or transform into incorporeal entities. However, they don't lose pleasant sensations, and transfer to Elysium, the Islands of Happy, where they lead a carefree life under the rule of Kronos [3, p. 76–77]. It also relates to the generation of "divine genus of heroes", participants in the Trojan War (Hes. Erga. 157–173).

But the most important function of the heroes is the geographical and colonial "oikism", the expansion of the oecumene. This side of mythology develops following the advance of the Greeks into foreign territories and the process of colonization, starting from the 9th century BC. It includes stories about Perseus (Medusa-Gorgo, the liberation of Andromeda on the Libyan, i.e. the African coast of the Mediterranean Sea), about Bellerophon and Chimera in Asia Minor, about the Argonauts' voyage to Colchis (the Black Sea coast of the North Caucasus), about Theseus' voyage to Crete, Spanish expeditions of Heracles for the bulls of Geryon and the golden apples of the Hesperides.

Another classification wording, "cultural hero", uniting mythological, semi-mythological and even historical personalities of Orpheus, Musaeus, Linus, Pythagoras, Plato on the basis of their contribution to the formation of Greek culture, has apparently a secondary, later, origin. Herodotus ranks Heracles among the "host of so-called second gods" (Herod. II 145).

The place of Heracles in the generation of heroes is special. He is the first among them, as the absolute "winner" in the conditional all-around, including all known types of confrontations. The compound name Ἡρακλῆς, etymologically related to the female theonym Ἥρα (goddess Hera, wife of Zeus) and the male Ἥρως (Eros), according to N.N. Kazansky, "conveying the idea of rage, passion, fullness of strength, anger," is of Mycenaean origin and possibly means "*well-known thanks to Hera*" [4, p. 54–58].

According to the myth's general version, Heracles was the son of Zeus and Alcmene, while she herself descended from the Perseus clan in the sixth generation. Her earthly husband was Amphitrion, king of Mycenae, grandson of Perseus [Apoll. II 4, 5; Diod. IV 9, 1). The mythological-chronological time of Heracles' life is in particular determined by his participation in the Argonauts' campaign, which took place one generation before the Trojan War. An eternal struggle with destined circumstances is the main fateful attribute of Heracles. In the process of education in his youth, and of performing labours-feats, the image of Heracles is overgrown with specific personal symbols. This is a lion's skin, "a sword received from Hermes, a bow and arrow from Apollo, a golden shell from Hephaestus, a cloak from Athena", as well as a club cut down in the Nemean forest (Apoll. 2, IV 11). These characteristic details, along with the recognizable athletic torso, became hallmarks of the Heracles' images in the places of his cult, and his colonial mythology was formed on the basis of the cult theology of "specific" sacred events.

3 Geographical Symbolism of the Tenth Labour in the Context of Ancient "methodologies"

Even before his birth, the goddess Hera, lawful wife of Zeus, had chased Heracles. It led him to the loss of royal status in Mycenae-Thebes-Tiryns-Argos (they appear in the myth as a single basileia [5, p. 214]) and the service (prophesied by Delphi) to the eccentric basileus Eurystheus, on whose orders Heracles had to perform ten labours/feats, impossible for an ordinary mortal. In the process of performing, their number grows to twelve due to the increasing demands of the "customer". The reasons and impulse of these events are the persecution of Hera, the atonement for the crime committed in a fit of madness, and the gain of immortality, which was promised to Heracles based on the results of the labours performed (Apoll. II 4, 12).

According to the canonical tradition Δωδέκαθλος (twelve feats), the tenth and eleventh labours of Heracles are associated with the southern part of Spain: "as the tenth labour, Eurystheus ordered to kidnap and bring the oxen/cows of Geryon from Erytheia" [Apoll. 2, X 5]. In their mythological geography, the Greeks called Erytheia as Iberia in general (Herod. I 163; Diod. IV 17, 1), 'Ερύθεια (Butler I, 667; Herod. IV 8; Pausan. IV 36, 4; V 10, 9; Apol. II 5, 10), Tartessos (Herod. I 163; Pausan. VI 2, 3), Γάδειρα (Herod. IV 8; Pausan. I 35, 8; X 4, 6; Diod IV 18, 2), Gades in Latinized form, and in different contexts it is a "country/area/island/river/city". By our time, the location of this country has been established around modern Cadiz, the Phoenician Gadir (Greek-speaking toponym Γάδειρα), to which in ancient times the island of Erytheia adjoined in the western part of the modern Continussa island (now they form a single island [6, p. 219–228; 7, p. 225–242; 8, p. 157]), and around the area of the indigenous civilization of Tartessos. The latter's population was Turduli and Turdetani surrounded by Oretani, Contestani, Bastetani, Mastiens, Lusitanians, Carpetani and other indigenous peoples [9, p. 691–734].

This labour of Heracles was already known in the oldest version of Hesiod (Theog. 289–293), where the canonical set of Dodekathlos was nevertheless missing. And the first detailed description of the plot belonged to the Sicilian Stesichorus (632\29 - 556/553 BC), described by him in the poem "Geryoneis" [10; 11, p. 217–237], which has come down to us in fragments. *Heroic-agonistic motivation and symbolism* dominate there: the battle with Geryon (as retribution for his wickedness towards the gods [13, p. 350–362], the murder of Orthrus as an anti-cosmic monster desecrating the Olympic world. Therefore, in the earliest versions - by Stesichorus and Hesiod - we see *a cosmological expansion of the myth*, which occurs mainly owing to the Geryon's image and develops through the geographical features' description of the extreme west of the oecumene. The latter was located at the limits of the Olympic space, where, according to all points of view, the power of Hades begins and Tartarus is placed. This is from where, according to Strabo, the name of the country Tartessos originated (Strab. II, II 12). This part of the world is painted in red color of sunset (ἐρυθρός – "red"). That interpretation of the myth is supported by A. Shulten, who considers Geryon a mythological avatar of the local Tarthessian king [14, p. 99]. J. Pointrenaud indicates a similar *cosmological expansion of the myth* in the story of the abduction of the Helios' cows by the satellites of Odysseus (Od. XII 353–365), when, according to the cited interpretation, "vandalism posed a threat to the world balance". The time itself and the cycle of its eternal renewal were violated, and the seasons were intermingled, because all oxen and sheep in this herd were three hundred and fifty according to the number of the lunar calendar days and without outside interference they were immortal [15, p. 16]). The person of Geryon, the herds' owner, requires even particular attention.

As the myth develops, this adventure story is saturated with details, which abound in the two most developed literary versions of Pseudo-Apollodorus and Diodorus Siculus.

In the Hellenistic version of Apollodorus (120–180 BC), *adventurous-geographical and oekistes symbolism* of action prevails: on the way to his goal, Heracles "passed through Europe, killing many wild animals, entered Libya, and when he came to Tartessos, he put up road signs - two identical columns" (Apollod. II 5, 10). It was said in a simple and straightforward way, but in fact the path was not close. According to the geography of mid. II - mid. 1st c. BC (the dating time of Apollodorus' "Library" [12, p. 119]), to get to Libya (following the story's logistics, this is the northern coast of Africa east of Egypt) Heracles should pass Europe through the northern coast of the Aegean Sea, straits, Asia Minor. The path of Heracles is overland until the moment of his crossing "across the Ocean" (in Greek geography it is a river that washes the land around) from Africa to Spain. There Heracles threatens the titan Helios, the god of the Sun, with a bow shot in order to borrow a means of crossing, namely his golden bowl, in which Helios makes his daily round of the world. This way Heracles was transported from the coast of Libya (Africa) to the island of Erytheia.

Then the story of Apollodorus turns into the genre of a police report, based on the rural theft of cattle: Heracles was staying for the night on Mount Abas, where he was smelled by the two-headed dog Orthrus. Heracles killed the monster with a club as well as the shepherd Eurytion who came running to the noise. Next, Meneteus appears, the boozer of Aid, whose plot was in the neighbourhood, and summons the herds owner, Geryon, who catches up with Heracles, at that very moment stealing the oxen like a thief, and they enter into battle. Heracles killed Geryon with an arrow from his bow (according to Stesichor, with a stone or a throwing projectile [13, p. 350]), then puts the oxen in the bowl of Helios and swims across to Tartess, that is, to the Iberian mainland. Finally, he returns the bowl to Helios (Apoll. II 5, 10).

Heracles' return journey is even more schematic: he crosses Abderia (the area around the Phoenician city in southern Spain). Here the Spanish path of Heracles breaks off, because he immediately arrives in Liguria (north-western region of Italy), where he kills the Poseidon's sons, who were trying to take away his cows. Then he drives the herd through Tirrenia and the entire western coast of Italy to the Greek colony of Rhegium, in search of the escaped bull he goes to Sicily, where he kills Eryx, the son of Poseidon and the king of Elymians. Having returned the bull with the help of Hephaestus, Heracles drove the herd to the Ionian Sea, then by land he turns out to be in Thrace (northeast of the European coast of the Aegean), loses half of the herd due to horsefly sent by the goddess Hera, then ferried across the Strimon River (located in modern Bulgaria, the north of Greece and flows into the Orfano Gulf of the Aegean Sea) and in a completely unclear way arrives in Tiryns, the eastern part of the Peloponnese, where he gives the bulls to basileus Eurystheus, who sacrificed them to Hera (Apollod. II 5, 10).

The *imperial euhemerist* Diodorus of Siculus (90–30 BC) gives a different version. In his story, Heracles prepares a *sea expedition for the voyage*: "a large number of ships and an army appropriate for such a campaign" (Diod. Cic. IV 17, 1). The king of Iberia is Chrysaor, who is incredibly rich according to his name (from the Greek word "gold") and relies on the support of three sons, who lead the "squads of warlike tribes" (IV 17, 2). Heracles, however, first went to Crete, where he simultaneously cleared the territory of wild animals (IV 17, 3), then arrives in Libya, where he killed Antaeus, again exterminated wild animals, after which it became possible to grow fruit trees, grapes and olives there (IV 17, 4). Heracles also killed "people who had violated the laws" there (IV 17, 5). The motive for the destruction of wild animals and the eradication of lawlessness was repeated many times.

Then Heracles arrived in Egypt, killed the king Busirid, passed again through the waterless part of Libya, founded the city of Hecatompylos, and went out to the Ocean, where he installed two stone pillars on both continents, the so-called Pillars of Heracles (IV 18, 1–2). Further, Heracles was transported by ships to Iberia, conquered it, killing all the sons of Chrysaor and, finally, reached the goal of his trip – to steal the cows of Geryon (IV 18, 1–2). It seems that Chrysaor and Geryon in this story are not related figures, but the rulers of neighboring areas, and the story of the abduction of the herd itself is devoid of details. In addition, there are no details of cattle theft at all.

On the way back, Heracles passed Iberia, "transferred the royal power over the Iberians to the best" of them, gave one of the kings a part of the herd and "these cows are considered sacred in Iberia up to our days (the time of Diodorus)" (IV 18, 3); through the mountainous region of the Alps, he passed Celtica (Gaul) and founded there the city of Alesia (near modern Dijon in western France). Then Heracles "made trip to Italy, passing the plain in the "present" (Cisalpine) Gaul (eastern coast of northern Italy), moved through Liguria (IV 19, 1–4), the country of Tyrrenia and reached the Tiber River, where, on the site of the future Rome, he founded his cult among the inhabitants of the Palatine (IV 21, 1–4); after that he moved along the western coast, where he destroyed the giants (IV 21, 5) and cultivated the lake with warm springs in Avernus, rid the area of cicadas in Regium and Locris, transported the herd to Sicily (IV 22, 1–6), defeated the local king and handed over the country to the indigenous inhabitants, near the city of Agria he built a sanctuary in honor of the hero Geryon (IV 24, 1–3). Then Heracles again crossed to Italy, passed the eastern coast, rounded the bay and descended into the area of the Greek Epirus, from where he arrived in Peloponnese to Eurystheus and handed cows to him. Eurystheus sacrificed cows to the goddess Hera. We have omitted some details, but the very fact of the "escalation" of events accompanying the livestock stealing should be clarified.

In the stories of Apollodorus and Diodorus, the Hellenistic *colonial interpretation of the event*, developing a cultural and heroic line, comes to the fore: Heracles cultivates the territory, brings olive and grapes, destroys creatures unsuitable for civilized life (monsters and animals), establishes cities, establishes a missing law, corrects society. Heracles here is an *oekist, archeget* (founder of power), *protagonist of the royal clans.* However, the prehistoric basis of myth in all cases is the *motive for the abduction of herds*, most likely its most archaic element dating back to the era when the Mycenaean basileis stole cattle from each other, which, along with land, was the most important component of property and well-being (Il. X 671–706; Il. XVIII 515–529; Paus. XXXVI 3–5; Od. XII 353–365; Paus. XXXVI 3–5; Apollod. III 10, 2).

4 Who is Geryon?

In the Hesiod-Stesichorus context, *Geryon is a titanium-pontid of the fourth generation*, i.e. pre-Olympic god, descendant of the pre-Olympic generation of gods, originating from the conjunction of Gaia and Pontus (barren sea). He is the son of Chrysaor and Callirrhoe, daughter of the titan Oceanus (Hesiod, Theogue 980–983). According to Stesichor, Geryon is the grandson of Poseidon [11, p. 228], that links his origin with sea divinity. His πατήρ Chrysaor, in turn, comes from Medusa the Gorgon, the daughter of exotic sea deities Phorcys and Ceto (Theog. 287–288), in whose offspring Hesiod collects numerous anti-cosmic monstrous and hybrid phenomena. He emerged from the severed head of Medusa: when Perseus decapitated the monster, "the horse Pegasus emerged from her and Chrysaor, whose name comes from the fact that he was born with a golden sword" (Hes. Theog. 280–283).

Undoubtedly, already in the early versions of the Heracles myth, Geryon was also included in the *solar symbolic circle of the titan Helios-Sun* (Heracles forced him to give the bowl for crossing) and in the *infernal circle of the underworld god, Hades*, since the "end of the world" in which he lives is consistent with the Greek concept of the entrance and exit of the day and night from the possessions of Hades. Meanwhile, Helios and Hades are at the same time the owners of herds (they graze in the immediate vicinity of the Geryon's herds); his two-headed guard dog Orthros is the brother of the three-headed dog Cerberus, guarding the entrance to Hades. In the interpretation of J. Pointerno, "the mythical residence of Geryon was another world outside the Ocean", Orthros was Cerberus, and Geryon himself is the avatar of Hades, from whom Eurystheus inherited the way to hide in the jug's depths. "The owner of the herd can be equated to Hades", and the tenth labour of Heracles itself is filled with the meaning of "victory over time and death" [15, p. 14–15]. A separate element of the catachthonic background of the myth in Stesichor was noted by M. Ercoles, who pointed to the alleged presence of Moira or Ker at the site of the battle between Heracles and Geryon, hovering in anticipation of the latter [13, p. 350–362]. These goddesses of death, carrying away the body of a killed warrior are a frequent element of the scene of the warriors' death in the Iliad (Il. II 301–302; VIII 526–527). N.N. Kazansky sees Erytheia, the daughter of Geryon, in the female figure [11, p. 217–237].

The question of the Geryon archetype's deep semantics is very complicated. To date, it includes a study of analogies with the Tartessian king Heron and the "castle of Heron" from the alleged Tartessian oral tradition [14, p. 93–94, 95], and (as an option) Theron, "the king of near Spain" (Theron rex Hispaniae citerioris, Macrob. Sat. I, 20, 12), with the trifunctional indigenous main local gods before the Roman conquest, for example, the Turmodigi [16, p. 159–179], or with the Turdetani's god Néit / Niethos / Netos [17, p. 37–70], as well as with the Gallic Cernunnos, the god of natural fertility [15, p. 24; 18, p. 63–85], with stories about the Brian brothers in Irish mythology [15, p. 14–15], with the "three-headed Kronos of the Orphic cosmogonies", and even with a triad of Samothrace Cabeiri [19, p. 34]. This question results in the problem of xenogeneity / indigenicity of the archetype of Geryon and his historical movement from his native Epirus (according to Arrian, Geryon is the king of Epirus, who owns rich herds (Arrian. Anab. II, XV, 16, 5–6)) to Arcadia, to Ithaca, or to Acarnania, as already indicated by L. Preller [20, p. 1344], to Sicily, or generally imaginative, metaphorical movement to the borders of the known world (Pausanias mentions Geryon's tomb in Lydia, M. Asia (Paus. I 2, 7–8).

With the general context of this image as a god-ruler of an alien country, who possesses innumerable riches, its *three-headed or the three-bodied* comes to the fore (analysis of variations in C.E. Gibbons [21, p. 290–297]), that reminds the Greek tricephals Hecate (PWBd VII, 2, 1912, s.2769 ff.) and Hermes, or Zeus-Triopus (three-eyed). At the same time, Spanish archaeologists most often deny the Spanish origin of the myth of Geryon's bulls [22, p. 228]. What actually happened and what did the Greek myth have to do with the historical events of the appearance of the Heracles religion in Spain?

5 Archaeology of Myth and Archaeology of Cult

By now, the Spanish archaeology has established the point of view that in the territory of ancient Spain there could be more than one deity named Heracles / Hércules of different origin [22, p. 219], whose cults appeared in Spain as a result of numerous historical events. The latter are included in the successive stages of the Phoenician-Carthaginian, Greek and Roman colonization of the peninsula.

The *Phoenician-Greek Heracles-Melqart* is to be considered first, known in the Romanesque era as Hercules Gaditanus, [22, p. 219] he appears on the Iberian Peninsula together with the first Phoenician colonists (Strab. III 5, 5) about 1100 BC [23, p. 279–317] as the *Tyrian Melqart* – "ruler of the city" [24, p. 27], archegetus and navigator, the leader of colonization [25, p. 263], of war like "Smiting god", worshipped by indigenous population in the vicinity of Gadir [26, p. 91–122] and near the northeastern Greek settlements [[27], p. 85–107)], an oekist and purifier of nature from the dark forces; the personification of the light principle (Non. Dionys. X, 369), the tamer of lions and other predatory animals [22, p. 219], the thief of herds, destroying the economic power of the most ancient gods [28, p. 57–104]; "lord of fire", giving an oracle about the foundation of Gadir (Strab. IIIV 5), about the beginning of the war to Hannibal at Celia Italic (III, 32–44), the solar god (Macrob. Satum. I, 20, 12; Non. Dionys. X, 370–374), dying and resurrected [24, p. 27 (Non. Dionys. X, 370–374)] as the patron of the transition from life to immortality through cremation (scene of Melqart's apotheosis, Sil. It. Ill, 42–44); the calendar sea god, whose cycle in Gadir opened and ended the season of wholesale tuna fishing [29, p. 91–93]; the guarantor of civil legislation, commercial shipping and commercial law [22, p. 219; thirty]. His Tyrian protagonism should testify to the self-sufficient antiquity of the cult, although in the general Phoenician theogony of Sanchuniathon-Philo of Byblos, the place of Melqart is far from the beginning of time, and he is the son of Demarount [25, p. 263].

According to Y.B. Tsirkin, the advancement of the cult was accomplished along the path of *two directions of Phoenician colonization at the end of the 12th - first half of the 11th century BC*: 1) from Rhodes along the western coast of Asia Minor to Thasos; 2) from Rhodes along the southern part of the Aegean archipelago to Sicily, Africa and southern Spain, rich in silver [31, p. 20], and more precisely to the coast, where the internal roads of Atlantic trade in metals and livestock went out. There were temples of the gods accompanying various aspects of the colonists' life, including trade deals, direct Greco-Phoenician contacts, and the associations of Tyrian merchants united around the cult of Heracles-Melqart [25, P. 264].

In the colonies on Thasos, Rhodes, Cyprus [22, p. 223], Malta [25, p. 262], Sicily, Sardinia [31, p. 85], Africa, and, finally, in the Spanish Gadir, the *original archaic aniconism* of the Phoenician God, who probably consisted of *Baetylus* [22, p. 232], perceived the counter iconography of essentially related gods. In Spain, these are indigenous "gerreros"- warriors- Smiting Gods, with whom the merchants dealt when the colonies were founded (Strab. IIIV 5), the gods-tamers of lions and other predatory animals [22, p. 219], the Greek Heracles-colonist on the islands, revered, for example, in the temple of Heracles on the island of Thasos (Herod. II, 44; Paus. V, 25, 12). The study of his traces suggests that we can talk about the Tyrian Heracles, i.e. the syncretized cult of Heracles-Melqart. I.S. Shifman convincingly shows the penetration of Greek culture also on the Phoenician coast of the Mediterranean Sea, especially in Sidon, that led to the "Greek rethinking" of Phoenician myths [25, p. 188]. The Boeotian sea hero Meliqert, who in addition to the consonance of names had a genetic connection with the Tyrian Cadmus (Odis. V, 333–334; Apoll. 1,9,1; III, 4,1 -3), could participate in the process of syncretisation [31, 29–30].

The completion of this process is attributed to approximately the VI century BC, which is shown in particular by coins from Cadiz [20, p. 223], literary evidence and important finds in Sancti Petri (Chiclana, in the southern outskirts of the former island of Cotinussa), that discovered the alleged structural remains of the famous temple of Heracles Gaditanus in Hades [22, p. 225]. In the Roman time, it retained the colossal importance as a religious centre of intercultural communication. In Cadiz, the so-called "el Circulo de Gadir" of the oral, literary and visual-theological tradition [29, p. 75–112] had formed, the most important functions of which being reflected in the "Visual theology" of the famous gates of Gaditan temple, and this theology is confirmed by other comparative data [28, p. 57–104]. Today, the ruins of a 16th century Spanish citadel in that place are a historical landmark of Cadiz.

6 Colonialism as an Impetus for Cultural Syncretisation

How does syncretisation of the cults of the two gods take place? It is primarily associated with their multifaceted functionality of militant heroism, "oikism" and navigation, and takes place in the course of sea trade expansions on the Mediterranean islands within Greco-Phoenician and Greco-Punic contacts. The reasons for the secret process could be both purely aesthetic and, first of all, of a practical-functional and "marketing" nature: the god accompanying the sacred legal support of commercial transactions [32, p. 61] and of promising trading activities should be accepted by all interested parties as their guarantor.

According to the 2002 map of finds in Spain, Heracles-Melqart owns a variety of objects, including epigraphy and archaeological artefacts, the first group of which can be a priori considered a "direct information source", while the second group is characterized by an indirect meaning and the lack of precise context. These are objects with abundant ornamentation of mythological themes, well-known cult images, etc. surrounded by common ritual objects such as altars, vessels, lamps, incense lamps, musical instruments, initiations, traces of sacrifices, etc. In general, there are 21 localizations mainly on the southern coast of the peninsula from Cabo de San Vicente (Portugal) to Isla de Escombreras (Murcia, south of Cartagena), of which 5 points are qualified as places of *worship of Heracles-Melqart* [22, p. 224]. Added to these are more ancient objects - bronze figurines imported from the Orient and appeared in places such as the vicinity of Sancti Petri (Cadiz) and La Barra de Huelva, close to the sanctuaries of Melqart, known from literary sources. They have a characteristic iconography of Oriental gods: static or in a position of attack (Smiting God), in Egyptian or Syrian clothing, but they gradually give way to the depiction of Melqart with the classic features of Heracles as a naked man with a club and a lion's skin; several bronze figures without exact southwestern origin (provinces of Seville and Extremadura), votive figurines; and above all, images of Cadiz origin, proving that at least from the 3rd century BC the assimilation of Melqart, Heracles and Roman Hercules was finally completed [22, p. 227].

7 Was the Cult of Heracles Spread in Spain by a Direct Greek Tradition?

The penetration of the Greek substrate of the cult of Heracles, together with Greek culture, into north-eastern Spain could begin no earlier than the second quarter of the 8th century BC during the sea "reconnaissance campaigns" of the *Rhodians*, which took place, according to A.H. Dominguez, without "any noticeable influence" on the cultural and economic life of the open territories of the north-east of Spain [33, 90–91 (Strab. III, IV 8; XIV, II 10)]. Another polis source of these campaigns was the *Samians*, who had appeared here from the end of the 7th century BC that is confirmed on Samos by "objects with a plot" of the battle of Heracles with Geryon [33, p. 91–92 (Herod. IV 152); 34, p. 53–90; 35, p. 207–221]. *Phocians* visited the territory of ancient Tartessos (modern. Huelva, ancient Onoba) [33, p. 92–93 (Herod. I 63)] from the end of VII - early VIth century BC and in result of the "regular activity" they gained a foothold on the territory from Hades to Malaga and La Fontena from the VI century BC. [33, p. 95]. From the second quarter of the 6th to the 5th centuries BC *emporia* (trading posts under the control of the local population) were developing, the largest of which *Emporion* (modern San Martí d'Ampurias, Catalonia) acquired the status of a polis, like *Roda* (modern Rosas, Catalonia) from the 4th century BC. They also remain the only settlements with the polis status during the entire period of the Greek presence [33, c.88]. As the dominant cults of Emporion are called the religion of Artemis of Ephesus, Dionysus, Asclepius, Serapis, Themis [30, p. 208–215], but not of Heracles.

In the same period from VI to III century BC, Greek emporias are moving along the coast of the modern provinces of Alicante and Murcia, and in the interior regions of Extremadura [33, 99–104]. And the last years of the V - the first half of the IV century

BC were called the "golden age in the trading activities of the emporias" [33, p. 105]. It was the period when the Greek cults and the associated ornamental culture imported by the Greek population should have left the greatest traces in archaeology, including manifestations of possible syncretisation with functionally similar indigenous cults, as happened with the cult of Heracles, for example, in the Northern Black Sea region. The character of the cult there retained the functions of salvation, protection and statehood, of fertility, chthonism, belligerence, and navigation [36]. In Spain, this situation could have continued at least until the beginning of the 2nd century BC [33, p. 106].

Discussing these issues, modern Spanish archaeology does not adhere to a completely definite view on the place of the Greek Heracles substratum in the religious realities of Spain in the 8th-5th centuries BC, when the Greek religious "picture of the world" came into contact with the aboriginal religions of Spain. One position is voiced by M. Oria Segura, who claims that the Greek colonies in the Spanish-speaking lands did not have Herculean legends, and the cities mentioned in these legends still have not provided reliable evidence of the cult [22, p. 234]. It is confirmed by a specialist in Greek epigraphy of Heracles, M. Pas de Os: there is no reliable evidence of a specific cult of the Greek Hercules, introduced by the Greeks on the peninsula [30, p. 208]. However, according to her, it was the Greeks who reached the shores of Andalusia in archaic times that undoubtedly played an important role in the arrival of the Greek god and his identification with Melqart of Gaditan, and in the fact that the Herculean tradition remained alive and rich in Roman Spain, although in literature and art it did more than in the cult itself, which was revived in Roman times with characteristics of the Roman religion [30, p. 208].

8 Roman Hercules

The cult of Hercules-Melqart in Spain perceived the "third wave of inspiration" from the classical Roman Hercules, starting from the 2nd century BC, with a predominance of imperial functions. At the same time, according to the conclusions of M. Oriya Segura, the Roman transformation of Hercules practically copies his Hellenistic iconography. However, the peculiarities of its "application" in the Romanesque environment are difficult to identify the object's very purpose. In this environment, gods and mythological scenes are used in a range of situations, from decorating temples and graves to works of art as an indicator of literary culture in the home, appearing in the decoration of public buildings. In such situations, an exact religious, household or political connotation cannot be appreciated [22, p. 228].

The distribution of inscriptions and images of the romanized classical Hercules spans the regions of the Levant, Tartessos, south-western Messeta, western Galicia and the Ebro river. The archaeological map of finds compiled by M. Oria Segura [22, p. 228] shows 80 objects of the Herculean symbolism of the Roman era, including inscriptions, sculptures, bronze, drawings, mosaics and ornamentals, concentrated mainly in Betica, Levant and in small quantities in western Galicia, Extremadura, in the areas of Indigenes upstream and downstream of the Ebro. The analysis of artefacts shows that the civil, imperial, funeral functions of the cult of Hercules, imported from the Latin environment, are brought to the fore in the Spanish-Latin environment, but its initial closeness to the popular Silvanus [37, 235–239] is apparently being lost, that also demonstrates another - political - side of Roman syncretism, which was considered by E.M. Shtaerman [37, 232].

References

1. Barth, J.: Chimera. Mariner, Boston (1972)
2. Zaitsev, A.I.: The wicked heroes. In: Almazova, N., Zhmud, J. (eds.) Zaitsev A.I. Selected Articles, vol. 2, pp. 462–466. Filologicheskij fakul'tet SPbGU, St. Petersburg (2003). (in Russian)
3. Rhode, E.: Psyche. Seelencult und Unsterblichkeitsglaube der Griechen, bd. I, s.155. Verlag von J. C. B. Mohr (Paul Siebeck), Tübingen und Leipzig (1903)
4. Kazanski, N.N.: To the etymology of the theonym HERA. In: Neroznak, V. (ed.) Paleobalkanistic and Antiquity, pp. 54–58. Nauka, Moscow (1989). (in Russian)
5. Zaitsev, A.I.: The ideology of the Homeric epic and the personal views of the author of the Iliad (Iliad XV, 638–651)]. In: Almazova, N., Zhmud, J. (eds.) Zajcev A.I. Selected Articles, vol. 2, pp. 462–466. Filologicheskij fakul'tet SPbGU, St. Peterburg (2003). (in Russian)
6. López Castro, J.L.: La territorialidad y los fenicios occidentales: estado actual de la investigación y perspectivas. In: Memorial Luis Siret. I Congreso de Prehistoria de Andalucía La tutela del patrimonio prehistórico, pp. 219–228. Junta de Andalucía, Sevilla (2011)
7. Niveau De Villedary y Mariñas, A.M.: La estructuración del espacio urbano y productivo de Gadir durante la Fase Urbana Clásica: cambios y perduraciones. Complutum **26**(1), 225–242 (2015). https://doi.org/10.5209/rev_CMPL.2015.v26.n1.49351
8. Domínguez Monedero, A. J.: Gadir. Mito y arqueología en el nacimiento de ciudadas legendarias de la Antigüedad / Fornis C. (coord.). Sevilla, pp. 153–197 (2012)
9. Fernández, F.J.G.: Tartesios, túrdulos, turdetanos. Realidad y ficción de la homogeneidad étnica de la Bética romana. In.: Yanguas, J.S., Gonzalo Cruz Andreotti, G.C. (eds.), Fernández Corral, M., Sánchez Voigt, L. (cols.) Romanización, fronteras y etnias en la Roma antigua: el caso hispano, pp. 691–734. Universidad del País Vasco I, Servicio Editorial, Vitoria-Gasteiz (2012)
10. Carvalho, S.D.G.: Stesichorean Journeys: tese de doutoramento em Estudos Clássicos. Universidade Coimbra, Coimbra (2017)
11. Stesichorus. Fragments. Translation and Commentary by N. N. Kazansky (Leningrad), M. L. Gasparov (Moskow) (eds.). Vestnik Drevney_Istorii, 2 (173), pp. 217–237 (1985) (in Russian)
12. Boruhovich, V.G.: Antique Mythography and «Apollodorus Bibliotheca». In: Apollodor. Mifologicheskaja biblioteka [Bibliotheca of Pseudo-Apollodorus], pp. 99–120. Nauka, Leningrad (1972). (in Russian)
13. Ercoles, M.: Stesichorus PMGF S21.1–3 (Geryoneis): A textual proposal. Greek Roman Byzantine Stud. **51**, 350–362 (2011)
14. Schulten, A.: Tartessos. Renacimiento, Sevilla (2006)
15. Poitrenaud, G.: Cernunnos et le tricéphale gaulois, Géryon et les triades celtiques. Academia.edu [Electronic resource] https://www.academia.edu/10619975/Cernunnos_et_le_tric%C3%A9phale_gaulois_G%C3%A9ryon_et_les_triades_celtiques. Accessed 18 May 2020
16. Carcedo de Andrés, B.P.: Religiones prerromanas en la Hispania antigua. Los turmogos. Norba. Revista de Historia **21**, 159–179 (2008)
17. Almagro-Gorbea, M.: Una probable divinidad tartésica. Identificada: niethos/netos. Palaeohispanica **2**, 37–70 (2002)

18. Blázquez Martínez, J.M.: Teónimos indígenas de Hispania Addenda y corrigenda. Palaeohispanica **1**, 63–85 (2001)
19. Bertrand, A.: L'autel de Saintes et les triades gauloises. Note lue a l'Académie dés inscriptions en décembre Didier, Paris (1880). http://bibnum.enc.sorbonne.fr/omeka/files/original/7b3e09a056635dba3a212cc968281816.pdf. Accessed 17 June 2020
20. Preller, L.: Griechische Mythologie. Drittes Buch die Grossen Heldenepen. Abteilung die Argonauten der Thebanische Kreis, vol. II. 4te auflage. Weidmannsche Buchhandlung, Berlin (1921)
21. Gibbons, S. E.: The labors of Heracles. A literary and artistic examination: thesis... for the degree of Doctor of Philosophy. London University, London (1975)
22. Üria Segura, M.: Religión, culto y arqueología: Hércules en la Península Ibérica. In: Ferrer Albelda, E. (ed.) Ex Oriente Lux: Las religiones orientales antiguas en la Península ibérica, pp. 219–244. Universidad de Sevilla, Fundación el monte, Sevilla (2002)
23. Ruiz Mata, D.: La fundación de Gádir y el Castillo de Doña Blanca: contrastación textual y arqueológica. Complutum **10**, 279–317 (1999)
24. Tsirkin, J.B.: Introduction. In: Gerasimov, N.K., Dovzhenko, J.S. (eds.) Phoenician mythology, pp. 5–36. Letniy Sad; Zhurnal «Neva», St. Petersburg (1999). (in Russian)
25. Shifman, I.S.: Ancient Phoenicia - mythology and history. In: Gerasimov, N.K., Dovzhenko, J.S. (eds.) Phoenician mythology, pp. 185–324. Letniy Sad; Zhurnal «Neva», St. Petersburg (1999). (in Russian)
26. Corzo Sánchez, R.: Sobre las primeras imágenes y la personalidad originaria de Hercules Gaditanus. SPAL **14**, 91–122 (2005)
27. Lorrio Alvarado, A.J., Royo Guillén, J.I.: El guerrero celtibérico de Mosqueruela (Teruel): una pintura rupestre excepcional de la Edad del Hierro en el Alto Maestrazgo turolense. Antiqvietas **25**, 85–107 (2013)
28. Almagro-Gorbea, M., Torres, M.: Escultura Fenicia en Hispania. In: Bibliotheca Archaeologica Hispana, vol. 32. Real Academia de la Historia, Madrid (2010)
29. Almagro-Gorbea, M.: El 'Círculo de Gadir' y el final de la literatura hispano-fenicia. In: Serrano,B., Cruz Andreotti, G. (eds.). La etapa neopúnica en Hispania y el Mediterráneo centro occidental: identidades compartidas, pp. 75–112. Ulzama Digital, Sevilla (2012)
30. Paz de Hoz, M.: Cultos griegos, cultos sincréticos y la inmigración griega y greco-oriental en la Península Ibérica. In: Paz de Hoz, M., Mora, G. El Oriente griego en la Península Ibérica, pp. 205–254. Imprenta Taravilla S.L. & Mesón de Paños, Madrid (2013)
31. Tsirkin, J.B.: History of ancient Spain. Filologicheskij fakul'tet Sankt-Peterburgskogo gosudarstvennogo universiteta, Nestor-Istorija Pabl., St. Petersburg (2011). (in Russian)
32. Almagro-Gorbea, M.: Una probable divinidad tartésica identificada: Niethos/Netos. Palaeohispanica **2**, 37–70 (2002)
33. Dominguez Monedero, A.J.: The Greeks in Iberia and their Connections with the native population. Quest. Hist. **4**(254), 88–107 (2005). (in Russian)
34. Brize, P.: Samos und Stesichoros: zu einem früharchaischen Bronzeblech. Mitteilungen des Deutschen Archäologischen Instituts, Athenische Abteilung, 100, 54–90 (1985)
35. Robertson, M.: Geryoneis: Stesichorus and the Vase-Painters. CQ, vol. 19, pp. 207–221 (1969)
36. Zaharova, E.A.: Cults of ancient Greek heroes Achilles and Heracles in the Northern Black Sea region, VI-I centuries BC. PhD Thesis. Moscow (2003). (in Russian). https://www.dissercat.com/content/kulty-drevnegrecheskikh-geroev-akhilla-i-gerakla-v-severnom-prichernomore-vi-i-vv-do-ne. Accessed 18 May 2020
37. Shtaerman, E.M.: Social Foundations of Ancient Rome Religion. Nauka Publ, Moscow (1987).(in Russian)

Geography of Russia's Foreign Trade in the 18th and Early 19th Centuries

Olga Pavlova(✉) and Anna Ryabova

Peter the Great St. Petersburg Polytechnic University, Saint-Petersburg, Russia

Abstract. The article deals with the features and traditions that have developed in the history of Russia's foreign trade relations. The purpose of the paper is to reveal and examine the role of the international trade for the Russian Empire in the XVIII and early XIX centuries, to analyze the special features and traditions of the state's trade relations. The authors discuss the historiography of Russian foreign trade problems, view the formation of the trade features and traditions specific for a given time period, identify the priority areas of Russian foreign trade and determine the reasons for particular directions of Russian trade. The geography of international trade relations is highlighted; the impact of foreign trade on the economic and political status of Russia in a given historic period is emphasized. The priority areas of international trade are identified, and the reasons and features of the selected areas are analyzed. The novelty of the research is emphasized by the necessity to consider the emergence of Russian new economic opportunities associated with the trade contacts and new trade agreements that led to the development of new diplomatic and cultural relations and the tremendous interest for the voyages of geographic discoveries in a given period. The political and economic background of Russia and its trade partners is discussed. The opportunities for economic relations, the existing peculiarities and their significance for the political and economic status of Russia in the historical process are examined. The analysis of works devoted to the international trade geography of the 18th–19th centuries is carried out.

Keywords: Foreign policy · International situation · International relations · Political situation · Political and cultural contacts · Trade agreements · Trade and economic relations

1 Introduction

Foreign trade took shape in the process of formation and development of the state. It was the driving force, and an important component of the entire economy of the country. The goal of the paper is to reveal the impact of foreign trade on the economic, political and cultural relations in a given period of Russian history. The tasks of the paper are: to analyze the historiography of the problem; to view the peculiarities of Russian foreign trade in a given period of time; to define the reasons for the geography of the trade relations; to discuss the historical and economic background of Russia and its trade

© The Author(s), under exclusive license to Springer Nature Switzerland AG 2021
R. Bolgov et al. (Eds.): *Proceedings of Topical Issues in International Political Geography*, SPRINGERGEOGR, pp. 44–57, 2021.
https://doi.org/10.1007/978-3-030-78690-8_5

partners in a given period of time. Foreign trade policy reflects the position of state power, as well as the relationship between government and private business. P. B. Struve [19, p. 4] expressed the opinion that class interests exist and affect foreign policy. The interests of the state and its relations with other countries rise above class interests. "This paints foreign trade policy in a national color, connects it especially closely with the very existence of the state, because external relations, foreign policy with particular vividness express the existence of the state as a separate state with its own face." Thus, foreign trade policy reflects various aspects of the internal and external position of the state. Studying geography, that is, the most important areas of Russia's foreign trade in the 18th and early 19th centuries, makes it possible, through external relations, to present not only the economic situation of the state, and to elicit the geography of trade routes, but also to expand knowledge about the internal state of society, the relationship between the emerging classes, the foreign policy of the state, and determine its priority areas. The accumulated experience, its historical analysis contributes to a better understanding of modern international relations, and possible options for the prospective development and strengthening of economic, trade, and political international ties.

2 Methods

The research is based on the principle of historicism, an interdisciplinary scientific research, using the methods of historical, historiographic, and source analysis. The principle of historicism is understood as a systematic approach to the analysis of historical events and facts, based on the scientific research studied by the authors. The object of the paper, the tasks and research methods are to be considered in relation to the impact of Russian foreign trade on the economy, international relations' development and political strategies in a given period.

3 Historiography

The history of Russia's foreign trade development aroused the interest of representatives of various scientific fields-historians, economists, sociologists, and cultural scientists. Among the economists of the early 20th century, the works of P. B. Struve [21] and I. M. Kulisher [11], devoted to the economic development of Russia at different historical stages, are widely known. These researchers paid special attention to Russia's foreign trade in the context of economic history and diplomacy.

At the present stage, the history of Russian foreign trade continues to interest researchers. Scientific interest in the problem can be divided into several aspects. The most common aspect of studying Russia's foreign trade is related to individual countries in a certain historical period. Considerable attention of researchers is attracted by the organization of the system of trade relations and the objects of exchange. An important aspect of the study is the state policy in the field of foreign trade. No less interesting for scientific developments is the study of the activities of specific personalities: statesmen and politicians, merchants, representatives of other estates who participated in the organization and implementation of Russia's foreign trade policy.

So, the work of A.V. Dyomkin [4] on the history of trade relations between Britain and Russia in the 18th century, and the activities of British merchants on the territory of the Russian Empire is of special importance.

Currently, China is one of Russia's active political and economic allies, and researchers' interest in Russian-Chinese trade in various historical periods is quite understandable. The history of the tea trade of the 19th and 20th centuries, based on the works of Chinese and Russian scientists, is considered by V. G. Sharonova [28]. For the study of the tea trade, the reference edition of I. A. Sokolov [20], that consists of two volumes has the special value.

Trade relations of the Russian Empire with Latin American countries are studied in the monograph by V.N. Shkunov [29].The series Academic Fundamental Studies: History reproduces the work by T. P. Yuzefovich [32] published in 1869 Treaties of Russia with the East: political and trade: Turkey, Persia, China, Japan, revealing the history of the formation of not only political but also trade relations between countries. V. N. Shkunov's monograph [30] is dedicated to state legislation in the field of foreign trade regulation in the 18th and19th centuries.

The article by B. V. Zmerzly [8] is dedicated to the state policy for the restoration of foreign trade in the Crimea at the end of the 18th century. The author analyzes the government's measures to restore trade relations after the annexation of the Crimea to Russia, special attention is paid to foreign trade agreements with the Ottoman Empire, Sicily, France and the Austrian Empire.

The history of Russia's foreign trade is reflected in scientific reports at conferences, such as the Lazarev Readings held in Sevastopol in 2019. Among other historical problems of the Black Sea coast, the conference addressed issues of Russia's foreign trade from the White to the Black Seas [6].There a lot of foreign authors who did numerous research targeted at the foreign trade study that discussed the foreign trade policy of the Netherlands, Great Britain, Germany, China, and etc. In those papers the researchers discussed the trade problems from the point of view of their national research schools and their national trade interests. As an example one can consider the work of Müller, L. devoted to the Swedish trade in the XVIII century [13], the investigation of the Chinese trade strategies of the XVIII century is described in the work of Van Dyke [33], the special features of Dutch trade relations with China are examined in the work of Yang [34]. The presented brief analysis of the historiography of recent publications dealing with various aspects of both Russian and its partners foreign trade in the 18th and early 19th centuries shows a great interest of researchers in the development of foreign trade and its wide-ranging issues.

4 Results and Discussion

The history of Russia's foreign trade development begins with the formation of statehood. The famous trade route "from the Varangians to the Greeks", known already in the 10th century, was a water communication connecting the Baltic Sea with Byzantium, passed through eastern Europe, where the process of creating the Old Russian state was going on. Along this route, the first cities appeared, the ethics of trade relations were formed, and internal and external political and economic ties developed. No less famous is the Silk

Road, so named by the German geographer Ferdinand von Richthofen (1877). It went from China, spanned many countries of the East, changed its directions, ran to the Black Sea coast, connected the East and Western Europe, spanned the lower Volga, reached to the southeastern Baltic. It contributed to the development of economic relations between the countries, and the expansion of political and cultural ties between the peoples of the East and West. Thus, foreign trade relations that originated during the formation of statehood in Russia went through several stages of development changing priority areas that largely depended on the external and internal state of the country.

Having travelled the path of the flourishing of the feudal state, the development of international political, dynastic and economic ties, Russia survived two hundred years of enslavement by the Golden Horde, which brought death of people, destruction of cities and villages, led to a decrease in the productivity of artisans and peasants, violated the established traditions of political and trade relations. By the 16th century, the Russian princes, who relied on the freedom-loving people, had formed a Moscow Kingdom free from external dependence. The process of strengthening the centralized state began, largely associated with the name of Tsar Ivan IV the Terrible.

At that time, the White Sea was in the center of European trade. English merchants in search of the shortest route to India explored the possibilities of trade through Russia. For this purpose, three ships were organized and sent, two of which sank, only the ship under the command of R. Chancellor reached the mouth of the Northern Dvina. R Chancellor was an English seaman who became famous for his voyages in the middle of the 16th century. According to the researcher V.N. Shkunov "The meeting of R. Chancellor with Ivan the Terrible laid the foundations for trade relations between Russia and England" [29, p. 23].

English ships reached the Dvina River basin; they contacted Scandinavian merchants, controlled maritime trade. Interested in strengthening international relations, Ivan IV issued a letter of duty-free trade to English merchants. The policy of strengthening and expanding trade relations was continued. In 1567, the Moscow Company, as the English merchants-entrepreneurs who traded in the Muscovite were called, received the right to establish trading yards, organize transit trips, and sail in the Russian seas. Export items were traditionally: hemp, flax, tar, wood and other goods necessary for the development of English industry. English merchants imported mainly manufactured goods. Relations between the two countries were uneven, which also affected trade relations. In 1587, the rights of the Moscow Company were restricted, merchants could only purchase goods through the mediation of the state.

In the 17th century, several priority areas of foreign trade were formed in Russia. Western European merchants controlled Russia's trade with the Netherlands, England, Hamburg, Lubeck, Bremen. At the beginning of the century, French ships exported leather, furs, hemp, tar, fish and other goods from Arkhangelsk and Kholmogory. Among the imported goods were woolen fabrics, wines, non-ferrous metals, luxury goods. The list of goods brought to Russia changed little throughout the century. In the catalogs of 1604 and 1654, about the same luxury items are mentioned, so in the Catalog of overseas goods brought to Kholmogory in 1604, among others, goods taken at the request of the king are indicated: large and medium pearls, emeralds, velvet [17]. In 1954 such goods as pearl, incense, dry fruit were delivered by the merchant from Bremen Ivan Bonmas [18].

With the preserved documents on the development of Russia's foreign trade with Western European countries, it can be inferred that reports on trade were received in two orders: the Novgorod Chet and the Ambassadorial Prikaz. According to those documents, it is obvious that Western European trade was most actively carried out through Kholmogory and Arkhangelsk. Then the goods were sent to Moscow and other cities. The state already in those days kept foreign trade under control, carefully treated tradespeople, using not only their trade ties, but also hoping for commercial ingenuity in trade and political affairs.

Among the most active merchants were residents of Moscow and Yaroslavl. Vologda residents were engaged in transportation and delivery of goods. The names of foreign merchants – Moscow Germans and Dutchmen – are mentioned in the documents, which testifies to Russia's active contacts with the German lands and the Netherlands [5]. The New Trade Charter (1667) was of great importance for the development of foreign trade, according to which foreign merchants received the right to trade within Russia, with the special permission of the sovereign.

In the 17th century, Russian-Dutch trade was actively developed. As with other countries, trade relations largely depended on the international situation. Contacts with the Netherlands have been known since the time of Ivan the Terrible (1533–1584), when Dutch merchants appeared on the Russian shores of the White Sea. In the 16th century, trade was conducted through Amsterdam, and the center of trade with the Netherlands in Russia was Arkhangelsk. Official contacts were developed under Tsar Mikhail Fedorovich. Political considerations were important. Having ascended the throne, the first of the Romanovs reported that to the European powers by sending out a 'promise'. The main purpose of the appeal was related to a request for a loan of funds, and support of the union against Poland and Sweden. An embassy headed by S.M. Ushakov was sent to Europe. An important role in preparing the arrival of the embassy belonged to Dutch merchants, who were r interested in trade with Russia.

In 1615, an embassy headed by I. Kondyrev and M. Neverov was sent to France. It was no coincidence that their route ran through the Netherlands, where they were instructed to submit a request to the Dutch authorities for assistance against Poland and Sweden. In 1616, the Netherlands mediated a truce between Russia and Sweden. Less successful was the embassy of A. Olyabyev and G. Larionov sent in 1631. The request to help in the fight against Poland, the States General of the Netherlands refused, but allowed to take some of the ammunition to Russia. In 1630–1631, the Dutch Embassy arrived in Moscow with the intention of establishing political and economic contacts. Further, in 1665, the Embassy arrived in Moscow from Holland and acted as an intermediary in organizing trade relations in Western Europe. In 1676, the embassy of the States General visited Moscow. The aim was not only to discuss the prospects of political relations, but also to negotiate a possible organization of trade in Siberia and China. A new stage in the development of the already established political and trade relations between Russia and the Netherlands is associated with the name of Peter I.

English merchants competed with the Netherlands in trade with Russia. As a result, the Dutch mostly used the ports of Riga, Revel, Dorpat for trade with Russia. The English, the cities of the Hanseatic League, traded through Arkhangelsk. At that time, merchants from Scandinavia intensified their trade exchange through Arkhangelsk. The ports of

Kola and Kem were transit points for European merchants. In Kola, fish was exchanged for cloth and metals, and in Kem, especially on Peter's day, a lively fair was organized, where the English, Norwegians, Danes, and Sami exchanged furs, fish, and blubber (liquid fat) [15]. Swedish merchants were attracted by Russian grain. Frequent crop failures created problems with bread in Sweden. The Swedes had their own courtyard in Novgorod where they supplied iron, copper, lead, and exchanged metal for Russian bread.

Having briefly considered the directions of European foreign trade of Russia in the 17th century, it should be noted that the countries of Europe throughout the history of Russia viewed the country as an important political and economic partner and competitor. By the end of the 17th century, economic ties developed, trade contacts were established, the market for exchanging goods and trade traditions were formed. These ties became the foundation for the development of trade and economic relations which influenced political and cultural contacts.

The beginning of the 18th century is associated with the global transformations of Peter I. Having achieved success in the Northern War and gaining access to the Baltic Sea, Peter I strengthened the country's international position and opened up new opportunities for the development of foreign trade. In 1703, the construction of St. Petersburg began, which was also considered in the future as a major trading port. Starting in the late 1920s, foreign merchants focused more on St. Petersburg. Despite the long war with Sweden, Peter I showed concern for the development of foreign trade. In 1719, he gave instructions to Russian ambassadors at foreign courts to offer "all nations trading in the St. Petersburg port" to trade freely in Russia. At the same time, the Arkhangelsk port continued to operate [27]. It is interesting to note that confirming his decree the tsar ordered to translate into Russian the "privileges" that foreign merchants who traded in Riga had during its Swedish state. So Peter practically confirmed the preservation of benefits for foreign merchants in the Baltic port.

The mercantile policy pursued by Peter I contributed to the development of the manufacturing industry, and positively influenced the expansion of foreign trade relations. The expansion of trade relations was influenced by the ongoing diplomacy aimed at deepening diplomatic ties. Representatives of various states were present at the royal court. At the same time, Russia's foreign trade was influenced by the Great Northern War. The instability of the situation, the threat of an attack by the Swedes on Arkhangelsk, forced the tsar to make a decision in 1720 to transport all goods to Kholmogory [27, p. 79]. English merchants, on behalf of the Commerce Collegium, were forbidden to enter Swedish ports with Russian goods. This caused significant damage to Sweden, which needed Russian goods. However, as V.N. Shkunov [29, p. 36] notes, the Northern War aroused great interest in the Scandinavian countries, trade relations practically did not stop, "Some Russian statesmen advocated the full development of trade and economic relations with Sweden, Norway, Denmark, Finland." The Nystad Peace Treaty strengthened trade relations between the two states. Trade with Denmark began to develop successfully. A number of Danish companies had a monopoly on maritime trade. The favorable development of trade relations between Russia and Denmark was facilitated by its political neutrality, the country was not drawn into military conflicts, which positively affected trade relations [9].

Despite the danger of merchant shipping, the Netherlands continued to trade with Russia through the Baltic Sea, equipping a military squadron to protect ships from attacks by Swedish privateers. Trade and economic relations with the Netherlands developed despite the instability caused by the period of palace coups. The trade turnover increased, and the need arose to regulate trade relations between the two countries. In 1741, negotiations were held on that matter, but they did not achieve any result.

England was of particular importance in Russia's European trade. The established trade ties and traditions developed, but foreign policy situations could not but affect economic relations. Tsar Alexei Mikhailovich broke off diplomatic relations with England, and Peter I in 1698 terminated the monopoly rights of the "Moscow company". In 1734 and 1741, trade and defense treaties were concluded, which somewhat improved international relations between the countries. The next trade agreement was concluded in 1766, which became a continuation of the agreement of 1734. After a long period of difficult international relations, only Alexander I, who came to power, restored diplomatic relations with Great Britain, having entered into an alliance with it against Napoleon. Thus, politics was closely linked to economic relations and adversely affected foreign trade relations. At the beginning of the 19th century, English merchants played a leading role in foreign trade with Russia. According to the researcher M.F. Zlotnikov [7], who provides English information, in 1802, exports to Russia from England amounted to 1,282 lb sterling, and imports from Russia – 2,182 lb sterling.

It is known that the policy of Peter I was not always consistent, this also concerned foreign trade. On the one hand, he in every possible way supported the development of foreign policy and trade relations, the private initiative of merchants, on the other hand, he interfered in trade relations, introduced bans and restrictions on imported and exported goods, and a strict customs policy was implemented. The authorities, through their representatives and diplomats, controlled information related to trade: prices of goods, harvests, exchange rates, and other information useful to Russian merchants. All information flowed to the Commerce Collegium in St. Petersburg. Thus, maintaining trade relations and merchants, the autocratic government pursued a protectionist policy, strictly controlling this process. The policy of protectionism was also pursued by the successors of Peter I.

The reform of Peter I, the ongoing policy of mercantilism and protectionism gave their positive results in the development of industry and trade. In the second half of the 18th century, Russia's foreign trade turnover increased, "the annual export of Russian goods abroad increased 5.6 times (from 12.0 to 67.7 million rubles), and the import of foreign goods – 4.5 times (from 9.3 to 41.9 million rubles); in 1796, export exceeded import by 25.8 million rubles" [16].

Pursuing a policy of physiocrats and supporting a relatively liberal customs policy, Empress Catherine II adopted customs tariffs for European trade in 1766 and 1782. However, the political situation and events in France influenced the decisions of the empress and intensified the protectionist policy. The decree of April 8, 1793 prohibited all trade with France in Russia, and also restricted any import of goods from "foreign lands" [23, 24].

Relying on the international situation and on his personal ambitions, as well as on the customs policy of the last years of his mother's reign, Paul I, who replaced Catherine

II, adopted a new customs tariff of 1797, which practically did not differ from the 1782 tariff adopted by Catherine II.

A new stage of Russia's economic development is associated with the reign of Alexander I, and has influenced the development and state of foreign trade. An important role in this process belonged to a member of the government N. P. Rumyantsev. Over the course of many years, he directly or indirectly contributed to the development of Russia's foreign trade occupying various positions in the government of Alexander I. In the years 1801–1809, he was Chief Director of the Department of Water Communications; he headed the Expedition of Road Construction; over the period of 1801–1810 he was Minister of Commerce 'with the retirement in their posts and titles", then during 1808–1814 – Minister of Foreign Affairs. Among the numerous merits of N.P. Rumyantsev in the development of foreign trade are: the development of communication lines system, without which any trade would be difficult; expansion of international relations, for instance, his support for the Russian-American company which was provided with preferential conditions "to use a loan from accounting offices for its industrial goods and for its own bills of exchange" [14].

The activities of N.P. Rumyantsev were aimed at the development of the Mariinsky water system that connected the Volga River with the Baltic Sea. It was built to provide St.Petersburg with bread, forest and other goods. They were brought to St.Petersburg via Ribinsk from the Volga River. Lately the export of wheat was provided to Europe via the Mariinsky water system. N.P. Rumyantsev took the initiative in creating the Belomorskaya joint-stock trading company, which contributed to the development of trade on the White sea. He supported the private initiative of domestic merchants. So, in 1806, the Arkhangelsk merchant K.A. Anfilatov organized the first commercial sea expeditions to the North American United States, while receiving permission for duty-free export and import of goods. That step contributed to the development of trade between the countries through the Arkhangelsk port. N.P. Rumyantsev was an opponent of the mercantilist policy, in that part of it, which concerned restricting and reducing imports. He proposed new principles of customs policy, with the establishment of a moderate tax.

Russia, due to its geographical location, could become the center of transit foreign trade, which required the construction of new ports that connected the waterways of the West and the East. The states that were washed by seas had a developed river system. N.P. Rumyantsev considered the transit of Asian goods to Europe by water to be promising for the development of trade.

The East was an important area of Russia's foreign trade. Along with the treaties that Russia concluded with Poland, Prussia, Denmark, the Hanseatic cities, Sweden, England, and Turkey, a 'special' trade Treaty with Persia was concluded in 1717, approved in 1720. Its peculiarity was that traditionally between states there were peace or union treaties which included the positions of trade agreements. A trade treaty of political significance was concluded with Persia [20, p. 70]. He had his own prehistory. The first trade agreement with Persia was concluded in 1567 under Ivan the Terrible, which helped to strengthen ties between the countries. This agreement was confirmed in 1673. Finally, in 1717, with difficult political relations, the Russian envoy to Persia, A.P. Volynsky, concluded a trade agreement that provided Russian merchants with freedom of trade in

Persia. The positive role of the treaty was not only of an economic nature but also of political significance. The agreement did not give Turkey an opportunity to strengthen its positions in the Eastern Transcaucasia. This was in line with Russia's interests in the Caucasus.

Trade relations between Russia and Persia were crucial in the Caspian region. Astrakhan was the center of trade. Silk and cotton fabrics, luxury goods, incense, and spices were brought from Persia. The list of exported goods included raw materials, items of handicraft and factory production. The turnover of trade with Persia at the beginning of the 19th century was constantly increasing, as evidenced by the customs books of Astrakhan. So, in 1805, the turnover was 984 thousand rubles, and by 1807, it amounted to 1550 thousand rubles [3, p. 210]. The turnover of trade is the important economic index of the trade organization, which reflects the volume of trade in currency during the certain period.

Baku, which had large customs warehouses and a commercial pier, occupied an important place in the intermediary trade in the Caspian Sea. The intensification of trade there was associated with political events. In 1803, Russia took under its patronage the Mangyshlak Turkmens of the Abdal branch. Alexander I considered this step important for the development of industrial activities of the Turkmen and the development of all Russian trade in the Caspian Sea.

At this time, trade through the Black Sea intensified, the Azov-Black Sea trade routes were gaining momentum due to the establishment of political and economic relations. According to A.V. Semenov, from 1792 to 1802, the total turnover of the Black Sea trade increased 18 times, but for a long time it was less than the White Sea trade and amounted to 4% of the country's foreign trade turnover. A noticeable growth was noted by the year 1805, and amounted to 9/4% of the total trade turnover in Russia [7, 13, 14].

The main role in the Black Sea-Azov basin belonged to Taganrog. Grain, which is in great demand abroad, was delivered here from the Volga, Don, and Azov regions. Besides bread, iron was also an important export item. At the beginning of the 19th century, the importance of the commercial port of Odessa increased. However, as the researchers rightly point out, southern trade, like the western one, largely depended on the political situation. Relations with Turkey, private wars created an unstable political and economic situation, and affected trade relations [12, 19].

China was Russia's most important partner in the Asian direction. According to historians (for example, N.Yu. Bolotina [2] and V.L. Uspensky [25]), the development of Russian-Chinese trade dates back to the 15th and early 16th centuries. This was facilitated by the development of Siberia and the Far East by the Russians. The peoples who settled on that land – Tajiks, Uzbeks, Uighurs – played a certain role in the development of trade relations between Russia and China, they were often called "Bukhara merchants". Representatives of the Cossacks took part in the development of these areas; they built several outposts along the banks of the Amur River. The city of Nerchinsk and the fortress of Albazin were of the greatest importance. Erofei Khabarov plays an important role in land development. The construction of fortresses and cities contributed to the settlement of these areas, the development of crafts and trade. The formation of diplomatic and trade relations between Russia and China dates back to the turn of the 16th and 17th centuries. In early 17th century, as part of the policy of Tsar Mikhail Romanov, one of the first diplomatic missions to China was carried out under the leadership of Ivan Petlin. One of the tasks of the mission was to establish trade relations. This was especially important as European merchants were already trying to use transit trade routes with China through Russia. The success of the mission is evidenced by the diploma brought by Petlin from the Chinese authorities, confirming the desire to develop trade relations. In 1654–1658, the Russian ambassador F.I. Baikov was sent to China, his task was to identify trade opportunities between the two countries. The importance of this direction in Russian diplomatic and trade relations is evidenced by the fact that it was controlled by the Governing Senate. Although the embassy of F.I. Baykov, as well as the embassy headed by Nikolai Gavrilovich Milescu-Spafari (Moldavian Greek) sent to China by Tsar Alexei Mikhailovich in 1675 had no success for the development of trade. To a greater extent, the last embassy was of scientific importance.

However, relations between the countries were tense. Frequent military clashes did not create stability. The ideology of the Qing Empire was aimed at a dominant position in the region. From 1685, in fact, with interruptions until 1688, military clashes arose between the Chinese and Russian troops, the Chinese repeatedly organized the siege of the Albazin fortress.

A certain stabilization of relations between the two countries took place in the late 80s of the 17th century, under Peter I. An important stage in the development of trade relations with China was the Treaty of Nerchinsk of August 27, 1689, which is considered the beginning of official trade between Russia and China. Tsar Peter I repeatedly sent caravans and ambassadors to China, including those from the merchant class, in an effort to strengthen trade relations, learn about the reaction of the Chinese side to the Treaty of Nerchinsk, and he also had a desire to open a Russian trading yard in Beijing. According to the researcher S.V. Bereznitsky [1, 7, 8], "Only the Kyakhta Treaty of 1727 between Russia and China confirmed the articles of the Nerchinsk Treaty and determined the further prospects for interstate, cultural, scientific and trade relations, and secured the consent of the Qing government to the presence of the Russian Spiritual Mission in Beijing."

The Nerchinsk prison was gradually gaining importance in Russian–Chinese trade in the 18th century. During that period, caravans from Russia through Tobolsk, Irkutsk, Selenginsk, Kyakhta and other settlements would reach Beijing.

Founded in 1727, the Kyakhta trading settlement was rapidly developing. In 1730, on the Mongolian territory, opposite Kyakhta, the Chinese "Trading Town" was created. At the end of the 18th century, Kyakhta became the center and the only legal point of Russian-Chinese trade that successfully developed at the turn of the century. During the reign of Catherine II, in order to strengthen the Russian merchants, it was allowed to conduct permanent trade in Kyakhta for six Central Russian trading companies and merchant tradesmen who were engaged in the supply of furs from Alaska and Kamchatka. At the beginning of the 19th century, the Kyakhta customs fees accounted for about 60% of all Asian trade [22, p. 214]. The Main export item of Kyakhta trade was traditionally fur. However, at the turn of the 18th-19th centuries, the share of furs decreased, while the export of cloth, leather, iron products and live cattle increased. Among the imported goods, silk and cotton fabrics prevailed, in addition, tea, sugar, and tobacco were in great demand. At the beginning of the 19th century, tea accounted for about 30% of the cost of all Russian imports from China [19, p. 163]. During that period, due to the Napoleonic wars, trade between European countries became more complicated, which was one of the reasons for the activation of trade between European countries with China. Russian fairs served as an intermediary for European trade with China. Transit trade (mainly European cloth) accounted for half of the value of exported goods (55%) [26, p. 136]. Thus, caravan trade with China was an important component of Russia's trade turnover.

Caravan trade was actively conducted with Turkey from Tiflis through Kars to Erzerum. According to S. M. Bronevsky [3, p. 211], the Tiflis customs collected duties in the amount of more than 96 thousand rubles in silver only in 1805–1807. The small trade turnover was attributed to the complexity and insecurity of caravan trips, as well as political instability in the region. It should be noted among the difficulties of developing trade relations with the East, the desire of eastern merchants not to use commodity–money relations, and instead to offer tea as a measure of value. In this regard, the aspirations of the Russian authorities to look for more reliable sea routes to expand trade relations with rich Asia, in particular with China, are clear.

Returning to the geography of Russia's trade relations in the Mediterranean basin, attention should be paid to the obviously not the most important trade direction, but showing the wide range of Russian trade relations in the 18th–19th centuries. The first political contacts of Russia with one of the states of the far north of the African continent, Tunisia, belong to the period of the reign of Catherine II and Tunisian Bey Hamuda Pasha (1782–1814). At that time, Russia was fighting to strengthen its positions in the Mediterranean and Black Seas. F. Ushakov's naval victories strengthened the country's position but did not make the area calm and stable for trade. Nevertheless, Russia's victories in the Russian-Turkish wars contributed to the growing interest of Russia in developing trade relations with African countries. The representative of Holland, A.G. Nissen, volunteered to mediate in establishing contacts with Tunisia. The desire to take a mediating role was not accidental, the position of a consul was prestigious and profitable. In addition, the practice of that time shows that it was possible to be a representative of several states. So, Nissen was Consul in Tunisia in addition to the Netherlands, Germany and Belgium. Nissen asked the envoy of the Russian Empire to the Ottoman port, A. Stakhiev, to support his initiative to represent the interests of Russia in Tunisia. His request was granted, moreover, the Dutch consul and his descendants continued to hold the post of Russian non-staff consuls in Tunisia from 1803 to 1912 [10].

5 Conclusion

The extensive historiography of Russia's foreign trade in the 18th and early 19th centuries testifies to the great scientific interest in the problems of the development of various areas of international trade, points to a wide range of scientific areas related to trade: state, legislative, customs policy, organization of the trade process, goods exchange items, trade routes, private business and the like. The Russian market played a prominent role in a world trade throughout the centuries. In Viking times Russia was a source of furs and slaves for the Arabian Caliphate. But later the trade with Novgorod came to be a vital element in the German merchants' activities, the Hanseatic League's counting-house in Novgorod was an important rallying-point.

The foreign trade policy of Russia during this period was regulated and directed by the state power represented by the emperors. The policy of mercantilism and protectionism promoted the expansion of the geography of foreign trade. The historically established trade relations with the countries of the West, Scandinavia, and the East were used, and new routes were opened up, to Africa, for instance. However, the development of foreign trade was greatly influenced by the international situation and foreign policy. Priority directions for Russia in the 18th and early 19th centuries were European countries, England, Holland, and German lands among them. The most active trade went through the Baltic and White Seas. Although, by the end of the 18th century, the trade turnover of the White Sea was declining. The instability of foreign trade relations due to military conflicts and foreign policy should be noted. A similar situation occurred eastward. The main partner of Russia was China; trade traditions were rooted in the early history of Russia. Caravan trade with Asian countries developed. At the turn of the century, there were searches for new opportunities in the development of trade infrastructure, construction of roads, and ports.

Both in the 18th century and the 19th century, the Baltic Sea was central to European trade. At the beginning of the 19th century, trade there accounted for 63.6% of all Russian foreign trade, and 88.5% of trade along the European border. The main place in that direction was occupied by the ports of St. Petersburg and Riga.

Thus, the geography of Russian foreign trade in the 18th and early 19th centuries covered almost all continents. Traditional historical ties with European and Asian countries were supplemented by the development of trade relations with the states of America and Africa. The strengthening of economic contacts testified to the expansion of Russia's international political and diplomatic ties, its active foreign policy, and the establishment of Russia's status as a great state.

It is obvious that foreign trade policy is an important part not only of the economic and foreign policy of the state. It also contributes to the development of cultural and scientific ties. The geography of Russia's trade routes, and the centuries-old history of their formation provide extensive material for scientific research. Along with traditional European and Asian trade routes, Russian merchants explored Africa and America, which is less studied and remains to be studied by representatives of various scientific disciplines.

References

1. Bereznitsky, S.V.: Caravan Trade of Russia with China and Domestic Science of the 18th Century, pp. 7–8. MAE RAS, St. Petersburg (2017).(in Russian)
2. Bolotina, N.Y.: I am going for my trade. Chinese Goods in Russia in the 18th Century, vol. 4, pp. 172–210. Historical Archive (2006). (in Russian)
3. Bronevsky, S.M.: The latest geographical and historical news about the Caucasus. S. Selivansky's Typogr. **1**, 210–211 (1823). (in Russian)
4. Dyomkin, A.V.: British Merchants in Russia in the 18th Century. Direct-Media, Moscow; Berlin (2019). (in Russian)
5. Western European merchants and their goods in Russia in the 17th century. In: Dyomkin A.V. (ed.) Collection of Documents, pp. 12–13. Institute of Russian History RAS, Moscow (1992) (in Russian)

6. Zakharov, V.N.: Three centuries of Russian foreign trade – from the White Sea to the Black Sea. In: Shpyrko, O.A., Khapayeva, V.V. (eds.) Lazarev Readings. Black Sea Region: History, Politics, Geography, Culture: Collection of Materials of the 17th International Scientific Conference "Lazorev Readings", pp. 12–16. Sevastopol (2019). (in Russian)
7. Zlotnikov, M.F.: Kontinentalnaya blokada i Rossiya [Continental blockade and Russia]. Nauka, Moscow-Leningrad, pp. 13–14 (1966). (in Russian)
8. Zmerzly, B.V.: State measures to restore the foreign trade of Crimea in the 1780s. In: Transactions of Vernadsky Crimean Federal University. Legal Sci. 2 (68), 4, 15–20 (2016). (in Russian)
9. History of Denmark. Busk, S., Poulsen, Kh. (eds.), Ves' Mir, Moscow, pp. 262–263 (2007). (in Russian)
10. Kazdagli, N.: Russian-Tunisian relations at the initial stage: the era of Catherine II and Hamuda Pasha in Tunisia. In: Bulletin of the Volgograd State University. Series 4: History. Regional studies. Int. Relat. 22(2), 119–127 (2017). (in Russian)
11. Kulisher, I.M.: History of Russian Trade and Industry. Sotsium, Chelyabinsk (2003). (in Russian)
12. Lobas, Y.V., Savchenko, M.M.: Foreign Trade and Customs Policy of the Russian Empire in the Early 19th Century. EPU of the Russian Customs Academy, Moscow (2017).(in Russian)
13. Müller, L.: Trading with Asia without a colonial empire in Asia: Swedish merchant networks and chartered company trade, 1760–1790. In: Antunes, C., Polónia, A. (eds.) Beyond Empires: Global, Self-organizing, Cross-imperial Networks, pp. 1500–1800. Brill, Leiden (2016)
14. Petrov, A.Y.: Russian-American Company: Activities in Domestic and Foreign Markets (1799–1867), pp. 59–70. Institute of General History RAS, Moscow (2006). (in Russian)
15. Pokrovsky, V.I.: Trade history in Russia. In: Brockhaus and Efron Encyclopedic Dictionary, 33a, pp. 561–588. Izdatelskoye delo, St. Petersburg (1901). (in Russian)
16. Pokrovsky, V.I.: Essays on the History of External Russia, p. 27. Frolov M.P., St. Petersburg (1902). (in Russian)
17. Russian State Archive of Ancient Acts. F.35. Op.1. D.41. L.47. (in Russian)
18. Russian State Archive of Ancient Acts. F.169. Op.3 D.2715. L7. (in Russian)
19. Romanova, G.N.: Trade relations between Russia and China in the late 17th and early 20th centuries (overview). In: Foreign East: Issues of the History of Trade with Russia, p. 163. Vostochnaya Literatura RAS, Moscow (2000). (in Russian)
20. Sokolov, I.A.: Tea, Sugar, and Coffee Merchants: Directory, vol. 2. Sputnik, Moscow (2017). (in Russian)
21. Struve, P.B.: Russia's Trade Policy. Sotsium, Chelyabinsk (2007). (in Russian)
22. Trusevich, K.I.: Russian Embassy and Trade Relations with China (Until the 19th Century). Malinski T, Moscow (1882).(in Russian)
23. Edict "On prohibition of imports of silk, wool, paper and other goods; about the reference to the court of customs officials and Ministers, contributed to the secret importation of prohibited goods evenly and those who brought them, and the impact perpetrated a denunciation of the prohibited import in the award all captured goods". Edict signed by the czar given to the Senate. June 26, 1789. In: Complete Collection of Laws, 23 (16781), 42–45. (in Russian)
24. Edict "On the suppression of the import into Russia from foreign lands, in the attached list of names". Edict signed by the czar given to the Senate. April 8, 1793. In: Complete Collection of Laws, 23(17111), 414–417. (in Russian)
25. Uspensky, V.L.: From the History of Russian-Chinese relations in the 17th Century (According to New Documents in Mongolian), 3(33), 10–17 (2012). (in Russian)
26. Khokhlov, A.N.: Kyakhta trade and its place in the politics of Russia and China (20s of the 18th century – 50s of the 19th century). In: Tikhvinsky, S.L. (ed.) Documents Refute: Against Falsification of the History of Russian-Chinese Relations: Collection of Articles. Mysl, Moscow (1982). (in Russian)

27. Chulkov, M.: Brief History of Russian Trade. Ponomaryov Printing House, Moscow (1788). (in Russian)
28. Sharonova, V.G.: Tea Story Box: Essays on the History of Russian-Chinese Tea Trade. Centre for Humanitarian Initiatives, Moscow-St. Petersburg (2017). (in Russian)
29. Shkunov, B.N.: Trade of the Russian Empire with Scandinavian Countries in the 18th and 19th Centuries. Ulyanovsk State University (2011). (in Russian)
30. Shkunov, B.N.: Trade Relations of the Russian Empire with the States and Peoples of Latin America in the 18th and 19th Centuries. RUSAINS, Moscow (2018).(in Russian)
31. Shkunov, B.N.: State and Legal Regulation of Foreign Trade of the Russian Empire in the 18th and 19th Centuries. IPR Media, Saratov (2018). (in Russian)
32. Yuzefovich, T.P.: Russia's Agreements with the East: Political and Trade: Turkey, Persia, China, Japan. LENAND, Moscow (2017).(in Russian)
33. Van Dyke, P.A.: Merchants of Canton and Macao: Politics and Strategies in Eighteenth-Century Chinese Trade. Hong Kong University Press, Hong Kong (2011)
34. Yang, L.: The Dutch East India Company's Tea Trade with China: 1757–1781. Brill, Leiden (2007)

The Role of Svalbard in the Policy of the Soviet Union and Norway Amid the Cold War

Sergei Nabok(✉)

St. Petersburg State University, St. Petersburg, Russia

Abstract. Nowadays the Arctic region is becoming an intersection point for many states' interest, with its strategic importance for the economic, transport, and military fields constantly increasing. The Svalbard archipelago plays a special role in the region, since for historical reasons it has a unique legal status. While maintaining the sovereignty of Norway, other countries enjoy considerable legal opportunities for performing economic activities, provided that the peaceful status of the archipelago is not violated. However, growing strategic importance of Svalbard forces Norway and other countries of the region to look for opportunities to use it, including for solving security problems, which gives a significant rise to regional tensions. This paper analyzes the historical dynamics of the Arctic policy of Norway, Russia, and other countries towards Svalbard, as well as present-day contradictions resulting from the specifics of its legal status and a new role it is playing in the modern international relations. The conclusion has been drawn about crucially important role played by reliable guarantees of presence on the archipelago in ensuring Russian strategic interests.

Keywords: Svalbard · Arctic · Arctic policy

1 Introduction

The Svalbard archipelago is of strategic importance for the Arctic countries – for Russia and Norway in the first instance, as well as for a wider range of states and international structures interested in both developing the region's economic resources and using it for security goals. The historical reasons, primarily the Svalbard Treaty of 1920 [1], predetermined the unique international legal status of the archipelago, implying the possibility of performing economic activity by the participating countries under maintained Norwegian sovereignty.

Historically, Demilitarization of the Svalbard, as the one of the central guidelines set in the Svalbard Treaty, preordained the current state of affairs in the region. The ninth article of the Treaty reads as follows: "Subject to the proviso related to the rights and obligations that may arise for Norway resulting from it joining the League of Nations, Norway undertakes not to establish or permit the establishment of any naval base in the areas referred to in Article 1, and not to build any fortifications in these areas, which should never be used for military purposes" [1].

© The Author(s), under exclusive license to Springer Nature Switzerland AG 2021
R. Bolgov et al. (Eds.): *Proceedings of Topical Issues in International Political Geography*, SPRINGERGEOGR, pp. 58–66, 2021.
https://doi.org/10.1007/978-3-030-78690-8_6

However, Norway was unable to comply with the article 9 of the Svalbard Treaty during the Second World War. Therefore, in 1944 the Soviet government raised the issue of revising the Treaty before the government of Norway. At the same time, shared governance of Svalbard (by creating a condominium) was proposed, as well as establishment of a Soviet military base on the archipelago, which would secure marine approaches to the Soviet territorial waters in the Western Arctic sector.

The Norwegian government agreed to revise the Treaty, which was put into writing in a joint statement in April 1945; however, due to V. M. Molotov being preoccupied, the convenient time for revising the Svalbard Treaty of 1920 was missed, and Norway flatly refused to consider this proposal. This issue was never raised again by the Soviet Union.

At the same time, Norway was preparing for the beginning of Svalbard militarization, which was introduced into practice in summer 1971, when construction of the Svalbard-Longyearbyen airport began on the archipelago.

The issue related to this construction was solved in full compliance with the fundamental articles of the Svalbard Treaty of 1920, since the Norwegian government informed all the states that had joined the Treaty by that time that it was planning to start construction of a civil airport on the archipelago. Now it seems that in response to this information, the Arktikugol Trust should have insisted on a participation interest in this construction and subsequent joint operating of the airport.

This option seems rather realistic, since at that time Norway was still quite a poor country, and Norwegians would have listened to the opinion of the Arktikugol Trust, since it was seen as the Soviet Union representative. Moreover, there were only three major land users in Svalbard – two Norwegian companies and the Trust. What's even more important, the relationship between the Trust and Norway was more than friendly. These relations level can be proven by the fact that in August 1981 Norwegians participated in the celebration events related to the 50th anniversary of the beginning of the economic activity of the Arktikugol Trust [2, p. 100].

The common knowledge is that the negotiations took place in the atmosphere of full mutual understanding and with the parties sincerely aiming at reaching mutually acceptable solutions. But it wasn't like that. Moreover, not only some issues remained open, but were obviously discriminatory.

For example, as early as when discussing the rules of operating the new airport, questions immediately arose about the number of Aeroflot representatives in Longyearbyen. Thus, according to E.G. Magerov, the representative office of Aeroflot was initially limited to 6 people only. However, the Norwegian side demanded that this number also included members of Soviet representatives' families, while the period of stay of family members in Longyearbyen was limited to 30 days during the calendar year, which automatically resulted in the requirement for monthly replacement of Soviet personnel at the airport [4].

Moreover, Soviet airliners could stay at the airport no more than 1 h and 20 min. This was not enough to unload the delivered cargo and passengers, collect luggage and register Soviet citizens departing from the archipelago, as well as to manage to perform an aircraft maintenance and issue all the necessary documents for departure. However, it is not surprising that these questions were not checked and clarified at that time, since the very possibility of direct delivery of Soviet miners and cargo along the route Moscow-Murmansk-Svalbard was extremely tempting.

Especially since before the airport was built, Soviet An-12 cargo aircraft could only fly to Svalbard during winter season. For that, a temporary airstrip in the Adventdalen area was used, which was every winter prepared by the Norwegian coal company Store Norske Spitsbergen Kulkompani specifically for receiving mail and cargo delivered to the archipelago by Brotenseif airline.

But even despite the above-mentioned inconveniences, and many of them being obviously discriminatory, Aeroflot aircraft delivered miners to their motherland and brought urgent cargo to the archipelago. After perestroika began, this airline was closed, and nowadays Russian citizens can get to the archipelago only by meeting traffic from Murmansk – on ships going to Svalbard or by Norwegian airlines via Norway [2, p. 149].

Thus, the history of international activity in the Svalbard region during the Soviet period was characterized by a rather tough struggle for national interests, primarily between the USSR and Norway. Changes of the international geopolitical situation that initially resulted from the collapse of the USSR, and further from other factors such as the rise of China, establishment of the European Union, which were greatly reinforced by valuable natural resources discovered in the Arctic zone and growing potential of the Northern sea route, made the transformation of the leading parties' approaches to the identification of the stature of Svalbard problem and to the policies of protecting their national interests in this area almost inevitable.

2 Research Objectives

The objectives of the current research are to analyze the interests of Russia and Norway on the Svalbard archipelago, retracing their evolution and development; characterize the policy of both countries in the region and find the grounds for its change; study the influence of the third parties in determining the current international significance of the archipelago.

3 Method of Research

To assess the content and results of the Russian, Norwegian, and other countries' policy in relation to the Svalbard archipelago and adjacent territories, methods of political analysis have been used aimed at identifying the goals and interests of the key parties that influence the regional international situation. Analysis of the documents content and expert opinion was used as direct research methods. The evidence base of the analysis is formed by official documents and statutory acts of the states, texts of international treaties, speeches of officials and policy makers, as well as expert appraisals by independent specialists and researchers.

4 Results of Research

The special character of the international situation and foreign policy in the archipelago region were largely determined by the potential and influence of Russia in the context of both the available economic resources and other countries' policies. At the initial stage, in the 1990s, the general weakness of the Russian government resulted in the intensification of Norwegian presence and control in the region.

The Norwegian authorities demonstrated the utmost activity on restricting the Russian rights on the archipelago and in this Arctic region in general became especially obvious since 1992, when the acting Head of the Russian government E.T. Gaidar officially declared: "The North is unprofitable!" Probably, Norwegians understood it as a clear sign that Russia either didn't need the unique Arctic regions at all, or very soon wouldn't use them.

However, E.T. Gaidar did not manage to consider that after the collapse of the USSR, the country's territory was significantly deformed, which couldn't but affected the direction of new Russia's national interests, which became even more northern country than it used to be before perestroika.

This happened against the background of the ongoing, almost systematic reduction of the Russian presence on the archipelago. The Norwegian government, on the contrary, made greatest efforts to strengthen and expand its positions and powers both on the territory of Svalbard itself and in the adjacent seas. Moreover, this happened not only in the Svalbard area.

For example, during 1997–2001, the Russian Arctic sector was visited by 52 foreign research vessels that worked there for 330 days. Rather plausible forecasts about the future of Russia in the Arctic region were made by experts at the first Russian Scientific and Historical Conference related to the 130th celebration of Arkhangelsk-Murmansk shipping line, which was held in Murmansk on May 27–28, 2005 [5, p. 66].

The tough idea of these predictions was published in Russkiy Vestnik (Russian Bulletin) weekly newspaper in materials by A.B. Rogozyansky "Russia may lose the Arctic". The author writes: "The territorial claims of other countries, the rapid outflow of residents from the Extreme North regions, and the loss of its own well-established infrastructure as a result of the shortfall policy – these are three problems that may kick Russia out of the Arctic countries list within the next 7–10 years" [6].

Meanwhile, the policy pursued by Norway is characterized by the fact that almost all anti-Russian activities are carried out behind the screen of Articles 2 and 6 of the Treaty, i.e. in the fishing waters of Svalbard covered by environmental slogans, and on the archipelago, militarization is undergoing under the aegis of establishing international research centers.

Thus, back in 1997, the Norwegian Parliament adopted a program for 30-year-long development of Svalbard, for which a special Center for environmental protection of the archipelago was established, which included six state and public organizations, which made it possible to perform all the Center's activities at the government level. Already in the late 1990s, the total amount spent on developing a scientific center in Svalbard exceeded NOK 500 million. Of course, these were not only Norwegian state investments, since within a year's time, about 130 research and development projects were registered on the archipelago, with 14 foreign countries participating in their development [7, p. 40].

Out of Norwegian international scientific activities on the archipelago the Russian government should have been particularly concerned about the obvious multi-purpose targeting of majority of the developed topics and the clear fact that Norwegian own funds were not enough to establish such undoubtedly dual-purpose scientific centers on the territory of Svalbard, which can be clearly seen even from a brief list of objects that are already operating on the archipelago:

1. In the New-Alesund area, there is a rocket range with a landing pad for heavy aviation, a space communication center, a radar system, missiles warehouse and an assembly building, there is also a mobile platform that enables to launch diverse kinds of geodesic missiles, as well as a telemetry station to control space targets. This station uses its own communication system to stay connected to similar facilities and a special NATO system;
2. Radar system ASAT located in Longyearbyen – the administrative center of Svalbard – and officially focused on studying anomalous atmospheric circumstances. However, capacities of these stations are far wider. They are capable to measure flight trajectories of land-based transcontinental ballistic rockets and ballistic missiles launched in the Arctic from rockets submarines of the Northern navy;
3. SvalSat satellite communications center (also located in Longyearbyen), which is capable of detecting and round-the-clock monitoring of all satellites in polar orbits. The center's premises are situated not only on the archipelago, but in Norway (Tromso) as well, and can receive information from satellites in polar orbits. The data from space objects can be received 14–15 times a day (every 1.5 h on average). SvalSat's particular advantage is its favorable geographical location, since it takes only one or two days to deliver personnel and cargo from the mainland, unlike the US stations in Antarctica, where it take up to six weeks to replenish the supplies. Moreover, it is one-of-a-kind object in the world, which according to military experts can influence natural atmospheric processes and in fact can be used as a meteorological weapon. SvalSat employs about 40 people [5, p. 67–68].

On January 31, 2004, the fiber-optic high-speed communication line between Longyearbyen (Svalbard) and mainland Norway was officially brought into use. The line consists of two 1,400 km-long cables. The project cost amounted to NOK 350 million (about USD 50 million). This work was mostly funded by NASA Space Agency and US weather service. By bringing this line in use, the importance of Svalbard was increased as a center for receiving information from satellites in polar geostationary orbits and its further transmission to consumers in the USA, Europe, and Japan. The dual purpose of the above-listed objects is obvious, so the issue of Norway complying with Article 9 of the Svalbard Treaty of 1920 is rather topical [8, p. 44].

The above-listed systems are capable of controlling the situation in the North-Western districts of the Russian Federation, including the White and Barents seas. Just as much as supervising the bases of the naval strategic nuclear forces of Northern fleet's and the space station in Plesetsk, following all the routes of Russian submarines and surface ships both in the Atlantic and Pacific oceans and even in the North pole areas.

It is also worth mentioning that the waters that wash Svalbard are the place of active performance of control systems, within which American nuclear submarines conduct "scientific" research supported by US Navy patrol airplanes Orion, based at the Longyearbyen airdrome. Moreover, one of the so-called submarine defense identification zones starts right off the North Cape. Further it passes along Medvezhy island shores and ends at the Svalbard archipelago. This zone is directly connected with Stave station located next to the Norwegian airbase Andoya.

In 2003, Norway also started preparing and launching missiles from the Svalbard proving ground New-Alesund. The first launch was performed back in January 2004. The launch was carried out along an optimal trajectory towards Greenland at a range of about 1 thousand km. According to Major General Vitaly Dubrovin – an expert in the field of space missile defense – this project aimed at "adjusting the complex military program of US radars in the Arctic." This assumption is based on the missile flight trajectory, whose "route" passed the responsibility area of American radars located in Northern Europe and Greenland. By including the latest radar system on Greenland into this system the USA managed to build up a solid radar "belt" in the North, with its capabilities being strongly supported by another powerful radar Globus-2 located in Vard (40 km from the Norwegian-Russian border) [5, p. 72].

Here, it would appear that the former headquarters commander of the Northern fleet, Vice Admiral M. Motsak was right when claiming that: "…Norway and its NATO allies strive… to limit the Russian presence in the Barents sea and in the Arctic in general as much as possible" [9]. For the first time, these Norwegian actions on militarization of the archipelago under the guise of scientific activities were assessed by V.V. Putin back in 2002 during his visit to Norway. Then, the Russian President clearly stated that our country allows rigid requirements for nature protection regarding the Arctic environment.

However, as far as activities on the archipelago in general are concerned, we are committed to a constructive dialogue between our countries, provided that Norwegian legislative activities concerning Svalbard are in full compliance with provisions of the Svalbard Treaty of 1920 [10].

Unfortunately, neither Russian Ministry of Foreign Affairs nor other state authorities took further stringent measures after this statement, with the active measures on militarization of the archipelago being continued, especially regarding various military maneuvers and training exercises, which are usually held in the immediate vicinity of the Russian borders.

The most illustrative examples of such exercises are the following large-scale military maneuvers: Strong resolve - 1998, Barrens peace – 1999 and Strong resolve - 2002. The operational background chosen for these exercises looks extremely interesting, since it implies that two bordering states – Blue land and Lime land – come into a military conflict. The Yellows are holding a demand against their opponents considering resources located on the contested territory of continental shelf that incorporates petroleum deposits and fishing lands. Withal, the Blues is a "respected state with modern democracy and developed interstate relations, including membership in NATO", while the Yellow is an aggressor, "an unstable state with a developing economy" [11, p. 125].

From a historical perspective, the "color" of the parties involved in these exercises looks provocative, because it is a common knowledge that since the Norwegian struggle for independence from Sweden, for the Norwegians blue has been a symbol of the national freedom, with yellow always being the color of their enemies. Therefore, the very version of the maneuvers performed next to the Russian borders implied a situation that was hardly acceptable in the context of good neighborhood with Russia.

According to the scenario, NATO protects its Blue ally and carries out a rapid military operation in full compliance with Article 5 of the North Atlantic Treaty "On collective defense" and the resolution adopted by the UN Security Council, which condemned intentions of the "Yellow" aggressor that is potentially threatening to the whole world.

After the end of the maneuvers, Norwegian Defense Minister K. Devold wrote in Aftenposten newspaper that "military training exercises in Norway, especially in the Northern part of the country, should be made more attractive to NATO member States", as this district "has all the prerequisites to test the real combat qualities of military personnel in poor weather and geographical conditions" [12].

In 2007, when Norway fulfilled a extensive NATO military operation Bold avenger – 2007 on its land, 13 NATO member countries such as Romania, Czech Republic and Poland took part in this exercise, with more than 100 units of flying machines were involved, in particular air tankers, long-range radar detection and electronic combat planes [13, p. 185].

In retaliation, on September 3–4, 2007, 12 strategic bombers of the Russian Federation flew to the northern latitudes. However, this was a single demonstration of force, but the Norwegians continue performing such exercises and maneuvers almost on a regular basis [14]. Thus, it appears that Svalbard is becoming NATO's outpost in the Barents Sea. Moreover, it should be considered that it is not only from a military point of view that the Russian Arctic has long been of interest to the USA. For example, the US National Strategy for the Arctic Region, passed in May 2013, implies maintaining the required level of economic and scientific activity in the Arctic region [15]. Moreover, the National Security Strategy adopted in the USA in 2010, directly indicates that there are good reasons to provide American business with an access primarily to the Russian resource reserves in its North and especially in the Arctic [16].

5 Discussion of Research

Therefore, despite the fact that the Svalbard Treaty of 1920 states that Norway undertakes not to use the specified territory for military purposes, the country failed to comply with this condition, which led to the issue of the necessary revision of the Treaty by the countries, with probable strengthening of the USSR's influence on the archipelago, but the convenient time was missed by the Soviet Union, and Norway refused to consider this proposal forever.

The militarization of Svalbard began with the construction of Svalbard-Longyerbyen airport: however, in legal terms Norway did not contravene the regulations of the Treaty. Later, this assumption led to the open militarization of the archipelago, which however is carried out guised by establishment of the international research centers, with financing of Norwegian projects involving participation of many foreign states.

The Russian short-fall policy regarding the Extreme North has made it possible for Norway to strengthen its positions to such an extent that currently there are many multipurpose facilities located on Svalbard that are capable of fully controlling the situation in the Russian border areas.

All this makes it clear that Norway and its NATO allies are trying to bound Russian attendance on Svalbard and in the Arctic region in the overall as much as possible. Rapid militarization activities continue, with various military maneuvers and exercises being conducted next to the Russian borders.

6 Conclusions

To sum up all the above-stated, it can be inferred that from an economic and especially a defense point of view, the loss of Svalbard will have disastrous consequences for Russia. The most unbiased opinion on this matter can be found in a published work by N.N. Yakush: "Voluntary and truly insane, from the point of view of geopolitics, retreat from Svalbard opens up unprecedented opportunities for NATO in terms of approaching the Russian borders. However, undoubtedly the most important thing, if this happens, is that the most significant Northern fleet will be blocked and partially cut off from the world ocean, since the only exit that will remain free will be under the ice of the North pole" [17, p. 68]. That is why today the issues of a reliably guaranteed Russian presence on Svalbard seem much more relevant than it was in 1912, when V.A. Rusanov managed to save the archipelago for Russia. It was hard to imagine that early in the 21st century, an era would begin when Russia could be displaced by Norway from the Russian lands of Svalbard, developed and inhabited by the coast-dwellers. Thus, the question is now more acute than ever: whether Russia will maintain its presence in Svalbard or be pushed knocked out by Norway and its allies.

References

1. Treaty between Norway, The United States of America, Denmark, France, Italy, Japan, the Netherlands, Great Britain and Ireland and the British overseas Dominions and Sweden concerning Spitsbergen 09.02.1920. http://www.lovdata.no/traktater/texte/tre-19200209-001.html. Accessed 29 May 2013
2. Mikhailov, I.A.: The Svalbard archipelago – conjunction of events and destinies, 1st edn. Nauchny Mir, Moscow (2004)
3. Ivanov, G.K.: Good traditions. In: Shirokov, V.L. (ed.) Half a Century at the Pole: Collection of Memoirs, Murmansk, pp. 144–156 (1983)
4. Magerov, E.G.: The mainland becoming closer. In.: Shirokov, V.L. (ed.) Half a Century at the Pole: Collection of Memoirs, Murmansk, pp. 136–139 (1983)
5. Fedorov, A.F., Zhbanov, A.V., Zhbanova, D.A.: What the Arctic ice is client about: the Arctic rush, Kharkov (2012)
6. Rogozjansky, A.B.: Russia may lose the Arctic. Russ. Bull. **3**(8), 7–8 (2005)
7. Hough, P.: International Politics of the Arctic: Coming in from the Cold, 1st edn. Routledge, New York (2013)
8. Umbreit, A.: Svalbard: Spitzbergen, Jan Mayen, Frank Josef Land, 3rd edn. Globe Pequot press, Guilford (2013)
9. Simonov, V.V.: A new redrawing. Our Power Doings Faces **2**, 12–14 (2006)
10. Scheglov, K.V.: In the Russian strategic interests zone. Russ. Fed. Today **5**, 7–10 (2003)
11. Archer, C., Joenniemi, P.: The Nordic Peace, 1st edn. Routledge, Aldershot (2003)
12. Magnus, O.: Flere «militærturister» til Norge. Aftenposten **5**(2) (2003)
13. Inzhiev, A.A.: Fighting for the Arctic. Will the North become Russian? 1st edn. Eksmo, Moscow (2010)
14. Aleksandrov, Y.V.: The Arctic triangle. Financial Control **6**(103), 14–16 (2010)

15. National Strategy for the Arctic region of the USA 10.05.2013. http://www.whitehouse.gov/sites/default/files/docs/nat_arctic_strategy.pdf. Accessed 20 June 2013
16. National Security Strategy of the USA 01.05.2010. http://www.whitehouse.gov/sites/default/files/rss_viewer/national_security_strategy.pdf. Accessed 20 June 2013
17. Yakush, N.N.: Russian national interests on the Svalbard archipelago. In: The Russian North in the System of Russian Geopolitical Interests, 1st edn. Russian Academy of Civil Service, Moscow, pp. 51–55 (2002)

Review of the Training of Grassroots Officials Since the CPC Became the Governing Party

Xing Xu(✉)

Zhou Enlai School of Government, Nankai University, Tianjin, China

Abstract. Since the Communist Party of China (CPC) became the ruling party in 1949, it has always attached great importance to the training of grassroots officials. From the very beginning of the founding of New China, the Party has been engaged in the training of officials at all levels. By 1966, the CPC's training of officials was aimed at taking over cities, establishing and improving government at all levels, and adapting to large-scale socialist construction. Entering a new era of reform and opening-up, the CPC has further strengthened its training of grassroots officials. With the development of the situation, the main content of training for grassroots officials has also been constantly updated, with the focus of training shifting to the training of new personnel in the "Four Standards" (revolutionary, youthful, intellectual and professional) and to improving the leadership, management and professional and technical skills of the majority of grassroots officials, and the training of officials has gradually become standardized and institutionalized. The two stages before and after the reform presented different functions and features depending on the background and content of the CPC's training of officials and achieved different results and experiences. It has important inspiration and guiding significance for the future training of CPC officials and the strengthening of ideological construction.

Keywords: Communist Party of China · Grassroots officials · New China · Government at all levels · Training of officials · Ideological education · Theoretical study · Operational training · Early years of the People's Republic of China · Reform and opening-up

1 Introduction

The education and training of officials is a preliminary, fundamental and strategic project for building the high-quality officials, and is also an important measure and a necessary long-term task for the CPC, as the ruling party, to strengthen its ideological construction. Since the founding of the People's Republic of China in 1949, the Communist Party of China has attached greater importance to the education and training of grassroots officials since becoming the ruling party. The training of grassroots officials in the Communist Party of China is roughly divided into two stages, the pre-Cultural Revolution period and the post-Reform and Opening-up period, with the objectives, content and effectiveness of the training varying at each stage depending on the background and tasks involved. This

© The Author(s), under exclusive license to Springer Nature Switzerland AG 2021
R. Bolgov et al. (Eds.): *Proceedings of Topical Issues in International Political Geography*, SPRINGERGEOGR, pp. 67–81, 2021.
https://doi.org/10.1007/978-3-030-78690-8_7

paper will examine and study the training of grassroots officials since the CPC became the ruling party, discuss its achievements and characteristics, and analyze its experience and inspiration for the training of officials in today's CPC and Chinese government. The grassroots officials in this paper refer to officials up to the levels of county and regiment within the Communist Party of China and the Government of China. This group is relatively large, and its work is directly related to the implementation of the various policies of the CPC and the Government of China. The performance of officials directly affects the image of the Party and the Government.

2 The Development of the CPC's Training of Grassroots Officials in the Pre-cultural Revolution Period

At the beginning of New China, when all kinds of work were waiting to be done, the CPC not only suffered from a shortage of grassroots officials, but also from a low level of leadership and education of them. The CPC, as the ruling party, felt an urgent need to train and rapidly improve the theoretical level, professional and technical skills and cultural qualities of most of the grassroots officials.

From the founding of New China to the pre-Cultural Revolution period, according to the different educational levels and job requirements of the grassroots officials, the CPC mainly conducted training in two ways: firstly, a large number of grassroots officials with a low level of education were mobilized and sent to literacy classes, evening schools, part-time schools, or part-time study and training in short-term training courses or accelerated education for workers and peasants; secondly, a number of grassroots officials were sent to primary Party schools, cadre schools, military schools and universities for off-the-job training, while higher-level professional and technical officials were selected to receive training in the Soviet Union and Eastern European countries. The duration of study for the various categories of officials varied from a few years to a few days, depending on the requirements of jobs and departments.

On 15 October 1950, the CPC Central Committee forwarded an instruction on the *Report of the 1st National Conference on Education for Workers and Peasants*, requesting that all localities should hold quick-entry secondary schools and remedial schools for workers and peasants, and that they should enroll workers and peasants who were over 18 and under 35 years of age, had a level of education equivalent to primary school, and had participated in the revolution for at least three years. The Report also criticizes the lack of attention to the training of officials in some regions, where only "surplus personnel and those officials who are difficult to manage are introduced to schools, while good officials are not sent to study [10, pp. 433–435]". To meet the needs of the country's modernization, in 1952, during the nationwide reorganization of higher education institutions, 12 new specialized industrial institutes for iron and steel, geology, water conservancy, mining and metallurgy, machinery, post and telecommunications, railways, and others were set up, and a group of industrial and agricultural officials and grassroots officials of the army were recruited to study in these institutes, so that they could acquire a certain degree of scientific and technological knowledge and, after graduation, assume important leadership positions in administration and technology.

To implement the First Five-Year Plan, the Second National Conference on Organizational Work in October 1953 pointed out that the cadres needed for Party and mass work in industrial construction should be trained accordingly. The existing industrial institutes of higher education and technical secondary schools throughout the country must be systematically expanded and strengthened in order to train many new technical officials [12, p. 521]. On 24 November 1953, the CPC Central issued *Decision on the Central Deployment of Officials, Unification and Reform of Pre-Existing Technical Personnel and Mass Training of Officials*. The Decision states: "If we do not take all possible measures to train a large number of officials for industrial construction and do not actively seek the training of new technicians and specialists among the workers and young intellectuals of the revolution, we will not be able to move forward [12, p. 571]". In July 1955, at the Second Session of the 1st National People's Congress, it was also proposed to train talents for various types of construction and to improve the political and theoretical skills, operational knowledge and technical level of in-service officials. On 1 February 1956, the CPC Central Committee issued *Directive on Strengthening the Work of Primary Party schools*. It affirmed that in the previous period, Party schools in provinces and cities "generally trained a large number of officials in the form of large-scale short-term training sessions". It was also decided that, in order to raise the work of the Party schools and gradually formalize them, all provincial and municipal Party schools should henceforth be renamed primary Party schools. Its task was mainly to train key members of the CPC junior leadership in rotation. "The primary Party school training is for all of the following kinds of Party members with a middle school education: (1) in the Party Committee system: members, department directors and deputy department directors of county Party Committees; secretaries, deputy secretaries and members of district Party Committees; (2) in the governmental system: deputy county chiefs, district chiefs and deputy district chiefs; (3) in the mass organizations: secretaries and deputy secretaries of county Committees of Communist Youth League; presidents and vice-presidents of Federations of Trade Unions; presidents and vice-presidents of Women's Federations; (4) officials equivalent to each of the above-mentioned positions in the various departments of the systems [13, p. 107]".

In accordance with the instructions of the CPC Central Committee, primary Party schools have been established in provinces, cities and counties since the mid-1950s. Various systems and industries, even some enterprises also set up training institutions for their own officials, and held short professional training sessions of varying lengths. At that time, education and training of officials was the joint responsibility of the Organization Department and Publicity Department of the Party Committees at all levels. Localities were responsible for organizing training for grassroots officials below the county level, while local Organization Department and Publicity Department jointly formulated rotational training and training plans for officials, and local Party schools and cadre schools were responsible for carrying out training tasks. It was clearly stated in 1957 that in order to build socialism at an early date, the working class must have its own technical officials, "this task should be largely resolved within the next 10–15 years [8, p. 550]". Entering the 1960s, the CPC continued to attach great importance to the training of grassroots officials. 23 August-16 September 1961 the CPC Central Committee held a working meeting at Lushan to discuss and adopt decisions such as

Decision of the CPC Central Committee on Rotational Training of Officials. "It was decided to conduct a general rotational training for leadership officials of the whole Party at all levels and in all areas by means of short-term training sessions [14, p. 608]". The purpose of rotational training is to help officials further understand and grasp the objective laws of socialist construction, to overcome the one-sided understanding and leftist or rightist errors among them, to raise their political and ideological level, and to build socialism with greater, faster, better, and more economical results.

In addition to centralized training on a full-time or part-time basis, another major method and approach used by the CPC to train grassroots officials at that time was to require them to participate in political campaigns, productions and various kinds of social works while undergoing training, and a large number of them were sent to the front line of productions and constructions to learn operational knowledge. In 1953, the CPC Central Committee stated in *Decision on the Central Deployment of Officials, Unification and Reform of Pre-Existing Technical Personnel and Mass Training of Officials*: "Each economic sector should, in accordance with the principles of tightening the upper management bodies and strengthening the lower production and basic construction units, with the exception of those who are indispensable personnel, the officials of the remaining organs should be sent to work in factories and mining enterprises as far as possible. This will not only solve the problem of the shortage of cadres in factories and mining enterprises but will also enable these officials to accumulate experience in leading basic constructions and productions. It will help to strengthen the economic leading authorities in the future [11, p. 569]".

Prior to 1966, the CPC trained grassroots officials in the following ways: firstly, it conducted political training, i.e., education and training in the guidelines, lines and policies of the CPC. In the light of the various political campaigns carried out at the time, the officials were organized to study the instructions and resolutions of the Party Central Committee and to grasp its spirit in order to lead the masses in implementing it. Secondly, theoretical training, i.e., instructing officials to selectively read the works of Marx, Engels, Lenin and Stalin, the history of the All-Union Communist Party and the Selected Works of Mao Tse-Tung, in order to improve their Marxist-Leninist theory level. Thirdly, educational training, i.e. illiteracy elimination and remedial training in general knowledge of science for grassroots officials with poor educational backgrounds, in order to improve their education and self-cultivation. Fourthly, training in professional skills, i.e. arranging for and requiring officials with a certain level of education to study modern science and technology in order to adapt to the needs of large-scale modernization construction, and to reserve cadres for the long-term development of the country. In addition, there was also education on revolutionary traditions and ideals, current affairs, etc.

The various systems also trained officials in professional skills to meet the requirements of their respective professions; for example, the industrial system at the time stipulated that officials with less than middle school educational level and technical officials who were workers, were to be raised to the level of middle school graduates in Chinese language, mathematics, physics and chemistry, and that officials who already had middle school education were to be raised to the level of high school or vocational school graduates.

In order to enhance the effectiveness of the training, an evaluation of the trained officials was conducted at the end of each course, and graduation examination was held at the end of the training session. Training institutions kept detailed records of each participant's learning and provided feedback on the assessment of the officials' training to the higher-level organizations and personnel departments as well as the former organizations of the participants. The document of the Central Committee stipulates that "the results of tests and examinations must be recorded and filed. Good or bad academic performance should be one of the important conditions for knowing and employing the officials [13, p. 110]".

3 Strengthening the Training of Grassroots Officials by the CPC Since the Reform and Opening-Up

Cultural Revolution of China, which caused a decade of turmoil, also dealt a severe blow to the training of grassroots officials, with a large number of grassroots officials being assaulted and criticized; Party schools, cadre schools and socialist institutes at all levels being closed down, causing heavy losses to the building of the CPC cadres.

Since the beginning of the reform and opening-up, the CPC has entered a new era in the training of officials at all levels. At the beginning of the reform process, Deng Xiaoping raised the issue of training new recruits in accordance with revolutionary, youthful, intellectual and professional standards, and accelerated the building of cadres with high level of leadership and governing ability. He stressed that in the new era, officials should be educated and trained to become "officials with ideals, morals, education and discipline [20, p. 205]". Chen Yun summed up the lessons from the history of selecting successors both in China and abroad and clarified the criteria for selecting and cultivating young and middle-aged officials, which are "both integrity and ability". Their correct ideology has become the guideline for the building and training of officials in the new era.

From the late 1970s onwards, Party schools at all levels were successively restored and newly built in various parts of the country, and the training of grassroots officials was intensified and increased. On 3 October 1982, the CPC Central Committee and the State Council issued *Decision on the Education Work of the Central Party and Government Organs*, which proposed that: "In order to adapt to the new situation, it is necessary to systematically reform the governing bodies and the cadre system, to strengthen the education and training of officials, to achieve revolutionization, youthfulness, intellectualization and professionalization of the officials [15, p. 106]". The Decision also proposed that education of officials should be included in the national education plan, and that in the future, all officials of the central organs should participate in rotational training in batches and phases, and to link the officials' training with the appointment.

Since the reform and opening-up, the Party has adopted a variety of methods and approaches for training officials, including training in Party schools, learning in central groups of Party Committees, tutorial by lecturer groups, and theoretical training seminars, and other effective forms such as newspapers, radio, television and the Internet. The main contents of the training of officials during this period were: basic Marxist theory education, Party spirit education, ideological and political education, and studies of the modern science, technology, and management. Deng Xiaoping made it clear: "The fundamental thing is to study Marxism-Leninism and Mao Zedong Thought, and to try to combine the universal principles of Marxism with The Four Modernizations of China have been integrated into concrete practice. At present, most officials should also focus on three areas of study: economics, science and technology, and management. Learn well so that to lead a high speed, high level of socialist modernization construction. Learn from practices, learn from books, learn from your own and others' lessons [19, p. 153]".

Training and Education of Young Officials. The Organization Department of the CPC Central Committee then issued the *Opinions on the Implementation of the Central Committee's Decision on Stepping up the Training and Education of Young Officials.* The Central Committee has decided that: "Over the next five years, theoretical training will be provided to all types of young officials in phases, in batches and at different levels [16, p. 247]". Young officials in organs at the provincial and prefectural levels shall be sent to Party schools at the same level for full-time rotational training; the duration of each rotation shall not be less than three months. Other young officials will be trained at cadre schools or training sessions organized by the respective systems. Young officials from enterprises, institutions and mass organizations will receive full-time or part-time training organized by the relevant systems.

At the turn of the century, the CPC launched a series of concentrated efforts among officials at all levels to study Deng Xiaoping theory, train them in professional skills and knowledge, and the important thought of Three Represents. On 24 June 1998, the CPC Central Committee issued the *Opinions on Deepening the Study of Deng Xiaoping Theory in the whole Party*, which requires "all Party members with a certain level of education and reading ability to work on the original works of Deng Xiaoping. Party members with low level of education should be given easy-to-understand explanations. Party schools at the county (city) level are responsible for the training of Party branch secretaries and major theoretical workers, and grassroots Party schools are responsible for the rotational training of Party members in rural areas, enterprises, institutions and neighborhoods". Under the unified planning of the Central Committee, hundreds of thousands of officials at and below the county level participated in Deng Xiaoping Theory training nationwide after the 15th National Congress.

Since the beginning of the new century, under the unified deployment of the CPC Central Committee, the officials at all levels throughout the country have made the study of important thought of Three Represents, the education of maintaining the advanced nature of Communist Party members, the implementation of the Scientific Outlook on Development, the building of learning party, and the study and implementation of Xi Jinping Thought on Socialism with Chinese Characteristics for a New Era the themes of the training and education work.

With the rapid development and changes of the world situation and the deepening of the great practices of China's reform and opening-up, the new situation requires building the CPC into a Marxist learning party, improve the Party's ideological and political level and comprehensive quality. In January 2007, the CPC Central Committee issued *National Education and Training Plan for Officials 2006–2010*, and issued a notification requesting all regions and organs to focus on Party and government officials, according to the principle of classification and full training, to ensure the implementation of education and training for Party and government officials, enterprise management personnel and professional and technical personnel [9]. The purpose of the training is to help officials improve the level of scientific, democratic and law-based governance as well as to develop the skills to promote the economy and society for better and faster growth. In 2009, the 4th Plenary Session of the 17th CPC Central Committee adopted *Decision of the CPC Central Committee on Major Issues Pertaining to Strengthen and Improve the Party Building in the New Situation.* The Plenary Session made specific plans for how to build the learning party organizations, proposing to create a strong atmosphere of learning throughout the Party, actively turning to books study, learning from practices and masses, optimizing the knowledge structure, improving the comprehensive quality, increasing innovation, so that the party organizations at all levels become the learning party organizations, and the leadership groups at all levels become the learning leadership groups. In the future, theoretical accomplishment and learning ability will be an important basis for the selection and appointment of leadership officials. Party schools, Academies of governance, cadre schools and the national education system will fully play important roles in building the Marxist learning party [3].

Since the 18th CPC National Congress, more emphasis has been placed on the training of grassroots officials from the central to local levels. In February 2010, the Central Committee convened the 11th National Joint Meeting on Education of Officials, to actively promote reform and innovation in the education and training of officials in accordance with the requirements of building the learning party, so as to enhance relevance and effectiveness and better serve scientific development and the healthy growth of officials [21]. The CPC Central Committee with Xi Jinping at the core, has also specially formulated and issued *Outline of the Reform of Education and Training of Officials 2010–2020*, which is an overall plan for the reform and development of education and training of officials in the next decade. General Secretary Xi stressed: "All regions and sectors should take into account the actual situation, conscientiously implement the work, enhance the relevance and effectiveness of the reform and innovation of the education and training of officials". In order to better promote the training of officials and to reform and innovate the form and content of training, in 2015 the CPC Central Committee issued Regulations on the Education and Training of officials, emphasizing that in cadre training, it is necessary to rule by law, strict management, conscientious implementation of "The Three Guidelines for Ethical Behavior and the Three Basic Rules of Conduct". The examination of officials' learning attitudes, performance, and results must be strengthened, and the training assessment must be used as important basis for officials' appointments, promotions and annual evaluations.

In the 19th CPC National Congress report, General Secretary Xi Jinping specifically addressed the "continuous improvement of the CPC governing ability and leadership level" and "building high-quality and professional officials". He particularly emphasized that "one must be politically competent as well as highly skilled [23]," and set stricter standards for leadership officials of the new era with both integrity and ability, also pointed the way forward for the future education and training of CPC and governmental officials at all levels.

4 Effectiveness and Characteristics of the Party's Training of Grassroots Officials Over the Past 70 Years

Since the CPC became the ruling party, the training of officials at the grassroots level over the past 70 years has made glorious achievements in terms of improving the quality and increasing the number of officials, as well as in terms of reserving talents and training young officials. At the beginning of the New China, thanks to the Party Central Committee's timely efforts to train a large number of officials with high political ideological integrity, with a certain degree of administrative talent and basic theoretical knowledge and education which enabled the rapid establishment and consolidation of people's power at all levels and the construction of socialism in China. From 1950 to 1955, more than 30,000 worker and peasant officials throughout the country attended remedial training courses, accelerated schools for workers and peasants, and various levels of Party schools where they have received more formal training and education, and more than 20,000 professional officials have been trained in institutions of higher education throughout the country. More than 1.2 million officials were educated in various types of specialized fields over the years, and by 1956, there were 3.84 million professional officials in the fields of science and research, education, engineering and technology, health, culture and arts [7]. They made an indelible contribution to the socialist revolution and the socialist construction. After 1957, due to the interference of ultra-leftist ideology, the training of officials in the CPC was affected to some extent, but until the Cultural Revolution was fully launched, the CPC continued to organize rotational training for officials at all levels; more than a decade before the Cultural Revolution, through the training of grassroots officials, the Party cultivated the officials of "red and expert" with a firm political stance and a certain level of education, which played an important role in China's major social transformations and large-scale modernization.

Most of the grassroots leadership officials trained before 1966 were strong in ideology and maintained a hard-working and honest tradition of excellence. Of course, the training of officials was influenced by the overall social environment of the time, and the training of officials showed the marks of the times. In many cases, the training of officials at that time revolved around political movements. The interference of ultra-leftist ideology, the singularity and deviation of training methods, and the lack of training in such subjects as laws and regulations, administrative management, professional skills and leadership abilities were clearly inadequate in the training of officials during this period. These problems have also provided the Party with valuable experience in defining the correct purpose of training and in improving the content and methods of training after the reform and opening-up.

In the new period of reform and opening-up, the Party has achieved remarkable results in training officials at the grassroots level throughout the country; according to incomplete statistics, from 1982 to 1987, 7.34 million officials were trained through various forms of training, of whom more than 1.15 million obtained college degree, more than 620,000 – vocational education and more than 560,000 – high school education [6, p. 266]. From 1993 to 1996, some 21 million officials nationwide participated in various forms of off-the-job learning. Since the end of the last century, the Party's cadre education and training system has been gradually taking shape, with an orderly turnover of new and senior officials. The cultivation of young and middle-aged officials who with both integrity and ability and who meet the "Four Standards" has kept the leadership of the CPC and the government of China at all levels continuously updated with new knowledge and vigor, it ensures China's political stability and sustained economic development.

In the new century, the CPC's training of grassroots officials has reached a new stage; according to preliminary statistics, from 2003 to 2006, some 19 million Party and government officials, 11 million enterprise management personnel and 38 million professional and technical personnel participated in various types of off-the-job training nationwide [1]. Through large-scale training, the officials have clearly enhanced their conscientiousness and firmness in implementing the CPC's routes, guidelines and policies, and have improved their theoretical and cultural qualities and working abilities, providing strong ideological and political guarantees, human resources and intellectual support for the great and fast development of the economy and society.

Since the 18th CPC National Congress, and especially since the 19th CPC National Congress, the awareness of the importance of officials' training, and the innovation of training methods and content have reached a new height. The training of officials has also entered a new phase. In June 2014, the CPC Central Committee issued *Opinions on Strengthening and Improving the Training and Selection of Outstanding Young Officials*, in order to optimize the path of officials' training, education and selection. Adhere to strict training and assessment and focus on cultivating and testing officials at the grassroots level and in areas of hardship. In the five years from the 18th to the 19th CPC National Congress, 974,000 civil servants were recruited through examinations nationwide, and more than 50,000 outstanding college graduates were selected and recruited. Publicly recruited 21,000 civil servants, of which more than 1,300 at the grassroots level were selected by the central organs. Many talented officials have entered the Party and government organs, greatly enhancing the vitality and vigor of the civil service. According to People's Daily report in 2017, "Since the 18th CPC National Congress, a total of more than 84 million officials at all levels, enterprise management personnel and other categories of officials have been trained nationwide [5]".

Comparing the two different historical stages before and after the reform and opening-up, it is not difficult to find that the CPC cadre education and training work has become increasingly standardized, institutionalized, legalized and scientific.

First of all, compared to the time before the reform and opening-up, the Party's training of grassroots officials has been moving in the direction of standardization and institutionalization, especially by entering the new century, it has gradually been incorporated into the legal system. Before the Cultural Revolution, the CPC Central Committee had issued some instructions on the training of officials and education in Party schools, but there was no formal planning or mechanism in place. After the reform and opening-up, the CPC Central Committee began to attach importance to this issue, and in 1983 it issued *Decision on the Formalization of Education in Party Schools*, and in 1984 the Publicity Department of the CPC Central Committee formulated *Regulations on the Formalization of Marxist-Leninist Theoretical Education for Officials*, proposing that the training and education of officials should gradually become regular, formalized and institutionalized. Since the 1990s, the Central Committee has attached greater importance to this issue, and has issued the *Key Points of the National Training Plan for Officials 1991–1995*, *National Education and Training Plan for Officials 1996–2000* and *National Education and Training Plan for Officials 2001–2005*, which set out detailed arrangements for the tasks, objectives, requirements and specific measures for the training of officials at all levels throughout the country. In July 2002, the CPC Central Committee promulgated the newly revised *Regulations on the Selection and Appointment of Party and Government Leadership Officials*. It further improves the necessary procedures for the selection and appointment of officials and specifies one of the qualifications for promotion to Party and government leadership positions: "Shall be trained by the Party schools, Academies of governance or other training institutions accredited by the organization (personnel) department for a cumulative period of three months within five years. If the above requirements for training are not met before promotion due to special circumstances, the training shall be completed within one year after promotion [17, p. 2449]".

Since the beginning of the new century, the CPC has incorporated the training of grassroots officials into the legal system in accordance with changes in the situation and the requirements of reform. On 27 April 2005, the 15th Session of the 10th National People's Congress adopted the *Civil Servant Law of the People's Republic of China*, which specifically stipulates that "civil servants who are promoted to leadership positions shall undergo training within one year before or after taking up such positions [2]". In January 2006, the CPC Central Committee promulgated the *Regulations on Education and Training of Officials (for Trial Implementation)*, which formally set out the guiding ideology, basic principles and main contents of officials' training work in the form of internal Party regulations. This is the first legislation on officials' education and training in the history of the CPC and is the basic regulation on officials' education and training in the new situation. At the end of 2006, the CPC Central Committee issued the *National Education and Training Plan for Officials 2006–2010*, which, in addition to making comprehensive arrangements for the training of senior and middle-level officials, also proposed that full use be made of channels such as local Party schools, Academies of governance, cadre schools and distance education, and that efforts be made to increase the education and training of rural grassroots officials (mainly township), urban grassroots officials (mainly neighborhoods), and grassroots officials in enterprises and institutions (mainly small and medium-sized enterprises and institutions). The content of the training of grassroots officials in rural areas, urban neighborhoods, and enterprises and institutions focuses mainly on education and training in the implementation of policies, accelerated development, serving the masses, acting in accordance with the law, strengthening social administration, carrying out community services, developing community undertakings, ensuring the reform, development and stability of enterprises and institutions, and giving full play to the role of the political core of Party organizations [9]. Since the 18th CPC National Congress, the promulgation and implementation of such regulations as the *Outline of the Reform of Education and Training of Officials 2010–2020* and the *Regulations on Education and Training of Officials* have highlighted the fact that the CPC has moved towards the legalization of the training of officials.

Secondly, compared to the period before the reform and opening-up, the institutional set-up and management system for officials' training has been more complete, and there has been a marked improvement in both the construction of training bases and the organization of teachers. In the 1950s and 1960s, there were not many training bases for CPC officials, but they were severely disrupted during the Cultural Revolution. After the reform and opening-up, Party organizations at all levels built a large number of new Party schools, cadre schools and training centers on the basis of the successive restoration of closed Party schools, and equipped them with the necessary facilities, such as classrooms, dormitories and canteens, and rebuilt the teaching staff. According to incomplete statistics, by 1987, more than 2,800 Party schools had been established at the central, provincial, municipal and county levels, and more than 2,200 cadre schools at the county level and above, with more than 90,000 full-time teachers [4, p. 204]. While the construction of training institutions for officials has begun to bear fruit, the management system for training of officials has also been gradually improved, with officials' education and training work leading groups or joint meetings of officials' education being established in various localities, forming a management system under the unified leadership of the Party Committees, with the Organization Departments in charge and the relevant departments dividing responsibilities among themselves. After the 16th CPC National Congress, in order to train officials on a large scale, the Central Committee decided to create new national training bases for officials, and three China Executive Leadership Academies were established in Pudong, Jinggangshan and Yan'an. In recent years, Party schools and cadre schools at all levels have been continuously recruiting newly graduated doctoral and master's degree holders to join the teaching staff, making the faculty of officials' training greatly enriched. At present, the system of officials' training management has taken shape in China, from central to local level, in which Party schools and cadre schools at all levels are basically rational in layout, with a clear division of tasks and respective responsibilities.

Lastly, compared to the time before reform and opening-up, training in the new era is more scientific, the content and methods of education are constantly being updated, and the forms of training are also more flexible, diverse and innovative. Before the Cultural Revolution, the training of officials in the CPC was relatively simple and inflexible in terms of both contents and forms. Since the 1980s, the CPC officials' training methods have been characterized by diversification and scientification. Party organizations at all levels, while rehabilitating Party schools and cadre schools, have made use of the various kinds of training methods that were then emerging, such as the Radio and TV Universities, evening universities, radio and correspondence courses, and self-study college entrance examination. After entering the new century, the CPC Central Committee paid timely attention to the issue of updating teaching contents. In 2002, the Organization Department, Publicity Department of the Central Committee and other 13 organizations established the National Steering Committee for the Compilation and Review of Officials' Training Materials. By 2006, two batches of 27 teaching materials had been produced, covering political theory, administrative management, leadership science, science and technology, business and finance, social security, law, history, literature and arts, playing an important role in training officials to adapt to new situations and solve new problems. With the development of society, the Party's approach to the training of officials has been constantly updated. In accordance with the requirements of the *Regulations on Education and Training of Officials (for Trial Implementation)* of the CPC Central Committee, the methods of education and training of officials should in future be based on the needs of economic and social development. It is necessary to "make comprehensive use of such method combinations as organizing training and self-selection of studies, off-the-job training and on-the-job self-study, in-country and out-of-country training, so as to comprehensively improve the quality and capacity of officials [18]". In recent years, the CPC Central Committee has adopted a variety of methods for training officials at all levels, including central group study, off-the-job centralized training, on-the-job self-study, special lectures, self-selected studies, television and telephone presentations, Internet seminars, field trips, MPA degree education, and sending officials abroad for professional training; these flexible and diverse methods have made training of officials more scientific.

5 Conclusions

Since becoming the ruling party, the CPC has attached great importance to the training of officials at all levels and has accumulated much experience in its 70 years of practice. At a symposium held at the China Executive Leadership Academy in Pudong in October 2010, General Secretary Xi Jinping made the following statement on how to further improve the education and training of officials under the new situation: "These successful practices should be adhered to for a long time in the light of the new reality. At the same time, it is necessary to adapt to new changes in world conditions, national conditions and Party conditions, and actively promote the reform and innovation of officials' education and training methods, so as to enhance the vitality of officials' education and training [22]". The report to the 19th CPC National Congress put forward new requirements for "building a high-quality and professional cadre force", which has become the guideline for future education and training of officials at all levels. If we carefully examine the history of the CPC's training of officials at the grassroots level over the 70 years since the founding of New China, we can sum up the following experiences and inspirations:

Firstly, the education and training of officials must be closely focused on the development needs and work priorities of the CPC Central Committee. This is the most fundamental requirement for the education and training of officials. The training of officials at the grassroots level must firmly grasp the correct political direction and guiding ideology, adhere to the direction of socialist school running, and always be subordinated to and serve the overall development of the CPC Central Committee and the Government of China. All trained officials should have a profound understanding of and implement the CPC Central Committee's lines, principles and policies, cultivate their firm political beliefs and correct political direction, and enhance their ability to distinguish between right and wrong. At present, the training of grassroots officials must comprehensively implement Xi Jinping Thought on Socialism with Chinese Characteristics for a New Era, take The Four Comprehensives of the Party Central Committee as the guide, focus on enhancing the consciousness and ability to govern, and provide a guarantee of grassroots management personnel for the acceleration of socialist modernization, to achieve the goal of training a large number of high-quality Party and government leadership personnel, enterprise management personnel, as well as professional and technical personnel at all levels through the raining of officials.

Secondly, the training of grassroots officials must pay close attention to theoretical learning. It is necessary to make the grassroots officials truly comprehend and understand Marxism-Leninism, Mao Zedong Thought, Deng Xiaoping Theory, the important thought of Three Represents, the Scientific Outlook on Development and Xi Jinping Thought on Socialism with Chinese Characteristics for a New Era, to help them establish the correct worldview and the idea of serving the people wholeheartedly, and to improve their theoretical cultivation and moral standards. "To do well in the education and training of officials under the new situation, it is necessary to highlight Marxist theoretical education, especially the education of theoretical system of socialism with Chinese characteristics and the Party spirit education, and focus on improving the ideological, political and moral quality of officials". It is necessary to learn to apply the Marxist positions, viewpoints and methods to guide practice and promote the Party's work. To cultivate the awareness of public servants and service consciousness among Party members and officials, to establish the correct vision of values, power and interests, and to enhance the cultivation of Party spirit among members of the Communist Party.

Thirdly, the training of officials in the Party must link theory with practice and apply what they have learned. Emphasis should be placed on the quality of training, and the relevance and effectiveness of training should be enhanced in order to improve the ability of grassroots officials to solve practical problems and deal with emergency situations. Practical training in front-line work is essential to the healthy growth of officials, and practice is the best training classroom for officials. At present, the focus should be on training and improving the leadership, management, professional and technical skills of the grassroots officials. The training of grassroots officials in operational knowledge should focus on strengthening the knowledge necessary for the performance of their duties and improving their ability to carry out those duties. Targeted training should be provided to improve officials' capacity to judge situations scientifically, to manage the market economy, to deal with complex situations and to govern in accordance with the law. At the same time, operational training should keep pace with the updating of knowledge, promote the updating of officials' knowledge structures, improve their knowledge systems, and improve their overall quality to adapt to the needs of the times. In the future, "We should make better use of the great practical classroom of reform and opening-up and socialist modernization construction, take the good experiences and approaches in the practice as the fresh teaching materials for officials' education and training, and organize officials-in-training to study at the front line of this great practice, at the most grassroots level of this great practice, and among the masses who are the main force of this great practice [22]".

Fourthly, the Party's training of officials should continue to move in the direction of legalization and formalization, further integrating training resources, optimizing training teams and improving the training system. At present, in China the main rules and regulations governing the training of official have been promulgated, but there is still a long way to go in terms of further improvement and implementation. In the future, it will be necessary not only to continuously improve the level of education and training of officials, but also to further improve the cadre management system and to implement the lawful administration. In order to strengthen the construction of educational and training institutions for officials, on the one hand, specific issues such as the types and status of training institutions, institutional settings, the scope of training, teachers, teaching materials, teaching methods and approaches, as well as the management of scientific research, should be clearly defined by law, so that the training of officials can enter a formalized and legalized track. On the other hand, it is necessary to continue to promote distance education, online learning and Internet education, actively create conditions for the training of officials at all levels and create a favorable learning environment. At the same time, it will further deepen the system of examinations and evaluations, as well as the combinative system of training and practice, so as to effectively carry out the tasks of training officials on a large scale.

Fifthly, "It is necessary to adapt to new changes in world conditions, national conditions and party conditions, and actively promote the reform and innovation of officials' education and training methods, so as to enhance the vitality of officials' education and training [22]". The education and training of officials should adhere to the innovative spirit of advancing with the times, the grassroots officials training methods, approaches and contents need to be based on the development and changes in the situation, to carry out continuous exploration, reform and innovation, so that officials in the ideological, theoretical, operational capacity, management level to keep pace with the development of the times. For example, how to learn from foreign advanced management models and training methods? How to combine the training of officials with the building of Marxist learning party? How can the inputs and benefits, quantity and quality of training be harmonized? How can training be made more scientific and modern? How to explore more effective ways to promote officials' awareness and capacity to govern? How can the relevance and effectiveness of education and training for officials be effectively improved? How to provide differentiated training for officials of different categories, levels, ages and experiences? How to enhance the training and education of grassroots officials so that they can better lead the masses in successfully achieving the Two Centenary Goals proposed by the CPC Central Committee? These are more pressing and urgent issues to consider in the future.

References

1. China's education and training of officials has risen to a new level since the 16th National Congress. (in Chinese). www.gov.cn/jrzg/2007-08/06/content_707931.htm
2. Civil Servant Law of the People's Republic of China. (in Chinese). http://www.npc.gov.cn/npc/c12488/201812/7a8739d2c6e146ddb3acf29b27336562.shtml
3. Decision of the CPC Central Committee on Major Issues Pertaining to Strengthen and Improve the Party Building in the New Situation. People's Daily, 28 September 2009. (in Chinese)
4. Deng, C., et al.: History of the Development of Deng Xiaoping's Thought on the Education of Officials. Party Building Books Publishing House, Beijing (2002). (in Chinese)
5. Education and training of officials have risen to a new level since the 18th CPC National Congress. People's Daily, 1 January 2017. (in Chinese)
6. Li, X. (ed.): A Brief History of the Education of Officials of the Communist Party of China. Communist Party of China History Publishing House (2009). (in Chinese)
7. Liu, W.: Historical experience in the building of cadres in the early years of the founding of New China. Contemp. China Hist. Stud. **2**, 38–45 (2006). (in Chinese)
8. Mao, Z.: The Situation in the Summer of 1957. In: Mao Zedong's Manuscripts Since the Founding of the Nation, vol. 6. Central Party Literature Press, Beijing (1992). (in Chinese)
9. National Education and Training Plan for Officials 2006–2010. (in Chinese). http://www.china.com.cn/policy/txt/2007-05/11/content_9252660.htm

10. Party Literature Research Center of the CPC Central Committee: Selected Important Documents Since the Founding of the Nation, vol. 1. Central Party Literature Press, Beijing (1992). (in Chinese)
11. Party Literature Research Center of the CPC Central Committee: Selected Important Documents Since the Founding of the Nation, vol. 3. Central Party Literature Press, Beijing (1992). (in Chinese)
12. Party Literature Research Center of the CPC Central Committee: Selected Important Documents Since the Founding of the Nation, vol. 4. Central Party Literature Press, Beijing (1993). (in Chinese)
13. Party Literature Research Center of the CPC Central Committee: Selected Important Documents Since the Founding of the Nation, vol. 8. Central Party Literature Press, Beijing (1994). (in Chinese)
14. Party Literature Research Center of the CPC Central Committee: Selected Important Documents Since the Founding of the Nation, vol. 14. Central Party Literature Press, Beijing (1997)
15. Party Literature Research Center of the CPC Central Committee: Selected Important Documents Since the 12th National Congress, vol. 1. People's Publishing House, Beijing (1986). (in Chinese)
16. Party Literature Research Center of the CPC Central Committee: Selected Important Documents Since the 13th National Congress, vol. 3. People's Publishing House, Beijing (1993). (in Chinese)
17. Party Literature Research Center of the CPC Central Committee: Selected Important Documents Since the 15th National Congress, vol. 3. People's Publishing House, Beijing (2003). (in Chinese)
18. Regulations on Education and Training of Officials (for Trial Implementation). People's Daily, 30 March 2006. (in Chinese)
19. Selected Works of Deng Xiaoping, vol. 2. People's Publishing House, Beijing (1994). (in Chinese)
20. Selected Works of Deng Xiaoping, vol. 3. People's Publishing House, Beijing (1994). (in Chinese)
21. The 11th National Joint Meeting on Education of Officials was held in Beijing on 27 February. (in Chinese). http://www.gov.cn/jrzg/2010-02/27/content_1543800.htm
22. Xi, J.: Do a good job of educating and training officials under the new situation. (in Chinese). http://www.ccps.gov.cn/xxsxk/xldxgz/201812/t20181223_126885.shtml
23. Xi, J.: Secure a Decisive Victory in Building a Moderately Prosperous Society in All Respects and Strive for the Great Success of Socialism with Chinese Characteristics for a New Era, Delivered at the 19th National Congress of the Communist Party of China. (in Chinese). http://www.xinhuanet.com/english/download/Xi_Jinping's_report_at_19th_CPC_National_Congress.pdf

National Policies and International Politics

Geography of International Activities of St. Petersburg Legislative Assembly of the Sixth Convocation

Anastasiia Zotova[1(✉)] and Sergei Poltorak[2]

[1] Saint-Petersburg State University, Universitetskaya nab., 7-9, 199034 Saint-Petersburg, Russian Federation
[2] Leningrad State University named after A. S. Pushkin, Petersburg Highway, 10, Pushkin, 196605 St. Petersburg, Russian Federation

Abstract. The article is dedicated to the study of political and geographical aspects of international activities of St. Petersburg Legislative Assembly of the sixth convocation, which operated between 2016 and 2020. The analysis is based on the documents kept in the Department of External Relations of the St. Petersburg Parliament. For the first time the scientific community is presented with statistical information about the delegations, which visited the parliament of St. Petersburg, discussion plans of deputies with foreign visitors and many other documents of international purpose. It is noted that during the deputies' activities of the Legislative Meetings of the sixth convocation, they had contacts with representatives of 52 countries. The article compares the priorities in international activities of modern parliamentarians with political and geographical priorities of the deputies of previous convocations. An attempt has been made to explain the reasons for the deterioration of parliamentary contacts with states in various regions of the world, including the Great Britain, Australia, Canada and Ukraine. Authors conclude that the deputies of the Legislative Assembly of St. Petersburg follow the official state policy of Russian Federation.

Keywords: Legislative Assembly of St. Petersburg · International relations · Political geography

1 Introduction

The Legislative Assembly of St. Petersburg is a city parliament. It dates back to December 1994. Its predecessor was the St. Petersburg City Council of People's deputies, which was formerly called the Leningrad City Council of people's deputies. The modern parliament of the city is an elected body of 50 deputies headed by the chairman, currently of the sixth convocation. This convocation was formed by the city residents' in September 2016. In addition to the legislative activities aimed at the development and functioning of St. Petersburg, the Legislative Assembly dedicates its efforts to international activities. These international activities of St. Petersburg parliament are based on the recommendations of the State Duma of the Russian Federation, Federation Council, Ministry of

© The Author(s), under exclusive license to Springer Nature Switzerland AG 2021
R. Bolgov et al. (Eds.): *Proceedings of Topical Issues in International Political Geography*, SPRINGERGEOGR, pp. 85–101, 2021.
https://doi.org/10.1007/978-3-030-78690-8_8

Foreign Affairs, Government of St. Petersburg, as well as at the initiatives of the deputies of the Legislative Assembly of St. Petersburg. Geography of international relations of the Parliament of St. Petersburg in September 2016–2020 allows to trace the dynamics of foreign policy priorities of both the government bodies of St. Petersburg as well as the Russian Federation as a whole.

The Legislative Assembly of St. Petersburg is located in the historical city-centre, in the building of the Mariinsky Palace. This palace was built in 1844 according to the blueprints of the outstanding architect A.I. Stackenschneider for the daughter of the Russian Emperor Nicholas I – Grand Duchess Maria Nikolaevna, Duchess of Leuchtenberg. The Mariinsky Palace is one of the dominants of St. Petersburg. Next to it is the fourth largest domed temple in the world – St. Isaac's Cathedral. The interiors of the palace are so unique that the state authorities of the Russian Federation sending foreign delegations, believe that the atmosphere of the palace is conducive for the emergence of trusting relationships and mutual understanding of politicians and representatives of different cultures, religions and countries.

In 25 years, the Legislative Assembly of St. Petersburg was attended by 405 foreign delegations representing parliaments and major inter-parliamentary organisations from 110 countries worldwide. During this period 16 international agreements, protocols, declarations and memorandums were signed. They regulate cooperation with foreign parliaments [1]. The city parliament is the initiator of meetings, excursions, conversations with a wide range of foreign visitors, including students, schoolchildren, businessmen, politicians, people of art and culture, and public organisations. There are 36 general consulates in St. Petersburg, 19 honorary consulates and one embassy branch. Deputies take part in protocol events of consular postings accredited in St. Petersburg, on occasions of national holidays and other memorable dates of various countries. One form of international cooperation is days, weeks and decades of culture of various countries and foreign cities in St. Petersburg, as well as days of St. Petersburg in other countries and foreign cities. These important political events are often attended by St. Petersburg parliamentarians. According to a long-standing tradition, the walls of the Mariinsky Palace hold international conferences, information meetings, round tables, which contribute to strengthening the relations between St. Petersburg with other countries.

2 Methods of Research

Thanks to the scientific methods such as analysis, synthesis, comparison and induction it was possible to process the main sources for the study of geography of international relations of the St. Petersburg parliament. Whilst writing this article, the authors used documents from the Office of External Relations of the Legislative Assembly of St. Petersburg, such as:

- statistical information about the delegations visiting the Legislative Assembly
- service letters and memos regulating the reception process of members of foreign delegations
- working correspondence between both Russian public authorities and its foreign partners on the goals and objectives of their visits

- visit programs
- plans of conversations of deputies with guests
- reports on receptions of delegations
- other documents of international character.

An important source for analysis of the deputies' actions in organising receptions with foreign partners is information about members of delegations, as well as about political, economic and cultural infrastructures of partner countries, prepared by specialists of the Committee for External Relations of the St. Petersburg Government.

One of the research methods was interviewing employees of the Legislative Assembly carrying out technical preparations of international meetings at the Mariinsky Palace. Also, these employees organise trips for the deputies to other countries with the purpose of strengthening international relations.

Using methods of historical periodisation as well as the comparative method, it was possible to trace the effectiveness of the deputies' work in establishing contacts with politicians of other countries, since the work of deputies of the sixth convocation was compared with the work of the deputies of the previous convocations. The geography of international contacts of the deputies of the study period and deputies of the Legislative Assembly of St. Petersburg, elected to the previous convocations was also analysed.

The methods of systematisation and generalisation made it possible to organise information about the international activities of St. Petersburg parliamentarians and identify the main trends in the development of national policy within the international sphere, as well as to identify the priorities of the international activities of politicians from other countries in the framework of relations with St. Petersburg and Russia as a whole.

3 Geography of International Contacts of the Legislative Assembly of St. Petersburg (2016–2020)

3.1 Priority Areas of International Activities of St. Petersburg Parliament of the Sixth Convocation

Throughout the period of the deputies' activity The Mariinsky Palace has been the centre of international events. The following events were held during this period:

- Informational meeting of the Chairman of the Legislative Assembly of St. Petersburg V.S. Makarov with heads of consular posts of the European Union member states accredited in St. Petersburg
- Signing a Memorandum of Understanding between the AP of St. Petersburg and Sukhumi City Assembly (Abkhazia)
- Information meeting of the Chairman of the Legislative Assembly of St. Petersburg V.S. Makarov with heads of consular offices of the CIS countries, accredited in St. Petersburg, and members of the Secretariat of the Council of the Interparliamentary CIS Assemblies
- Events to mark the 200th anniversary of the birth of the Duke Maximilian of Leuchtenberg de Beauharnais

- Meeting of the Council of the Parliamentary Assembly of the Treaty Organisation on collective security and the plenary session of the Parliamentary CSTO Assembly
- Meeting of the Association of Secretaries General of Parliaments Participants of the Interparliamentary Union at the 137th meeting Interparliamentary Union
- Meeting of the Chairman of the Legislative Assembly of St. Petersburg. V.S. Makarov with newly appointed consuls of foreign states accredited in St. Petersburg
- Signing a Memorandum of Understanding between the AP of St. Petersburg and Maslikhat of the city of Alma-Aty (Republic of Kazakhstan)
- Signing a Memorandum of Understanding between the AP of St. Petersburg and Bishkek City Kenesh (Kyrgyz Republic)
- Round table on the development of the urban environment with the participation of delegates to the Legislative Assembly of St. Petersburg and the delegation of Maslikhat of the city of Alma-Aty
- Round table on urban planning and innovation with the participation of delegates of the Legislative Assembly of St. Petersburg and the delegation of the Council city commissioners of Turku
- Round table on urban transport development with participation of delegates of the Legislative Assembly of St. Petersburg and the delegation of Kenesh of Bishkek City
- An extended meeting of the standing committee on ecology and environmental management of the Legislative Assembly of St. Petersburg with the participation of the delegation of the Council of City Commissioners of Turku
- Informational meeting of the Chairman of the Legislative Assembly of St. Petersburg V.S. Makarov with 11 general councillors of foreign states, accredited in St. Petersburg
- Meeting with a delegation led by the Director General of the Permanent International Secretariat of the Council of the Baltic Sea States Maira Mora. Reception took place in preparation for the anniversary events of the Council of the Baltic Sea States. In the spring of 2017, the Council was 25 years old.

It is of interest to analyse the geography of states with which the deputies collaborated during the sixth convocation. Based on the analysis of information about the meetings of delegations of the Legislative Assembly of St. Petersburg with delegations of other countries a rating of the highest and the lowest activity of international relations has been compiled (Table 1).

Thus, based on the authors' calculations presented in the table, People's Republic of China is the most active partner of the Legislative Assembly of St. Petersburg. Relations of China and St. Petersburg are truly diverse. Among them are meetings of St. Petersburg deputies with delegations from various provinces of the PRC, with high-ranking officials of China, including the deputies of the National People's Congress of Representatives, with functionaries of the Committee on Financial and Economic issues of the National People's Congress, as well as leaders of large joint projects. Among them is a meeting with the General Director of Baltic Pearl CJSC Mr. Hemin. In a short time, the St. Petersburg parliamentarians met with representatives of the leadership of this project three times. According to experts, the Baltic Pearl project is the largest Chinese investment project in the world [2]. It includes shopping, entertainment and business complexes of both urban and international level, hotels, restaurants, sports,

St. Petersburg is one of the largest centres of Korean studies in Russia. The Centre for Korean Language and Culture was opened at the Faculty of Oriental Studies of St. Petersburg State University in 1995, which is supported by government organisations of the Republic of Korea. The Centre is currently implementing research programs in the creation of new textbooks related to Korea, descriptions of ancient Korean manuscripts and woodcut prints held in St. Petersburg. The publication of the "Bulletin of the Centre for Korean Language and Culture" is being carried out in St. Petersburg. This centre holds scientific conferences and seminars, evenings of Korean culture, and organises annual examinations in the Korean language. The Centre has a rich reference library with a collection of audio, video and multimedia materials about Korea.

On June 14, 2016, the Second Political and Economic Forum was held at the St. Petersburg State University as part of the events of the Russian- Korean Dialogue Forum "Russia and the Republic of Korea in a Changing World".

Over a hundred cooperation agreements have been signed between the universities of St. Petersburg and Korea. In the 2017/2018 academic year, 50 Korean students studied at the universities of St. Petersburg.

At present, more than 4,000 Koreans live in St. Petersburg, there are two national public associations whose activities are aimed at reviving Korean culture and education of Koreans living in the city.

The development of international relations with two more partners, Azerbaijan and Germany, is encouraging. Relations between Russia and Azerbaijan have a long history; since several decades ago they were subjects of one state - the Soviet Union. Strong business and cultural relations developed between them still during the Soviet era. The Azerbaijani diaspora in St. Petersburg is currently one of the largest and most active. In addition, St. Petersburg and the capital of Azerbaijan, Baku, are linked by strong industrial and trade ties. Over the past four years, the parliamentarians of St. Petersburg have met 12 times with their counterparts from Azerbaijan: 7 times Azerbaijani parliamentarians were guests at the Mariinsky Palace and 5 times the deputies of the St. Petersburg Legislative Assembly have visited Azerbaijan. In Baku, they met with parliamentarians, representatives of the business community of Azerbaijan - heads of industrial, oil refining and agricultural enterprises, as well as with representatives of the Russian community of Azerbaijan [6]. St. Petersburg deputies visited the Baku Orthodox Cathedral of the Holy Myrrh-Bearing Women. In September 2017, deputies from St. Petersburg laid flowers on the grave of the President of the Republic of Azerbaijan, national leader Heydar Aliyev and visited the Heydar Aliyev Centre - a cultural complex opened in Baku in 2012. The deputies of the Legislative Assembly of St. Petersburg have developed good business and friendly relations with the chairman of the Milli Mejlis (parliament) of the Republic of Azerbaijan Oktay Asadov and with members of the working group on inter-parliamentary relations "Azerbaijan-Russia".

The quintessence of relations between St. Petersburg and Azerbaijan is expressed in the words of the Chairman of the St. Petersburg parliament V.S. Makarov: "Azerbaijan is one of the key partners of Russia in the Transcaucasus. This region is of great importance for the foreign policy of the Russian Federation. Only the best diplomats are capable of successfully defending the interests of our country here, finding the most successful, compromise solutions to the most difficult foreign policy problems. Employees of our embassy have more than a quarter of a century since the beginning of diplomatic relations between the two countries and have demonstrated the highest professionalism, wisdom and endurance. I am sure that thanks to their efforts, the relations between the two states will continue to develop, becoming ever closer" [7].

With all the reliability and stability of relations between the parliamentarians of St. Petersburg and Azerbaijan, one should not forget about several difficult moments in real politics. In particular, it is important that, unfortunately, an extremely tense situation has developed in relations between Azerbaijan and Armenia. Of course, the modern conflict has long historical roots. The specificity of the activities of the St. Petersburg Parliament is that they no less successfully cooperate with their colleagues from the Republic of Armenia. This is confirmed by the fact that only between 2017 and 2019, Armenian parliamentarians visited the Legislative Assembly of St. Petersburg 8 times. The St. Petersburg deputies have developed strong relations with the Council of Elders of Yerevan, which is the highest representative body of power in Armenia.

It is fair to say that a common historical fate lies at the heart of strong relations between the St. Petersburg parliament and the Council of Elders of Yerevan. In particular, in modern Armenia they still remember that the Russians and Armenians fought together during the Second World War against Nazism. On January 26, 2018, a monument to the Courage of Leningraders was unveiled in St. Petersburg. The opening ceremony was attended not only by the Governor, Chairman of the Legislative Assembly of St. Petersburg, city residents, but also by members of the delegation of the National Assembly of the Republic of Armenia headed by the Deputy Chairman of the National Assembly Eduard Sharmazanov [8]. The delegation included deputies of the National Assembly Hayk Babukhanyan and Romik Manukyan. The head of the Armenian delegation, which was in St. Petersburg at the invitation of the Speaker of the Legislative Assembly of the city, made a speech to the participants at the opening of the monument [9].

The creation of the monument became possible thanks to the financing of the famous Armenian benefactor G.M. Poghosyan [10]. The Armenian philanthropist, speaking at the opening of the monument, said: "It is not necessary to be born in St. Petersburg to love it with all your heart. I am incredibly happy to be connected to the history of this city and doubly happy because I had the opportunity to contribute to the chronicle of its life. <…> Thousands of children of the Armenia also fought against the enemy in the battle of Leningrad, and we still, once again, express our resolute negative attitude towards fascism and all evil on Earth. The defence of Leningrad is a common victory for our peoples; none of us have a right to forget this heroic feat" [11].

Receiving the Armenian delegation, the chairman of the St. Parliament V.S. Makarov drew attention to an important circumstance: "Our countries are linked by the closest friendly and allied relations. During the World War II, we fought shoulder to shoulder against the fascist invaders. The memory of the sons of the Armenian people who died in the battles for Leningrad will live forever in St. Petersburg. And we are glad that the deputies of the National Assembly of Armenia will participate in the celebrations on the occasion of our victory in Leningrad" [12]. Vyacheslav Serafimovich supported the initiative of the Armenian delegation to erect memorial plaques in St. Petersburg in honour of famous Armenian scientists, politicians and artists, whose life and work was strongly associated with St. Petersburg. In particular, the conversation turned to the establishment of a memorial plaque in honour of the prominent military leader and statesman Count M.T. Loris-Melikov - a hero of the Russian-Turkish war of 1877–1878. The head of the delegation of the Republic of Armenia invited the Chairman of the Legislative Assembly of St. Petersburg to come on an official visit to Yerevan on the occasion of the celebration of the 190th anniversary of the Turkmanchay peace treaty, as a result of which Eastern Armenia, having come out of the influence of Persia, had become a part of Russia.

Of course, our common historical past influences the relations between Armenia and St. Petersburg. The Soviet Union collapsed a long time ago, but many people, especially the older generation, continue to feel their commonality regardless of which country they live in now.

Thus, it can be argued that the St. Petersburg parliamentarians, while strengthening relations with Azerbaijan and Armenia, are doing everything possible to reconcile these two states. And this international mission in the modern complex conditions of the development of world civilization is extremely important.

Germany is known to be the most successful state in the European Union. St Petersburg has traditionally established strong international relations with this country. This is confirmed by the fact that the deputies of the 6th convocation at the official level met with their colleagues 12 times. It is impossible to disregard the fact that the relations between the parliamentarians of St. Petersburg and their German counterparts have recently been of one-way direction. Guests from various German states have visited The Mariinsky Palace 11 times, while the deputies of the Legislative Assembly of St. Petersburg have visited Germany only once. Recently, the St. Petersburg deputies met with delegations from Bavaria, Dresden and Hamburg. The deputies of the Free and Hanseatic City of Hamburg are the most frequent guests of the St. Petersburg parliament. Since 1995, cooperation between the Legislative Assembly of St. Petersburg and the Burgerschaft of the Free and Hanseatic City of Hamburg began. St. Petersburg and Hamburg are both port cities and major cultural and economic centres of the Russian Federation and Germany. Their direct contacts make a significant contribution to the development of state relations between Russia and Germany. In 2007, the Legislative Assembly of St. Petersburg and the Burgerschaft of the Free and Hanseatic City of Hamburg signed a new Cooperation Agreement between these two parliamentary institutions. The document was an updated version of a similar Agreement, adopted in 1996.

At the end of 2017, Germany became one of the three leading trading partners of St. Petersburg. Commodity turnover between St. Petersburg and Germany increased by 14% compared to 2016 and amounted to $ 3.7 billion. Export from St. Petersburg to Germany in 2017 compared to the previous year increased by 26% and amounted to $ 1.67 billion. Mineral fuels account for 87% of exports to Germany. Imports in 2017 increased by 6% and reached $ 2 billion. Germany supplies products to St. Petersburg for mechanical engineering, optical devices, electrical equipment, polymer materials and plastics, vehicles [13].

At the end of 2017, the balance of German investments in the economy of St. Petersburg amounted to $ 1.17 billion. Among the most reputable FRG businesses operating in the St. Petersburg are companies such as Siemens, which took an active part in the creation of the North-West CHP. She also implemented a traffic control system on Nevsky

Prospect, provided several medical institutions of the city with the latest equipment, opened a refrigerator factory. Production of metal structures for electrical panels, consoles and systems is successfully undertaken by the company RITTAL. Back in August 2006, in Kolpino, the Knauf industrial group built the Knauf Gypsum St. Petersburg plant, which produces, in addition to gypsum, other construction materials based on it. For many years the company Metro Cash & Cary has been operating in St. Petersburg, its products highly valued by consumers, especially trades people.

It has become a good tradition, with the active assistance of the St. Petersburg Legislative Assembly and the General Consulate of the Federal Republic of Germany, to hold the "Week of Germany" in St. Petersburg every April. Each such "Week" is dedicated to one of the federal states of this country. For example, in 2017 it was dedicated to Hamburg, and in 2018 to Bavaria.

Friendly ties between St. Petersburg and Germany are developing effectively thanks to tourism. In 2017 alone, 115.5 thousand Petersburgians visited Germany. The flow of tourists from Germany was even greater. The city was visited by 220 thousand German tourists.

In the 2017/2018 academic year, 104 students from the Federal Republic of Germany studied at the universities of St. Petersburg. 59 universities and scientific institutions of St. Petersburg cooperate with educational and scientific establishments in Germany. As a consequence of that cooperation - over 350 bilateral agreements were signed in 2017.

The partners continue to strengthen mutually beneficial cooperation. Both sides see the prospect primarily in the development of mutual tourism, in the exchange of experience in organising social and labour policies. Joint work in the field of health care and science seems promising. In particular, a programme for the exchange of specialists in the field of cardiac surgery, neurosurgery, rehabilitation, and anaesthesiology is successfully operating between the healthcare institutions of St. Petersburg and Hamburg. German and the St. Petersburg parties see great potential in humanitarian cooperation, as well as in supporting the programme for the exchange of young journalists [14].

Often, the Legislative Assembly of St. Petersburg becomes a link in the establishment of international relations, including cases where the initiative is shown by ordinary citizens and representatives of St. Petersburg and foreign business. An example of this is the appeal to the St. Petersburg parliament of Elena Meringova, originally from St. Petersburg, who now lives in Germany. She, being the director of the tourist office in Bonn, specialises in organizing trips for German citizens to Russia, and helps to ensure that as many people as possible show interest in our country, in its history and culture. Elena Meringova suggested organising for a group of German tourists, including members of the society of family friends Leuchtenberg, a short tour of the halls of the Mariinsky Palace. The director of the tourist bureau quite rightly emphasised that such visits "give much more than even a visit to our amazing museums, for mutual understanding and friendly relations" [15]. In accordance with the decision taken by the leadership of the Legislative Assembly, 12 German citizens visited the building of the Mariinsky Palace, including the Information and Educational Centre of the Legislative Assembly of St. Petersburg [16].

The breadth of international relations of the Legislative Assembly of St. Petersburg is eloquently indicated by contacts with parliamentarians of African states. The deputies of the 6th convocation strengthened business contacts with representatives of such countries as Burundi, Morocco, Nigeria, Ethiopia and South Africa.

In the autumn of 2016, a delegation of the National Council of Provinces of the Parliament of the South African Republic, headed by the Chairman of the Council, Ms. Tandi Modise, visited the Mariinsky Palace. In her opinion, South Africa is considering Russia as one of the most promising partners. For this country, economic and humanitarian ties with St. Petersburg are especially important. In particular, we are talking about medical education. At the same time, Russian-speaking audience in South Africa can expand not only on the account of students. Russian-language educational programs can be extended to a wider audience of South African residents. Currently, three St. Petersburg universities are closely cooperating with universities of the South African Republic. This is one of the promising areas of educational and scientific activities.

Delegates of the National Council of Provinces of South Africa expressed a unanimous opinion on the establishment of direct contacts between the parliaments of South Africa and St. Petersburg. The port of Cape Town, the economic centre of Johannesburg and the Western Cape province are already actively working with St. Petersburg. Entrepreneurs of St. Petersburg are interested in developing relations with the port of South Africa Durban; cooperation in the field of shipbuilding, power engineering and pharmaceuticals is of mutual interest. The St. Petersburg business has also expressed interest in supplying medical equipment from the BAR that meets the level of world standards. In turn, business partners from the South African Republic expect to attract representatives of such industries as the enrichment of natural minerals to their economy, since this segment is very well developed in St. Petersburg.

South Africa, due to the specifics of its geographic location, is known to have unique climatic conditions. The South African sun has such a beneficial effect on local vineyards that wine made from local grapes, according to experts, often has its own qualities and surpasses the famous French wine. Tandi Modise, that the reason for her special pride is the fact that the Russians, including St. Petersburgians, have already appreciated the quality of South African wine and a number of other types of agricultural products [17].

On October 1, 2017, Russia and Burundi celebrated the 55th anniversary of the establishment of diplomatic relations. In their congratulatory messages, the Foreign Ministers expressed their intention to continue strengthening cooperation for the benefit of the peoples of the two countries. On May 24, 2018, Deputy Chairman of the Legislative Assembly of St. Petersburg S. Soloviev met with the Chairman of the Senate of the Republic of Burundi Reverien Ndikuriyo, one of the most authoritative politicians your country. He has extensive life and political experience. Studied at the University of Burundi in Bujumbara. During the Civil War from 1995 to 2004, he participated in the Hutu guerrilla movement, of which he is. For several years he was the governor of one of the provinces of the country, a member of the National Assembly of Burundi, the chairman of the commission for political, administrative and diplomatic affairs. Since 2010 he was a member of the Senate of the Republic Burundi, and since 2015 as its chairman. A lot speaks about the diversity of this man's creative abilities, including the fact that he absolutely loves football, being at the same time the founder of the football club Aigle Noir ("Black Eagle") and the president of the Football Federation of Burundi [18]. Mr. Reverien Ndikuriyo spoke about the political and economic situation in his country, and also expressed confidence that contacts between the Republic of Burundi and the Legislative Assembly of St. Petersburg will strengthen from year to year [19].

The prospects for cooperation between St. Petersburg and the Republic of Burundi are great. The Norilsk Nickel Company and the Gipronickel Research Institute located in St. Petersburg have expressed their desire to start working on the issue of mining rare earth metals in Burundi. Petersburg is ready to supply equipment to Burundi and participate in projects related to the development of mineral resources. In June 2017, the second vice-president of Burundi, Joseph Butori, visited St. Petersburg and held talks with potential Russian partners. Petersburg businessmen are interested in purchasing in Burundi, their coffee, which is famous for its high quality.

On May 4, 2018, in the Red Hall of the Mariinsky Palace, a meeting was held between the deputy, chairman of the profile commission on physical culture and sports, Y. Avdeev with a delegation from the Federal Republic of Nigeria. The Nigerian delegation was headed by the Extraordinary Plenipotentiary Ambassador of the Federal Republic of Nigeria to the Russian Federation and concurrently to the Republic of Belarus S.D. Ugba. The delegation also included Minister-Counselor of the Embassy of Nigeria in the Russian Federation N. Dankadai, experts P.A. Gbadamosi and D. Andura [20].

During the meeting, the St. Petersburg parliamentarian noted that St. Petersburg is ready to develop relations with Nigeria in the field of science, higher education and tourism. It was noted that by the time of the meeting, 76 students from Nigeria were studying at St. Petersburg universities, although their number could be significant more. At the Faculty of Oriental Studies of St. Petersburg State University, the languages Hausa, Yoruba, and a number of others, which are the most widespread in Nigeria, are studied [21].

The hope was expressed for the growth of trade turnover between St. Petersburg and the Federal Republic of Nigeria. In 2017, the trade turnover of St. Petersburg and Nigeria increased by 86% compared to 2016 and reached its historical maximum, amounting to $ 13.1 million. Thus, Nigeria took 97th place among the countries – trade partners of St. Petersburg. Trade is characterized by a significant predominance of exports. Exports from St. Petersburg to Nigeria doubled in 2017. This increase is due to the supply of grain, which accounted for 70% of total exports. In addition, mineral fuel, paper and cardboard are sent from St. Petersburg to this African country. Imports in 2017 decreased by 87% and amounted to $ 59 thousand. For import to St. Petersburg Nigeria ranks 136th. Basically, this country supplies coffee, tea and spices to St. Petersburg.

On the eve of the 2018 FIFA World Cup, Y. Avdeev wished the guests success to the Nigerian national team, which was to play in St. Petersburg with the Argentine national team on June 26, 2018. The participants of the meeting discussed the course of preparation for the championship. The Ambassador of the Federal Republic of Nigeria noted that he planned to get acquainted with the St. Petersburg sports facilities, including the new stadium on Krestovsky Island. A special place in the conversation was occupied by issues of economic and political cooperation. Mr S.D. Ugba noted: "We need Russian assistance in the development of the steel and aluminium industries, the energy industry, including the construction of nuclear power plants. We are also ready to undertake joint efforts in the fight against international terrorism" [22].

2018 marks 120 years since the establishment of diplomatic relations between Russia and Ethiopia. Cooperation between St. Petersburg and Ethiopia is also developing. In 2017, the trade between the two countries amounted to $ 3.86 million. At the same time, exports are $ 0.72 million, and imports are $ 3.14 million. Import supplies are dominated by vegetables, root crops and coffee. Petersburg business is interested in additional supplies of agricultural products from Ethiopia, as well as in the sale of equipment and participation in various projects for the development of minerals [13]. It seems that cooperation between St. Petersburg and Ethiopia has a great future. The languages of the peoples living in Ethiopia are studied at the Faculty of Oriental Studies of St. Petersburg State University. Medieval Ethiopian manuscripts are in the collections of the Russian National Library, the Institute of Oriental Manuscripts of the Russian Academy of Sciences and the Museum of Anthropology and Ethnography named after A. Peter the Great.

In March 2011, a delegation from the Ethiopian capital Addis Ababa, led by Mayor Kuma Demexa, who later became the Ambassador of Ethiopia to Germany, visited St. Petersburg. The Ethiopian delegation held a number of important meetings, including with deputies of the St. Petersburg parliament, heads of a number of committees of the City Administration, with the leadership of the Oriental Faculty of St. Petersburg State University and the State University of Railways.

3.2 Changes in the Course of Development of Political and Geographical Relations of the St. Petersburg Parliament in Recent Times

Strengthening and weakening of political contacts is a common occurrence in the life of any political, diplomatic and legislative institution. In recent years, the priorities of international communication have also changed in the St. Petersburg Legislative Assembly. But if we look at the emerging tendencies of the international dialogue of St. Petersburg parliamentarians from political and geographical positions, then the emerging picture makes it possible to better understand the objective successes and failures of the international activities of St. Petersburg parliamentarians in recent years. Since 2016, i.e. since the beginning of the work of the deputies of the sixth convocation, contacts with a number of states located on all continents have practically been lost.

Also, the deputies of the St. Petersburg parliament of previous convocations have repeatedly met with delegations of the parliaments of Ireland, Malta, Mali and Guatemala.

As can be seen from Table 2, among the most tangible losses is the loss of contacts with parliamentarians from Great Britain, Australia, Canada and Ukraine.

The loss of parliamentary contacts with Great Britain was primarily a consequence of the Skripals case. The situation was significantly influenced by the fact that in 2018 the British Consulate General in St. Petersburg was closed.

It is possible that the deteriorating relations with the UK affected the political, diplomatic, economic and cultural relations between Russia and Australia. One of the mistakes of parliamentary activity St. Petersburg is, in our opinion, little attention to maintaining and development of relations with this state. An indirect example of this inattention is, for example, the fact that in the city with five million population only one scientist is engaged in research in the history of Australia.

Table 2. International contacts of parliamentarians of St. Petersburg with delegations of foreign countries, lost by the deputies of the sixth convocation [23]

Activity rating of international relations (in descending order)	Number of meetings	Country
1	7	UK
2	5	Australia, Pakistan, Canada, Ukraine, Mongolia
3	4	Uruguay
4	3	Peru, Chilli
5	2	Lithuania, Aland Islands, Brazil, Romania, Mexico, Angola, Gabonese Republic, Jordan, Colombia, Luxembourg, Argentina, Syria, El Salvador, Laos, Kenya
6	1	Taiwan, Portugal, Venezuela, Bolivia, Bangladesh, Cambodia, Cameroon, New Zealand, Ghana, Madagascar, Botswana, Slovenia, Montenegro, Yemen, Sri Lanka, Iceland, Algeria, Saudi Arabia, Albania, Myanmar, Afghanistan, Ecuador, Paraguay

The problems associated with a sharp deterioration in contacts between Russia and Ukraine are well known. For all their importance, they are not the subject of this article. But we consider it necessary to note that, in our opinion, the loss of all-round contacts between these two countries negatively affects the development of relations between Russia and Canada, as well as between St. Petersburg and those Canadian cities with which relations were previously very positive. One of the important reasons for the cooling of relations between St. Petersburg and Canadian parliamentarians, it seems, was the fact that the influence of the Ukrainian diaspora in Canada is quite strong. 1 million 359 thousand citizens of this country, i.e. 4% of the population are Ukrainians. Many of them have an impact on the politics and economy of the state, as well as on the process of forming public opinion [24].

It is not difficult to notice that the Legislative Assembly of St. Petersburg is following the official state policy. The MPs are a parliamentary body consisting of five party factions. They represent parties such as United Russia, Fair Russia, Communist Party of the Russian Federation, Liberal Democratic Party, Party of Growth and Yabloko Party. But this circumstance does not qualitatively affect the course pursued by the Petersburg parliamentarians aimed at implementing a single nationwide line in building bilateral relations with other countries.

4 Conclusions

Looking at the example of the work of the Legislative Assembly of St. Petersburg of the sixth convocation, one can see the main trends in the development of political and geographical contacts between Russia in general and the countries of the world. It is easy to notice that relations with the United States of America have deteriorated significantly, a large number of countries in South America, Africa and, especially, Europe. The efforts of the Russian Federation to compensate for these political and geographical losses by strengthening ties, primarily with such states as the People's Republic of China, the Republic of Kazakhstan, Republic of Azerbaijan, Republic of Korea, are obvious. The parliamentarians of St. Petersburg are trying to solve the problem of political and geographical changes in implementing comprehensive contacts by emphasasing all-round interaction with a number of European states and, first of all, the Federal Republic of Germany, Finland, Turkey and the State of Israel. Since the bilateral relations of all these countries are not without many problems, the deputies from St. Petersburg focus on strengthening not all the contacts at the same time, but on deepening the interest of these countries precisely in bilateral contacts. This increases the room for manoeuvre. St. Petersburg and, in particular, its parliament, is an important tool for establishing new contacts and breathing life into old ones, since the history of the city, its geographical position, economic potential and logistic perspective are attractive aspects for other states, including their parliaments. A feature of modern political and geographical contacts is that St. Petersburg develops contacts, as a rule, neither with one or another state in general, but with certain geographical entities of different countries, for example, with such states of the Federal Republic of Germany as Bavaria, Free and Hanseatic City of Hamburg and Dresden. This also applies to contacts with other countries, including those with whom relationships are built on a very delicate basis.

References

1. Zotova, A.V., Vezhbitskite, V.: On the results of 25 years of international activity of the Legislative Assembly of St. Petersburg. History of St. Petersburg **4**(77), 48–51 (2019)
2. Yagya, V.S.: Thoughts wide open. St Peterssburg, Evropeyskiy dom (2008). (in Russian)
3. Vyacheslav Makarov: in St. Petersburg, Chinese entrepreneurs will find understanding and support. Bezformata. https://sanktpeterburg.bezformata.com/listnews/peterburge-kitajskie-predprinimateli/52604818/. Accessed 18 May 2020
4. A delegation from the People's Government of Hong Kong visited the Mariinsky Palace. https://www.assembly.spb.ru/article/633200002/103547/Mariinskiy-dvorec-posetila-delegaciya-Narodnogo-pravitelstva-Gonkonga. Accessed 18 May 2020
5. Foreign economic activity of regions. Customs. Logistics. https://imexp.ru/blog/import-export-regionov-rf/import-spb/. Accessed 18 May 2020
6. Vyacheslav Makarov met with the Chairman of the Milli Mejlis of Azerbaijan. St. Petersburg Diary, 18 May 2018
7. Vyacheslav Makarov met with the Charge d'Affaires of the Russian Federation in Azerbaijan. https://www.assembly.spb.ru/article/633200002/99609/Vyacheslav-Makarov-vstretilsya-s-vremennym-poverennym-v-delah-RF-v-Azerbaydzhane. Accessed 18 May 2020

8. Foreign parliamentary delegations adopted by the Legislative Assembly of St. Petersburg. in 2018. From the funds of the Foreign Relations Department of the Legislative Assembly of St. Petersburg
9. https://www.azatutyun.am/a/28999788.html. Accessed 06 Apr 2020
10. Nikiforova, D.: Dedicated to the courage of Leningraders. Noah's Ark **2**(301) (2018)
11. Azg Armenian Daily NewsPaper, 15 February 2018. https://www.azg.am/AM/2018021631. Accessed 06 Apr 2020
12. Vyacheslav Makarov met with the delegation of the National Assembly of Armenia. Legislative Assembly of St. Petersburg. https://www.assembly.spb.ru/article/633200002/95202/VyacheslavMakarov-vstretilsya-s-delegaciey-Nacionalnogo-Sobraniya-Armenii. Accessed 12 May 2020
13. Foreign economic activity of regions. Customs. Logistics. https://imexp.ru/blog/import-export-regionov-rf/import-spb/. Accessed 12 May 2020
14. Materials for the conversation of the chairman of the profile commission for the environmental protection of the population of St. Petersburg of the Legislative Assembly of St. Petersburg Tikhonova N.G. with a delegation of the heads of the Hamburg metalworkers' trade union "IG metal", May 24, 2018, 11.00, Mariinsky Palace, Red Hall. From the funds of the Foreign Relations Department of the Legislative Assembly of St. Petersburg. Folder "Conversations" (2018)
15. Letter to the Chairman of the Legislative Assembly of St. Petersburg V.S. Makarov from the director of AM Kontinent GmbH Konsular - Travelservice Elena Meringova dated 05/03/2018. From the funds of the external relations department of the Legislative Assembly of St. Petersburg. Folder "Conversations" (2018)
16. Memo to the Chief of Staff of the Legislative Assembly of St. Petersburg M.V. Subbotin from the head of the department of external relations of the Legislative Assembly of St. Petersburg N.A. Akhadova dated May 21, 2018 No. 804053-2. From the funds of the Foreign Relations Department of the Legislative Assembly of St. Petersburg. Folder "Conversations" (2018)
17. Petersburg will arrange a creative cultural exchange with South Africa. Petersburg diary, 29 September 2016
18. Reverien Ndikuriyo President of the Senate of the Republic of Burundi. Reference material. From the funds of the Foreign Relations Department of the Legislative Assembly of St. Petersburg. Folder "Conversations" (2018)
19. The Mariinsky Palace was visited by the Chairman of the Senate of the Republic of Burundi. https://www.assembly.spb.ru/article/633200002/99946/Mariinskiy-dvorec-posetil-Predsedatel-Senata-Respubliki-Burundi. Accessed 12 May 2020
20. List of participants in the meeting of the chairman of the profile commission on physical culture and sports of the Legislative Assembly of Y.V. Avdeeva with Ambassador Extraordinary and Plenipotentiary of the Federal Republic of Nigeria to the Russian Federation Steve Davis Ugba, May 4, 2018, 15.00, Mariinsky Palace, Red Hall. From the funds of the Foreign Relations Department of the Legislative Assembly of St. Petersburg. Conversations folder" (2018)
21. Materials for the conversation of the chairman of the profile commission on physical culture and sports of the Legislative Assembly of St. Petersburg Y.V. Avdeev with Ambassador Extraordinary and Plenipotentiary of the Federal Republic of Nigeria to the Russian Federation Steve Davis Ugba, May 4, 2018, 15.00, Mariinsky Palace, Red Hall. From the funds of the Foreign Relations Department of the Legislative Assembly of St. Petersburg. Folder "Conversations" (2018)
22. The Ambassador of the Federal Republic of Nigeria visited the Mariinsky Palace. https://www.assembly.spb.ru/article/633200002/99202/Mariinskiy-dvorec-posetil-Posol-Federativnoy-Respubliki-Nigeriya. Accessed 12 May 2020

23. List of heads of official delegations who attended the Legislative Assembly of St. Petersburg I and II convocations in 1995–2002. year 2000; Foreign delegations adopted by the Legislative Assembly of St. Petersburg. in 2003; Foreign delegations adopted by the Legislative Assembly of St. Petersburg. in 2004; Foreign delegations adopted by the Legislative Assembly of St. Petersburg. in 2005; Foreign delegations adopted by the Legislative Assembly of St. Petersburg. IV convocation in 2007–2011; Foreign delegations adopted by the Legislative Assembly of St. Petersburg. V convocation in 2011-2016. From the funds of the Foreign Relations Department of the Legislative Assembly of St. Petersburg
24. Oleksandra Khichiy, President of the Congress of Ukrainians of Canada. Ukrinform. https://www.ukrinform.ru/rubric-society/2676441-aleksandra-hicij-prezident-kongressa-ukraincev-kanady.html. Accessed 11 June 2020

Problems and Prospects of Promoting Digital Trade in Russia

Anastasia Osypa, Sergey Pogodin(✉), and Anna Matveevskaya

Peter the Great St. Petersburg Polytechnic University, St. Petersburg, Russia

Abstract. Recently, Internet commerce has developed in world practice. It is carried out through the online platform, on the basis of which it is possible to buy freely goods and services, place orders and pay for them. In recent years, the Internet market has been successfully developing in the Russian Federation. The number of users in the country is more than 90 million people, 9 out of 10 residents have already made at least one purchase via the Internet. Internet commerce in Russia appeared in 1996 and is increasing every year. The following e-commerce models can be distinguished: B2B Business to Business; B2C Business to Consumer; B2G Business to Government) - commercial relations between business and government; C2C Consumer to Consumer. Internet commerce has got a positive economic impact, promotes competition, and creates new jobs. E-commerce in Russia is developing rapidly and in the near future may reach the level of the world leading countries. But there are a few obstacles: the lack of clear terminological formulations and official statistics connected with e-commerce, not a clear legislative framework. Some work should be done to study the global experience of e-commerce and its use in Russia.

Keywords: Russia · E-commerce · Digital economy · Smart community · Logistics

1 Introduction

Relevance of research. In modern conditions Internet trading is widespread. This is a new form of selling goods and services through online platforms, where you can freely choose the desired product, place an order for it and make payment. In Russia, the Internet market is successfully developing. The agency IT-analyst Radius Group has published data on the volume of Internet trade in Russia, for 2019 sales amounted to 1.7 trillion rubles, compared to 2018, sales have increased by almost 60%, this is the best figure for 10 years [3].

Internet users currently prefer to make online purchases. Germany's largest market research institute, GfK, has published data on Russia: the country has the largest number of registered network users in Europe. The number of users is more than 90 million people, 75% of the country's population have access to the Internet. Data from the U.S. market research company, Nielsen Media Research, notes that out of 9 out of 10 people in Russia have ever made purchases over the Internet. These data indicate a high potential

© The Author(s), under exclusive license to Springer Nature Switzerland AG 2021
R. Bolgov et al. (Eds.): *Proceedings of Topical Issues in International Political Geography*, SPRINGERGEOGR, pp. 102–111, 2021.
https://doi.org/10.1007/978-3-030-78690-8_9

for the development of online commerce in Russia. TAdviser predicts that the corporate information portal is next: "If in 2017 sales of physical goods via the Internet were measured at $18 billion, in 2020 and 2023 the turnover may reach $31 billion and $52 billion respectively [3].

Based on the above, the following conclusion should be drawn: Internet market is an active developing trend in trade, which provides unlimited opportunities for the buyer and has a great impact on the development of the Russian economy.

The aim of the work is to identify the main problems and prospects in the development of e-commerce in the Russian Federation.

The following tasks have been defined to achieve the objective:

1. to characterize the specifics of the emergence and condition of modern e-commerce in the Russian Federation.
2. identify the main problems in e-commerce for the Russian Federation and determine the ways of their implementation.

Research on e-commerce in the Russian Federation has become widespread. It is worth mentioning works of Savinov [16], Novikov [11], Shaydullin [17], Rossoha [15], Tsenina and Zou Tung [19]. At the same time, digital trade/e-commerce issues are in the scope of studies over the globe [22, 23]. For instance, Biagi and Falk [20] focus on the European experience regarding the impact of e-commerce activities on employment. Ferracane and colleagues [21] attempt to measure the restrictions to e-commerce with the cover of 64 countries. The International Journal of Electronic Commerce focuses on e-commerce issues.

2 Method of Research

The study consists of 5 steps:

1. We analyze the concept of "e-commerce", as well as the main criteria and indicators of e-commerce.
2. Then we evaluate individual analysis results to form an overall picture of e-commerce.
3. We determine the conclusions on e-commerce development in the Russian Federation.
4. We study private cases from e-commerce sales.
5. We study specific e-commerce participants.

Moreover, we use the elements of institutional analysis to analyze the organizational structure of e-commerce in Russia, as well as the policy paper analysis regarding to laws and policy papers which regulate e-commerce in Russia.

3 Results of Research

E-commerce is an activity aimed at "selling goods and services using information technology based on network interactions between buyer and seller [5]". In other words, it is the process of buying and selling over the Internet.

It is possible to allocate the following models of electronic commerce which will be similar to traditional forms of sales:

- B2B (Business to Business) - business for business. With this model, the seller and buyer are an organization, an example of this form can be the following fact: one company produces colorful children's balls and sells in bulk to another company, which in turn is the organizer of children's holidays.
- B2C (Business to Consumer) - a business for the consumer. With this model, the buyer is a specific person, he does not buy a party of colorful children's balls, and a specific amount of not large, for example - 5. To celebrate the birthday of your child.

According to TAdviser research, these two models "provide the majority of transactions on the Internet [3]. The following e-commerce models, but less so in online commerce:

- B2G (Business to Government) - commercial relationship between business and government.
- C2C (Consumer to Consumer) - a form of trade in which the consumer and the seller are not legally entrepreneurs.

The following segments of the Internet market are highlighted:

1. Online retail (trade in goods);
2. Trade in services;
3. Content trading;
4. Electronic payments.

Internet commerce appeared in 1979, thanks to Michael Audrich, who connected cable television with a computer using telephone lines, which allowed viewers to shop online [8]. In this way, e-commerce appeared. In the eighties, the first forms of B2B and B2C online shopping appeared.

The first browser was created in 1990, the first online store was opened in 1992. Beginning in the mid-nineties, the giants Amazon and eBay appeared. At the end of the twentieth century there was a payment system PayPal. A big role in the development of e-commerce was played by Google. According to N. Omelchuk in 1998: "annual volume of online sales reached $8 billion [12]". The first conference on online sales was held in 2005 in Chicago, and there was also an exhibition of samples of online trade.

The creation of Internet commerce has had a major impact on economic and lifestyle changes. The emergence of online stores has contributed to competition, which has affected the emergence of new jobs, stimulated the mastery of new professions: content managers, administrators of online stores, digital marketing professionals, etc. All this created new opportunities for the population, contributed to the development of business and ultimately the sale of goods.

In the Russian Federation Internet commerce appeared August 30, 1996, when first appeared in Runet online store - books.ru. It was an online bookstore; it exists to this day and enjoys great popularity. The first online banking system, appeared in Russia in 1998, its organizer was "Internet Service Bank" founder Autobank. At the same time, the population of Russia got acquainted with the IMTB virtual bank and Cyber Plat payment system. According to N.A. Novikov, a Russian researcher of e-commerce market in Russia, "in April 2005, the total volume of payments through CyberPlat from subscribers of mobile operators, satellite and cable television and Internet service providers was one billion dollars [11]".

The processing of a large number of applications for the purchase and sale of securities required the creation of an Internet gateway in 1999. The advent of this system for processing requests has led to the fact that they can now be processed in seconds. Since then, Russian Internet trading has been developing. According to Novikov's research, "in the first two years of the gateway's existence alone, the market share of Internet brokers on MICEX grew to 50%, and the share of transactions concluded through the gateway - to 63% [11]". The creation of the Internet Exchange turned out to be a change in the balance of power in banking circles. The advantage was gained by companies that managed to reorient themselves towards Internet transactions and apply trading in their activities.

According to the article "History of online shopping development [6]". The main products bought in the network are clothes and shoes. One of the first successfully developing clothing stores was Ozon. According to the number of orders and estimates for 2019, the online app stores AppStore and Google Play, was recognized as the largest online clothing store on the Wildberries [7]. The e-commerce research agency "Data Insight" published data for 2018: total site sales amounted to RUB 1,200 million, while in 2019 the share of orders in the top 3 online stores (Wildberries.ru, Citilink.ru and Mvideo.ru) increased by 90% [7].

According to the same publication, the share of Internet sales and GDP increased by 1.3% in 2019, while the volume of B2C amounted to $30.6 billion by the volume of Internet trade [7]. The index of Internet inclusiveness has reached the mark of 19. Russia has become a rapidly growing country in terms of e-commerce markets, this indicator was 23%. This enabled the Russian Federation to enter the top 5 fastest growing e-commerce markets, with a per capita value of $170.

Forbes published a rating of "10 top sellers of Runet" on February 20, 2020 [9]. The rating includes the largest Internet retailers, their goods and services are sold offline, but sold through online channels. The first place is occupied by the online hypermarket Wildberries with a turnover of 223.5 billion, while the second - "Aeroflot" with a turnover of Br221 billion from online sales, and the third place belongs to the store AliExpress Russia, with online revenues reaching Br201.4 billion. It is interesting that AliExpress Russia store appeared only in September 2019 and managed to become one of the three leaders in online sales in Runet in such a short time. The oldest Hypermarket of clothing Ozon, one of the oldest online stores, takes only seventh place, its revenue being Br80.5 billion. In tenth place is the construction shop "Petrovich" with a turnover of 60 billion rubles per year.

The Audience of Internet Users in Russia. According to Data Insight, 78% of the population of the Russian Federation are Internet users, of which 90% use the network every day, the monthly audience of Internet users is 95.7 million people [7].

Online B2C in Russia. B2C called "market of final consumption" (from English business to consumer, literally "business to consumer") is described in detail in A. Ovsyannikov's work "Modern marketing": "the term referring to the commercial relationship between the organization (business) and a private, so-called "final" consumer (consumer). B2C market is a set of buyers who make purchases for personal, family, home use [14]".

According to the retail trade "Data Insight", in 2019, the number of orders amounted to 425 million rubles, the average purchase price was 3,800 rubles, total revenue was 1.6 trillion rubles, which is 6.7 times more than in 2011, when the sales volume was equal to 240 billion rubles. For comparison: the highest growth in sales volume was observed in 2014 - 35%, the lowest in 2017 - 20%. Last year, the growth in sales volume was 24%. At the same time, we can state the significant growth of orders: in 2015 it was 7%, and in 2019 it grew by 41% [7]. However, the average purchase receipt has been reduced to 14%. In 2015, the highest growth rate of 18% was observed. From all of the above we can conclude there was an increase in the total volume of online sales, which is associated with the growth in the number of orders, while reducing the average purchase receipt. The number of orders is growing, due to an increase in purchases. Consequently, we observe an increase in the number of people switching to online shopping, with an increase in confidence in this form of trading.

C2C Market in Russia. It is a form of trade that provides that subjects are not legally registered entrepreneurs. According to the analytical company "Data Insight"; "The number of sellers involved in C2C trading has almost doubled in two years and reached 13.9 million sellers. The number of buyers has increased to 11.8 million [7]". At the same time, the total amount of sales amounted to 177 million rubles, and the average check was equal to 3,210 rubles, with annual revenues of 568 billion rubles. In 2017, annual revenue was only 295 billion rubles, doubling compared to 2019.

Logistics. Domestic analyst D. Rossokha in his work "Logistics in Russia: problems, opportunities, solutions", notes: "In the concept of modern entrepreneurs, logistics means optimizing the shipment of products from one place to another [15]. The chosen rational method of delivery of goods guarantees several times, to increase the competitiveness of the firm-supplier. It is possible to define several delivery options. The first one, with the help of special logistics companies. The second, by the selling company's own forces.

The following Internet order delivery channels are available in the Russian Federation:

- "Russian Post";
- Logistics companies, to the door;
- Logistics companies, in self-delivery points (PVZ) and postamats (automated postal station);
- Own store services.

In the analysis of Internet order delivery services, we can state a significant growth of own services at large electronic commercial companies. This situation is related to the concept of "last mile problem", which means the final stage of delivery of goods to the customer. In this mode, work the largest online stores runet: Wildberries, Ozon and Apteka.ru [7].

By the number of self-delivery points (PVZ) and postamats (automated postal station), the first place is occupied by Svyaznoy Euroset, it owns more than five thousand points. According to "Data Insight": "as of spring 2019, the total number of unique points of order and postal orders from 27 major market players exceeds 35 thousand, while the growth of large networks by the number of points in the second half of 2019 was from 15% to 30% [7].

However, not all companies resort to logistics companies, courier services and PVZ organizations. From the TOP-100 Runet online shops, only 23% offer all four types of delivery: mail, to PVZ, post or to the door.

Based on the above data, we can conclude that the logistics delivery system is not very well developed in the Russian Federation. According to the data of the conducted surveys, 56% of sellers sell goods only in their settlement, 30% - within their region, 13% - send goods across Russia and only 1% abroad. This situation is explained by the lack of an established and reliable transportation system. It should be noted an interesting fact, for 65% of buyers the cost of delivery is more important than its timing. Consequently, 32% of customers prefer to receive goods from PVZ logistics companies for greater savings.

When analyzing a brief overview of e-commerce in the Russian Federation, we can draw the following conclusion: the market is constantly growing and developing. We can highlight the giants of the industry, which are successfully developing their activities: Wildberries and Ozon. With the existence of a well-established system of ordering and delivery of goods, there are major shortcomings and a number of unsolved problems.

4 Discussions of Research

According to V. Savinov, "Internet commerce accounts for only 3–4% of the volume of retail trade turnover in Russia. In western countries this indicator reaches 10–20%" [16]. For wide development of e-commerce in the Russian Federation there are a number of problems and unsolved issues. This section is devoted to their analysis and ways to overcome them.

When assessing objectively the state of electronic commerce in the Russian Federation [10], there are no clear terminological formulations, such as: "e-commerce", "e-commerce", "online trade", "e-business" and "internet trade". There are no official statistics on e-commerce in Russia, so it is possible to give an objective assessment only if there are expert opinions or research works. This approach completely excludes the possibility of a comprehensive, objective analysis.

The Russian Federation does not have a clear legislative base on the problem of e-commerce development in the country. E-commerce is subject to general provisions of regulatory documents: Civil Code, Civil Procedure Code, Arbitration Procedure Code and regulations of the Central Bank [4].

Transport transactions are regulated by legislation of the Customs Code and the Customs Union. The Federal Law "On Information, Information Technology and Information Protection" of 27.07.2006 № 149-FZ [1] contains general provisions on electronic commerce. There are a number of special laws regulating hotel elements of e-commerce: Federal Law "On Electronic Signature" dated 06.04.2011 № 63-FZ [13]. "Retail e-commerce is additionally regulated by legislation in the field of consumer protection (Federal Law "On Protection of Consumer Rights" of 07.02.1992, № 2300-1, Rules for the sale of goods by remote means, approved by the Government of the Russian Federation of 27.09.2007 № 612) [4]".

The above legislation does not reflect the specifics of regulating commercial transactions carried out online, but aims to regulate traditional trade relations. There are no specific laws regulating consumer protection on the Internet. The absence of such special laws leads to the fact that the buyer is exposed to potential risks: deception, provision of low-quality goods, etc.

There is no fair competition between foreign and domestic online retailers in the Russian Federation. In order to attract new entrepreneurs and stimulate Internet retailing, the Russian government has lowered the duty-free import threshold. From 1 January 2020 it was 200 euros per parcel up to 31 kilograms, in 2019 the duty rate was up to 500 euros [2]. According to the Russian scientist V. Shaydullin, author of the study "The state and prospects of electronic commerce in Russia", "import of goods from foreign online stores weighing up to 31 kg and costing up to 1 thousand euros is not subject to duties at all [17]". Foreign online shops do not pay VAT. Therefore, the average check when buying from abroad, is cheaper for Russian citizens than at home. As V. Shaydudulina writes, in 2018 the Russian Federation was ranked second in the world by the number of imports in e-commerce [17]. This situation significantly hinders the development of the domestic economy.

One of the main problems associated with the online store is public distrust of them due to the high risks of fraud on the Internet and the insecurity of personal data of customers. A separate group of problems is distrust and prejudice to the payment system itself through the Internet. In Western countries the share of cashless payment is about 90% of all payments made by the population, in the Russian Federation the main method of payment is cash [17].

Actively realized delivery of goods when buying through online stores, still has many unresolved issues. Leading retailers are investing heavily in logistics development, renting warehouses, developing a network of post offices, developing new ways to deliver goods, but the system itself is still very far from improving. An example of unsolved problems is the case of Aliexpress, which wanted to deliver goods from the companies "Five" and "Cross", due to the complexity of the cashiers of these companies, which led to queues in shops.

Measures for Solving the Above Problems and Effective Development of Economic Commerce in the Russian Federation. In September 2019, the draft "Strategy for Trade Development in the Russian Federation until 2025 [18]" was published. In the project a whole section was devoted to electronic commerce. The annex to the project defined electronic commerce as "a form of trade carried out using information systems, information and communication network Internet or other communication procedures through electronic transactions on the Internet or other communication networks. In this case, payments, delivery of goods or provision of services may be made either online or offline [18]". Separately, the concept of e-commerce as a retail e-commerce has been highlighted: "a form of e-commerce in the field of retail (B2C), in which the buyer is acquainted with the goods and the terms of sale, as well as informing the seller of the intention to buy the goods occurs through an Internet information and communication network [18]".

The draft strategy notes that the lack of a regulatory framework to regulate this type of trade, as well as the lack of comprehensive data evaluation, are key challenges to e-commerce development. This prevents accurate forecasting of e-commerce development and hinders the development of entrepreneurial activity in this direction. It should be noted the following fact, in the absence of strict legal and regulatory framework for e-commerce, leads to the growth of Internet retail, as in the Internet space seller can become almost anyone. This situation impedes objective data collection, and Internet commerce itself can become part of the shadow economy.

In the authors' opinion, the following measures are necessary to address this situation:

1. promoting the initiative of small and medium entrepreneurs;
2. the elimination of bans on the sale of certain categories of goods online;
3. introduction of new developments for data collection and analysis, promotion of data openness for objective evaluation of e-commerce state;
4. Improving payment systems, ensuring cyber security and protecting the rights of buyers and sellers in online transactions.

In 2017 a document on e-commerce development in the Russian Federation was adopted.

"Strategy for the Development of Electronic Commerce until 2025 [4]". The strategy aims at achieving the following indicators:

- 20% share of electronic commerce in total trade volume;
- at least 70% of retail stores that use the Internet channel in their sales;
- at least 80% of the population over 12 years of age who use the Internet to make purchases;
- increase of Russia's share in the global e-commerce market to 10%;
- increase in exports through e-commerce channels, by at least 5%;
- at least 100,000 people are Internet sellers engaged in electronic commerce in Russia.

5 Conclusions

According to V. Shaydullin, "prospects for further development directly depend on the purchasing power of citizens, the state of the Russian economy and other equally important factors [17]". To this statement should be added the necessity of development and correction of legislation in e-commerce. However, it should be noted that the rapid growth of e-commerce is largely due to the absence of strict measures and restrictions for entrepreneurs. Such areas as logistics, protection of consumer and buyer rights, improvement of cross-border online commerce system, etc. require detailed study, development and improvement.

In general, e-commerce in the Russian Federation is rapidly developing. World experience should be actively used for effective adaptation to Russian reality. At the same time, there are still many unresolved and open issues in e-commerce in the Russian Federation.

References

1. About information, information technologies and about information protection: Federal law from 27.07.2006 № 149-FZ. Administration of the President of Russia. http://kremlin.ru/acts/bank/24157/page/1
2. Changes in the duty-free quota of import of goods for personal use, delivered from abroad in international mail or by carrier from January 1, 2020. Federal Customs Service. http://customs.ru/fiz/mezhdunarodnye-pochtovye-otpravleniya/vnimanie!c-1-yanvarya-2020-goda-izmenenyayutsya-normy-besposhlinnogo-vvoza-tovarov-dlya-lichnogo-pol-zovaniya,-dostavlyaemyx-iz-za-rubezha-v-mezhdunarodnye-pochtovyx-otpravleniyax-ili-perevozchikom
3. Electronic commerce. TAdviser. 05 July 2019. http://www.tadviser.ru/index.php/%D0%A1%D1%82%D0%B0%D1%82%D1%8C%D1%8F%D0%AD%D0%BB%D0%B5%D0%BA%D1%82%D1%80%D0%BE%D0%BD%D0%BD%D0%B0%D1%8F_%D0%BA%D0%BE%D0%BC%D0%BC%D0%B5%D1%80
4. Expert Council on Digital Economy and Blockchain Technologies at the State Duma Committee on Economic Policy, Industry, Innovation Development and Entrepreneurship Development of Electronic Commerce in the Russian Federation. State Duma, September 2018. http://esgosduma.ru/static/pdf/GD1.pdf
5. Gryaznova, A.G.: Financial and Credit Encyclopedic Dictionary. Finance and Statistics, Moscow (2002)
6. History of development of internet-shops. Arguments and facts, November 13 2015. https://oren.aif.ru/dosug/purpose/istoriya_razvitiya_internet-magazinov
7. Internet trading in Russia 2019. Data Insight, 18 December 2019. http://datainsight.ru/sites/default/files/DI_Ecommerce2019.pdf
8. Jahar, P.: Fact-checking: four great non-Chinese inventions. BBC, 04 April (2018). https://www.bbc.com/russian/features-43626963
9. Kazmin, I.: 10 main sellers of Runet. Routing Forbes. AC RUS MEDIA, 02 February 2020. https://www.forbes.ru/biznes-photogallery/393349-10-glavnyh-prodavcov-runeta-reyting-forbes
10. Konkov, A.E.: Digital politics vs political digitalization. Vestnik of Saint Petersburg University. Int. Relat. **13**(1), 47–68 (2020). https://doi.org/10.21638/spbu06.2020.104
11. Novikov, N.A.: Research of the electronic commerce market in Russia. South Ural State University (2017). (in Russian). https://dspace.Susu.ru/xmlui/bitstream/handle/0001.74/19259/2017_421_novikovua.pdf?sequence=1
12. Omelchuk, N.: How electronic commerce (infographics) was developing. PaySpace Mag. 15 March 2017. https://psm7.com/special-projects/infographics/kak-razvivalas-elektronnaya-kommerciya-infografika.html
13. On electronic signature: Federal Law of 06.04.2011 № 63-FZ. Administration of the President of Russia. http://www.kremlin.ru/acts/bank/32938
14. Ovsyannikov, A.A.: Modern Marketing. At 2 pm. Part 1: Textbook and Workshop for Bachelor's and Master's Degree. Moscow, Yurait Publishing House (2017)

15. Rossoha, D.Y.: Logistics in Russia: problems, opportunities, solutions. Young Sci. **13**(1), 94–96 (2016). (in Russian)
16. Savinov, A., Gavryushin, O., Taranovskaya, E.: About strategy of the electronic trade development. Int. Econ. **8**, 16–26 (2019). (in Russian)
17. Shaydullin, V.K.: State and prospects of electronic commerce in Russia. Univ. Bull. **4**, 118–122 (2019). (in Russian)
18. Strategy of trade development in the Russian Federation till 2025. State automated information system "Management". http://gasu.gov.ru/stratpassport
19. Tsenina, T.T., Zou, T.: Information logistics role in the world electronic trade development. NarodRossii **04**(12) (2019). (in Russian). https://narodirossii.ru/?p=69656
20. Biagi, F., Falk, M.: The impact of ICT and e-commerce on employment in Europe. J. Policy Model. **39**(1) (2017). https://doi.org/10.1016/j.jpolmod.2016.12.004
21. Ferracane, M.F., Lee-Makiyama, H., van der Marel, E.: Digital trade restrictiveness index. European Center for International Political Economy (ECIPE) (2018)
22. Ferencz, J., Lopez Gonzalez, J.L.: Digital Trade and Market Openness. OECD Trade Policy Papers, 217. OECD Publishing, Paris (2018). https://doi.org/10.1787/1bd89c9a-en
23. Meltzer, J.P.: Governing digital trade. World Trade Rev. **18**(1), 23–48 (2019). https://doi.org/10.1017/S1474745618000502

Institutional Analysis of eHealth Development in EAEU Countries

Vitalina Karachay[1], Gennadii Orlov[2], and Radomir Bolgov[3](✉)

[1] ITMO University, Saint Petersburg, Russia
[2] NP PRIOR North-West, Saint Petersburg, Russia
[3] Saint Petersburg State University, Saint Petersburg, Russia
r.bolgov@spbu.ru

Abstract. Recently, there is a growing trend of expanding multilateral cooperation within the economic and geographical zones in the post-Soviet space (Customs Union, Eurasian Economic Union). In context of the regional integration, the task of ensuring the common standards for cross-border electronic interaction of the state information systems becomes urgent. This research is focused on studying the eHealth development in the countries of the Eurasian Economic Union (EAEU). The factors affecting the quality of eHealth information systems, as well as the problems and perspectives of the digital healthcare in EAEU countries in order to facilitate the Eurasian integration processes are analyzed in the paper. The analytical overview of the main institutions in eHealth domain of the EAEU member states is conducted. Such research directions as monitoring the eHealth systems development in the EAEU countries by using open data from both the EAEU countries and the European Union, the issue of common standards development in EAEU healthcare are planned on exploring in further works. Moreover, development of the effective tools of measuring the eHealth impact on the quality and effectiveness of the health services provided in EAEU countries is also the subject of future research.

Keywords: Eurasian integration · EAEU · eHealth · e-Governance · e-Government · ICTs · Healthcare · Health information systems · Institutional analysis

1 Introduction

Recently, there is a growing trend of expanding multilateral cooperation within the economic and geographical zones in the post-Soviet space. The creation of the Customs Union (CU), the Eurasian Economic Union (EAEU), and others can be considered as the confirmation of this statement. In context of the development of the integration processes, the task of ensuring the cross-border electronic interaction of the state information systems becomes urgent. Cooperation between the states in the development and establishment of the e-governance can facilitate the integration of the countries of common economic zone. In order to solve this problem, it is necessary to develop common

© The Author(s), under exclusive license to Springer Nature Switzerland AG 2021
R. Bolgov et al. (Eds.): *Proceedings of Topical Issues in International Political Geography*, SPRINGERGEOGR, pp. 112–124, 2021.
https://doi.org/10.1007/978-3-030-78690-8_10

standards for electronic interaction at the inter-country, regional, and global levels and to tackle key problems in their implementation [1, 2].

The Eurasian Economic Union is a dynamically developing integration. During the five years, Eurasian integration transformed from the Customs Union in 2010 to the Eurasian Economic Space in 2012 and to the origin of the Eurasian Economic Union in 2015. The EAEU members – Armenia, Belarus, Kazakhstan, the Kyrgyz Republic, and the Russian Federation share a commitment to improve the effectiveness of national socio-economic policy for growth quality of life and well-being of the population.

In the long term, the goals of the association can be reduced to a single global objective: to contribute to the achievement and maintenance of sustainable economic growth of the member countries and the Union as a whole by realizing their competitive advantages [3].

To achieve the goals of the EAEU, the most important tasks are the economic integration, the increasing of the level of interaction between the participants. This article reviews the state of these issues in context of healthcare digitalization, or eHealth.

It should be noted that the strategic document regulating the basic directions of digitalization in the EAEU, "Digital Agenda of EAEU" [4], does not include the eHealth direction, which, in our opinion, is a serious omission, which is particularly evident in the crisis caused by the COVID-19 pandemic. One of the UN Sustainable Development Goals is to ensure healthy lives and promote well-being for all at all ages. Moreover, the statement of the need to provide support for the integration of economic systems, including the ensuring the legitimate electronic interstate interaction of eHealth systems, which encourages the activation of the movement of a significant number of citizens from one country to another does not require special justifications.

Analyzing the existing publications on EAEU eHealth topic it can concluded that it is necessary to equalize the levels of eHealth development. In particular, there are information systems of individual medical organizations in the Republic of Belarus; however, the idea to create a Centralized Health Information System is still in plans [5]. In Russian Federation, there is a long history of eHealth development and the creation of a Unified State Information System in the field of Health [6]. Medical information systems have been developed and are successfully functioning in a number of cities of the country, telemedicine technologies, cloud solutions are being improved, and an intelligent business environment in the field of digital health is being formed. In general, the market for software products for medicine and healthcare has been formed in Russia. State requirements for the development of information technologies for medicine are constantly growing [7]. Besides, the issue on use the Internet technologies for healthcare ensuring the mass involvement of citizens in electronic interaction with doctors and the health care system as a whole is developing in Russia. The digital technologies can potentially be crucial in changing the architecture of health promotion and medical care in the next few years [8]. In other members states of the EAEU projects in the field of eHealth are also implemented [9–11]. In more detail, the state of eHealth in each member of the EAEU is described in the Sect. 2.

The research project presented in this article is aimed at studying the institutional development of eHealth in the member states of the Eurasian Economic Union. The eHealth processes in EAEU countries are examined including the exploring the factors affecting the quality of eHealth information systems, i.e. its effectiveness and compliance with international standards, the identification of barriers and problems, in order to facilitate integration processes taking into consideration the interests of citizens of these countries.

2 eHealth in Eurasian Economic Union

The study was conducted using a neoinstitutional approach. The methodological basis was the concept proposed by Scott [12], and further developed in the works of Kostova [13], Baba et al. [14], Friel [15], and Thoenig [16]. According to this approach, there are regulatory, normative, and cognitive sources of institutions. These sources, together with related activities and resources, provide stability and relevance to social life. From their point of view, we can understand the complexity of the development of eHealth and evaluate the experience of EAEU countries.

The method of research is the case study which allows analyzing practice of eHealth in EAEU countries. The study was conducted based on the information of "Atlas of eHealth country profiles: the use of eHealth in support of universal health coverage: based on the findings of the third global survey on eHealth 2015" [17], published in 2016 by World Health Organization, updated on the materials of the official bodies, online press and publications on the research topic available in the public access.

2.1 Armenia

After the post-Soviet transition period, characterized by strong economic growth, public spending on health in Armenia dropped sharply. Health care use rates declined partly because of formal and informal out-of-pocket expenditures had to compensate for the decrease in public spending. Facilities were not maintained and informal payments for health services were widespread due to the extremely low wages of health personnel. Hospital use and efficiency remained extremely low [18].

The health reforms, implemented by the Armenian government, were aimed at improving the cost efficiency and included a shift in health care financing, reorienting the system toward primary health care (PHC). The Ministry of Health became a policymaking and supervisory body. The State Health Agency (SHA) was established in 1998 as a purchaser of publicly financed health care services. Health care providers became managerially and financially autonomous and derived their income from annual contracts with SHA and private out-of-pocket payments. As the system decentralized and public service provision was reconfigured, operation and ownership of health services was devolved to provincial governments (hospitals) and local governments [18].

National health information system (HIS) policy was adopted in 2010. Two years later national health coverage policy was adopted. Funding sources for eHealth are private or commercial. Public funding is absent. Armenian law protects the privacy of individuals' health-related data held in electronic format in an Electronic Health Records (EHR). The legislation governs the sharing of personal and health data between research entities, civil registration and vital statistics, as well as national identification management systems. The introduction of new legislation was the important contribution.

As for the national EHR system, it is absent now. The only component of this system represented in Armenia is medical e-billing systems. At the moment there is no use of eLearning in health sciences. mHealth programmes such as toll-free emergency, health call centers, appointment reminders etc., are informal or pilot and represented mainly at local and intermediate levels. Some healthcare organizations provide social media to promote health messages as a part of health promotion campaigns, make general health and emergency announcements. The use of big data in the health sector is not provided yet [17].

The gaps in financial and human resources complicate the implementation of the Electronic Health Information Management System (e-HIMS), however, there are the solutions, including better collaboration and exchange of experience within the field and improved training, as well as better strategic planning in eHealth domain [9].

2.2 Belarus

The eHealth development in Belarus is one of the priorities of the state policy reflected in the Strategy for the Development of the Informatization in the Republic of Belarus for 2016–2022 and in the State Program for the Development of the Digital Economy and Information Society for 2016–2020 [5].

The main directions of the health care informatization in Belarus are the following: automatization of healthcare institutions; creation of automated disease registers determining the demographic security of the country; creation of information diagnostic systems, including those using artificial intelligence elements; development of information and analytical systems for specialized healthcare institutions; telemedicine development; entry into international registers and interstate information interaction to improve national and international treatment protocols etc. [5].

However, the correlated health information system is absent. Funding sources for eHealth are public or donor. Commercial funding is absent. As for the legal frameworks for eHealth, Belarusian law protects the privacy of personally identifiable data of individuals irrespective of whether it is in paper or digital format, as well as governs civil registration and vital statistics. The only telehealth programme provided is in teleradiology at the intermediate level, which is informal.

Policy development and priority setting are centralized processes where the Ministry of Health is the key actor. The Ministry of Health is overall responsible for the health system, although the funding and purchasing of primary and secondary care is devolved to the regional level. Highly specialized tertiary care hospitals are funded directly from the Ministry of Health budget. Healthcare providers manage healthcare delivery under the supervision of regional healthcare departments and local government, but the system is in essence hierarchical. Thus, the Ministry of Health plays the main regulatory role at all levels of the health system, although regional and district governments are also key stakeholders as they are responsible for financing the system at their level [19].

National EHR system was introduced in 2005. However, there is no legislation governing the use of the national EHR system. The system provides primary, secondary and tertiary health facilities with EHR. However, laboratory, pathology, pharmacy, and automatic vaccination alerting information systems are absent. There are some eLearning programmes in health sciences such as medicine, public health, biomedical/life sciences.

mHealth programmes such as toll-free emergency, health call centers, patient monitoring, are informal and represented mainly at local and intermediate levels. Social media are used only by individuals and communities for learning about health issues and providing feedback to health facilities or health professionals. The use of big data in the health sector is not provided yet [17].

In Belarus, participation of the public in health policy-making is rather limited. Officially, citizens are given the right to participate in public discussions on proposed policy, but whether their concerns are actually considered by decision-makers when formulating and promulgating policy is unknown due to lack of transparency in policy-making processes.

The main challenges faced in emergency care are therefore: the shortage of medical staff; the absence of a single, unified system of monitoring and management of ambulance services; inadequate coordination of pre-hospital and inpatient stages of emergency care provision; insufficient system of coordination for all emergency services (Ministry of Internal Affairs with Ministry of Health structures) in emergency situations; the absence of a single set of technological standards in the provision of emergency care and continuity of medical care; the inefficient system of financing for emergency care services; the lack of a telephone triage and consultation system etc. [19].

2.3 Kazakhstan

Kazakhstan national health strategy, eHealth policy, and health information system (HIS) were adopted in 2013. Funding sources for eHealth are public, public-private or donor. Commercial funding is absent that's similar with Belarus. The legal framework is quite mature. The national law defines medical jurisdiction, liability or reimbursement of eHealth services such as telehealth, addresses patient safety and quality of care based on data quality, data transmission standards or clinical competency criteria, governs the sharing of digital data between health professionals in other health services in the same country through the use of an EHR, etc. However, the legislation does not regulate demand of the deletion of health-related data from individuals EHR. The law does not allow individuals to specify which health-related data from their EHR can be shared with health professionals of their choice. Telehealth programmes are represented in teleradiology, teledermatology, and telepathology at intermediate and national levels.

National EHR system was established in 2003. However, there is no legislation governing the use of the national EHR system. The situation related to health facilities with EHR is same as in Belarus. Medical e-billing systems are introduced in Kazakhstan. There are some eLearning programmes in health sciences such as public health and biomedical/life sciences. The wide range of mHealth programmes is established (or launched as pilot) and represented mainly at all the levels. Despite the absence of national strategy on the use of social media by government organizations, social media in eHealth are widely used, for instance, to promote health messages as a part of health promotion campaigns, help manage patient appointments or learn about health issues. The big data is not engaged in the health sector yet.

According to National eHealth Strategy 2013, the factors hindering the development are the following: unsatisfactory material and technical base of healthcare organizations; lack of joint responsibility of the employer and citizens for healthcare; difference in the quality of medical services between regions and cities; insufficient quality of training of medical personnel; poorly developed institute for health managers training; shortage of personnel in areas of narrow specialization; low provision of qualified personnel for the healthcare system; poorly developed system for protecting the rights of patients and medical workers; low availability of medical services.

The factors promoting the development are the following: clearly defined development priorities; significant increase in government healthcare funding; stabilization and improvement of basic medical and demographic indicators of the country's population, including a decrease in the incidence of socially significant diseases; restoration and construction of new healthcare facilities; introduction of new medical technologies in medical and diagnostic process; availability of telemedicine points in medical institutions of rural terrain; the presence of a republican center for health development with branches acting as medical information and analytical centers in all regions of the country.

Support for the development and dissemination of ICT is also carried out within the framework of the project of the World Bank and the Government of the Republic of Kazakhstan "Technology transfer and institutional reform in health care of the Republic of Kazakhstan". From the funds of the WB project, the IT infrastructure of the UHMIS of Akmola and Karaganda oblasts was equipped, as well as retrofitting in Astana.

The issue of creating an information system for the implementation of the compulsory social health insurance system, which has been introduced since 2017, is relevant.

Ultimately, eHealth is evolving through centralization, led by the Ministry of Health, with insufficient involvement of business and citizens in the decision-making.

2.4 Kyrgyzstan

National Health Strategy was adopted in 2012. The E-Health Center was established in 2016 by the Ministry of Health with the aim to speed up implementation of e-services in healthcare and pharm market. The Center is the Strategy's main administrative body. The Government approved the E-Health Program of the Kyrgyz Republic for 2016–2020 and its Action Plan aimed to create E-Health Technical Architecture and a Unified Health Integrated Information System.

At the moment, there is neither national eHealth policy nor national health information system (HIS). The situation with funding sources for eHealth is similar to Armenia: funding is donor or commercial. Public funding is poor. As for the legal frameworks for eHealth, the national law protects the privacy of personally identifiable data of individuals irrespective of whether it is in paper or digital format, as well as governs civil registration and vital statistics and national identification management systems.

The telehealth programs provided are teleradiology and telepsychiatry programs at the intermediate and local levels, and the programs are informal.

National EHR system was not introduced yet. There are no health facilities with EHR. eLearning programs in medicine and public health cover health professionals only. mHealth programs are absent. The big data is not engaged in the health sector yet.

Ultimately, eHealth is led by the Ministry of Health, with significant involvement of business, NGOs, and international organizations (UNDP, World Bank, WHO) in the decision-making.

2.5 Russia

The Ministry of Health is the main policy maker in health care sector in Russia. National health strategy, eHealth policy, and health information system (HIS) were adopted respectively in 1991, 2013 and 2011. All the types of funding are used (public, private, public-private and donor). As for the legal frameworks for eHealth, the national law protects the privacy of personally identifiable data of individuals irrespective of whether it is in paper or digital format, as well as governs civil registration and vital statistics and national identification management systems. All types of telehealth programmes are represented at national and intermediate levels.

The National EHR system was introduced in 2013. However, there is no legislation governing the use of the national EHR system. The system provides primary, secondary and tertiary health facilities with EHR. However automatic vaccination alerting information system is absent.

Increasing the availability of medical care to the population is one of the priorities in the framework of the main direction of the strategic development of the Russian Federation "Healthcare" for the period up to 2025, approved by the Protocol of the meeting of the presidential Council for strategic development and priority projects dated March 21, 2017 No. 2. The most important service that increases the availability of medical care to the population is an appointment with a doctor. Over the past few years, this service has been converted to electronic form in most regions of Russia and has become the first successful step towards the digital transformation of healthcare – doctor appointments have become faster, easier and available online. Globally the importance of eHealth services is rapidly increasing. They may reduce costs, improve the access to services and patient self-management [20].

The Single State Information System of Health Care (SSISHC) approved by the Order of the Ministry of Health and Social Development of the Russian Federation of April 28, 2011 No. 364 [21]. The Concept of Creating a Single State Information System in Healthcare defines the purpose, principles, general architecture, the main stages of creating the information system in the health sector, the governance mechanisms and resource support for its creation and maintenance, as well as the expected socio-economic effect. The SSISHC was created for ensuring effective information support of bodies and organizations of a health care system and also citizens within processes of management of medical care and its direct receiving.

It is planned that in 2024 all the state and municipal medical organizations will provide access to citizens to electronic medical documents in the Personal account "My Health" on the State Services portal; will use the medical information systems to organize and provide medical care to citizens and provide information interaction with the SSISHC; will ensure continuity of medical care to citizens by organizing the information interaction with centralized subsystems of the state information systems in health care sector in Russia.

eLearning programmes in biomedical/life sciences cover health professionals only. The wide range of mHealth programmes is established (or launched as pilot) and represented mainly at all the levels. Social media in eHealth are widely used, for instance, to help manage patient appointments or learn about health issues. The use of big data in the health sector is not provided yet.

The initiatives in the public-private partnership (PPP) format in eHealth sector have not yet become widespread. There are only exceptional examples in the regions [22]. The PPP mechanisms facilitate the improving of the quality of medical services. Activation of theoretical and practical developments in this direction will serve as a balanced cooperation between the state and business in the field of health care [23].

Studying the information presented above it can be summarized that managing the eHealth development in EAEU is impossible without a common system for measuring the level of eHealth services and an objective comparison of different regions with each other. However, it is important to ensure the appropriate infrastructure and Internet access to all the health institutions.

3 eHealth in the European Union

At the present stage, expanding the focus of research on eHealth development requires the use of complex tools and is shifting towards expert assessments that can best reflect the current state of the problems and build predictive models of development based on it. In this regard, the experience of colleagues implementing research projects that are similar in goals and objectives is of interest. The EU's experience shows the importance of integrating the eHealth as a basis for achieving seamless interaction of public and private information systems of the European Union [24].

The Eurasian Union has individual features, its own identity, and a unique legal nature, but the concept of its construction cannot but rely on international practice and, above all, on the rich experience of the European Union. In this regard, it is necessary to study the European integration processes in order to draw a rational grain.

The important role of eHealth in health and social care systems has been widely recognized in the EU Member States. The European Commission is supporting them in implementing eHealth in their healthcare systems through different activities including the work towards the deployment of electronic health records (EHR). The Directive on the application of patients' rights in cross-border health, the eHealth Task Force Report 'Redesigning health in Europe for 2020', the eHealth Action Plan 2012–2020: 'Innovative healthcare for the 21st century', projects assessing the impact of eHealth tools on patient empowerment – these are few examples of documents and EU-wide projects that aim at enabling fast access to vital information and sharing of information among health professionals, improving access to quality and efficiency of care [25].

An overview of the current national laws on electronic health records (EHRs) in the EU Member States and their interaction with the provision of cross-border eHealth services is discussed in [26].

The European Commission's eHealth Action Plan 2012–2020 [27] provides a roadmap to empower patients and healthcare workers, to link up devices and technologies, and to invest in research towards the personalized medicine of the future.

The eHealth Digital Service Infrastructure (eHDSI) facilitates the cross-border exchange of health data including patient summaries and e-prescription. Through 'core services', the European Commission is providing a common ICT infrastructure and crosscutting services (terminology, interoperability etc.) to EU countries. They can then set up ‘generic services’ to connect national eHealth systems through ‘National Contact Points for eHealth (eHealth NCPs)’, with financial assistance from the Connecting Europe Facility (CEF) [28].

The electronic cross-border health service is an infrastructure ensuring the continuity of care for European citizens while they are travelling abroad in the EU. This gives EU countries the possibility to exchange health data in a secure, efficient and interoperable way.

The following two electronic cross-border health services are currently progressively introduced in all EU countries:

- ePrescription (and eDispensation) allows EU citizens to obtain their medication in a pharmacy located in another EU country, thanks to the online transfer of their electronic prescription from their country of residence where they are affiliated, to their country of travel.
- Patient Summary provides information on important health related aspects such as allergies, current medication, previous illness, surgeries, etc. It is part of a larger collection of health data called electronic Health Record. The digital Patient Summary is meant to provide doctors with essential information in their own language concerning the patient, when the patient comes from another EU country and there may be a linguistic barrier. On a longer term, not only the basic medical information of the Patient Summary, but the full Health Record should become available across the EU. The exchange of ePrescriptions and Patient Summaries is open to all the Member States [29].

4 Discussion

Digital economy development index lets to conduct the cross-country comparisons of the degree of readiness, use and impact of digital technologies on socio-economic development and contains the tools for assessing the level of development of the digital economy in the country [30].

Currently, a system of indicators for monitoring the digital economy at the international level is being developed, in particular, within the framework of the Group of twenty (G20). The organization for economic cooperation and development has updated the standards for statistical monitoring of the use of digital technologies by people and businesses, and a number of international organizations and analytical agencies have made efforts to develop composite indices for the development of the digital economy and society. In countries implementing digital economy strategies and programs, monitoring systems are being formed at the national level.

The calculation of the digital economy development index makes it possible to conduct a comparative analysis of the current situation in Digital Economy Program’s directions and assess the digitalization processes in various sectors of the Russian economy, based on the international comparisons [3].

However, use of the digital solutions in health care system become an increasingly significant part of the country’s critical infrastructure and play a vital role in digital economy development. The digital technologies increase the efficiency and productivity of the health care system, attract the investments, and improve the treatments.

The values of integrated indicators for Russia are close to or higher than the average for the countries included in the rating in the areas of cyber security, digital government, and digital health. Therefore, there is a good basis for using the potential of digital healthcare in Russia as a basis for integration with other EAEU member states.

The issue of the digitalization of healthcare consider in Digital Agenda of EAEU in the member states without set development indicators and assessment of development dynamics [4], in each country independently, without discussing integration issues. However, the success of the implementation of common single eHealth space in EAEU mostly depends on the base state in both ICT and healthcare sectors.

Table 1 demonstrates the general context of healthcare development in EAEU member states. The information, presented in table, is retrieved from the World Health Organization official materials, ICT Development Index rank 2017, and Networked Readiness Index 2016 [31].

Table 1. Country context indicators

Indicator\Country	ARM	BLR	KAZ	KGZ	RUS
Life expectancy at birth (years)	71	72	68	69	69
Total health expenditure (% GDP)	4.5	6.1	4.3	6.7	6.5
ICT Development Index rank 2017	5.76 (75)	7.55 (32)	6.79 (52)	4.37 (109)	7.07 (45)
Internet users (% population)	39.2	46.9	53.3	21.7	53.3
Impact of ICTs on access to basic services (e.g., health), WEF Networked Readiness Index 2016, indicator 10.01	4.3 (63)	–	4.5 (53)	3.2 (122)	3.9 (88)

Based on the analysis of the data presented in the table, we can summarize that there are significant differences in the basic conditions of medical development of EAEU countries. It can be stated that the average life expectancy is about the same in all member states. Besides, the spending on healthcare in Belarus, Russia, Kyrgyzstan and in Armenia and Kazakhstan are almost the same. However, high expenses for healthcare (as in Kyrgyzstan) must be accompanied by ICT development and Internet availability; otherwise, the eHealth implementation will not be technically possible. Besides, implementing the common eHealth space and information integration in EAEU, the technical, semantic, organizational and legal conditions should be provided, including such parameters as unification, standardization, and harmonization [6].

Therefore, despite the availability of eHealth services in the EAEU countries, they are not yet widely used in the real medical practice. Thereby, it is important to promote the successful solutions and develop of the assessments methodologies evaluating and demonstrating the eHealth systems benefits.

However, the main problem that is still under the discussion is the inability to make an adequate objective assessment of eHealth development. The clear quantitative outcome indicator to reflect the effectiveness of institutional strengthening interventions is missed.

To achieve the objectives of improved access, quality, efficiency, and governance, it is proposed a multilevel (national and regional), multisectoral (primary and secondary health care), and multi-intervention approach that focused on implementation of the family medicine model, the optimization of the hospital networks, and strengthening government capacity for health sector policymaking and monitoring. Besides, the institutional strengthening in eHealthcare sector will facilitate the policy making, planning, regulation, human resources development, for more effective governance system, better managing of public health threats. The health system modernization includes not only the investments in infrastructure, but also the acceptance of the proposed organizational changes involving strong stakeholders in the hospital sector [18].

5 Conclusion and Future Work

Based on the analysis of the eHealth development in EAEU countries, it can be concluded that the absence of a targeted policy for the eHealth development within the framework of the Digital Agenda of EAEU may cause the major challenges and barriers in integration processes related to the increase in the flow of labor migration, scientific, technical and educational cooperation of the EAEU countries. Significant restrictions imposed by the government and health systems on cross-border movements due to the epidemiological situation require the development of special procedures and regulations for the interaction of medical information systems of the EAEU countries.

eHealth development in the member states of the Eurasian Economic Union contributes to the fundamental improvement in the quality and availability of health services, while stimulating the growth in this new promising sector. Based on the fact that EAEU member states do not conduct regular assessments on citizens' use of eHealth features and services and have not established indicators to measure the level of eHealth development, the issues on elaborating the common standards in eHealth, monitoring the eHealth development systems in the EAEU countries, using the experience of both the EAEU countries and the EU are perspective and can be viewed as the directions of further investigations. In this regard, the researches based on big data analysis using artificial intelligence technologies can be promising in further work.

The introduction of the information technologies in various spheres of public life poses new challenges to the society and requires the development of new approaches to ensuring the information security. In this regard, an important component is to provide conditions for the protection of private information at all stages of the introduction of the digital technologies in public relations and, mainly, at the stage of their design and development.

The success of digital technologies implementation in healthcare sector depends on the real participation of all the stakeholders at all levels of care involved in the development and not just implementation of future reforms [18]. Moreover, without strategic planning at the governmental level, it is impossible to achieve serious success in this area. The state has not only financial resources, but also administrative levers that can be used to create infrastructure conditions for the introduction of digital technologies. Therefore, a competent digital policy [32] may facilitate the creation of the effective eHealth system.

The focus for future reform is on strengthening preventive services and improving the quality and efficiency of specialist services. The key challenges in achieving this involve reducing excess hospital capacity, strengthening health care management, use of evidence-based treatment and diagnostic procedures, and the development of more efficient financing mechanisms [19]. However, the digital transformations [33] must be provided in the entire healthcare ecosystem, encompass all its aspects and innovations.

References

1. Bolgov, R., Karachay, V., Zinovieva, E.: Information society development in Eurasian economic union countries: legal aspects. In: ACM International Conference Proceeding Series. 8th International Conference on Theory and Practice of Electronic Governance, ICEGOV 2014, pp. 387–390 (2014). https://doi.org/10.1145/2691195.2691278
2. Bolgov, R., Chugunov, A., Filatova, O., Misnikov, Y.: Electronic identification of citizens: comparing perspectives and approaches. In: ACM International Conference Proceeding Series. 8th International Conference on Theory and Practice of Electronic Governance, ICEGOV 2014, pp. 484–485 (2014). https://doi.org/10.1145/2691195.2691245
3. Meshkova, T. (ed.): Eurasian economic integration: prospects of development and strategic goals for Russia. In: XX April International Scientific Conference on Economic and Society Development Issues, Moscow, 9–12 April 2019. HSE University (2019). https://www.hse.ru/data/2019/04/12/1178006152/11%20Евразийская_интеграция.pdf. (in Russian)
4. Digital Agenda of EAEU. 2016-2019-2025. Eurasian Economic Commission. M. (2019). http://www.eurasiancommission.org/ru/Documents/digital_agenda_eaeu.pdf. (in Russian)
5. Lapicky, V.A., Tom, I.E.: E-health of Belarus: current state and perspective. Informatics **15**(4), 63–71 (2018). https://inf.grid.by/jour/article/view/468/639. (in Russian)
6. Rusanova, N.E.: History and issues of digital healthcare in Russia. Popul. Econ. **2**(2), 5–40 (2018). https://doi.org/10.3897/popecon.2.e36046
7. Gusev, A.V., Pliss, M.A., Levin, M.B., Novitsky, R.E.: Trends and forecasts for the development of medical information systems in Russia. Doct. Inf. Technol. **2**, 38–49 (2019)
8. Lebedev, G., Zimina, E., Korotkova, A., Shaderkin, I., Kirsanova, E.: Development of Internet technologies for health care in the Russian Federation. Public Health Panor. **05**(01), 103–111 (2019). https://www.euro.who.int/__data/assets/pdf_file/0003/398172/PHP-vol5-issue1-eng.pdf?ua=1. Accessed 30 June 2020
9. Davtyan, K., Davtyan, H., Patel, N., Sargsyan, V., et al.: Electronic health information system implementation in health-care facilities in Armenia. Public Health Panor. **05**(01), 44–53 (2019). https://www.euro.who.int/__data/assets/pdf_file/0003/398172/PHP-vol5-issue1-eng.pdf?ua=1
10. Ministry of Health of the Kyrgyz Republic: eHealth in the Kyrgyz Republic. Strategy and Action Plan 2015–2020 (2015)
11. Sharman, A.: A new paradigm of primary health care in Kazakhstan: personalized, community-based, standardized, and technology-driven. Cent. Asian J. Glob. Health **3**(1), 186 (2014). https://doi.org/10.5195/cajgh.2014.186
12. Scott, J.: Institutions and Organizations. Sage Publications, Thousand Oaks (1995)
13. Kostova, T.: Country institutional profiles concept and measurement. Acad. Manage. Proc. **1**, 180–184 (1997)
14. Baba, M.L., Blomberg, J., LaBond, C., Adams, I.: Approaches to Formal Organizations. A Companion to Organizational Anthropology. Blackwell Publishing Ltd., Hoboken (2013). https://msu.edu/~mbaba/documents/New_Institutional_Approaches_to_Formal_Organizations.pdf

15. Friel, D.: Understanding institutions: different paradigms, different conclusions. Revista de Administracao **25**(2), 212–214 (2017)
16. Thoenig, J.-C.: Institutional theories and public institutions: new agendas and appropriateness. In: The Handbook of Public Administration, pp. 185–201 (2011)
17. Atlas of eHealth country profiles: the use of eHealth in support of universal health coverage: based on the findings of the third global survey on eHealth 2015. World Health Organization (2016). https://apps.who.int/iris/rest/bitstreams/908392/retrieve
18. Armenia. Achievements and Challenges in Improving Health Care Utilization. A Multiproject Evaluation of the World Bank Support to the Health System Modernization (2004–2016). Report No. 134584, 20 March 2019. http://ieg.worldbankgroup.org/sites/default/files/Data/reports/ppar_armeniaimprovehealthcare.pdf
19. Richardson, E., Malakhova, I., Novik, I., Famenka, A.: Belarus: health system review. Health Syst. Transit. **15**(5), 1–118 (2013)
20. Wynn, R., Gabarron, E., Johnsen, J.-A.K., Traver, V.: Special Issue on e-Health services. Int. J. Environ. Res. Public Health **17**(8), 2885 (2020). https://doi.org/10.3390/ijerph17082885
21. Order of the Ministry of Health and Social Development of Russia, 28.04.2011 No. 364 "On Approval of the Concept of Creating a Single State Information System in Healthcare"
22. The Ministry of Health proposes to establish the responsibility of regions for the introduction of electronic prescriptions. https://medvestnik.ru/content/articles/Minzdrav-predlagaet-ustanovit-otvetstvennost-regionov-za-vnedrenie-elektronnyh-receptov.html. (in Russian)
23. Skryl, T.V.: Public-private partnership in health care system: case study in Russia. CITISE **4**, 459–466 (2019). https://doi.org/10.15350/24097616.2019.4.43
24. EU Health Programme. https://ec.europa.eu/health/funding/programme_en
25. Peetso, T.: Addressing eHealth at the EU level. In: Rinaldi, G. (ed.) New Perspectives in Medical Records. T, pp. 115–122. Springer, Cham (2017). https://doi.org/10.1007/978-3-319-28661-7_9
26. Overview of the national laws on electronic health records in the EU Member States and their interaction with the provision of cross-border eHealth services Final report and recommendations, 23 July 2014. https://ec.europa.eu/health//sites/health/files/ehealth/docs/laws_report_recommendations_en.pdf
27. eHealth Action Plan 2012–2020 - Innovative healthcare for the 21st century. https://ec.europa.eu/newsroom/dae/document.cfm?doc_id=4188
28. eHealth: Digital health and care. EU cooperation. https://ec.europa.eu/health/ehealth/cooperation_en
29. Electronic cross-border health services. https://ec.europa.eu/health/ehealth/electronic_crossborder_healthservices_en
30. N 35 National index of the development of digital economy: Pilot realization. Moscow, Rosatom (2018). (in Russian)
31. ICT Development Index 2017. https://www.itu.int/net4/ITU-D/idi/2017/index.html
32. Konkov, A.: Digital politics vs political digitalization. Vestnik of Saint Petersburg University. Int. Relat. **13**(1), 47–68 (2020). https://doi.org/10.21638/spbu06.2020.104
33. Chugunov, A., Bolgov, R., Kabanov, Y., Kampis, G., Wimmer, M., et al.: Preface. In: 1st International Conference on Digital Transformation and Global Society, DTGS 2016. Communications in Computer and Information Science (CCIS), vol. 674, pp. III–VI (2016). https://doi.org/10.1007/978-3-319-49700-6

National Policy Against Corruption in the Framework of Global Integration Processes

Anna Mokhorova(✉), Dmitriy Mokhorov, Bella Kerefova, and Sergey Kosarev

Peter the Great St. Petersburg Polytechnic University, St. Petersburg, Russia

Abstract. Combating corruption is an important part of the national policy of states. However, measures taken at the state level are no longer sufficient to fight corruption, which is a new challenge for states and governments. As a result, international anti-corruption cooperation is strengthening, attempts are made to develop a unified strategy to combat corruption at the regional and interstate levels, new methods of combating corruption are formed, and a regulatory framework for combating corruption is created.

The research methodology is based on a systematic approach to the phenomenon under study that examines the fight against corruption from the political and legal, socio-economic, and moral perspectives; the institutional method that made it possible to consider the fight against corruption as a special political and legal institution of the modern society; and the concept of a philosophical worldview that allows an in-depth analysis of integration anti-corruption processes.

The result of the work is related to determining the need for a clearer legal regulation of the nomenclature in the field of combating corruption at the international level, as well as the need to coordinate anti-corruption strategies and methods to in different countries.

Keywords: Corruption · National policy · International cooperation · Combating corruption

1 Introduction

Corruption is a multidimensional and multilayered phenomenon. As history shows, it is inherent in any society, and its structure, scale, and dynamics are the result of political, social, and economic problems. The level of corruption increases when a society is on the verge of social and political changes and economic reforms.

Corruption is always of social importance and is an activity aimed at distorting the norms of social relations and criminalizing their content by means of individuals using their powers for lucrative purposes contrary to their professional (official) interests.

Imperfection of legislation, ugly forms of social and economic transformation, poverty, stratification of society, inadequate activity of state structures, legal nihilism,

© The Author(s), under exclusive license to Springer Nature Switzerland AG 2021
R. Bolgov et al. (Eds.): *Proceedings of Topical Issues in International Political Geography*, SPRINGERGEOGR, pp. 125–132, 2021.
https://doi.org/10.1007/978-3-030-78690-8_11

people's disbelief in the punishment of corrupt officials lead to the growth of corruption in society. The determinants of corruption may vary, with corruption itself preserving and aggravating social and economic problems, creating difficulties in eliminating factors that help contain it. The root cause of corruption, however, always lies in the very essence of the state and is created by the problems that occur in it.

The legislative power does not always keep up with the changes which take place in society. There may be "confidential jurisprudence", legal norms may be introduced that are obviously not applicable in practice; lobbyism may arise; bills may be "stuck" and their essence may be diluted, including through malicious intent. Corruption may arise both because of gaps in the legal framework and because of the adoption of complex rules hardly suitable for application.

The growth of corruption is also facilitated by the instability of the political structure of society, substitution of political will with slogans, imperfection of legal methods of economic management, reduction of the moral potential of society when corruption manifestations are not regarded as something shameful.

The presence of organized crime in the country, whose main task is to redistribute the financial and material resources of the state, significantly contributes to the growth of corruption. New forms of organized crime have emerged with the creation of their own commercial structures, banks, and financial-industrial companies (groups) with the establishment of control over the major equity stake in enterprises, with penetration into the banking and financial systems of the state, with the establishment of financial and economic control over profitable enterprises defining the personnel policy in them, with submission of traditional spheres of economy (oil and gas, non-ferrous metals, etc.) with infighting and elimination of undesirable officials and journalists, with protection racket of structures, with bribery of law enforcement officers, with transfer of money to foreign banks to the accounts of rogue firms, with the influence on the conduct of elections at all levels by bribing voters and exerting pressure on election commissions with promoting their own deputies to power structures and their subsequent involvement in liquidating economic, administrative, and political competitors.

In fact, organized crime is an active stimulator of all corruption processes in society; it is difficult to prove the guilt of organized crime leaders, there are problems with return of money received as a result of such an activity.

Corruption is constantly growing and its forms are evolving and changing, which is a serious challenge to states to create efficient measures to combat it. Combating corruption implemented within the framework of national policies can no longer reduce the level of corruption, and therefore, international cooperation in this area should be developed.

2 Literature Review

Issues of combating corruption in the context of international cooperation have received considerable attention in both foreign and Russian literature.

A number of authors directed their research to develop a typology of anti-corruption national policies in order to implement interdisciplinary approaches to the study of this phenomenon [1]. The influence of the norms of national law on international law is considered with the purpose of forming new approaches in the fight against corruption [2]. The transformation of global norms of the fight against corruption and changes in their scope are studied [3].

In Russia, anti-corruption cooperation between states is considered in the following aspects. The possibility of transferring some functions of combating corruption to international or regional organizations in a globalizing world is analyzed. The reverse side of such actions is permeation of corruption into international relations, which leads to such consequences as transnationality and supranationality of corruption [4, 5]. The influence of international law on the mechanisms of combating corruption in Russia is considered [6–9]. The analysis of international and Russian anti-corruption principles is conducted, as well as their effectiveness in international and Russian anti-corruption experience [10].

3 Methodology

A national anti-corruption policy is seen as a system that consists of a number of interrelated elements that interact with each other and with the environment (regional and international organizations, international law). The systematic approach to studying corruption and combating it provides an opportunity to identify the basis for corruption and important elements of corruption offences that need to be influenced when implementing anti-corruption policies. The main elements of the fight against corruption are identified - official corruption, which is associated with the use of "administrative resources", and organized crime, which covers a significant number of individuals and broad areas of economic activity; the main directions of anti-corruption national policy and international cooperation are identified. The systematic approach allows creating a coherent mechanism for combating corruption, determining the degree of importance of combating one or another manifestation of corruption and eventually developing an effective anti-corruption policy, which is impossible without systematized legislation aimed at combating corruption.

Application of the institutional approach allowed describing the impact of anti-corruption processes on various institutions of society in the social, political, and economic spheres.

The use of the concept of a philosophical worldview in relation to the institutions and mechanisms of anti-corruption policy implementation made it possible to identify the main trends in the development of anti-corruption activities, peculiarities of the national anti-corruption policy in the Russian Federation, and the level of interaction with international institutions to combat this phenomenon [11–13].

4 Findings

The problem of corruption is recognized by all governments, and transnational corruption infringes on normal relations. Many authors share the view that it is necessary to form a ground for combating corruption in the form of a clear and balanced system of anti-corruption legislation and subordinate legal acts [11, 14]. However, the abundance of legal documents somehow affecting the fight against corruption in Russia, unfortunately, today does not mean that there is a comprehensive system of legal regulation of this issue. For the first time in our country, the first National Anti-Corruption Plan was approved only in 2008, and a serious legislative framework to combat corruption began to be created, organizational measures were taken to prevent corruption, and the activities of law enforcement agencies to combat corruption intensified [15].

However, despite such measures of the state and society, corruption continues to seriously hinder the normal functioning of all social mechanisms, hinders social transformation and modernization of the national economy, causes serious concern and distrust in state institutions, creates a negative image of Russia in the international arena, and is rightfully considered as one of the threats to the security of the Russian Federation [16, 17].

These provisions demonstrated the necessity to adopt the National Anti-Corruption Strategy, which is a constantly improving system of measures of organizational, economic, legal, informational, and personnel nature, taking into consideration the federal structure of the state, covering all levels of government, aimed at eliminating the root causes of corruption in society and consistently implemented by all subjects of state and power activities, as well as legal entities and citizens.

The National Anti-Corruption Strategy allows for a certain system of measures aimed at combating corruption, which includes legal and organizational frameworks to combat corruption, creation of barriers to the implementation of corrupt behavior, development of coercive measures against persons who violate anti-corruption requirements [4, 18].

A national anti-corruption policy presupposes a wide participation of civil society institutions in combating corruption; expanding the system of legal education of the population; improving the efficiency of activities of state authorities and local government bodies to combat corruption; improving law enforcement practice of law enforcement bodies and courts in the cases related to corruption; improving the efficiency of execution of judicial decisions; improving the current legislation in state service, state procurement, civil, labor, and pension legislation; forming an organizational framework of anti-corruption activity.

On the basis of the principle of reciprocity, the Russian Federation cooperates with foreign states, their law enforcement agencies and special services, as well as with international organizations in the field of combating corruption. The main purpose of such cooperation is investigation and solving corruption crimes of international level, information exchange, and coordination of anti-corruption activities [19, 20].

Currently, the following forms of international cooperation in the fight against corruption exist: 1) based on international treaties; 2) based on the principle of reciprocity.

International treaties are understood as interstate, intergovernmental, and interagency treaties regardless of their type and name.

Russia's international cooperation in this area is also based on various international legal acts on anti-corruption [21–23].

Within the CIS, the legal basis for this activity is the Treaty of the States Parties of the Commonwealth of Independent States on Combating Money Laundering and Financing of Terrorism, adopted in Dushanbe on October 5, 2007, aimed at combating various types of crimes, including corruption.

In accordance with the Agreement on Partnership and Cooperation establishing a partnership between the Russian Federation and the European Communities and their Member States, the parties agreed in 1994 on cooperation to prevent illegal activities in the economic sphere, including corruption problems.

In 2006, during the G-8 Summit in St. Petersburg, Russia joined the Anti-Corruption Initiative to combat corruption among top officials.

On February 1, 2007, Russia joined GRECO, the Group of States against Corruption, an international organization founded by the Council of Europe in 1999, whose main goal is to help member countries fight corruption. GRECO sets anti-corruption standards (requirements) for government activities and monitors compliance of practices with these standards, helps to identify shortcomings in national anti-corruption policies, and proposes the necessary legislative, institutional, or operational measures, provides a platform to share best solutions to detect and prevent corruption [24].

Criminal liability for corruption comes under the norms established by international treaties or under national legislation.

Thus, the Palermo UN Convention against Transnational Organized Crime of November 15, 2000 allows for unification of legislation in this sphere, and corruption is attributed to organized crime, it is recognized that honest business is made unprofitable, its moral principles are reduced.

This Convention, the Council of Europe Convention on Criminal Responsibility for Corruption of January 27, 2006 and the OECD (Organization for Economic Co-operation and Development) Convention on Combating Bribery of Foreign Public Officials of December 17, 1997, propose to criminalize corruption for legal entities, to recognize non-material benefits (benefits, services, etc.) as objects of bribery, and to make foreign public officials and international civil servants liable for corruption.

Part of these recommendations have already been taken into consideration in the latest Russian legislation.

Currently under consideration is the issue of Russia's signing of the Council of Europe Convention on Civil Liability for Corruption, which defines various civil law aspects of protecting individuals from corruption, in particular, if a defendant has committed an act of corruption or has not prevented it, a plaintiff has suffered damage and there is a causal link between these facts.

The Palermo UN Convention against Transnational Organized Crime of November 15, 2000 proposes preventive measures to focus on strengthening cooperation between law enforcement agencies, to develop a standard procedure designed for the conscientious work of lawyers, notaries, accountants, and managers of commercial enterprises to create a public registry of legal entities and individuals involved in corrupt practices, of persons deprived of the right to hold the posts of heads of legal entities with an exchange of such information on the international level, including independent media that have a possibility of investigation which has nothing to do with bribery of journalists who can create biased materials [25].

The issues of corruption prosecution are roughly the same in all countries.

This is the high latency of corruption and the complexity of proving in corruption-related cases, which is greatly facilitated by the connections and position of corrupt officials.

However, it should be noted that criminal cases of corruption can also be initiated by political opponents.

Besides, unification of legislation, mutual assistance of states on extradition of criminals plays a huge role in the fight against corruption. The legal framework for this process is developed in sufficient detail at the international level [26, 27]. There are also international treaties on joint operational and investigative activities in corruption cases. These documents are developed in bilateral and multilateral acts.

Thus, Russia participates in many international anti-corruption projects: it has ratified certain international conventions, concluded about 100 bilateral treaties on strategic and operational cooperation with foreign countries in the fight against corruption, and adopted international anti-corruption standards [28].

5 Conclusions

Every year, the problem of combating corruption is becoming more and more global. Today, the process of globalization involves a huge number of actors, the relations between states are developing at a rapid pace, and corruption is becoming cross-border. This process leads to "diffusion of corruption" [28], that is, easy penetration of such a phenomenon into any country and spread to a significant circle of social relations. Consequently, the fight against corruption within one state becomes obviously insufficient and does not imply achieving an unambiguously positive result. Therefore, there is a need to join efforts of the international community to implement anti-corruption policy.

The following can be highlighted as the main directions of joint activities of the states. One of the most important is the law-making activity aimed at creation of international convention norms and their implementation in the national legislation of states. And the important issue here is to define the nomenclature. Unambiguously, today, the concept of corruption is not defined either at the legislative level (neither international, nor regional, nor national), as a result, no derivatives of it can be defined. Discussion of the problem has been reflected in the works of many authors [29], but no solution has been found so far.

The law enforcement activities of states should ensure all established universal principles of international cooperation and respect the implemented norms aimed at combating corruption.

From a political standpoint, states must maintain an appropriate level of competition in the management of state affairs. The correlation between the level of corruption and competition in the state has been studied by a number of authors [22], who concluded that the less democratic the electoral procedures are, the less protected human and civil rights and freedoms are, the higher the level of corruption in the state, the greater the danger in the international arena in terms of corruption processes is posed by this actor. Depending on the level of competition in public administration, it is necessary to develop a set of measures to combat corruption. In other words, measures effective for combating this phenomenon in democratic states may "not work" in authoritarian states.

From the point of view of social processes taking place in states, it is necessary to take into account the location of the country in the period of socio-economic or social political transformation. In conditions of instability, corruption processes increase and intensify. Russia is one of the transforming societies where institutional systems are weakened, resulting in deviations of social practices from the existing formal norms, when they acquire non-legal nature. Practice shows that the reliance of the state on the bureaucracy, with little effective control over the performance of functions defined by the corresponding legislative mechanisms, provokes increased corruption, just as corruption in such conditions increases bureaucracy. In this case, corruption is a consequence of the blurred "rules of the game" established by the state. That is, in the existence of laws that do not clearly and unambiguously formulate conditions for a civil servant to perform their official duties, the limits of authority, "can" and "cannot", there are prerequisites for corrupt behavior.

Only those countries where anti-corruption activities have been introduced into the rank of state policy, where work is carried out in a systematic, transparent, and monitoring manner, are the most successful. Nevertheless, it can be concluded that the fight against corruption is not only on a solid international legal basis, but is already having a fairly tangible effect in many areas.

A thoughtful balance between Russian and international anti-corruption principles is an opportunity for prospective minimization of acts of corruption and a basis for effective fight against corruption.

References

1. Villeneuve, J.-P., Mugellini, G., Heide, M.: International anti-corruption initiatives: a classification of policy interventions. Eur. J. Crim. Policy Res. **26**(4), 431–455 (2019). https://doi.org/10.1007/s10610-019-09410-w
2. Lang, B.: China's anti-graft campaign and international anti-corruption norms: towards a "new international anti-corruption order"? Crime Law Soc. Chang. **70**(3), 331–347 (2017). https://doi.org/10.1007/s10611-017-9742-y
3. Jakobi, A.: The changing global norm of anti-corruption: from bad business to bad government. Z. Vgl. Polit. Wiss. **7**, 243–264 (2013). https://doi.org/10.1007/s12286-013-0160-y
4. Gurzhy, T.A.: Corruption as a phenomenon of globalization. Actual Probl. Econ. Law, 24–31 (2014)
5. Rudenko, V.V.: Anti-corruption policy in the context of globalization. Current issues of scientific support for the state anti-corruption policy in the Russian Federation, 61–67 (2019)
6. Shtatina, M.A.: Effect of globalization on formation of anti-corruption administrative and legal mechanisms. Globalization Publ. Law 211–221 (2016)
7. Polyakov, M.M.: Anti-corruption administrative and legal regulation in public administration. Actual Probl. Russian Law **9**, 109–115 (2017). https://doi.org/10.17803/1994-1471.2017.82.9.109-115
8. Nechevin, D. K.: Anti-corruption problems. Publ. Priv. Law, 200–211 (2015)
9. Pashchenko, F.F.: Corruption as a sign of system crisis in Russian statehood. Problem analysis and public administration design, 2–4 (2011)
10. Tsurikov, V.I.: Corruption and corrupter in modern Russia. Sociodynamics **4**, 1–24 (2017). https://doi.org/10.7256/2409-7144.2017.4.20868
11. Nisnevich, Y.A.: Public power and corruption: socio-anthropological approach. Polis **6**, 89–96 (2014)
12. Baranova, M.V., Chikireva, I.P.: International and Russian anti-corruption principles. J. Nizhny Novgorod Acad. Ministry Intern. Affairs Russia **4**(44), 56–59 (2018). https://doi.org/10.24411/2078-5356-2018-10417

13. Bukharina, N.P.: Concept and indications of corruption in acts of international law. law. J. High. School Econ. **1**, 166–176 (2016). https://doi.org/10.17323/2072-8166.2016.1.166.176
14. Kuznetsova, O.A.: Several features of corruption determination in modern Russia. Tambov Univ. Rev. **12**, 128–129 (2013)
15. Sokolov, M.S.: Theoretical and applied aspects of anti-corruption policy: Russian and international experience. Secur. Issues **1**, 15-26. (2016). https://doi.org/10.7256/2409-7543.2016.1.18350
16. Levakin, I.V., Shishova, Z.A.: Anti-corruption policy: development of legislation and law enforcement practices in Russian Federation. Monograph. Ed. S.M. Shakhray. Moscow, NII SP, vol. 29 (2012)
17. Okhotsky, E.V.: International anti-corruption legal standards and their implementation into Russian practice. Publ. Priv. Law **3**, 4–77 (2015)
18. Aust, A.: Handbook of International Law, 2 edn, vol. 68. Cambridge University Press, Cambridge (2010)
19. Golovanova, S.V., Meleshkina, A.I.: Assessment of mutual influence between corruption and competition. Econ. Policy, 11–15 (2016)
20. Gravina, A.A.: Transnational corruption as a component of the international crime, 16–21 (2015). https://doi.org/10.12737/16640
21. Sukhanov, V.A., Dulova, I.E.: Legal aspects of activities of international anti-corruption organizations. Moscow J. Int. Law, **106**(2), 128–135 (2017). https://doi.org/10.24833/0869-0049-2017-106-2-128-135
22. Golubykh, N.V., Lepikhin, M.O.: Improvement of anti-corruption laws. Actual Probl. Econ. Law **2**, 102–109 (2019). https://doi.org/10.17803/1994-1471.2019.99.2
23. Rumyantseva, E.E.: Breaches of Russian anti-corruption laws and unwritten rules of the anti-corruption policy. Monit. Publ. Opin. Econ. Soc. Changes **3**, 116–127 (2017). https://doi.org/10.14515/monitoring.2017.3.08
24. Trubnikova, E.I., Trubnikov, D.A.: The problem of institutional corruption in the system of higher education. Vysshee obrazovanie v Rossii - Higher Educ. Russia **27**(12), 29–38 (2018). https://doi.org/10.31992/0869-3617-2018-27-12-29-38
25. Babeluk, E., Savin, S., Shchepelkov, V.: On the national system of monitoring the effectiveness of corruption counteraction. Russian J. Criminol. **12**(5), 613–621 (2018). https://doi.org/10.17150/2500-4255.2018.12(5)
26. European convention on the supervision of conditionally sentenced or conditionally released offenders (Strasbourg, 30 November 1964)
27. European convention on the international validity of criminal judgments (along with «The list of other offences, except for offences under criminal law»), signed in the Hague on 28 May 1970
28. Convention on the transfer of sentenced persons (Strasbourg, 21 March 1983)
29. European Convention on Extradition (Paris, December 13, 1957); European Convention on Mutual Assistance in Criminal Matters dated April 20, 1959; European Convention on the Transfer of Proceedings in Criminal Matters (Strasbourg, 15 May 1972)
30. Lipinsky, D.A., Musatkina, A.A.: Problems of compliance of Russian anti-corruption legislation with international corruption standards. Vestnik of Saint Petersburg University. Law **4**, 673–690 (2019). https://doi.org/10.21638/spbu14.2019.404

Russia-China Cooperation: Linking the Eurasian Economic Union and Belt and Road Initiative

Darya Railian, Jingcheng Li(✉), and Sergey Pogodin

Peter the Great St. Petersburg Polytechnic University, St. Petersburg, Russia

Abstract. Bilateral relations between Russia and China are among the most significant in today's rapidly changing world. Reliability, stability and scale of the partnership define various spheres of the tendency of economic development. So far, relations between the two countries are supported by participation in major international organizations and integration projects – Belt and Road Initiative (BRI) and the Eurasian Economic Union (EAEU). Both countries have strategic interests in such projects. In this regard, the review of status and prospects of cooperation between Russia and China in promoting the "two initiatives" (One Belt - One Union) is a relevant measure to predict the development of the Eurasian political and economic integration in general. This article analyzes the cooperation mechanism and forms of the EAEU and BRI in the framework of Russia-China relations. The research revealed the basic strategic interests of Russia and China in the integration process, as well as the role of the EAEU and BRI in bilateral trade and economic cooperation. In conclusion, it is noted that with the expansion of comprehensive cooperation, the Russia-China relations are deepening and bringing benefits to the world.

Keywords: Russia · China · Russia-China cooperation · Eurasian economic union · Belt and road initiative · Silk road economic belt · One Belt - One Union

1 Introduction

The relevance of this study is explained by the fact that the world is constantly undergoing profound transformations in which modern international relations are not to defend or defeat the enemy, but to find allies and strengthen their positions, primarily in regions. The economic growth of emerging countries in recent decades has far exceeded that of traditionally developed countries. This situation is unprecedented in modern history.

In the newest geopolitical conditions, such cooperative efforts as the Belt and Road Initiative (BRI) and the Eurasian Economic Union (EAEU) are prospective multilateral platforms for cooperation between Asian, European and African countries. Relations between Russia and China are the good example of mutually beneficial strategic cooperation. This can be seen both at the level of the two countries and at the level of international economic integration. The two countries have reached an important consensus on interfacing the construction of the EAEU and the Silk Road Economic Belt

© The Author(s), under exclusive license to Springer Nature Switzerland AG 2021
R. Bolgov et al. (Eds.): *Proceedings of Topical Issues in International Political Geography*, SPRINGERGEOGR, pp. 133–139, 2021.
https://doi.org/10.1007/978-3-030-78690-8_12

(part of BRI) in order to implement major economic, political and cultural projects. In 2019, Russia and China celebrated the 70th anniversary of establishing diplomatic relations. The development of cooperation between the two countries has had its ups and downs, but within the framework of the research it would be more practical to consider the current stage and prospects of their interaction.

The purpose of this article is to identify the mechanism of interaction between the EAEU and BRI in the framework of Russia-China cooperation. This article analyzes the main interests of Russia and China in the course of cooperation between the EAEU and BRI; defines the role of the EAEU and BRI in promoting trade and economic cooperation between Russia and China, as well as the prospects of EAEU and BRI development in the future.

Russian scholars A. Holkina [5], M. Korotov and V. Muntiyan [9], A. Tsoi [13], Y. Glittova and A. Toropygin [4], P. Kadochnikov [6] wrote about creation and development of the Eurasian Economic Union. The current stage of development, along with the problems and achievements of the Belt and Road Initiative are highlighted in studies by D. Kozlov [8], S. Luzyanin and A. Afonasyeva [12]. Bilateral relations between Russia and China in the framework of the EAEU and BRI were covered in the studies of Russian and Chinese researchers: E. Davydenko and T. Kolesnikova [3], Li N. [10], Kong Q. [7].

2 Method of Research

The scientific and objective approach to the research of the cooperation mechanism of the Eurasian Economic Union and Belt and Road Initiative in the framework of the Russia-China cooperation provided the theoretical basis for this work. The paper mainly applied a systematic analysis, which allowed to identify the main interests of Russia and China in the linking of EAEU and BRI; the historical method was important in the study of information sources (regulations, journals, etc.); the prognostic method was used to determine the prospects for the development of EAEU and BRI.

3 Results of Research

3.1 Strategic Interests of Russia and China in the EAEU

Since its establishment, the Eurasian Economic Union has aimed to support its member countries (Armenia, Belarus, Kazakhstan, Kyrgyzstan, Russia) in the field of economy. Even though Russia dominates the Eurasian Economic Union economically (87% of total GDP), demographically (80% of total population) and geographically (85% of total territory), while decisions on EAEU policy should be made jointly and each member state has voting rights [9]. That is why there are discussions in the scientific and economic community about the relevance of Russia's accession to the EAEU, arguing that this is due to the huge money spent on modernization of infrastructure projects, development of business and production, as well as support for the EAEU economy. Apparently, Russia is doing all this with the expectation of further economic and social results.

Undoubtedly, there are also positive reasons for Russia's participation in the EAEU, such as strengthening its borders and ties with neighboring countries – Belarus, Kazakhstan, Kyrgyzstan and Armenia. The Union has a positive impact on Russia's geopolitics; moreover, for Russia, the EAEU market is an investment-attractive market, where the presence of Russian industry can be expanded. Support and status in the world economy are also a big plus for Russia. Acting under the auspices of such a large association is much easier to promote its interests on the world arena, as well as to create partnerships with other countries [5].

Thanks to cooperation with different countries within the framework of the EAEU, Russia has received the possibility to develop both traditional directions (for example, trade communications with the non-member countries, development of agrarian and industrial complex), science and technology, innovative projects, education, sports and tourism. But the most important issue has become the creation of digital area shared by all the EAEU member countries.

The development of environmental and energy efficiency projects should be considered as an advantage of Russia's entry into the EAEU. The problem of excessive use of resources in Russia in 2020 is seen as being under control, because the country has huge reserves of mineral resources. However, as many natural resources are not renewable, this problem may arise in the foreseeable future. Researchers Bordachev T., Vishnevsky K. and Glazatova M. state that "the energy intensity of Russia's GDP significantly exceeds that of developed countries. Reducing energy intensity will increase the competitiveness of domestic industry and improve the environmental situation [1, p. 34]".

By participating in the EAEU Russia is modernizing its transport and logistics system. The need to create a coordinated transport policy within the EAEU has pushed Russia to implement long planned changes in the transport infrastructure [6]. Scientific and innovative projects are also the main reason why Russia joined the EAEU. If earlier it was said that the future lies in digital technologies, today it can be confirmed that the future has already come and now there is progress in the field of Internet technology – e-commerce, digital economy, big data and so on. It should be noted that Russia has made huge investments in the EAEU development and it is obvious that the results leave much to be desired. However, by adjusting some sectors of the economy, industry and customs regulations to the EAEU norms, Russia has improved a lot of domestic social and economic structures.

In May 2018, China and the EAEU signed an agreement on trade and economic cooperation. In the agreement, the parties settled issues related to transparency, trade protection measures, technical barriers to trade, customs cooperation and trade facilitation, sanitary and phytosanitary measures, intellectual property, competition, electronic commerce and public procurement. As former Chairman of the Eurasian Economic Commission Board V.B. Khristenko noted, "this agreement marks an important stage in the development of economic cooperation, which regulates the entire structure of relations and creates a basis for further movement, including for a possible agreement on a free trade area" [1, p. 117].

The economic partnership between China and the EAEU is a well-thought-out and mutually beneficial step for the parties. The EAEU needs partners and foreign investments [14]. For China, such cooperation is beneficial for several reasons. First, with the creation of a transport corridor from the East China to the EU through Russia and Kazakhstan, China would significantly reduce the time of transporting goods, and it is the fastest way from China to European countries. Kazakhstan and Russia are members of the EAEU, therefore there is a common customs border between them, which makes this option of transportation more logical and cost-effective for China. It should be noted that under current conditions the transportation of goods from China to Europe by sea (mostly via South China Sea) takes 45 days or more, and 18–20 days by the Trans-Siberian Railway, while the Lianyungang - Hamburg international railway (through Kazakhstan) takes about 11 days [13].

Secondly, in case of effective cooperation between the EAEU and China regarding the unified transport route, it is planned to develop cooperation in all areas of the economy and to create a common economic area between China and the EAEU – a free trade area [4]. For Russia, this kind of integration with China also has its advantages. The Chinese market is suited for the export of Russian agricultural products and minerals. It is also a positive thing that China could invest in infrastructure and energy projects in Russia.

3.2 Interaction Forms Between Russia and China in the BRI

In 2013, during his visit to Kazakhstan and Indonesia, President Xi Jinping announced the idea of "Silk Road Economic Belt" and "21st Century Maritime Silk Road" (Belt and Road Initiative, BRI). The given initiative at the present stage became the largest international initiative of China, therefore many forces and hopes have been invested in its realization and advancement.

In 40 years of dynamic economic development, China has managed to become the second economy in the world. This is an unprecedented indicator, given that active development was not so long, and the country had insufficient resources and support from other countries to make such a dramatic leap. As a fact, it should be noted that from an agricultural and relatively closed state with the largest population, China has turned into one of the most technologically advanced, trade-attractive and economically powerful countries in the world [7]. However, over the last decade the pace of China's economic development has slowed down. This has become one of the substantiations explaining the necessity of industrial transformations in the country.

China has consistently called on the international community to cooperate in economic and social development and to change the inequitable world order. China supports the idea of global governance with the participation of developing countries. Legal instruments on the BRI have increasingly included the word "balance" as one of the main definitions of the desired outcomes [8].

China's opinion on the world stage is becoming more often considered and perceived, but this is mainly because of its global economic ties. Seeing this as a sustainability issue, China has focused on working on the foreign policy image of the country. Realizing this, China comes in search of methods to position itself as a peaceful, friendly and responsible international actor through the principle of soft power. The essence of soft power is to use social, economic and political spheres to achieve long-term foreign policy goals. The idea of reviving the Silk Road was preceded by other Chinese projects such as the China Dream, the Community of shared future for mankind [12]. But the most ambitious and significant is the Belt and Road Initiative.

China is aware that the support of its developmental environment is necessary for the successful implementation of all plans. The next important point is to strengthen the non-violent approach in China's foreign policy. Disseminating information about Chinese history and culture, social life and philosophical beliefs through various sources is an important stage in building the international image of major country diplomacy with Chinese characteristics.

As for bilateral relations with Russia, it should be noted that in 2018 the trade turnover between China and Russia reached a record level of over $100 billion. China is Russia's largest trade partner for many years running. Thanks to the visible success of the Belt and Road Initiative, Russia has chances to promote the ideology of the Greater Eurasian Partnership based on the Eurasian Economic Union. For Russia, participation in BRI is of strategic importance, as is cooperation with China as a whole. On the eve of 70th anniversary of establishment of diplomatic relations in 2019, the two countries once again confirmed the high level of cooperation, agreed to adhere to the concept of good neighborliness, friendship, cooperation and mutual benefit, to develop comprehensive relations of strategic interaction and partnership in the new era.

In 2016, during the St. Petersburg International Economic Forum, Russian President Vladimir Putin announced Russia's intention to continue its Eurasian-directed policy. This fact determines the change of Russia's course in the world, focusing on the integration process of the Eurasian continent. At the forum, President Putin urged European countries to participate in the Greater Eurasian Partnership project. The guidelines of the program proposed by Russia are similar in many respects to the principles and provisions of China's Silk Road Economic Belt, but this does not cause much dissonance on the Chinese side, and even on the contrary – China has officially supported the development of this concept through cooperation of the Silk Road Economic Belt and the EAEU into a joint platform – One Belt, One Union or Belt and Union.

It should be noted that back in 2015, Russia and China signed a historically important cooperation agreement to connect the construction of the Eurasian Economic Union and the Silk Road Economic Belt. After all, many earlier assumed that these two projects would become rivals on the world stage. However, Russia and China aim at close cooperation. For example, Russia is attracting Chinese investments in projects to modernize infrastructure facilities in the Far East, Siberia, Central and Northwest [3]. Cooperation in the energy sector is also actively developing: a number of large-scale contracts have been concluded between gas and oil companies of Russia and China for oil supply and processing (Power of Siberia, Yamal LNG project) and joint construction of nuclear power plant units in China.

In implementing the Silk Road Economic Belt, China sees Russia as a strategic reliable partner. For Russia, in addition to implementing its economic projects based on the EAEU, the revival of the Silk Road promises opportunities to modernize the remote regions of the country, especially in infrastructure and agrarian complex [11]. The Trans-Siberian Railway plays a particularly important role in the project. Nowadays it is a connecting link for routes from China to the main logistical centers of Europe, e.g., Hamburg and Warsaw.

The importance that China attaches to cooperation with Russia is evidenced by The Second Belt and Road Forum for International Cooperation (2019, Beijing), where the

Russian leader was given the floor after the Chinese leader, as the first among all foreign guests, while in Chinese media the Russian President was positioned as the main guest of the Forum. In addition, the Russian side paid special attention to the Northern Sea Route (NSR) as a promising area of maritime trade and a variant of cooperation between the NSR and the 21st Century Maritime Silk Road. Based on this idea, China proposed the joint development of the Ice Silk Road. This is of great importance for China, since about 90% of Chinese international trade is carried out by sea transport, the development of a new transport route will provide new opportunities and prospects.

After a long period of economic sanctions imposed by Western countries on Russia, participation in the Chinese project became a progressive push in the international economic and foreign policy of Russia. Besides, it may become a step towards successful development of the ideology of the Greater Eurasian Partnership. After all, the main directions and principles of the Chinese and Russian vision coincide: the development of Eurasian transport corridors, the creation of free trade zones, the integration of economies and close interaction between regions. These recent events give high expectations for friendly and mutually beneficial relations between China and Russia within the framework of the BRI and integration with the EAEU.

4 Discussions of Research

According to various estimates, the EAEU will probably be a potential competitor to the Silk Road Economic Belt, as Russia and China seek to expand their influence in Central Asian countries. Crossing interests at the level of transport routes is also a contentious point. Such inconsistencies may lead to the growth of tension in the region, and to avoid this, new pragmatic plans for the deepening of trust and interaction of projects are needed [2]. But it is worth noting that many Russian and Chinese experts claim that there is no conflict of interests between the countries and both BRI and EAEU may develop simultaneously in the same region [10]. This is explained by the different realizations and interaction forms between the countries in this region.

The best way to solve the above problems is to continue cooperation on connecting the Silk Road Economic Belt and the EAEU. China sees the possibilities of such integration as strengthening regional trade and economic cooperation, complete trade and investment throughout the Eurasian space, which will allow the countries of the region to develop more intensively [15]. Also, Russia and China are constantly exchanging experience, implementing joint innovation projects, concluding a large number of deals and agreements, and at every meeting invariably pledge to strengthen the comprehensive strategic partnership, supporting close cooperation in various areas.

5 Conclusion

To sum up, it should be noted that relations between Russia and China are mutually beneficial. Both countries are implementing major integration, developing international cooperation and investing in internal modernization. Cooperation on One Belt, One Union is a great opportunity for Russia to strengthen its geopolitical borders and influence, expand the export area of its goods and raw materials, conclude new mutually beneficial agreements with other countries, exchange experience and technologies. For China, it is of interest to improve its international image, create strong partnerships with different countries, expand the presence of its capital abroad and international influence. Having analyzed the strategic interests and interaction forms of bilateral cooperation within the framework of the Eurasian Economic Union and Belt and Road Initiative, it can be concluded that economic, political and cultural relations between China and Russia are stable and strong, and the partnership will be continuously broadened and strengthened.

References

1. Bordachev, T.V., Vishnevsky, K.O., Glazatova, M.K., et al: Eurasian economic integration: development prospects and strategic tasks for Russia. Moscow, 123 p. (2019). (in Russian)
2. Cheng, G., Chen, L., Degterev, D., Zhao, J.: Implications of "One Belt, One Road" strategy for China and Eurasia. Bull. Russian Univ. Peoples' Friendship Ser. Int. Relations **1**, 77–88 (2019)
3. Davydenko, E.V., Kolesnikova, T.V.: "One Belt - One Road" as a catalyst of investment cooperation between Russia and China. J. Eurasian Sci. **5**, 11 (2018). (in Russian)
4. Glittova, Y., Toropygin, A.: Political and legal basics of the conjugation of the Eurasian economic union and the silk road economic belt: formation process and perspectives of cooperation. Admin. Consult. **12**, 33–47 (2018). (in Russian)
5. Holkina, A.A.: Advantages and disadvantages of joining the Eurasian economic union. Actual Issues Econ. Sci. **46**, 12–21 (2015). (in Russian)
6. Kadochnikov, P.A.: On prospects of development of the Eurasian economic union. Econ. Dev. Russia **3**, 13–21 (2020). (in Russian)
7. Kong, Q.: Prospects for the "One Belt - One Road" initiative in the development of russian-chinese economic relations . Theory Pract. Soc. Dev. **5**, 83–86 (2017). (in Russian)
8. Kozlov, D.I.: The first results and the nearest prospects of realization of the Chinese Initiative – "One Belt, One Road" . PolitBook **2**, 127–156 (2018). (in Russian)
9. Krotov, M.I., Muntiyan, V.I.: Eurasian economic union: history, features, prospects. Admin. Consult. **11**, 33–47 (2015). (in Russian)
10. Li, N.: "One Belt, One Road" Initiative as a new model of cooperation between China and Russia and Central Asian Countries. Bull. Russian Univ. Peoples' Friendship. Ser. General History **4**, 382–392 (2018). (in Russian)
11. Lo, S.A.K.T: The linking of the belt and road initiative and the Eurasian economic union and its impacts on Sino-Russian relations. Russian Politol. **1**, 65–71 (2019)
12. Luzyanin, S.G., Afonasyeva, A.V.: "One Belt, One Road" - political and economic dimensions. Bull. Tomsk State Univ. Econ. **40**, 5–14 (2017). (in Russian)
13. Tsoi, A.V.: Problems of the Eurasian economic union development. Region. Econ. Theory Pract. **7**, 1223–1234 (2018). (in Russian)
14. Wang, S., Wan, Q.: The silk road economic belt and the EAEU-rivals or partners? Central Asia Caucasus **3**, 7–16 (2014)
15. Zhang, J.: How eurasian economic union and silk road can coexist and prosper. Int. Trade Trade Policy **1**, 27–32 (2016)

The Continuity of the US Policy Towards Central Asia

Aigerim Ospanova[1(✉)], Andrei Shenin[2], Aiym Shukyzhanova[1], and Kilybayeva Banu[1]

[1] L.N. Gumilyov Eurasian National University, Nur-Sultan, Kazakhstan
[2] Narxoz University, Almaty, Kazakhstan
andrey.shenin@narxoz.kz

Abstract. The article represents the retrospect of U.S. foreign policy towards Central Asia since the collapse of the Soviet Union. Authors compare various geopolitical doctrines of Bill Clinton, George Bush Jr., Barack Obama and Donald Trump administrations in order to evaluate the U.S. impact on economy and political landscape in Central Asia. Considering strategies of all administrations, authors conclude that despite different approaches each strategy was based on the same principles – supporting the sovereignty and independence of Central Asian countries through economic and democratic reforms, creating an environment for penetrating U.S. companies at local markets, resisting to Russian and Chinese influence in the region, and combating terrorism after 9/11. In sum, this article demonstrates that the of U.S. foreign policy towards Central Asia have been always the same just changing priorities in places. The research results provide us with the ground for future research on U.S. plans regarding the region.

Keywords: U.S. foreign policy · Central Asia · Bill Clinton · George Bush jr. · Barack Obama · Donald Trump · Afghanistan

1 Introduction

Since the collapse of the Soviet Union, the United States have been showing their interest to Central Asia. A huge territory with colossal hydrocarbon reserves is extremely important for each component of the world development – economy, energy, security, logistic; therefore, if Washington wants to keep the global dominance, it must take the region under control. However, any practical attempts to exploit regional wealth encountered with numerous problems ranging from geographical distance and the poor understanding of domestic processes in five Central Asian countries (Kazakhstan, Uzbekistan, Turkmenistan, Tadzhikistan, and Kyrgyzstan), to the resistance from Russia and China.

According to this, each U.S. administration has faced the problem of forming an appropriate foreign policy approach towards Central Asia. A strategic approach has to follow specific points such as pursuing U.S. interests in the region and keeping a balance between regional actors, lobbyists that champion interests of U.S. trans-national

© The Author(s), under exclusive license to Springer Nature Switzerland AG 2021
R. Bolgov et al. (Eds.): *Proceedings of Topical Issues in International Political Geography*, SPRINGERGEOGR, pp. 140–148, 2021.
https://doi.org/10.1007/978-3-030-78690-8_13

corporations at Congress, and an interest group that holds power, whether it was "liberal-progressists" of Bill Clinton, "neoconservatives" of George Bush Jr., "wilsonians" of Barack Obama or "realists" of Trump. However, despite any ideological distinctions among U.S. administrations, it can be said that the general trend has been remained unchanged and is based on the same geopolitical principals: supporting the sovereignty and independence of Central Asian republics, creating favorable conditions for penetrating U.S. companies at local markets, resisting to Russian and Chinese influence in the region, and combating terrorism after 9/11. For the last 25 years, an implementation of these goals has been providing over the support at building strong borders, economic and democratic reforms, and development of an energetic cluster in Central Asia. The Deputy Assistant Secretary of State for Central Asia Daniel Rosenblum stresses that regardless of situation at hand, the U.S. policy towards Central Asia will keep the continuity of previous strategies without significant changes.

An analysis of U.S. policy towards Central Asia has a significant value for the region, since it indicates an impact of the global superpower on the development, security and political landscape of the region. The research question of the paper is to find out priorities of U.S. foreign policy to Central Asia and how focusing on one of them changes the situation on the ground.

The primary hypothesis of this article is that the regional policy has been always based on the same ideas over the last four administrations from Clinton to Trump. However, each administration re-evaluated a list of priorities, concentrating on democratic reforms, security or economy. This can be proved over scrutinizing years of U.S. activity on regional projects ranging from signing of the U.S.-Kazakhstan "Charter on Democratic Partnership" during Clinton administration, to the war against terror in Afghanistan under Bush Jr. presidency and launch of "The New Silk Road" at Obama administration.

2 Methods and Sources

This research deals with historical and comparative methods. Under the historical approach, U.S. policy towards Central Asia is arranged in chronological order from Clinton to Trump administration, while their distinctions are analyzed within the comparative approach. Additionally, important data was collected from regular and systematic observation of academic journals, news articles, and content analysis of official governmental resources of Kazakhstan, the U.S. and Russia.

Within the research the authors used official statements and doctrines developed by U.S. officials. First of all, the speech of the deputy State Secretary Strobe Talbott "Farewell to Flashman" that defined the key points of U.S. foreign policy towards Central Asia in the end of 1990s [6]. Second, the document titled "Reliable, Affordable, and Environmentally Sound Energy for America's Future" by U.S. vice-president Dick Cheyney in 2001 that put energetic security at the top of the U.S. priorities in the region (after 9/11 the White House added new "pillar" to the list of strategic goals in Central Asia – the war against terrorism) [12]. Third, an important contribution to this work have brought numerous debates in Congress and "think tanks" during Obama and Trump presidencies. For instance, the discussion on U.S. policy towards Central Asia at "Atlantic Council" in 2017 [2].

Also, especially valuable to this research were articles written by top-U.S. experts such as Jeffrey Mankoff [11], Eugene Rumer, Richard Sokolsky or Paul Stronski [13].

Among Russian and Kazakh researchers should be mentioned Murat Laumulin, Timur Shaimergenov, Konstantin Syroezhkin, Evgeniy Troitsky and Andrey Kazantsev. Laumulin, for example, in his article "U.S. and Central Asia: in search of the doctrine" describes the variety of views on Central Asia within the American establishment [10]. Shaimergenov, in his turn, analyzes the state and prospective of Kazakhstan-NATO relations and its impact on the region [14]. Russian researcher Kazantsev in the article "U.S. in Central Asia" provides us with a ground for more detailed analysis of U.S. policy towards the region, describing key foreign policy paradigms of the White House [8]. Additional information came from articles by Evgeniy Troitsky, who researched U.S. presence in Central Asia from an angle of energetic cooperation between U.S. and local governments [17].

Providing this research, the authors also analyzed press releases and official statements from the U.S. State Department archive. As a result of careful analysis of mentioned above sources, the authors have summed up various points of view on advantages and disadvantages of U.S doctrines on Central Asia and their transformations over the years.

3 Results of Research

3.1 Discovering Central Asia

In the first half of 1990s, Clinton administration did not have a geopolitical strategy towards the region, focusing on three specific tasks that worried White House most at that moment. At first, U.S. helped Kazakhstan to export and disposal the weapons of mass destruction (nuclear and biological weapons) that had left on Kazakh territory after the collapse of USSR. Second, Washington put efforts to support independence, sovereignty and territorial integrity of the newly formed republics over economic and political reforms. Third, U.S. officials championed creating conditions for the penetration of American trans-national companies at local markets. Forth, it was seemed extremely important to reduce the influence of Moscow on Central Asian governments over the building of new pipelines in avoidance of Russia [3].

However, after the U.S. had fulfilled their key goal – signed contracts for exploiting local oil\gas deposits and moved the nuclear weapons out of Kazakhstan – the interest towards the region has decreased rapidly. Moreover, in 1997 the deputy state Secretary Strobe Talbott officially claimed at his speech "Farewell to Flashman" that U.S. has no strategic interests in Central Asia [6]. The new Central Asian strategy, he said, should be based on two simple ideas: the U.S. should keep providing very limited support to Central Asian republics in terms of democratic and economic reforms, and politically prevent the region from falling under control of neighboring superpowers. These ideas were unofficially named the "Talbott's doctrine."

Following that strategy, officials stressed that Central Asia is an area of rivalry of such powerful players as Russia, China, India, Pakistan, Iran and Turkey. Backing one of them by Washington could result in disbalance and tensions, while U.S. unilateral interference in Central Asian affairs was seriously hindered by geographical distance and cultural distinctions. To reverse the situation, Washington has offered to turn the region into a zone free from any foreign influence, hoping that Central Asian wealth under U.S. advisory support would help to promote sustainable development in local republics and create societies loyal to West.

Over the last years of Clinton's presidency, the administration has been trying to follow the strategy presenting Central Asian republics to the world as democratic states, but high level of corruption and a lack of progress in democratic reforms has ruined their initially positive image and cumbered funding of international governmental programs in the U.S. These factors meant the failure of Talbot's doctrine, therefore the next administration decided to shift emphasizes of U.S. policy towards Central Asia.

3.2 Energy is a Key

It is well-known today that the Central Asian policy of Bush Jr. administration was initially based on the document titled "Reliable, Affordable, and Environmentally Sound Energy for America's Future" by vice-president Dick Cheney. The new national energy policy emphasized that the U.S. have to diversify gas and oil flows from Central Asia to the West in order to reduce dependence on Persian Gulf monarchies and fluctuations of global markets. To do this, the Cheney's team has offered plans to build Turkmenistan-India (TAPI), Baku-Ceyhan (BTC), Trans-Caspian pipelines, etc [12].

Also, the new strategy insists that developing democracy and maintaining state independence must remain as a basis for Central Asian prosperity but means towards this end should be revised. Experts saw authoritarian regimes as a key barrier towards regional democratization but warned that any attempts to stimulate the transit of power could led to political instability (as it was in Georgia in 2003, Ukraine in 2004 and Kyrgyzstan in 2005). Fearing to shake the situation, Washington has reasoned that it cannot rely entirely on stagnant political regimes or non-profit organizations, and the only power able to become a guarantor of maintaining order in case of instability is the military. That is why American analysts have advised to the administration to promote the creation of local military elites loyal to Washington.

On the one hand, the September 11 attacks contributed this idea strengthening ties between U.S. and Central Asian militaries, but on the other hand, the anti-terrorist campaign shifted the focus of White House from energetic to security issues. Nevertheless, energy was still a one of the top priorities, so that the combination of security and energetic issues has transformed Central Asia from just a source of energy to a geopolitical hub that must be taken under military and political control.

As a first step to American dominance in the region and response on terrorist attacks, in 2001 the U.S. initiated the full-scale military campaign in Afghanistan using logistical support from Central Asian countries. Local rulers have agreed to back U.S. troops, since it promised them not just generous economic aid, but also political stability for ruling elites that stop fearing ideological pressure from the U.S. As it turned out, in vain.

Against the background of the anti-terrorist campaign, Washington kept intensity on supporting independence and sovereignty of Central Asian countries, respect for human rights, resisting influence of other superpowers, and exploiting new oil and gas deposits. The most remarkable actions in these fields were signing the Declaration on strategic partnership with Uzbekistan in 2002, launch of BTC pipeline, opening USAID offices in Kazakhstan, broad criticism of Russia-led integrations (The Custom Union, EurAsEC, and EEU), etc. However, in 2002-2005Washington pushed its own policy ignoring interests of local authorities, that is known as an "aggressive realism". An aggressive incorporation of Western values led to negative consequences in Eurasia, such as a set of "color revolutions" in Georgia, Ukraine and Kyrgyzstan. Central Asian elites felt unsecure and started countering to American NGOs and political initiatives in the republics (Uzbekistan even went further moving U.S. troops out of Karshi-Khanabad base), that ended as a crisis of Bush's Central Asian policy.

After the U.S. presidential elections in 2005, the balance of power in Bush administration has changed shifting from neoconservatives to realists that prefer building energetic cooperation instead of military supremacy in Central Asia. Nevertheless, at that moment, the U.S. was facing with numerous problems in domestic economy, bogged down in Iraq and Afghanistan, could not compete with Russia and China in the region, and had no clear plan on improving relations with Central Asian republics. Washington has attempted to quit and fill the vacuum by European forces, but Brussels did not want to clean the room after Americans and took a very moderate part in Central Asian processes [15].

Today it can be said with certainty that Bush Jr. administration was following to the same ideas and motives as the Clinton administration, but relied more on the brute force and strength of the dollar, ignoring the interests of local authorities. If the Clinton period could be characterized as "we will show how to do it," then Bush's policy was more like "we will come and do whatever we consider the best" that faced resistance from local elites. Shared objectives, carried out by different means, brought the modest achievements of Clinton administration to nothing and left the next president, Barack Obama, with a bunch of unresolved problems.

3.3 Obama's Revision

The third round of U.S. policy towards Central Asia has started right after Barack Obama won the U.S. presidential elections in 2008. The new administration did not change the set of U.S. priorities, but each component of the regional policy was revised. For example, taking into account the need to withdraw troops from Afghanistan, military-political influence and infrastructure (the lost bases in Uzbekistan and Kyrgyzstan) lost their critical importance for Washington. Moreover, the growing role of Russia and China in the region, in fact, forced the US to admit defeat in the "big game" and abandon the "Talbot doctrine". Central Asian republics have demonstrated an inability to adopt democracy, while lowering oil/gas prices decreased importance of TAPI-pipeline [15]. After the revision of priorities, American analysts and officials concluded that a new basis for U.S. strategy should be strengthening of independence and sovereignty of Central Asian republics over the development of economic spheres, instead of taking control using brutal force or financial aid.

Following this idea, in 2011 then Secretary of State Hillary Clinton announced the launch of U.S. project the "New Silk Road" that is going to unite Central Asia with India, Pakistan and Afghanistan into a single transport corridor. The economic cooperation opens new spheres for investments that pleased U.S. business. Thus, Obama administration made a bid for a middle and long-term economic cooperation as a basis for keeping influence in the region and further integration of Central Asian republics into global markets [7].

As far as the process of democratization, the experience of the Bush administration had shown that pressure on human rights issues and stimulating democratic reforms from outside threaten to destabilize the region. State department officials admitted that the process of democratization has reached its limit and stopped, but they also stressed that halting work on this direction could create a favorable ground for spreading terrorism and religious extremism across the region. In testimony in May 2011, then-Assistant Secretary Blake stated that leaders in Central Asia "are suspicious of democratic reforms, and with some exceptions have maintained tight restrictions on political, social, religious, and economic life in their countries" [4]. However, Obama administration was not going to spend money for projects in corrupt and unstable republics of Central Asia. Washington remained the spread of democratic values as one of the key elements of long-term political stability but did not insist on intensification of work in this area. In addition, a bunch of problems that left after the withdrawal of U.S. troops, lack of funding, chaos in Afghanistan, failure of TAPI-pipeline, and inability to compete with neighboring superpowers forced Democrats to admit that "The Great Game" is up and "Talbot's Doctrine" does not work anymore. However, their policy towards Central Asia was one among many vulnerabilities that White House was criticized before U.S. presidential elections in 2016. A decade later their vision was approved by scholars [9].

To reverse the situation, in 2014–2016 Democrats began to call for a new round of the "Great Game" for the predominance in the region [1, 8]. The development of the strategy was given to an authoritative analytical think-tank "The Carnegie Endowment for International Peace", however their outcomes did not match with expectations of politicians. In the research paper "U.S. Policy Towards Central Asia 3.0" experts stressed that U.S. have fulfilled its mission in the region, since they supported the young post-Soviet republics in their striving for independence and sovereignty at a critical moment, and also laid the foundation for long-term liberal-democratic development [13]. Moreover, the strengthening of Beijing and Moscow has significantly limited the U.S.'s capabilities in the region.

Experts emphasized that the decline of U.S. political activity and loss of influence in Central Asia would not be a big problem, because Washington does not need it today. The key task is to keep presence in the region without being involved at competition with major players. As a solution, U.S. Secretary of State John Kerry offered a new format of interaction between U.S. and five Central Asian republics "C5 + 1," through which Washington planned to coordinate common projects. Unfortunately, the sum of money allocated by the U.S. turned out to be very tiny - just $ 15 million that is clearly not enough to solve a whole range of regional problems from building trade infrastructure to combating terrorism [9]. Thus, the new idea did not work out.

All after all, it can be concluded that the second term of Obama administration has ended with an ill-defined approach to Central Asia. Despite the flashy launch of TAPI-pipeline in 2015 and stable work of CASA-1000, other infrastructural projects of "The New Silk Road" were unable to compete with Chinese "One Belt – One Road" investments or be implemented under such a little funding [17]. In addition, the withdrawal of

American troops from Afghanistan has significantly reduced the U.S. political capabilities in the region, while "C5 + 1" failed because of lack of money and support within American establishment. As a result, after more than 25 years of U.S.-Central Asia interactions it seems as a paradox that the U.S. had reached all the goals on supporting sovereignty and independence of local republics, helped building pipelines, stimulated democratic and economic reforms, but lost the struggle for Central Asia.

3.4 Trump's Uncertainty

For the moment, Trump administration does not present a single document on the future of U.S. policy towards Central Asia. The fundamental reason of that is an ongoing political struggle between supporters and opponents of Trump himself, political parties and interest groups. The polarization of U.S. society and political establishment has a negative impact on forming foreign policy doctrines.

Unlike Clinton, Bush and Obama administrations, Trump's team does not have a basic idea for Central Asia. For example, Clinton based its Central Asian policy on democratic values, Bush's "energetic policy" was grounded on building new pipelines to the West and combating terrorism, and Obama was focused on development of local economies to bring stability in the region [18].

Regarding Trump, there are grounds for believing that the process of developing a policy towards Central Asia will be intensified in the near future as it evidenced by the president's idea of turning U.S. policy towards Afghanistan. In addition, the balance of power across the Eurasian space is changing rapidly that will have an influence on the whole world. If U.S. want to keep the global leadership, they are obliged to react to all these changes.

Today, experts and officials in Washington have initiated a vast discussion on new approaches to U.S. policy in Central Asia. It should be noted that American ruling elite has a range of different opinions about the region and its importance for U.S. geopolitical plans, and which of them will ultimately prevail is still a big question.

The only thing that can be said for sure is that the Central Asian policy will be based on the same "pillars": maintaining security and stability in the region, developing energy resources, supporting sovereignty, etc. However, it is quite possible that Trump administration will revise accents of its policy. Many U.S. officials believe that the current situation is unfavorable for the expansion of capital -- the demand for hydrocarbons and investments in regional economies are declining. The political stability of these states is also getting worse, while the proximity of the turbulent Middle East makes the future virtually unpredictable. So, risks for Western corporations, working in Central Asia, are growing up from day by day. It seems that hydrocarbons can no longer be a source of prosperity for the region, so local elites have to think about the diversification of economies based on the development of trade and transit routes under Chinese BRI projects. This will make the countries of the region more independent economically, support the sovereignty, increase their ability to resist the pressure of neighboring powers, and create opportunities to help Afghanistan. But the strategy of trade and transit development is quite expensive, therefore some expert's advice Washington to join Beijing's project. However, the dispute on this issue is not over yet [19].

4 Conclusion

To sum up the U.S. policy towards Central Asia for the last 26 years since the collapse of the USSR, we can stress that despite radical changes of the political landscape, Washington still holds the basic ideas presented by Strobe Talbot in 1990s. Each administration from Clinton to Obama grounded its regional policies on three pillars: supporting independence of Central Asian republics over political and democratic reforms, developing energetic sector over building pipelines, and resisting to attempts of any local superpowers to take the region under control (plus combating terrorism after 9/11). However, leaving the priorities unchanged, each administration has placed emphasizes in its own way.

During the Clinton period, the main idea was to limit the influence of Russia and China in the region and help American corporations to penetrate into local markets. The Bush administration, in turn, preferred to focus on combating terrorism and actively developing new hydrocarbon transportation routes, which should reduce the dependence of the West on the countries of the Middle East. However, the brutal incorporation of Western values, within the framework of "aggressive realism," set Central Asian republics against the U.S., leaving many problems for the next administration. To solve the U.S.-Central Asian contradictions, Obama administration decided to focus on developing local economics and sovereignty, but the withdrawal of troops from Afghanistan and unwillingness to invest significant funds under "The New Silk Road" project have reduced U.S. influence in the region. The vacuum of power was immediately filled by neighboring superpowers via Eurasian Economic Union and Shanghai Cooperation Organization.

Now, Donald Trump should create a new approach to Central Asia. Officials of the administration officially declare that the foreign policy concept will not undergo significant changes and will be based on the same "pillars" as before. In view of this, it can be safely assumed that Trump is likely to continue Obama's policy of development local economies but giving more attention to the issues of fighting terrorism over returning a small military contingent to Afghanistan. Also, Washington will focus on energy issues and the resistance to Russia and China in the region, while incorporating democratic values is left on the third roles.

References

1. An Address by Deputy Secretary of State Antony Blinken. "The United States and Central Asia: An Enduring Vision for Partnership and Connectivity in the 21st Century". Event at the Brookings Institution (2015)
2. Blond, S.: "Central Asia: US Foreign Policy at a Great Power Crossroads" Discussion at the Atlantic Council (2017)
3. Cohen, A.: U.S. Policy in the Caucasus and Central Asia: Building A New "Silk Road" to Economic Prosperity. The Heritage Foundation (1997)
4. Commission on security and cooperation in Europe. "Hearing on Central Asia and the Arab Spring: Growing Pressure for Human Rights." U.S. Congress (2011)
5. Cornell, S., Starr, F.: Modernization and regional cooperation in Central Asia: a new spring? Silk Road Papers (2018)

6. Deputy Secretary Talbott. A Farewell to Flashman: American Policy in the Caucasus and Central Asia. U.S. state archive (1997)
7. Hillary Clinton's Remark on India and the United States: A Vision for the 21st Century. U.S. Department of State (2011)
8. Kazantsev, A.: The U.S. policy in post-Soviet Central Asia: character and prospects. Bulletin MGIMO **4**, 155–164 (2012)
9. Laruelle, M. Royce, D.: No Great Game: Central Asia's Public Opinions on Russia, China, and the U.S. The Kennan Cable (2020)
10. Laumulin, M.: The U.S. and Central Asia: in search of doctrine. Continent **7**(94), 35–50 (2016)
11. Mankoff, J.: The US policy in Central Asia after 2014. Pro et Contra **17**(1–2), 41–57 (2013)
12. National Energy Policy - Reliable, Affordable, and Environmentally Sound Energy for America's Future. Report of the National Energy Development Group (2001)
13. Rumer, E., Sokolsky, R., Stronski, P.: U.S. Policy Towards Central Asia 3.0. Carnegie Endowment for International Peace (2016)
14. Shaimergenov, T., Biekenov, M.: Kazakhstan and NATO: assessment of likely prospects for cooperation. Central Asia Caucasus **1**, 40–59 (2010)
15. The Development of Energy Resources in Central Asia. Hearing Before the Subcommittee on Europe, Eurasia, and Emerging Threats of the Committee on Foreign Affairs. U.S. House of Representatives (2014)
16. Troitskiy, E.: The U.S. policy in Central Asia: the approaches of the second administration of G. Bush (2005–2009) and B. Obama (2009–2010). Comparat. Polit. **4**(6), 65–74 (2011)
17. Troitskiy, E.: U.S.-Kazakhstan relations in energy. Bull. Tomsk State Univ. (313) 101–106 (2017)
18. United States Strategy for Central Asia 2019–2025: Advancing Sovereignty and Economic Prosperity (Overview). U.S. Department of State (2020)
19. Wilson, W.: China's Huge 'One Belt, One Road' Initiative Is Sweeping Central Asia. The Heritage Foundation (2016)

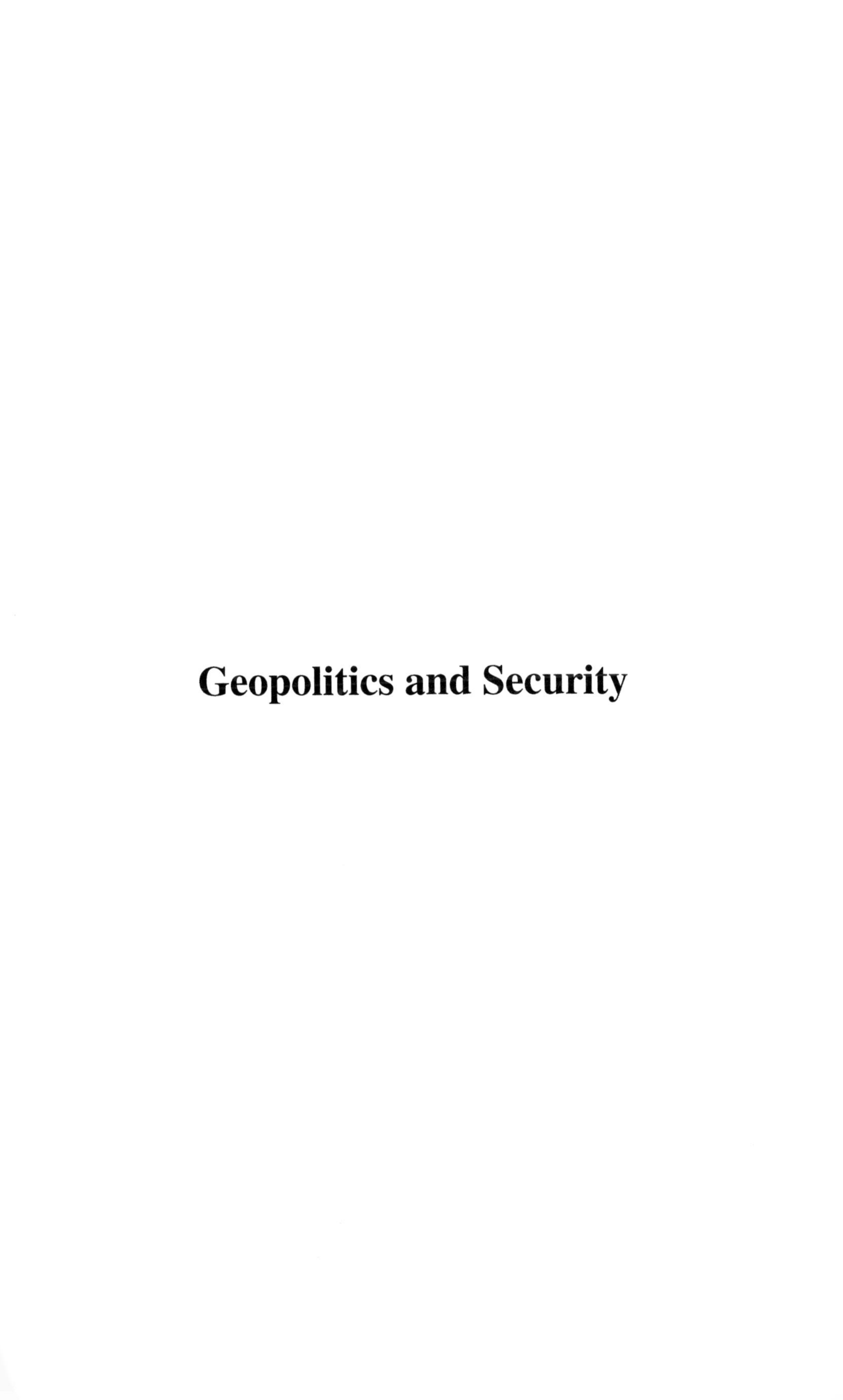

Geopolitics and Security

Xinjiang. The New Development Trends

Svetlana Kozhirova(✉)

International Science Complex "Astana", Nur-Sultan, Kazakhstan
office@isca.kz

Abstract. Over the last several years XUAR, being a part of China, became an important participant in international relations. Mostly the region has been deprived of adequate attention in the shadow of China's Economic Rise. Most commonly the economic rise used to go with East Chinese regions development, not West. The new development trends in Xinjiang presented in the paper prove that the region is playing increasing significant role in Chinese foreign trade relations at the whole. The research is based on review the economic news released by Chinese, Russian and Kazakh official agencies, as well as the analytical papers released by ISC "Astana", Institute of World Economics and Politics, Kazakhstan, and Boğaziçi University, Turkey. The paper concludes, that Xinjiang is assigned a role of the "window" to the Western land within the implementation of the Belt and Road initiative, which, on the other hand, gives a new impetus to the region development.

Keywords: XUAR · Xinjiang · Belt and road · Silk road economic belt · China · Central Asia

1 Introduction

Over the past decades, the Xinjiang Uygur Autonomous Region (XUAR) has been developing at a rapid pace. Thanks to the support from other regions of China, XUAR has experienced economic growth in recent years. Infrastructure projects the process of modernization have brought the region closer to the rest of the country. XUAR is significant for Chinese national interest due to the following reasons: it provides country with energy resources; it has strategic importance in Chinese foreign trade; it is considered as China's gateway to Central Asia; it is assigned a role of the "window" to Europe, etc.

In this regard, the study of strategic goals, main directions and priorities of China's policy in Xinjiang, as well as the key tools and mechanisms for their implementation are altogether becoming a subject of special importance.

The purpose of this research is to do a comprehensive study and systematic analysis of the new development trends in Xinjiang. To achieve this purpose the following objectives have been identified:

- to determine new approaches of Chinese interests and policies in Xinjiang in the energy sector and infrastructure;

© The Author(s), under exclusive license to Springer Nature Switzerland AG 2021
R. Bolgov et al. (Eds.): *Proceedings of Topical Issues in International Political Geography*, SPRINGERGEOGR, pp. 151–160, 2021.
https://doi.org/10.1007/978-3-030-78690-8_14

- to identify the process of modernization and innovations in XUAR;
- to examine the role of XUAR in Chinese foreign trade relations at the whole;
- to reveal the content of the development of Kazakhstan-Xinjiang relations in the sphere of trade and economy as the basis of formation of mutual interest;
- to substantiate the main aspects of the interaction in car industry and textile manufacture;
- to figure out new tendencies in labour market of XUAR.

Evolution and dynamics of changes in Xinjiang on holistic way are one of the unexplored areas of modern science about the regions. Such research is not completely absent, but it is limited to a very small number of foreign and domestic authors. Therefore, our research is concentrated mostly on review of Chinese, Russian and Kazakh official news agencies, in addition to the analytical papers released by ISC "Astana", Institute of World Economics and Politics, Kazakhstan, and Boğaziçi University, Turkey.

The method of system analysis in its conjunction with other main approaches or methods of scientific analysis in political science (institutional, sociological, comparative and historical) have been applied in this paper. The combination of these methods, as well as consideration of facts and events have provided a high degree of objectivity and validity of assessments and conclusions, allowed fully and comprehensively reveal the new trends of development in XUAR.

The novelty of the study is determined by providing evidences of the importance and strategic role of Xinjiang in China's economic security, transit route of energy resources, foreign trade and infrastructure network through Belt and Road Initiative, which gives a new impetus to the region development.

2 Xinjiang as Important Energy Base and Travel Corridor of China

Xinjiang Uygur Autonomous Region (XUAR, PRC) is an important participant in international relations and hold a special place in the cross-border cooperation of China with neighboring states. That is because of geographic proximity with 8 neighboring states at once – Afghanistan, India, Kazakhstan, Kyrgyzstan, Mongolia, Pakistan, Russia and Tajikistan, which is 40% of all countries bordering China.

This area of cooperation is closely going with solving two key tasks:

firstly, ensuring border security, including border protection;
secondly, the creation of cross-border trade-economic networks, which subsequently should be transformed into border free trade zones.

The XUAR geographic location causes the following greatest borderline risks:

- neighborhood with unstable states as Afghanistan, Kyrgyzstan, Pakistan and Tajikistan;
- the presence of the US military (as well as the private companies) in Afghanistan;
- illegal traffic, arms traffic, drug trafficking and illegal migration;

- cross-border activities of the extremist and terrorist organizations, the fighters-immigrants from Xinjiang and Central Asia in the "gray zones" of neighboring countries;
- the threat of the irredentist and separatist movements in certain neighboring states of Central Asia.

Xinjiang has always played a special role in China's relations with the Central Asia. For Kazakhstan, Xinjiang is the "Southeast Gate" to China. In the current context there is a need for new approaches to regional cooperation focused on ensuring national interests and active Belt and Road promotion. China's economic reforms pursuing to transform Xinjiang and Western territories into the Central Asian "industrial workshop" are focused primarily on neighboring states. This implies the need of the implementation the economic transformations in the XUAR with respect to both the advantageous geographical position of the region, its economic and geographical indicators, and the geographical features and economic opportunities of neighboring countries. This is a fundamental truth, because regional economic integration is always based on the geographical factor – sharing space, common frontiers, approximately similar climatic conditions, ethnic and historical ties, transport and logistics components, natural resources and more.

All transport and logistics arteries connecting China with the countries of Central Asia, Russia and Europe pass through Xinjiang. It is the main transit corridor for the export of Central Asian gas and Kazakhstan's oil. Oil and gas significant reserves are concentrated here, their exploitation is developing rapidly. Kazakhstan-China oil pipeline is located through the province territory, as well as the gas pipeline connecting Central Asia to China. According to the Chinese leadership estimates, a new integration network of internal and external pipelines will be formed here with respect to the mutual benefits of all parties. Xinjiang is assigned the main operational role resulting in creating, in the medium term, a "safety bag" of energy for all of China.

The region is the main center of concentration for all Eurasian transit routes of China passing through the territory of Central Asian countries. The autonomous region has 29 checkpoints, including 17 checkpoints of the first category and 12 checkpoints of the second category. This creates a promising opportunity for its successful industrial development. The Urumqi oil refinery plant is one of China's 7 largest refineries. Oil and petroleum products account for 22.2% of Xinjiang's total industrial production. This is important in the energy supply of the Chinese economy. The textile and food industries are also developed in the region.

The Government of China has decided to create a $ 40 billion Fund in cooperation with leading Chinese commercial banks. The investments will primarily be aimed at improving the XUAR transport network, as well as developing the infrastructure of Central Asian countries and parts of European countries through which transit routes – Eurasian and Pan-Asian – will pass. In recent years, Xinjiang has begun to show activity as a potential investor in projects implemented by China in Central Asia.

On January 16, 2014, the 12th People's Congress in Xinjiang set a task to create a "road map" on developing the XUAR into a key region within the implementation of the Silk Road Economic Belt (SREB). In a long term, it is planned to create the processing industry cluster, a resource cluster, a trade cluster, a service delivery cluster, all focused on foreign countries, as well as to stimulate the advanced equipment, technologies, and services. It is worth to note that it is Xinjiang companies that plan to expand cooperation with the countries of Central Asia in the field of scientific and technological innovations, as well as in such an unfortunate bilateral issue as agriculture. Chinese experts believe that Xinjiang should make efforts to speedily implement the construction of indicative agricultural cooperation zones as China – Kazakhstan, China – Turkmenistan and China – Tajikistan, as well as to intensify cooperation in the distribution of agricultural machinery.

In the future, thanks to the revival of the Silk Road and Northwestern China development, Xinjiang will transform into a large transport and energy hub, an economically developed externally oriented region, which will spur the strengthening cooperation between China and Central Asia.

3 Modernization and Innovations in XUAR

XUAR has been developing rapidly over the past decades: a comprehensive industrial system including metallurgy, oil refining, textiles, engineering, electronics, and coal has been formed here. According to the National Development and Reform Commission, since the end of 2012, 19 provinces and cities have allocated a total of more than 13 billion yuan (about $ 1.91 billion) for industrial development and created more than 500 thousand jobs. Thanks to the support of other regions of the country, XUAR had a good economic growth in recent years, and the region's infrastructure made it closer to the rest of the country. According to official data, in 2016 XUAR economy grew by 7.6%, which is 0.9% higher than the national indicator. Per capita income grew by 8.9% reaching 18 355 yuan, which also turned out to be above the average country income. The White Paper released by China's State Council Information Office shows that the percent of poverty in XUAR fell to less than 10% by the end of 2016.

XUAR rapid development is facilitated by the infrastructure modernization in course of which greater focus placed on the overproduction control, reforms extension and Belt and Road initiative support. According to XUAR governmental report, in 2017 Xinjiang planned to allocate 1.5 billion yuan for infrastructure development, more specifically, 200 billion for the new roads construction, 34.7 billion – railway network construction, 14.4 billion – Urumqi airport modernization. The White Paper also notes that the length of high-speed roads in XUAR reached 4395 km, high-speed rail lines – 717 km. In addition, the region has 18 civilian airports. In 2017, 2,400 companies were registered in Xinjiang, the city's GDP reached 5.12 billion yuan with an increase of 278% over the previous year. In 2017, industrial investment in XUAR reached 461 billion yuan, in particular, investment in production amounts to 280 billion yuan, thereby increasing by 51% in annual terms [1].

The XUAR Manufacturing Program 2025 envisages the creation of an industrial hub, 6 districts and 8 centers focused on the production and processing of raw materials. The main goal is export to Western countries and expanding presence in European markets. According to experts, the creation of production infrastructure will take about 10 years. At the initial stage, it is planned to create an industrial base as part of the Belt and Road.

Six industrial areas, based on the availability of natural resources and geographical advantages, have been identified:

1. Urumchi-Changji-Shixjeczy;

2. Kuitun- Dushanzi-Karamay;
3. Hami;
4. Korla;
5. Kashgar;
6. Ili-Kazakh Autonomous Prefecture

The XUAR Manufacturing Program 2025 includes the development of elaborations and researches, as well as the intellectual equipment products invention in such areas as robotics, electric power, construction equipment, agricultural equipment, equipment of new energy sources. By 2025, it is planned to introduce elaborations into production everywhere, to establish a special industrial base. It is planned to create eight industrial clusters with a potential profit of 100 billion yuan. It is expected that the level of production by 2025 will rise by 60% in relation to current indicators.

Chinese expert Jia Yu Chong explains that the SREB projects and the XUAR Manufacturing Program 2025 are complementary to each other. Without the implementation of the XUAR Manufacturing Program 2025, it is impossible to implement the SREB initiative, since the region is not able to cope with the expected tasks in view of weak industrial development. In addition, this will reduce the dependence of the Western provinces of China on the Eastern.

Fan Gang, a Member of the Monetary Policy Committee of the People's Bank of China and a Director of the National Economic Research Institute, notes that in 2015, China's foreign investment exceeded foreign investment in China for the first time, which marked a new era for the economy. In this regard, the investment of Chinese enterprises should be directed to Central Asian countries, as the main tool for implementing the SREB initiative.

Liu Yilei, the Xinjiang Production and Construction Corps Deputy Secretary General stressed, that it is necessary to create an open economic system in XUAR, which will allow it to integrate with the international economic system. In addition, it is necessary to organize multilevel cooperation with Central Asian countries [2].

4 XUAR's Foreign Trade

During the Belts and Roads implementation, XUAR is constantly increasing the import-export transactions with five countries of Central Asia, Russia and Mongolia. As the result the cross-border trade has developed rapidly. Xinjiang Government places a high priority on the cross-border e-business development. On the report of the XUAR Development Program for the 13th Five-Year Plan released by the Xinjiang Economic and Information Commission in December 2016, by 2020, Xinjiang will become a large-scale regional center for cross-border e-commerce and electronic trade in specific goods. This will bring even more development opportunities to local foreign trade enterprises.

In recent years, additionally to the traditional trade development, Xinjiang has continuously increased steps towards economic and technical cooperation. This is done, first of all, by encouraging the internationalization of the enterprises, especially due to participation in infrastructure construction, resource development and contracted works in neighboring countries. The export of sophisticated equipment, technologies and services is stimulated along with export credit, external aid and other supporting methods. This is becoming one of the main specifics of Xinjiang's export trade. As the result, a number of internationally competitive enterprises have emerged, including TBEA. Tebian Electric Apparatus Stock Co., Ltd specializes in energy projects. It is known for the construction of power lines. The Chinese company operates in 70 countries. The number of the company's employees reaches 20 thousand. TBEA received government support in Kazakhstan and Tajikistan, it is implementing social projects in Tajikistan. It built roads and secondary schools in Dushanbe and Vahdat, as well as Rudaki and Dangar regions. The construction of a highway in Dangarinsky district, funded by TBEA, consists of three components and includes the restoration of the highway in various sections of the Dangarinsky district.

Xinjiang Sanbao Industry Group has taken the first place among foreign trade enterprises in the XUAR over the past three years. It comprises 10 member enterprises, successfully cooperates with many countries of the world. The company has set up branches or offices in Kazakhstan, Russia, Tajikistan, Uzbekistan, and Afghanistan. Sanbao is one of 200 top enterprises in foreign trade of China. It exported more than 2000 units of various special equipment to Kazakhstan from China, while a lot of after-sale service centers have been set up successively. In recent years, the Group has reached internationally advanced level and made up 2/5 of domestic sales market share.

Oil equipment is exported to the Russia and Kazakhstan. The company enjoys overseas engineering contracting license approved by the Ministry of Commerce, it has undertaken 19 overseas large-scale projects. The polypropylene projects have filled up the gap in petrochemical field of Kazakhstan. The ongoing cement plant of 2 million tons per annum in Kazakhstan is a cement plant with the largest scale and highest technology content in Central Asian countries [3].

Yema Group International ("Yema" – Wild Horse) is the regional leader in Internet trading. It has been engaged in traditional foreign trade for more than 20 years and its annual foreign trade turnover is approximately $800 million. The company occupies a leading position among export-oriented XUAR companies. In 2015 the leadership of Yema decided to establish a cross-border electronic commerce unit. The employees of the group independently developed an online store focused on Russian-speaking countries. The business model was created in the format of "Internet + foreign trade", matching traditional foreign trade with E-commerce.

Chen Qiang, president of Yema Group Company Limited noted: Our advantage is in professional teams in foreign trade. Yema has branches in Russia, Kazakhstan, Kyrgyzstan and Uzbekistan. Combining this advantage with Internet trading, our company has gained enormous opportunities, which can hardly be achieved by those who are just starting to engage in trade in this area. However, the spread of cross-border e-business in Russian-speaking countries is relatively worse than in China, this gave us some time to develop activities [4].

Unlike Yema, which specializes in E-business, Zhongxin Silu, a Xinjiang Internet technology joint-stock company, is making efforts to create an e-business platform that provides comprehensive cross-border trade services, such as procurement, marketing, payment, logistics, warehousing, procedures at customs, etc. The company wants to succeed in e-business in the market of five Central Asian countries. The company has a significant advantage due to the independently developed payment tool Zhongxing-fu. Large banks of five Central Asian countries have been paid a security deposit, in order to control the exchange rate, a special company has been identified that carries out conversion and payments in real time. Payments are made in national currencies; the company receives them in CNY [5].

5 Kazakhstan-Xinjiang Trade and Economic Relations

Today, Kazakhstan is the main trade and economic partner of Xinjiang among all 8 countries with which this region borders. XUAR accounts for more than 75% of the trade turnover between Kazakhstan and the PRC. In this regard, Xinhua News Agency reports that in 2017, Xinjiang's trade with Kazakhstan amounted to $ 9.42 billion: Xinjiang's trade with Kazakhstan increased by 49.3%, amounting to $ 9.42 billion, with Pakistan – by 60.4% to $ 460 million and with Mongolia – 84.3% to $ 149 million. [6]. Experts associate such dynamic growth with the advancement in China-Kazakhstan cooperation on production capacities, the construction of China-Pakistan and China-Mongolia-Russia economic corridors within the Belt and Road. In 2018, XUAR imported $ 480 million worth of agricultural products from Europe and Central Asia. Urumqi Customs reported that Xinjiang's foreign trade increased 17.1% from the previous year to $ 20.66 billion, including the number of exports – by 13.8% to $ 17.73 billion, imports by 42.6% to $ 2.93 billion. The bulk of exports, as agency reported, were mechanical engineering and electronics products. The labor-intensive traditional export products, and resource and agricultural products were predominated in imports. According to agency, further progress in these areas will give Xinjiang the opportunity to more fully demonstrate its advantages as a "key zone" of SREB. In this regard, it is worth to mark out such factors as the recent decision of China's State Council on the next reduction in import tariffs; the using of CNY in trade between China and Pakistan; the launch of a new trade route from China to Europe - Urumqi – Kuryk – Baku, which is recognized as a very convenient and highly efficient logistics corridor connecting China with Georgia, Iran, Azerbaijan, Turkey and European countries.

On December 3, 2019, Xinhua Agency reported that for the first 10 months of 2019 XUAR foreign trade amounted to 131.5 billion yuan (about $ 18.7 billion). This indicator exceeded by 28% the indicators of the first 10 months of 2018. The region's exports amounted to 100.2 billion yuan, up 19.9% year on year, while imports soared 63.1% to 31.3 billion yuan, Urumqi customs office said. Kazakhstan topped the list of Xinjiang's largest trading partners for the indicated period. The trade with Kazakhstan increased by 28.2% to 60.2 billion yuan [7].

Among other major partners of XUAR in the region are Kyrgyzstan, Australia, Pakistan, the United Kingdom, Argentina and Vietnam.

6 Car Industry. Textile Manufacture

The car industry is forming in Xinjiang. Currently, the production of SAIC Volkswagen plant in Xinjiang is 20 thousand cars per year. Some of them are sold on the local market, and the rest are delivered to other provinces in China. The local government has provided preferential taxation and land acquisition policies. The first car has been produced in this plant in 2013, when China put forward the Silk Road Economic Belt initiative. In 2016, another car manufacturing company – GAC Motors – opened a factory in Urumqi and hopes to occupy its niche in the promising market [8].

The Vice President of the China Textiles Import and Export Chamber of Commerce Zhang Xian made a note that Five Central Asian countries expected to be the largest Xinjiang textile market. In 2017, clothing exports from China to Central Asia reached $ 7.2 billion, 80% of which supplied by Xinjiang. Among all the countries and regions of the SREB and the Sea Silk Road, Eastern Europe has become the second largest market for Xinjiang's light industry. From 2014 to June 2018, the number of sewing enterprises in Xinjiang increased by 2,200 and the total number of employees in this industry reached 540 thousand people. Many textile enterprises in the economically developed East of China are setting up their new manufacturing facilities in Xinjiang. This region is considered one of the leaders in cotton manufacture in northwest China. According to the results of 2016, 112,300 workers were employed in the textile industry of the region. This figure exceeds 50% of all Xinjiang's industrial workers hired last year. In 2016, investments in textile projects in the region amounted to 65 billion yuan ($ 9.39 billion). The annual capacity of the cotton mills in Xinjiang has grown by 150% compared to 2015. According to the results of January–October 2016, the exports of textile products and clothing in XUAR reached 43 billion yuan ($ 6.2 billion). This is 49% more than a year earlier. Kazakhstan and Kyrgyzstan are the main targeted markets for these goods. Russia is in third place.

7 Labour Market

The Silk Road Economic Belt implementation caused a revival in XUAR labor market. Xinjiang's public and private enterprises are expanding their international activities. Meng Fengchao, President of China Railway Construction Co., said Xinjiang is considered one of China's most future-proof regions for developing rail services. He also noted that it is planned to open rail routes from Urumqi to Mersin in southern Turkey, which will pass through Kazakhstan, Turkmenistan and Iran. Thus, a regular railway links to connect Urumqi with Moscow will be established. China Railway Construction Co. decided to recruit more than 600 local residents with academic degrees to develop the marketing research in the region.

Kang Ning, Vice President at China Post Group in an interview with Chinese media said that China Post is interested in developing and improving the mail service established in 2016 between China, Russia and Kazakhstan. Local post offices recruit new employees from Xinjiang residents, with preference being given to ethnic Russians and Kazakhs. The e-commerce development has also influenced the revitalization of the local labor market. By the end of July 2017, about 24.59 million parcels were sent from China to Russia and Kazakhstan. This explains the extreme interest of China Post in increasing the number of employees not only from local residents, but from those who speak Russian or Kazakh. It is expected that within the Belt and Road, the need for such employees will increase at enterprises of other sectors of the economy [9].

Xinjiang becomes the center of building the Silk Road new Economic Belt, which creates great opportunities for development in areas as transparency, deepening reforms, optimizing the production structure, coordinated development of regions, enhancing the soft power of culture and ensuring social stability. In Xinjiang is planned to create "three corridors" of transport, energy and communications, "three bases" related to the (a) production, processing and storage of oil and gas, (b) the development of the coal industry, (c) the development of wind energy and solar energy. The region also plans to create "five centers": transport, trade, culture, finance and health. The strategy of "going beyond" makes Xinjiang a center for regional development.

8 Conclusion

The Belt and Road identifies economic and geographical contours and detailed indications of the tasks for various regions of China in realizing the large-scale aim of the initiative. Xinjiang is assigned a role of the "window" to the Western lands.

The socio-economic and national model of development implemented in XUAR, as well as ethnic separatism problems has a great importance. Despite the positive dynamics in the development Xinjiang has socio-economic, political, cultural and national problems. It is extremely important to understand these issues. The processes taking place in XUAR have an impact on the Central Asian states. And the deal with not only economic expansion or demographic pressure. The deal is, that is most important, with transferring the Xinjiang conflicts to the Central Asia. In the Central Asian policy of China, the decisive role of Xinjiang has a security factor. And it is not only about the Central Asian processes' negative impacts to XUAR, but also about the involvement of the region states in China's anti-separatism efforts.

China is becoming an active participant in Central Asian large-scale transport, logistics and infrastructure projects, as well as a major lender to the economies of the region. China gradually displaces Russia from its leadership position as a trading partner of the Central Asian states. This happens not only because PRC acquires goods and raw materials traditionally exported to the Russian Federation, but also because of increased competition with Russian enterprises for sales markets. The Chinese concept of turning the country into a Global Trade Power, along with the country's production and investment capabilities, leads to a decrease of the Russian importance as a trade and economic partner for the countries of Central Asia.

First, Xinjiang fulfills a vital and strategic role in China's economic security: the region provides country with natural gas, oil, and coal.

Secondly, XUAR is the transit route of energy resources imported from abroad, which will play an increasingly crucial role in China's supply security. Most of the main communications go through Xinjiang territory.

Xinjiang's strategic importance in Chinese foreign trade is growing, as China's gateway to Central Asia and beyond. The region is the largest hub for foreign trade with reduced logistics costs and the single largest raw material supplier of China's booming textile industry.

Third, although Xinjiang is behind most of the rest of China in terms of its share from the national income (and likely to remain so for years to come) the rest of China owes a great deal for its economic growth to Xinjiang. Apparently, Xinjiang's economic structure fits the picture of the periphery, of which the main function is to provide the core with energy supplies and raw materials. This is precisely why the Chinese government focuses on the economic development of the region. A developed and stable Xinjiang is one of the keys of ensuring China's economic security in the long term. Beijing cannot take the risk of instability in a region, which produces its energy supplies and cotton, serves as a major foreign trade hub, and hosts important transportation routes connecting China with the rest of the world are passing. [10].

Generally speaking, the Belt and Road Initiative, aimed at creating a trade and infrastructure network connecting Asia with Europe and Africa along the ancient trade routes of the Silk Road, gives a new impetus to the Xinjiang development.

References

1. Net, X.: Ekonomicheskoye obozreniye: ekonomika Sin'tszyana na puti bystrogo razvitiya. http://russian.news.cn/2017-07/13/c_136440871.htm
2. «Polit-asia.kz» online news outlet, Programma razvitiya Sin'tszyana. https://polit-asia.kz/programma-razvitiya-sintszyana/
3. Sanbao, X.: Industry Group. http://www.xjsanbao.com/english/index.html
4. YEMA Group Co., LTD. http://yema.uz/golovnoy-offis-knr
5. SILUXGC New Silk Road Observation, SUAR stanovitsya kommerchesko-logisticheskim tsentrom v ekonomicheskom poyase Shelkovogo puti. http://ru.siluxgc.com/html/R1683/201711/6659457273051802597.shtml
6. Zona KZ online newspaper, $9,42 mlrd sostavil ob"yem torgovli kitayskogo Sin'tszyana s Kazakhstanom v (2017). https://zonakz.net/2018/01/23/942-mlrd-sostavil-obem-torgovli-kitajskogo-sinczyana-s-kazaxstanom-v-2017-g/
7. Russian.people.cn, Vneshnyaya torgovlya Sin'tszyana na fone initsiativy «Poyasa i Puti» vyrosla boleye chem na 17 protsentov po itogam 2017 goda. http://russian.people.com.cn/n3/2018/0124/c31518-9419024.html
8. Net, X.: Sin'tszyan stanovitsya novym promyshlennym tsentrom na Shelkovom puti. http://russian.news.cn/2018-09/14/c_137468404.htm
9. Kozhirova, S.: Sin'tszyan v proyekte Ekonomicheskogo poyasa Shelkovogo puti, International Science Complex "Astana". https://isca.kz/ru/analytics-ru/2760
10. Atli, A.: Role Of XinjiangUyghur Autonomous Region In Economic Security Of China – Analysis, Eurasia Review. https://www.eurasiareview.com/03012016-role-of-xinjiang-uyghur-autonomous-region-in-economic-security-of-china-analysis/

The State Policy of the Republic of Kazakhstan in the Field of Combating Religious Extremism

Li Yang[1(✉)] and Svetlana Kozhirova[2]

[1] Kazakh Humanitarian Law Innovative University, Semey, Kazakhstan
[2] International Science Complex "Astana", Nur-Sultan, Kazakhstan
office@isca.kz

Abstract. The preeminence of the national policy of Kazakhstan in prevention and hindrance of religious extremism is considered as one of the most significant courses. The main achievements are the creation of various state and public bodies responsible for the above issues. Broad system of rules and obligations, civil programs are being enhanced. Present religious position in Kazakhstan has been thoroughly investigated by the author. The paper discusses that there has been consolidation and the emergence of radical religious organizations and outlines a huge expansion in the framework of contemporary beliefs. The article works on analysis of the impact and enlargement of foreign preacher movements, particularly effecting on youth due to the widening of the public base of sectarian communities. The primary purpose of this article is to survey all of the above-mentioned issues and examine their relation to extremism in Kazakhstan. The paper aims to examine the aspect of spiritual feature in the mode of life and everyday activities of ordinary citizens of Kazakhstan. Furthermore, it investigates the rapid growth of the influence of non-traditional religious movements. This paper observes anti-extremist legacy and strategy of Kazakhstan. Much attention in this study has been drawn to disseminated examination of the ongoing appearance of extremism in the country considering priority parts of pertinent papers. Analysis has proven improvements of foreign contribution in anti-terrorist missions in Kazakhstan, which in turn influenced on improvement of national domestic law.

Keywords: Extremism · Terrorism · Non-traditional religious movements · Republic of Kazakhstan · Legal framework · State policy

1 Introduction

In the modern world, religious extremism has transformed to one of the most difficult social problems, becoming a part of domestic politics and an instrument for conducting foreign policy activities to realize the goals of individual states, forces and movements on the world stage. In this case, devotional concepts are used to justify political actions or to mobilize people for non-religious purposes. Such manipulations are often the cause of the radical position on social issues and consciousness of citizens and the commission of offenses of the person who is fanatical in pursuit of religious ideals and owns radical

© The Author(s), under exclusive license to Springer Nature Switzerland AG 2021
R. Bolgov et al. (Eds.): *Proceedings of Topical Issues in International Political Geography*, SPRINGERGEOGR, pp. 161–170, 2021.
https://doi.org/10.1007/978-3-030-78690-8_15

character, which have recently turned into a universal human problem. Mode of living is not build on above mentioned circumstance; therefore, it exists both in developed and in poor countries.

Problems in the process of countering religious extremism today are relevant in the country. All national laws qualify this type of act as the most dangerous type of crime that poses a threat to national security, sovereignty, territorial integrity and the foundations of the constitutional order. Security and the cohesion of people entirely rely upon successful functioning of the devotional field of the government.

Therefore, the purpose of this research is to present a comprehensive study of the state policy of the Republic of Kazakhstan in the field of combating religious extremism. To achieve this purpose the following objectives have been identified:

- to substantiate appraisal of the present state of religion in the country;
- to examine the legal framework for the fight against extremism;
- to identify the effectiveness of state bodies;
- to substantiate the main aspects of verification of the information space.

In addition, this research develops increment of preacher actions, widening of futuristic beliefs, uprising and intensification of the radical devotional entities, growth of dissident communities mostly involving younger generation describe the current situation in the country. Also, it investigates the rapid growth of the influence of non-traditional religious movements. The article examines the state policies for solving the above mentioned issues. The advancement of domestic state laws and external assistance with the view of overcoming the raise of extremist-related terrorist actions has been carefully analyzed.

Today there is a sufficient amount of literature devoted to the problem of extremism in Kazakhstan, but most of them are confidential. Therefore, one of the main sources that have been studied with the aim to achieve the purpose and objectives of this research, is the group of state and governmental documents, which determine the main directions of domestic and foreign policy. In addition, analytical materials and scientific works of foreign, Russian and Kazakh researchers have made a significant contribution to the study of various trends and mechanisms of the manifestation of extremism in the country.

The method of system analysis in addition to institutional, sociological, comparative and historical methods and approaches have been applied in this research. Assembling and researching the significant state records taking notice to special consideration of factors is one of the main methods of investigating the situation in Kazakhstan.

2 Glimpse at the Religious Situation in Kazakhstan

Extremism in any form of its manifestation has become one of the most dangerous in its scale, unpredictability and consequences of socio-political and moral problems. The origins of extremism are rooted in ancient times. Its first manifestations should be sought in the beginnings of the emergence of the state in the stratification of society into classes.

At different times, quite a lot of different legal and scientific definitions of the concept of "extremism" were given, but nevertheless, there is no single definition today.

As defined in the Oxford Dictionary, the word "extremism" means political, religious and social ideas or activities that are extreme, inappropriate, unreasonable, and unacceptable to most people [1].

Describing the first steps towards radicalization, R. Cottone uses the term "Consensuality" [2]. He claims that a person who constantly interacting with extremists, with people who have been radicalized or marginalized, begins to perceive a certain point of view as the absolute truth. In his opinion, religion is more inclined towards extremism and radicalization, since social constructivism is much stronger in religious groups. This theory provides an understanding of the process within the personality that leads to extremism.

Poor socio-economic conditions are the "root causes" of political violence. However, history shows that only poverty can serve as neither a valid reason nor a reliable indicator of tension in society [3]. Studies also don't support the view that discontent is a sufficient factor to inspire collective political violence [4]. The social injustice encourages vulnerable segments of the population, especially young people, to resort to slogans of social justice and to express political protest in a religious form. All this has a negative effect on the institutions of power and is a fertile ground for the intervention of external forces to destabilize the situation in the region by activating the "sleeping" cells of terrorist organizations.

According to E. Sergun, extremism is loyalty to views, concepts, ideas, which is built on the basis of political, racial, national and religious hostility to an individual or certain groups, to various ethnic representatives or the state. In accordance with this, religious extremism is usually understood as a rejection of the system of traditional religious values of society and the propaganda of "ideas" that refute their idea [5]. At the same time, A. Nurullayev and A. Nurullayev determine the definition of religious and political extremism as motivated religious or hidden religious activity, which is aimed at changing the state regime or usurping power by violent means. It is also worth noting the violation of the independence and integrity of the territory of the state, the creation of illegal armed formations, the incitement of religious and national differences and hostility [6].

The state policy of the Republic of Kazakhstan in relation to religion has changed radically during the period of gaining independence. According to the Constitution, Kazakhstan proclaimed itself as a secular state in which religion is separated from the state. The forbiddance on religious extremism, terrorism demonstrations and prohibitions on them are specified in Article 5.20 of the Constitution [7].

The Law "On Countering Extremism" contained precise characteristics of the ideas of "political", "national" and "religious" extremism. Religious extremism is determined as instigation of devotional hate that contain those related to ferocity or calls for violence [8].

A distinctive feature of the development of the religious situation in Kazakhstan has been the diversity of religious associations, which resulted a logical expansion of the religious space and diminution of the commonwealth aspects along with regulation of the processes. It created the ground for the rapid growth of the influence of non-traditional religious movements.

Firstly, religion begins to occupy absolute influential provision in the socio-political life of Kazakhstan. Concurrently the religious factor begins to play an increasingly important role in the life and everyday activities of ordinary citizens. Accordingly, religion gradually started to influence the decision-making processes of both the individual and society as a whole, becoming a new paradigm in the public and social relations of people.

Secondly, the activities of non-traditional religious associations in the republic are increasing.

Kazakh scholars Z. Shaukenova, E. Burova [9] have developed the significance of the revival of internal family traditional values as a guarantee of stabilization of the internal political situation in the country. The Kazakh society has not lost its traditional values and spiritual roots, which can be a force in the fight against extremism. The authors also points out the need to revive the tradition of religiosity while maintaining a secular orientation.

According to a forecast the socio-economic situation in the country will further worsen, and, subsequently, the social situation of ordinary citizens. One should expect only an increase in the pace of activity of non-traditional religious associations, as they mainly attract people of low social status who have material difficulties into their ranks. At the same time, B. Bekturganova [10] draws a parallel of conflict cases between representatives of different confessions, arguing that an important component in the process of countering religious extremism is the work carried out directly with the younger generation within the framework of state youth policy.

3 Legal Framework for the Fight Against Extremism

The position of religion on certain issues has led to the ambiguity of state policy. In particular, we can highlight the regulation of the activities of foreign missionary religious and charitable organizations in the country, in the life of society and the state. Consequently, present situation has naturally demanded from the Kazakh authorities to adopt a new package of laws in the framework of adjustment of spiritual performances.

Due to a need to strengthen state policy regarding non-traditional religions, on February 10, 2000, Decree No. 3332 of the President of the Republic of Kazakhstan "On measures to prevent and suppress manifestations of terrorism and extremism" was adopted. Despite this, as shown by the Council for Relations with Religious Associations, by 2001 the number of religious organizations was observed to increase to 2252, and almost half of them, 1080, were not registered with the justice authorities, being essentially violators of the act [11].

The Law of the Republic of Kazakhstan "On Counteracting Extremism", adopted in 2005, can be noted as one of the most decisive laws in the struggle against extremism in this framework [8]. This law demonstrates the juridical and organizational basis for fighting against extremism, aimed to advocate the rights and liberty of individuals and nation.

Kazakhstan has not faced any problem with extremism and terrorism during the two decades of its independence. The beginning of terrorist attacks was indicated in Kazakhstan since 2011 (Aktobe case). After that the state policy in this direction was activated by adopting stricter laws regulating religious groups in the country [12].

In 2011, to replace the Law of the Republic of Kazakhstan "On Freedom of Religion and Religious Associations", adopted in 1992, the Law of the Republic of Kazakhstan "On Religious Activities and Religious Associations", introduced by the Agency of the Republic of Kazakhstan for Religious Affairs, was adopted. State-confessional policy made the evolution of a new phase after the endorsement of the law. This indicates the dynamic engagement of state authorities in regulating the work of religious entities, maintaining equilibration of interests of the country and community as a whole [13].

During the implementation of the Law, the joint efforts of central and local administrative institution in spiritual and religious spheres streamlined the issues of missionary activity, construction and location of places of worship, established places, prayer rooms for religious ceremonies outside religious buildings, streamlined the distribution of literature, data and evidence of religious scope aim and purpose, and the procedure for conducting religious studies examination has been established. The most important achievement of the implementation of the Law and the work of the devotional state associations were reflected in the awareness of citizens that religious activity is necessary only according to the rules defined by the state.

There was an increase in crimes linked to extremism and terrorism, the evidence that a new law was adopted and a special body was created to combat the above problem did not change the situation at all. There was a rapid increment in the number of criminals prosecuted under this law in 2016–2017. Only in 2018, there is a significant decrease in numbers by 19 people registered.

There was a huge increase in the dissemination of information that contains open propaganda of extremism in the period of 2013–2016. These materials called for joining religious organizations, which ultimately led to significant activity of a larger population in the religious sphere.

However, since 2016, there has been a decrease in the number of criminal cases related to the circulation of extremist materials such as leaflets, books and other media. The Government took timely measures on this issue. Thus, the rules for conducting religious studies examination were approved; budgetary funds were allocated to pay for the work of religious experts. In the Law of the Republic of Kazakhstan dated October 11, 2011 "On Religious Activities and Religious Associations", a separate article is reserved for religious studies, and on its basis the legislative act "Rules for conducting a religious studies examination" was developed, which was endorsed by Kazakhstani authorities.

The legislative consolidation of the legal and regulatory basics of the fight against manifestations of political, national and religious extremism was also facilitated by the ratification of the Law of the Republic of Kazakhstan "On Amendments and Additions to Some Legislative Acts of the Republic of Kazakhstan on Countering Extremism and Terrorism", which established a ban on the creation, registration and operation of terrorist organizations as well as illegal paramilitaries.

A clear position of the state in this course was indicated by the country in a number of strategic documents. The first President of the Republic of Kazakhstan Nursultan Nazarbayev identified the task of counteraction to all forms and manifestations of extremism as a priority in his Address to the nation "Strategy "Kazakhstan 2050": a new political course of the established state" [14]. He pointed out the urgency to extend the avoidance of religious extremism in society, specifically among youth, to educate the population of religious self-awareness that meets folk traditions and cultural values of a secular rule of law.

Following the above mentioned Address on Combating Religious Extremism and Terrorism for 2013–2017, the Program of government that counts preclusive remedies as the main way to fight was approved by the first President in September 24, 2013 [15]. The program sets itself the manifestation of extremism and the prevention of terrorism threats to the security of man and community as the most important goal.

The system "countering extremism" is completely built-up and has three main spheres of activity of government organizations: combating extremism, repression of terrorist and extremist actions, the detection of terrorist crimes and liquidating the outcomes of terrorism and extremism.

At the end of 2016, after a comprehensive discussion of the draft laws "On Countering Extremism" [8] and "On Countering Terrorism" [16] with international experts and representatives of civil society, the First President signed a law amending these legislative acts of fundamental importance. In particular, the national security bodies were charged with the obligation to conduct operational investigative, counter-intelligence remedies likewise the regulations of the Republic of Kazakhstan on motivated conclusions of state bodies, take measures to prevent the entry into the Republic of Kazakhstan of foreigners and stateless persons who by their actions pose a threat or damage the security of society or state. The next step towards changes is that the state bodies of the Republic of Kazakhstan to counter extremism collaborate with the competent authorities of foreign states. In turn, they use law enforcement organization, carry out operational-search counterintelligence activities in the country within the framework of an international treaty.

The Concept of State Policy in the Religious Sphere of the Republic of Kazakhstan for 2017–2020, adopted in 2017, made it possible to develop priority areas in the religious sphere [17]. The secular grounds of the progress of community were strengthened in all areas related to the competence of the government. The state's positions on topical issues of society, such as wearing hijabs in schools, refusing to vaccinate children for religious beliefs, entering into spiritual marriage "Neke", receiving religious education abroad, and others, have been developed.

As indicated in the Concept, the key direction of the work of state bodies and institutions in the religious sphere is considered outreach and educational work. The main goal of outreach is to create a public understanding of religion policy aimed at strengthening the secular grounds of the state, maintaining stable state-confessional and interfaith relations, discrediting the activities of organizations harmful to public, personal and state security under the guise of religious, educational, charitable, commercial and other goals. The present aim is to preclude any intent of extremists, reducing the number of adherents of destructive movements in community.

Building sustainable immunity from radical extremist ideas is a key challenge. To implement these tasks, a program was developed and adopted by the President's Decree "State Program on Counteracting Religious Extremism and Terrorism in the Republic of Kazakhstan for 2018–2022" [18]. This state program indicates the main ways to achieve the goals and the corresponding measures. In turn, this will ensure the safety of individuals, society and the state from violent manifestations of religious extremism.

4 Effective State Bodies

In order to ensure a scientific approach to the analysis and forecasting of the religious situation, in January 2007, a Scientific and Research Center of the Ministry of Justice of the Republic of Kazakhstan for Religious Affairs was established. During this period, the problem of victims who suffered from the activities of religious entities of a destructive nature is openly posed. In connection with this situation, the first non-governmental organization appears that provide psychological and legal assistance to people affected by enterprises of devastating devotional objects.

In this regard, one of the significant events of 2011 was the reorganization of the Ministry of Culture and the establishment of a new wing of the Agency of the Republic of Kazakhstan for Religious Affairs with the relocation of functions and powers engaged with interfaith consent, providing authority of civilians to liberty of devotional entities and interaction with religious associations [19].

By the order of the Minister of Culture and Sports of Kazakhstan, a new text "Rules for conducting religious studies examination" was approved in December 30, 2014 [20]. Currently, there are two types of religious studies carried out in the country: judicial religious studies supervised by experts of the Center for Forensic Expertise of the Ministry of Justice of the Republic of Kazakhstan; and the religious examination carried out by an authorized body-Committee of public consent of the Ministry of Information and Social Development of the Republic of Kazakhstan.

The functions of coordinating the activities of state bodies in the fight against terrorism were divided into 2 main levels: central and regional. This structuring became possible after adopting the state program in 2013–2017 with the aim of countering terrorism and religious extremism. The National Security Committee of the Republic of Kazakhstan, together with the Anti-Terrorist Center of the CIS countries, coordinates the activities of all state bodies at the central level. While the Akimat of the regions, the cities of Almaty and Astana (now Nur-Sultan), as part of the anti-terrorist commissions, coordinates at the regional level.

The next stage of the Center's practice was the prohibition of organizations whose activities are extremist in nature. The list of such organizations includes 18 foreign associations. These include: Al-Qaeda, Azbad al-Ansar, Muslim Brotherhood, Islamic Movement of Uzbekistan, Islamic Party of Turkestan, Hizb ut-Tahrir and several others. In 2015, ISIS was banned in Kazakhstan. Article 5 "On the National Security of the Republic of Kazakhstan" states that terrorist activities of organizations of foreign states and individuals, as well as extremism in any form, primarily refers to the group of risks to the national security [21].

Central and local executive bodies have established local awareness groups within the Concept of State Policy in the Religious Sphere of the Republic of Kazakhstan for 2017–2020. Experts-religious scholars, theologians, scholars, psychologists, specialists from the central and local executive bodies for religious affairs, leaders and representatives of traditional religions, youth and veteran associations are involved in the work. In total, 211 regional groups have been created in the country, of which 2,623 people are working. According to the Committee, in 2017, with the participation of members of regional Advocacy groups, 35 thousand various-format events of a general preventive orientation were conducted with coverage of more than 2 million 450 thousand people [22].

Religious problems research centers, which were created in each region, carry out targeted work with adherents of destructive religious movements, mosque zhamagati, a believing part of the female population, market traders, unemployed and self-employed citizens.

5 Verification of the Information Space

Studies have shown that the main direction of the content of document is the ideology of the power of extremism in radical Islamic literature. It is impossible to exclude operating mode of conventional structures, criticized in the texts, intolerance, animosity, devotional hate towards representatives of other religious entities.

815 titles of religious literature and information materials were considered prohibited to be imported into the country and distributed after a court decision from 2009 to 2019. In obedience to the accomplishment of the project "Automated Monitoring of the National Information Space" the work is being carried out to increase the effectiveness of monitoring the information space. Since 2017, as part of the Safe Internet pilot project, a draft of the Cybernadzor information system has been prepared to identify sites with illegal content. In 2016 alone, more than 7000 sites were analyzed, where 177 resources with the aforementioned content were identified. Whereas in 2017, the content of more than 10174 Internet resources was analyzed, of which 3555 Internet resources containing materials of an illegal nature were identified.

At the time of the implementation of these projects, access to materials from 1958 was limited, indicating the propaganda of the idea of terrorism and religious extremism for consumers by means of the Law "On Communications" adopted in 2018. In addition, 241 court decisions were issued on 3235 Internet resources and links promoting terrorism and religious extremism. The owners and administrators of Internet resources seized more than 250 thousand illegal propaganda materials [23].

Annually, methodological recommendations are developed for organizing informational work on religions among target groups of the population in the regions, and training seminars are held for Advocacy group members on religious issues.

We can observe the promotion of values and the prevention of extremism in 5 documentaries (Media-Astana LLP) released by the state order for 2017–2020. These documentaries were released as part of the fulfillment of the State Policy Concept and aimed at clarifying policy in devotional sphere of Kazakhstan by promoting traditional values, and preventing extremism. Maintenance and the leadership of the Committee has led to the creation of the cultural and educational Internet resource "E-islam" containing objective information about Islam and the traditional spiritual values of the Kazakh people.

Separately, it is necessary to note such an important direction in the prevention and counteraction to the spread of extremism as works in prisons. It is difficult for people who do not have stable views, psychologically broken imprisonment, to adapt to life behind bars. It is here that the prisoner easily falls into the conviction of the injustice of laws and the entire system, looking for any reason to justify his crime. Another prerequisite for the radicalism of prisoners is the religious illiteracy of those who end up in prison. Therefore, ideological and propaganda work is being strengthened in the penitentiary system with the involvement of psychologists, theologians, religious scholars, and representatives of the clergy. Non-governmental organizations are also involved in this work. For example, the potential of the Public Foundation "Akniet" Information and Advocacy and Rehabilitation Center, the main activity of the company is deradicalization of those convicted of crimes of an extremist and terrorist nature.

6 Conclusion

It can be concluded that the prevention of extremism and terrorist manifestations in the framework of countering religious extremism, as well as that the creation of a national process is sufficiently confirmed by a thorough analysis of state policy. The most prosperous accomplishment in determining the tough issues of the religious sphere are the observance of the current legislation, which is evidenced in the presence of a constitutive interlocution among the government and devotional community, the reciprocal believe, frankness of conception and reverence of importance.

It is very important to note that active work has been done in the anti-terrorist structure of the Commonwealth of Independent States. Decisive measures are being taken to combat extremism and consistent efforts are being made to counter terrorism. The country's activity in the Collective Security Treaty Organization, the Conference on Interaction and Confidence Building Measures in Asia, the OSCE and the UN, as well as in the Shanghai Cooperation Organization was noticed. Kazakhstan has acceded to all fourteen international universal conventions to overcome terrorism, extremism and manslaughter. The above measures will lead to the improvement of the legal framework together with international organizations in order to combat extremism, as well as contribute to the strengthening of legislation. The devotional understanding of nationals that leads to the manifestation of extremism and terrorism has a huge risk from the point of view of a radical position in the medium and long term for all countries without exception. Therefore, following main tasks of the state should be implemented on an ongoing basis:

- to overcome extremism and terrorism by enhancing structure countrywide;
- liquidate the presence of contradictions in legal system;
- corrections in the actions of government organizations in the fields of combating terrorism and extremism, particularly total lawmaking adjustment of the arranging and regulating actions;
- strike a blow against extremism and terrorism by means of dynamic delimitation, specification of authority and capability of the government entities;
- monitoring of destructive resources, which is the part of the comprehensive system of measures for providing protection of data of the government;
- high-quality informational, scientific and methodological support of measures to prevent extremist activities in society.

Moreover, only the joint work of state bodies and civil society institutions will be able to raise to a qualitatively different level the protection of the devotion of community, human being and the nation from any kind of manifestations of extremism and the emergence of terrorist threats.

References

1. Oxford dictionaries. http://www.askoxford.com. Accessed 13 Jun 2020
2. Cottone, R.: Toward a Positive Psychology of Religion: Belief Science in the Postmodern Era. John Hunt Publishing, Winchester (2011)
3. Webber, C.: Revaluating relative deprivation theory. Theoret. Criminol. **11**(1), 97–120 (2007)
4. Brush, S.: Dynamics of theory change in the social sciences: relative deprivation and collective violence. J. Confl. Resolut. **40**(4), 523–545 (1996)
5. Sergun, E.: Legal support of extremist activity in the Russian Federation. Legal Policy Legal Life **2**, 197–199 (2006)
6. Nurullayev, A., Nurullayev, A.: Religion and Politics, pp. 124–125. KMK Publishing, Moscow (2006)
7. Constitution of the Republic of Kazakhstan (1995)
8. Law of the Republic of Kazakhstan "On Counteracting Extremism", No. 31-III, 18 February 2005
9. Shaukenova, Z., Burova, E.: Religious identity in family strategies. Kazakhstan Spectr. **2**(76), 7–18 (2016)
10. Bekturganova, B.: Peers of independence: ethno-religious identifications. Kazakhstan Spectr. **1**(79), 7–19 (2017)
11. Decree of the President of the Republic of Kazakhstan "On measures to prevent and suppress manifestations of terrorism and extremism", No. 332, 10 February 2000
12. Cornell, E., Starr, F., Tucker, J.: Religion and the Secular State in Kazakhstan, pp. 54–65. Central Asia-Caucasus Institute, Washington (2018)
13. Law of the Republic of Kazakhstan "On Religious Activity and Religious Associations", No. 483-IV, 11 October 2011
14. Address of the President of the Republic of Kazakhstan - Leader of the Nation Nursultan Nazarbayev to the People of Kazakhstan "Strategy "Kazakhstan-2050": a new political course of the established state", 14 December 2012
15. State Program on Countering Religious Extremism in the Republic of Kazakhstan for 2013–2017, No. 648, 24 September 2013
16. Law of the Republic of Kazakhstan "On Countering Terrorism", No. 416, 13 July 1999
17. Concept of state policy in the religious sphere in the Republic of Kazakhstan for 2017–2020, 20 June 2017
18. State program on counteracting religious extremism and terrorism in the Republic of Kazakhstan for 2018–2022, 15 March 2018
19. Izbairov, A.: Islam in Kazakhstan, pp. 95–101, Almaty (2013)
20. Resolution of the Government of the Republic of Kazakhstan "Rules for Conducting a Religious Expertise", No. 209, 7 February 2012
21. Law of the Republic of Kazakhstan "On National Security", No. 527-IV, 6 January 2012
22. Report on the implementation of the strategic plan of the Ministry of Religious Affairs and Civil Society of the Republic of Kazakhstan for 2017–2021. https://www.evaluation.egov.kz/sites/default/files/pages/otchet_na_rus13022018121532.docx. Accessed 15 Mar 2020
23. Law of the Republic of Kazakhstan "On Communications", No. 567-II, 5 July 2004

Forming the Package of Framework Documents and Laws of Azerbaijan in the Sphere of Military Security in 1994–2019

Niyazi Niyazov(✉) and Galina Niyazova

Saint Petersburg State University, St. Petersburg, Russian Federation
{n.niyazov,g.niyazova}@spb.ru

Abstract. After the collapse of the USSR the conflict because of Nagorno-Karabakh between Azerbaijan and Armenia turned to be a largescale war. As a result of military actions Armenia wrested control of not only disputable Nagorno-Karabakh but also seven neighbor regions. In spring 1994 parties to the conflict came to the cease-fire agreement. By that time the UN adopted four resolutions demanding withdrawal of Armenian military forces from Azerbaijani territory. As Armenia didn't follow the UN resolutions Azerbaijan proceeded to reforms in the sphere of national security. Taking advantage of concluding "The Contract of the Century" Azerbaijan gained financial opportunity to improve its military forces. Besides, the process of promoting military and technical cooperation with European and Asian states was started. All this was accompanied by the creation of conceptual and legal framework for military forces advancing, including interaction with NATO. The objective of this article is to determine the peculiarities of transformation process of conceptual and legal framework approaches to military security challenges. Stating this objective defined the methods applied in the research. Taking into consideration that we focus on long duration process, historic and genetic method is used in the research. Moreover, the necessity to work with legal documents made it essential to use the method of legal hermeneutics. Furthermore, researchers used the legal compativistics method. The results obtained due to the research reveal that not only the conflict with Armenia influences the process of conceptual documents and laws passing but also geopolitical terms of Azerbaijan and viewpoint of political and military leaders on perspective of national security system development.

Keywords: Military security legislation · National security · Military doctrine · IPAP · The concept of national security · The Nagorno-Karabakh conflict

1 Introduction

Despite the fact that the states appeared after the collapse of the USSR with varying degree of success are engaged in nation-building, the global society is likely to evaluate the level of their development in a perfunctory manner, believing that these geopolitical entities are not able to ward off their Union Republic past. According to the authors of this article, this viewpoint is linked to certain factors:

© The Author(s), under exclusive license to Springer Nature Switzerland AG 2021
R. Bolgov et al. (Eds.): *Proceedings of Topical Issues in International Political Geography*, SPRINGERGEOGR, pp. 171–177, 2021.
https://doi.org/10.1007/978-3-030-78690-8_16

- Firstly, after the collapse of the USSR scientists from ex-Soviet state to the great extend focused on studying western countries, predominantly the USA and Western Europe;
- Secondly, in their turn western theorists often studied post-Soviet space usually focusing on Russia.

Concurrently, contemporary processes on the international arena, including uncertainty, set a problem of providing national security, and specifically in military sphere, for majority of post-Soviet states.

This research aims at revealing political and legal aspect of military security policy of Azerbaijan, one of the most dynamically developing ex-Soviet states.

The objective of the research is to determine the peculiarities of shaping conceptual and legal approaches to military security policy of the state [1] and adopting laws and regulations, legally providing the activity of defense and law enforcement agencies including military forces.

We studied historical and political factors, contributing to implementing and accepting new legal acts, aiming at providing military security of Azerbaijan.

Hypothesis. Shaping conceptual approaches to providing military security in Azerbaijan started with it being a Union Republic of the USSR and was closely connected with the outbreak of Nagorno-Karabakh conflict. Having obtained independence, Azerbaijani leaders changed their views on providing military security and only after Heydar Aliyev came to power conceptual approaches to military security challenges started to be maintained by passing necessary legislative acts. Cooperation of Azerbaijan with international organizations including the UN, Council of Europe, NATO etc., influenced greatly the shaping legal framework of Azerbaijan in military sphere.

2 Methods and Materials of Research

To understand the essence of the phenomena under study we refer to different historical events, thus historic and genetic method is used in the research. It helps to study cause and effect links and laws of sequences of events or phenomena and gives an opportunity to notice peculiarities of objective and subjective factors, which influence the events under study.

Taking into consideration that the research focuses on analyses of the content of Azerbaijani laws, the authors apply the method of **legal hermeneutics**, which is "essential in experiencing legal reality and helps understand the meaning and essence of legal regulations" [2, p. 63].

Furthermore, the researchers used the method of legal compativistics, which gives an opportunity to compare close legal systems of different countries and the legal norms established to outline the perspective of the further cooperation or rivalry.

The empirical base of the research is presented by a wide range of references being in the public domain.

3 Results of Research

After the collapse of the Soviet Union a largescale war caused by the desire to gain control over Nagorno Karabakh by Armenia was launched. In spring 1994 a ceasefire agreement was signed by the parties to the conflict. By that time Armenian troops had occupied 20% of Azerbaijani territory. Despite four UN resolutions [3], Armenia didn't leave the occupied territory and insisted on acknowledgement of Nagorno Karabakh as an independent state [4]. In its turn, Azerbaijan was not going to give up struggle for its territory.

Since 1994 Azerbaijani political leaders, headed by H. Aliyev, started to deal with army build-up. Moreover, they tackled multiple problems linked to searching for political and military allies, organizing military and technical cooperation with foreign states and improving work of inner military-industrial sector [5].

At the same time specialists were working out a package of laws, conceptual documents and legal acts in the sphere of military security. One of the first laws was pushed forward by H. Aliyev. It was a defense act passed on November, 26 1993. According to the document, the Republic of Azerbaijan "relying on the Constitutional act "On state independence of the Republic of Azerbaijan" and international legal norms, takes principles of friendly relations with all states as a framework and maintains its defense capability at the level necessary to provide protection against hostile acts [6].

The very fact of joining NATO program "Partnership for Peace" on May, 4 1994 had a certain impact on shaping conceptual framework of military security policy. The framework Document is known to construct formal basis for "Partnership for Peace" program [7].

As financial capability of Azerbaijan was growing, the state started to invest greater sums of money in modernization of military forces and security sector as a whole. Besides, the development of legislative structure in the sphere of military security was taking place.

In June 2004 President of Azerbaijan Ilham Aliyev signed Azerbaijan's "National Security Act" [8]. In fact, this act was close to Russian Federation National Security Concept, as in force in 2000 [9]. At the same time Azerbaijan made a decision that this document shouldn't be just articulation of the concept, which is not of compulsory nature, but following this law must be obligatory.

Meanwhile, processes of military forces development made policy-makers face the problem of choice of army construction pattern, that might be either "Soviet" or "NATO" ("western").

The main advantage of "Soviet pattern" was presented by its state of exploration. Moreover, the state had access to officer material, taught according to "the Soviet pattern". Furthermore, this model looked less expensive. And there was one more detail connected with the "Soviet pattern" - its engaging contributed to strengthening Russia-Azerbaijani relations both in political and military spheres.

However, despite all these "advantages" Azerbaijani leaders decided to choose "western pattern" of military forces development, although we should acknowledge that a number of Soviet model elements, especially those concerning military sector managing, stayed unchanged. There were various factors contributing to such decision-making: political, economic, geopolitical and, finally, military.

Political Factor was linked to the stand Azerbaijan took in the system of international relations after the Cold War. After the conclusion of truce in 1994 Azerbaijani leaders focused on upward mobility of the state in the opinion of the global society. Thus, it was vitally important for Azerbaijan to be integrated into international organizations, where western states played the leading role.

The fact that Russian leaders in 1991–1999 tended to support Armenia in its struggle against Azerbaijan, demanding Azerbaijan to conduct its foreign policy according to Russian needs, gave an additional impetus to this tendency. As a result, Azerbaijan started to look for counterbalance to pressure of Moscow and it found it cooperating with western states. Step by step the idea of leaving Russia became one of the central concepts of Azerbaijani foreign policy and stayed the same up to 2000s, when V. Putin came to power in Russia and Russia and Azerbaijan started to reshape their relations.

Economic Factor was tightly connected with political one. By the time the ceasefire agreement was signed Azerbaijan, as an economic partner, was not attractive for leading global politics actors. Majority thought that Azerbaijan depleted its resources and oil production failed to meet its own demands. Moreover, Azerbaijan was thought to lose its opportunity to sell oil globally, because even if it managed to find oil, it might fail to deliver it to global markets.

Azerbaijani leaders realized this reality. Thus, Azerbaijan, being in need to rise finds both for social problems tackling and army construction development, had to choose the West because it was ready to invest money in new geopolitical entities, which appeared after the collapse of the USSR.

Geopolitical situation in the region was different. Existence of neighbors, who backed various, sometimes contrary, viewpoints on global politics development, made Azerbaijan be careful on the international arena and even in domestic policy, correlating its actions with interests of global leaders and regional actors.

Military aspect was linked to two details. Firstly, Russia, under the pretext of prevention of Karabakh conflict escalation put embargo on equipment and armament delivery to Baku and Erevan, continuing to provide Armenia darkly with free combat systems [10]. Secondly, at the turn of the XXI century it was obvious that NATO troops were better educated and equipped than Soviet army. We should state that even not having access to western markets of equipment and weapons Azerbaijan didn't broke with NATO, but on the contrary, continued to strengthen relations with it.

Furthermore, having started "Partnership for Peace" program, NATO managed to attract post-Soviet states, presenting predominance of "western" pattern of military structure of state system over "Soviet" one.

Azerbaijan was satisfied with the fact that NATO paid attention to the possibility of cooperation with Baku, Azerbaijan's interests were taken into account, studied and even adopted. Thus, after joining Individual Partnership Action Plan (IPAP) in 2002 Azerbaijan insisted it to be adopted by the Euro-Atlantic Partnership Council, the prior body of NATO [11]. Besides, Azerbaijan intensified work on shaping package of necessary conceptual documents in the sphere of national and military security. According to

Armenian political theorist I. Muradyan "following the executive order from President H. Aliyev a group of people including officers, lawyers and political scientists started to work upon the Azerbaijani Defense Doctrine. Officially this document was called "Concept of the National Security". Simultaneously a group of independent experts, whose ideas were to be taken into consideration in creating the official document, faced the same task" [12].

On their part, NATO representatives were talking openly about the necessity to bring forward the adoption of the Military Doctrine and Concept of the National Security of the state. In September 2006 visiting Azerbaijan deputy commander-in-chief of European Command of US Armed Forces William Ward stated that he expected the adoption of the Military Doctrine and the Concept of the National Security, changing and adaptation of modern standards of military legislation and regulations" [13].

However, the process of accepting the Military Doctrine and the Concept of the National Security turned out to be protracted. One of the reasons was that NATO representatives tried to avoid including in the text of the Concept of the National Security and the Military Doctrine of Azerbaijan articles, where Armenia was named as an external threat to the system of national security of Azerbaijan. From the point of view of Azerbaijan, it was absolutely impossible to delete these clauses, but for NATO officials it was essential to keep friendly relations with Armenia, which despite it is a member state of the CSTO, actively cooperates with NATO in military sphere. Furthermore, NATO officials prefer to ignore clauses where Azerbaijan was called an enemy in doctrinal documents of Armenia. Thus, in the beginning of 2007 the National Security Strategy was adopted in Armenia. It tells that "aggressive policy of Azerbaijan conducted regarding Nagorno Karabakh problem is direct military threat to Armenia" [14]. The same year Armenia adopted the Military doctrine where Azerbaijan was mentions as an enemy [15]. Nevertheless, we should acknowledge that interaction with NATO in the sphere of shaping conceptual documents, linked to various aspects of national security, contributed to adoption of the Concept of the National security of Azerbaijan on May, 23 2007 [16]. In this document, in the part named "Europe and integration into Euro-Atlantic bodies" it is noted that Azerbaijan together with NATO is settling different problems in Europe and on Euro-Atlantic space. Moreover, it was mentioned that hostile Armenian policy is "the main barrier on the way of comprehensive regional cooperation in South Caucasus" [16].

NATO officials positively estimated the adoption of the Concept of the National security of Azerbaijan. Then-Special Representative of the UN Secretary General for the Caucasus and Central Asia R. Simmons, commenting on the adoption of the document, stated that it was a great achievement for the state, noting that "the document paves way for more expanded conceptual documents which would allow the government and the parliament of Azerbaijan to manage reforms in institutions of the security sector" [17]. The adoption of the Concept of the National security made Azerbaijan deal with the doctrine of Military Security (Military Doctrine). In the beginning of spring 2010 Azerbaijan was adopting the Military Doctrine of the Republic of Azerbaijan, but suddenly there was offered an idea to make some changes in the document since NATO was going to adopt a new Strategic Concept [18]. In this case Azerbaijan had to stop calling Armenia an enemy. By that time, it was known that experts from the USA, Great

Britain and Germany, engaged in formulating the document, "opposed the decision of Azerbaijan to call Armenia an enemy directly" [19].

In the end of May 2010, the Military Doctrine was passed to the National Assembly where it was revealed that, despite all the efforts of western partners, Armenia is called the key enemy of Azerbaijan. On June, 8 2010 the National Assembly adopted the document. Adoption of the Military Doctrine allowed Azerbaijan to start working upon another conceptual document – the Strategic Defense Review. In August 2010, Azerbaijani mass media wrote about the document supposing it to be created by the end of the year. [20] Nevertheless, by the moment this article was written there was no official information on the matter. It is likely to be connected with the desire of Azerbaijan to follow its own interest concerning the issues of army shaping while continuing collaboration with NATO.

4 Conclusion

To sum up we should state that the process of shaping legal and conceptual basis of military security policy in Azerbaijan was long and complicated. The key factors influencing this process were and still are geopolitical terms of Azerbaijani survival, viewpoints of political and military leaders of the state regarding the prospects of both national security system of the state and its part – military security system.

Up to settling the conflict Azerbaijan is not likely to rewrite its conceptual documents, connected with military security provision where Armenia is pointed as a key source of threat in military sphere. It won't change it even under the political pressure of superpowers who want to remove this idea from official documents.

In its turn, Azerbaijan will consider such pressure as double standard of the actor articulating this approach. Consequently, it may lead to step backward, with all visible channels of communication being preserved. Moreover, the position of Azerbaijan is high due to the fact that the state is building up its military, political and economic capital on the international arena.

References

1. Niyazov, N.: Mutual interference and strategic concepts of international relations in the interwar period. Int. Relat. **11**(2), 146–170 (2018). https://doi.org/10.21638/11701/spbu06.2018.203. Vestnik of Saint Petersburg University
2. Selyutina, E.N., Holodov, V.A.: Methodological problems of legal hermeneutics. Vestnik Povolzhskogo instituta upravleniya **1**, 60–67 (2016). (in Russian)
3. Resolution 822 (1993). https://undocs.org/en/S/RES/822(1993). Accessed 15 Aug 2020; Resolution 853 (1993). https://undocs.org/en/S/RES/853(1993). Accessed 15 Aug 2020; Resolution 874 (1993). https://undocs.org/S/RES/874(1993). Accessed 15 Aug 2020; Resolution 884 (1993). https://undocs.org/S/RES/884(1993). Accessed 15 Aug 2020
4. Mutagirov, D.: Recognition of as a factor of state legitimacy and stability. Int. Relat. **10**(4), 310–319 (2017). https://doi.org/10.21638/11701/spbu06.2017.403. Vestnik of Saint Petersburg University
5. Niyazov, N.S.: Main Vectors of the Military Security Policy of the Republic of Azerbaijan in 1994–2010. SPbGU Publ, Saint Petersburg (2010). (in Russian)

6. Law of the Azerbaijan republic about defense. https://mod.gov.az/az/qanunvericilik-015/. Accessed 05 Jul 2020
7. Partnership for Peace. https://www.nato.int/cps/ru/natohq/topics_50349.htm. Accessed 12 Jul 2020
8. The law of the republic of Azerbaijan on National Security. Azerbajdzhan. 181 (2004/08/06). http://www.dtx.gov.az/pdf/qanunlar/1.pdf. Accessed 11 Jul 2020
9. National security concept of the Russian Federations. https://www.mid.ru/en/foreign_policy/official_documents/-/asset_publisher/CptICkB6BZ29/content/id/589768. Accessed 12 Jul 2020
10. Baranec, V.N.: General Staff Without Secrets. Vagrius Publication, Moscow (1999). (in Russian) http://militera.lib.ru/research/baranets1/02.html. Accessed 15 Jun 2020
11. Gadzhiev, H.: Elements of an individual partnership. Zerkalo. 26 July (2008). (in Russian)
12. Muradyan, I.: Military doctrine of Azerbaijan. (in Russian) http://www.lragir.am/russrc/comments13951.html3. Accessed 12 Jun 2020
13. Mamedov, D.: The pentagon declared war on Soviet Standards. Voenno-promyshlennyj kur'er, 39(155) (2006). (in Russian)
14. Republic of Armenia National Security Strategy. https://www.mfa.am/filemanager/Statics/Doctrineeng.pdf. Accessed 18 Jul 2020
15. Military Doctrine of the Republic of Armenia. http://www.mil.am/files/LIBRARY/Hayecakargayin/825.pdf. Accessed 18 Jul 2020
16. The concept of national security of the Republic of Azerbaijan. http://www.mns.gov.az/img/3766779-_5me02.%20Milli_Tehlukesizlik_Konsepsiyasi.pdf. Accessed 19 Jul 2020
17. Sumerenli, D.: Azerbaijan and NATO. Voenno-promyshlennyj kur'er, 24(190) (2007). (in Russian)
18. Bajramova, D.: Great hopes of little Azerbaijan. Zerkalo, 27 February (2010). (in Russian). http://www.zerkalo.az/2010-02-27/politics/7446-nato-odkb-azerbaijan/print. Accessed 18 Feb 2011
19. Bajramova, D.: The military doctrine was submitted to the parliament. Zerkalo, 27 May (2010). (in Russian). http://www.zerkalo.az/2010-05-27/politics/9785-Doktrina-parlament-ildirim. Accessed 11 Feb 2011
20. Hanmamedov, G.: Azerbaijan's strategic defense review will be ready by the end of the year. (in Russian). http://www.aze.az/news_strateqicheskoe_oboronnoe__40513.html. Accessed 11 Apr 2011

Development Trends of the Strait of Malacca Countries in the Field of Countering Non-traditional Threats to National Security in the Context of the Indo-Pacific Region in 2010s–2020s

Adam Titovich[1] and Vadim Atnashev[1,2](✉)

[1] North West Institute of Management, RANEPA, St. Petersburg, Russia
[2] Saint Petersburg State University, St. Petersburg, Russia

Abstract. The Strait of Malacca is an important sea axis connecting the Indian Ocean with the South China Sea, it is providing a link between the Middle East, China, Japan, South Korea and the markets of Western countries. More than 65,000 ships cross the Strait of Malacca every year. Due to the establishment and development of the processes of globalization and the general trend towards democratization in the early 90s of XX century in the countries of the Strait of Malacca there was a trend towards a pragmatic perception of regional processes. ASEAN is gradually stabilizing the conflict factors of the Strait of Malacca. According to the International Maritime Bureau of Piracy, piracy in the Straits of Malacca increased significantly in the 1990s, but the number of reported incidents has declined markedly over the past decade. It is possible to designate different relationship to the trend in the development of military spending in the countries of the Malacca Strait in relation to the trend in military spending in Southeast Asia. The exclusion of the Strait of Malacca countries from the general trend of development of the countries of Southeast Asia is proof of the construction of a sub-regional model of security architecture.

Keywords: Strait of Malacca · Non-traditional threats · National security · Indo-Pacific region

1 Introduction

The Strait of Malacca countries are bridge-builders, which are seeking to secure objective national interests through the erosion of old geographic boundaries due to increased competition between the People's Republic of China and the United States of America. The Association of Southeast Asian Nations is shaping a new vision of the geographic and functional boundaries of the region through ASEAN Outlook on the Indo-Pacific 2019. Understanding the above trend and vital national interests of the Strait of Malacca countries seems to be even more important in expanding the international agenda of the

© The Author(s), under exclusive license to Springer Nature Switzerland AG 2021
R. Bolgov et al. (Eds.): *Proceedings of Topical Issues in International Political Geography*, SPRINGERGEOGR, pp. 178–196, 2021.
https://doi.org/10.1007/978-3-030-78690-8_17

Russian Federation in the context of bilateral and interregional cooperation, especially in view of establishing the free trade zone between the EAEU and the Republic of Singapore.

The Indo-Pacific is a neutral name for a new strategic political concept, which is centered on the maritime nations of East and Southeast Asia. The Pacific and Indian Ocean states are bridge-builders, which are seeking to secure objective national interests through the erosion of old geographic boundaries due to the rise of the People's Republic of China. The Indo-Pacific concept claims a multipolar regional order.

The Strait of Malacca is an important sea axis connecting the Indian Ocean with the South China Sea, is providing a link between the Middle East, China, Japan, South Korea and the markets of Western countries. More than 65,000 ships cross the Strait of Malacca every year. Due to the establishment and development of the processes of globalization and the general trend towards democratization in the early 90s of XX century in the countries of the Strait of Malacca there was a trend towards a pragmatic perception of regional processes. ASEAN is gradually stabilizing the conflict factors of the Strait of Malacca. According to the International Maritime Bureau of Piracy, piracy in the Straits of Malacca increased significantly in the 1990s, but the number of reported incidents has declined markedly over the past decade.

It is stressed that there is no relationship between the trend in the development of military spending in the Strait of Malacca and the trend in military spending in Southeast Asia. In terms of seasonal fluctuations, it is possible to note the impact of the Asian financial crisis of 1997–1998 and the World financial crisis of 2008.

According to the analysis of the cases of multilateral cooperation in the fight against non-traditional national security challenges based on the activities of the Strait of Malacca countries in the 2000s and 2010s, it can be seen that the periodically manifested tendencies of the military spending of the Republic of Singapore to increase correspond to the security situation in the territorial waters of the Republic of Indonesia, the deterioration of the situation occurs during periods of economic crises and the instability of the political situation associated with bursts of activity by extremist organizations and the role of the religious factor in the political process of the Republic of Indonesia.

The exclusion of the Strait of Malacca countries from the general development trend of the countries of Southeast Asia is proof of the construction of a separate subregional model of security architecture. The analysis carried out by the authors makes it possible to designate a hypothesis about the emergence of a fundamentally new model of regionalism in the political space of the Strait of Malacca countries. The Association of Southeast Asian Nations is shaping a new vision of the geographic and functional boundaries of the region through ASEAN Outlook on the Indo-Pacific 2019.

Understanding the trends and vital national interests of the Strait of Malacca countries seems to be even more important in expanding the international agenda of the Russian Federation in the context of bilateral and interregional cooperation, especially by reason of the creation of a free trade zone between the EAEU and the Republic of Singapore.

2 Materials and Methods

The purpose of the study is to reveal the relationship in the military spending of the Republic of Indonesia, Federation of Malaysia, and the Republic of Singapore on the trend of Southeast Asia. The object of the research is the foreign policy of the Republic of Singapore, the Republic of Indonesia, and the Federation of Malaysia in the second half of the 2000s, 2010s, 2020s in the field of regional security of Southeast Asia in the context of countering non-traditional threats to national security.

The subject of the research is the regularities of the time series of the security architecture development model of the Strait of Malacca countries up to 2025 with the involvement of the following indicators: (1) Gross domestic product (in accordance with purchasing power parity) of the Republic of Indonesia, Federation of Malaysia, Republic of Singapore, (2) relative values of military expenditures of the Republic of Indonesia, Federation of Malaysia, and the Republic of Singapore (3) absolute values of military expenditures of the Republic of Indonesia, Federation of Malaysia, and the Republic of Singapore (4) relative values of military expenditures of the countries of South-East Asia.

Methods used are exponential smoothing method, building a time series sequence diagram, constructing linear regression. Open databases of the Stockholm Peace Research Institute (SIPRI) [1] and the World Bank (World Bank Group) [2] were also used. Authors prepared a model for the development of the security architecture of the Strait of Malacca countries until 2025 using the IBM SPSP Statistics ver. 22. The research hypothesis is that the development trends of the security architecture of the Strait of Malacca are not in line with the trend of militarization of Southeast Asia.

3 Problem Statement

The Strait of Malacca plays an important role in the Indo-Pacific concept. The potential of the Strait of Malacca and its historical, political role has been discussed by many groups of researchers, the authors considers it possible to single out the following four areas of research. The first group focuses on the manifested role of ports and trade routes of the 6th-18th centuries AD. As a rule, researches are focused on Srivijaya, an ancient Malay kingdom centered on the island of Sumatra (Indonesia), Melaka (Malacca, Indonesia), Aceh (Ace, Indonesia), Johor (Malaysia), among many authors would like to highlight the Russian orientalist, historian M.Yu. Ulyanov.

The second group of studies focuses on the analysis of the competition of Western powers for control of trade routes in the 16th-19th centuries AD, among researchers there are N. Tarling [3, 4], D.K. Bassett [5] should be singled out separately, an important aspect of the process is the London Treaty of 1824, the consequences of which, according to H.R.S. Wright [6, c. 234], paved the way for the formation of modern nations.

The third group of studies is associated with the development of the national historiography of the Strait of Malacca countries, which made the nation the focus of research, ensured the tracking of the historical heritage of the colonial era, the process becoming part of the modern historical process, in this group it is impossible not to note the contribution of K.S. Sandhu, P. Wheatley [7], K.K. Khoo [8], the center of the historical process was transferred to the region of the Malacca Strait, along with this, the later archaeological research of N.H. Shuhaimi and N.A. Rahman [9], Z. Majid [10] made significant adjustments to the above hypothesis.

The fourth group of studies focuses on the growth and development of modern port cities along the historical trade routes of the Strait of Malacca, taking into account social and demographic factors and their role in the process of state building [11]. The complex aspect of identity, built on the reflection of historical experience, is touched upon in the works of many Malay researchers, identity, in accordance with the conclusions of R. Zainuddin [12], W.Z.A. Abdul [13], S.A. Baharuddin [14], A.H. Ramli [15], in a joint monograph of the National Institute of Public Administration of the Federation of Malaysia (INTAN) [16], was formed as a result of complex interaction and collision of values, the nature of thinking between emerging nations and the colonial government, which led to the modern nature of social groups exploiting regional sea routes at the present stage and financial, migration flows.

4 Research Questions

The Strait of Malacca is an important sea axis connecting the Indian Ocean with the South China Sea, is providing a link between the Middle East, China, Japan, South Korea and the markets of Western countries. The parameters of the strait are determined by the width of 65 km and the length of 800 km, they also determine the world transport highway through which more than 200 ships pass daily, including more than 40 oil tankers, which is 15–16 million barrels of oil [17, p. 256]. The Strait of Malacca accounts for a third of the world's trade and half of the energy supply [18]. The straits are vital in regional economic development, as they provide more than 60 and 80% of the oil supplied to the People's Republic of China and Japan, respectively [19]. The countries of the Malacca Strait have great potential in socio-economic development, due to the possibility of creating many jobs, developing infrastructure focused on seaports along the straits [20, p. 21]. The nodal infrastructure points are the regions associated with seaports: Selangor (Klang port [21]), Penang (Penang port), Johor (Pelepas port and Johor port), Malacca (Tanjung Langsat port [21] Langsat and Kuala Linggi port), etc. - these seaports [22] located along the straits seem to be extremely important for shipping, labor migration, knowledge exchange, for the formation of a stable system of international relations between the countries of the region. The Strait of Malacca, even if the construction of a major infrastructure project, the Thai Canal, is successful, will remain strategically important for trade between the Persian Gulf, Indonesia or Australia, especially for container ships [23].

The uncertainty that arose after the final departure of the Western colonial powers and Japanese imperialism was caused by the emerging independence of the players of the Malacca Strait, the potential dominance in the region of a single country led to an increase in conflict potential, entrenched in the competitive race of Indonesia, Malaysia and Singapore for dependence/independence of sea routes. A competitive race has certainly been escalated by ethnic disputes. In the first half of the 60s of XX century, there was a non-illusory potential for resolving disputes by actual combat. In an effort to ease tensions, the political elite of the straits of Malacca have taken important based on consensus and the priority of socio-economic development steps. An important role in this process was played by both the restrictions of opposition parties (Federation of Malaysia, the Republic of Singapore), the dominance of power structures in the political space (the Republic of Indonesia), and the Association of Southeast Asian Nations, which had appeared in 1967, and provided an opportunity for an equal political dialogue.

With the establishment and development of the processes of globalization and the general trend towards democratization in the early 90s of XX century in the countries of the Strait of Malacca there is a steady tendency towards a pragmatic perception of regional processes and diligent simplification towards economic benefits. ASEAN is gradually stabilizing the conflict-generating factors of the Strait of Malacca, however, the problem of the South China Sea is not able to solve, as A. Acharya noted in his work [24]. The democratization processes are gradually eroding the stakeholders of the national systems of public administration. In the context of globalization at the beginning of the 21st century, the modernization of the functional approaches of the political departments of the nation state is becoming more and more in demand for the sake of the general aspiration of management systems towards optimization, economization, decentralization, and adaptability. It should be noted that state authorities represent in the administrative culture the most conservative and reactionary institution in the field of public administration instruments [25, p. 128]. At the same time, the objective conditions of regional integration groupings, international organizations and world political processes deform the traditionalist approach and cause the need for innovation in the administration of the political department. The legitimacy of the decisions made, or rather the decision-making mechanism itself, directly depends on the transparency of the system of political institutions of the nation state. Information and communication technologies of the 21st century have made it possible to achieve a previously inaccessible level of interdepartmental communication and communication in the line of human interaction with state bodies, in a broader sense, with the state. The processes of democratization create conditions under which public authorities are more transparent and, therefore, more susceptible to public opinion and the influence of interest groups both within the country and beyond national borders. The above trend raises a serious question before the governments of the Republic of Singapore, to a lesser extent, the Federation of Malaysia and the Republic of Indonesia, to a greater extent, a pragmatic tendency in resolving emerging disputes.

There was clearly perceived need for ensuring national security against non-traditional types of threats, however, still traditional for sea straits. Securitization of the problem of piracy in the Strait of Malacca is an important aspect of the case study of the political security structure of the South-East Asia region.

International law, in accordance with Article 101 "Definition of Piracy" of the 1982 United Nations Convention on the Law of the Sea [26], defines piracy as an unlawful act of violence, detention, robbery, committed for personal purposes by the crew or passengers of a ship (or aircraft apparatus), directed against persons, property on board, against persons of property located in a place outside the jurisdiction of any state. Piracy means an act of voluntary participation, with knowledge of the circumstances, as well as any act that is incitement or deliberate assistance in the commission of the aforesaid actions. The International Maritime Organization (IMO) definition covers an attack carried out in territorial waters (including internal waters) and on ships in a port, with a view to pursuing personal interests. However, as it was noted by S. Mair [27, p. 5], cases of sea piracy have mixed personal and political interests, the narrow definition of piracy offered by international law restricts and prevents the international community from the right to intervene in conditions where the country cannot resist piracy within their territorial waters. Anti-piracy restrictions outside the 12 mile zone (national territorial waters), the existing principle of non-interference defines the framework for the effectiveness of the anti-piracy system.

The flourishing of piracy in the Strait of Malacca is usually associated with the temporary weakening of the Republic of Indonesia at the end of the 20th century and the restrictions on countries' spending due to the Asian economic crisis of 1997–1998. The interest of the Strait of Malacca countries (and a number of sympathetic countries) made it possible to ensure the comparative security of the Strait of Malacca and shift the center of the piracy problem from Southeast Asia to the Gulf of Aden and the Somali Basin. However, the threat of piracy and illegal fishing remains significant in the Straits of Malacca, which causes significant financial costs for the insurance of valuable cargo and ships, an additional concern is the question of where the profits obtained from piracy are spent, the manifested supplies of weapons and equipment to continue criminal activities, which are already a real danger of the emergence of both economic (and its subsequent effects) and an objective danger to national security, in the case of the integration of organizations into large transnational networks, for example, JemaaIslamia, a radical Islamist organization in Southeast Asia.

Effective counteraction to piracy in the Malacca Strait, ensuring maritime security is caused by the stabilization of the Indonesian state, which has become a central factor [27, p. 7], and trilateral cooperation of the Strait of Malacca countries, which has become an important aspect of multilateral programs to strengthen security, improve the economic and social conditions of coastal regions.

According to the International Maritime Bureau of Piracy (ICC International Maritime Bureau (IMB)), piracy cases in the Straits of Malacca increased significantly in the 1990s, but the number of reported incidents has declined markedly over the past decade. In 2004, the Straits of Malacca had 38 cases with signs of piracy, 11 cases in 2006, 7 cases in 2007, two cases in 2008 and 2009. Lloyd's Market Association, Lloyd's Market Association, representing the interests of persons operating in the insurance market and providing information and technical assistance, declared the Malacca Strait zone a war zone in 2005, which significantly increased the financial costs of insurance, next year, due to the normalization of the situation criterion was removed.

The socio-economic circumstances of the communities of Southeast Asia (its insular part) is an important factor in the development of piracy, for example [28], the fishermen communities of the Batam island of the Riau province often resort to piracy, due to the fact that the legal income in the region in the framework of their professional activities is 3/4 of a dollar a day, while engaging in piracy brings in $ 7,000 to $ 10,000 in case of success. The socio-economic condition of a large part of the population of the Republic of Indonesia deteriorated significantly, due to the Asian financial crisis of 1997–1998, the financial crisis affected all three states of the strait, however, the number of cases of piracy increased significantly only in 1999, from 47 to 115 compared to 1998 year.

Issues of state sovereignty over territorial waters and the nature of local government in certain regions of the Republic of Indonesia indicate rather unclear boundaries of competence. The Republic of Singapore and the Federation of Malaysia have relatively stable authoritarian or semi-authoritarian systems of government, while the end of Suharto's authoritarian rule in 1998 marked a political turning point towards democratization amid the financial crisis, and therefore political instability, corruption and weak power structures. Violent clashes between the Republic of Indonesia and Timor Leste (East Timor) during the rule of Yusuf Habibi resulted in the emergence of an independent state, the collapse of the tourism industry associated with the explosion in Bali in October 2002, a general decline in the socio-economic status of citizens, pressure on political transformation ensured an increase in cases of piracy, by 2003 the number had reached 121 reported cases.

Piracy issues were not included in national security threats until 2004, the year when the United States of America proposed [18] the formation of a Regional Maritime Security Initiative (RMSI), the plan called for cooperation between the United States and the Strait of Malacca countries, however, ambiguity of the wording did not allow to take the initiative for implementation, however, it caused a positive effect, making the fight against piracy a priority in ensuring security. Subsequent initiatives include one-, three-, multilateral platforms. Unilateral initiatives depend on the objective national interests of the Strait of Malacca countries; differences in geographic location affect the intensity of the development of initiatives. The Strait of Malacca in the political space of the Republic of Singapore is the only maritime space that it needs to be protected and controlled [29, p. 35], which determines its further understanding of the concept of the Indo-Pacific region. In April 2005, the Government of the Republic of Singapore launched the Accompanying Sea Security Teams (ASSeTs), a group of specially trained military personnel deployed on merchant ships carrying valuable cargo. The Republic of Singapore, on the basis of the Changi group, its control center [30], has developed an information-analytical and monitoring center that ensures coordination of various national security initiatives. Of key importance is the fact that the National Security Strategy of the Republic of Singapore traces a clear link between piracy and transnational terrorism [31, p. 98].

The unilateral initiatives of the Federation of Malaysia in the mid-2000s are driven by the interests of historically established territorial self-determination and the protection of sovereignty. During the reign of Abdullah Badawi in 2005, a single national coast guard Malaysian Maritime Enforcement Agency (MMEA) [32] was created, which included eleven disparate state bodies, with the aim of ensuring effective control over territorial waters and demonstrating state sovereignty in foreign policy discourse. At the same time, there is a tightening of national legislation [33, p. 74].

The consolidation of the statehood of the Republic of Indonesia came during the reign of Susilo Bambang Yudhoyono, which began in 2004. The government provided training for the coast guard, a modern navigation system, and modern patrol boats. Interdepartmental cooperation between the Coast Guard and local state law enforcement agencies was ensured, and information and analytical work was ensured in the direction of preventing corruption. Separately, it is necessary to note the comprehensive approach of President Yudhoyono, who ensured systematic patrols by the fleet [34, p. 91] of the Malacca Strait and the fight against ground causes of sea piracy: the growth of economic prosperity, strengthening of government bodies, the fight against corruption, for example, it was launched a comprehensive national action plan against corruption to improve the transparency and efficiency of government institutions. Through a systematic approach and combined strategy, the ICC International Maritime Bureau records a significant drop in piracy cases in the Straits of Malacca from 121 cases in 2003 to 15 reported cases in 2009.

Within the framework of the tripartite initiative, in July 2004, a regulatory agreement was signed, the MALSINDO program, in particular, due to the desire to avoid the limitation of sovereignty through the intervention of external actors in the emerging security architecture of the Strait of Malacca. The MALSINDO program provided for coordinated (but not joint) patrolling of territorial waters, the pursuit of pirates or ships engaged in illegal fishing in foreign territorial waters was not allowed, in accordance with the fundamental principle of the Association of Southeast Asian Nations, the principle of non-interference. The program, despite the limitations, has shown its effectiveness, which can be evidenced by the subsequent introduction of standards for the procedure for border actions and the inclusion of new formats of interaction. In April 2006, the program was renamed Malacca Straits Patrols (MSP). To date, the program is being implemented through three elements: MSSP (augmented form of the aforementioned MSP), EIS (Eyes in the Sky), MPO (Intelligence Exchange Group). As part of the improvement and intensification of data exchange by the Intelligence Exchange Group, USA, the MSP-IS, Malacca Straits Patrols - Information System software is being developed, to which four countries of Southeast Asia currently have access: the Republic of Indonesia, the Federation of Malaysia, the Republic of Singapore and Kingdom of Thailand (since October 2008).

Multilateral cooperation at the regional level within ASEAN is limited by the principle of non-interference, it is not yet possible to speak of a formed strategy of joint fight against piracy. However, the previously indicated trend towards the development of maritime cooperation within the ASEAN Indo-Pacific Outlook 2019 may play a role in establishing multilateral approach. The geostrategic situation and the significance of the Strait of Malacca indicated a great interest of regional economic powers in ensuring security measures. Defense ministers from seventeen countries, including the United States of America, the People's Republic of China, Japan, and the Republic of India, as part of the fourth Shangri-La Dialogue [35] held in the Republic of Singapore in 2005, agreed on the basic principles of security cooperation in the Strait of Malacca, in some places different from those stipulated in the 1982 UN Convention on the Law of the Sea. Responsibility for the security of the Strait of Malacca lies with the Republic of Singapore, the Republic of Indonesia, and the Federation of Malaysia, which reinforces their key status in the security architecture of the Strait of Malacca. The international community provides financial, material and informational advisory support, conducts joint exercises, but does not interfere with the implementation of security measures; this, in turn, reinforces the sovereignty of the Strait of Malacca countries. Multilateral cooperation in the Malacca Strait is subject to international maritime law and the sovereignty of the Strait of Malacca countries, which provides countries with ambivalent opportunities to resolve conflict situations, for example, with the People's Republic of China.

The United States of America, Japan and the Republic of India are of the greatest importance in ensuring the security of regional sea straits, providing regular financial, logistical and advisory assistance to the Federation of Malaysia and the Republic of Indonesia, for example, joint exercises and the establishment of a satellite video surveillance and communication system.

In November 2006, 15 countries ratified the Agreement on Combating Piracy and Armed Robbery against Ships in Asia (ReCAAP), which became the first international agreement on multilateral cooperation to combat piracy in Asia. In the same year, a Data Clearinghouse was opened in the Republic of Singapore, which improved communication and coordination. The role of the United States of America in the foreign policy discourse of the Republic of Indonesia and the Federation of Malaysia has been assessed ambiguously, as M.Valencia noted in his work [36, p. 92], the importance of the United States as a trading partner and source of investment is important, however, the US regional presence contributes to radical Islamism in Southeast Asia, the religious factor has a growing influence on the foreign policy of the Federation of Malaysia and the Republic of Indonesia.

The 2009 ReCAAP Clearinghouse (ISC) Annual Report highlighted new areas of vulnerability in the Malacca Strait security architecture. The incidents of piracy changed their spatial extent from the northern part of the strait to the Singapore Strait and the southern part of the South China Sea, while in 2009 the number of reported cases of piracy increased 5 times compared to 2005, the following areas can be distinguished: PulauDamar, PulauMangkai, Natuna Islands. Such existing problems as the limited number of patrol vessels, the length of the coastline, the need to expand routes to the southern part of the South China Sea cause the requirement for pooling assets and joint patrolling of vulnerable areas [37, c. 2]. Moreover, the need to create a single information center for the exchange of information at the points of contact between the Straits of Malacca and Singapore is identified, now the system is, in general, decentralized in the Port of Klang, Malaysia (Traffic Service, VTS), and in Singapore (Port Operations Control Center (POCC)), the existing RSN (IFC) currently does not provide accurate monitoring of passing vessels, despite advanced integration with the AIS satellite tracking system (providing an identification system). Within the framework of the AIS system, it seems possible to provide monitoring of vessels with a displacement of less than 300 gross register tons, however, the initiative of the Republic of Singapore did not receive support from other countries of the Strait of Malacca.

The successes achieved in building a regional security system for the Strait of Malacca in the 2000s and 2010s certainly have their drawbacks, along with the increased transparency of the naval forces of the Republic of Indonesia, the Federation of Malaysia and the Republic of Singapore, the introduction of new technological communication systems and satellite communications, the use of modern weapons systems, including submarines, and an irreconcilable attitude towards violations of territorial waters, perceived as a direct threat to national sovereignty, should build a clear tendency to militarize the Strait of Malacca, however, in the security architecture of the Malacca Strait in the context of the Indo-Pacific region, the dynamics of the region Southeast Asia is not traced.

5 Findings

Authors prepared a model for the development of the security architecture of the Strait of Malacca countries up to 2025 using the IBM SPSP Statistics ver. 22. A model was created for the development of the security architecture of the Strait of Malacca countries by 2025 (statistical data were built in accordance with the 1975–2018/2019 time series and the predicted values obtained by the exponential smoothing method by 2025). In order to construct the time series, the above-mentioned indicators were used (see Methods). Using the method of exponential smoothing (see Figs. 1, 2 and 3), the analysis of the behavior of the above time series, the patterns of their time series was carried out.

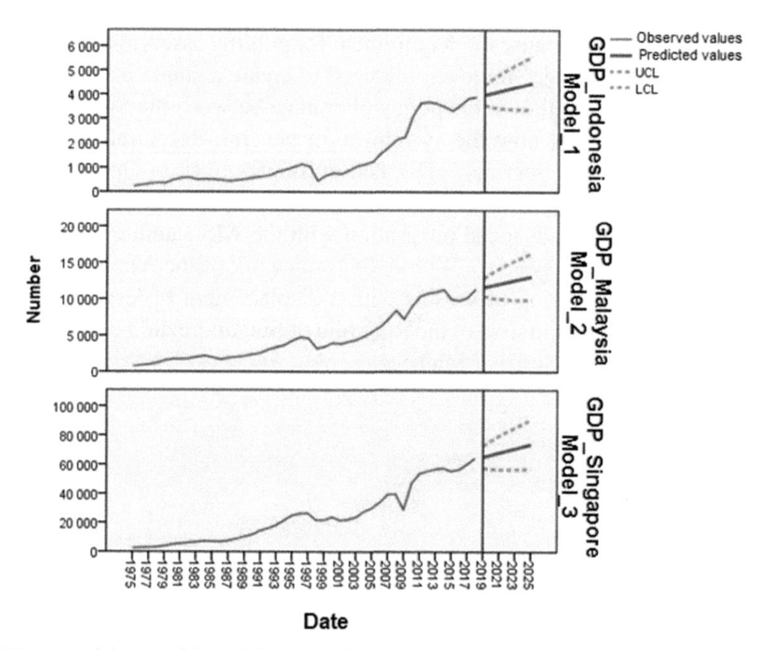

Fig. 1. Exponential smoothing of the GDP time series (in accordance with PPP) of the Republic of Indonesia, Federation of Malaysia, and the Republic of Singapore.

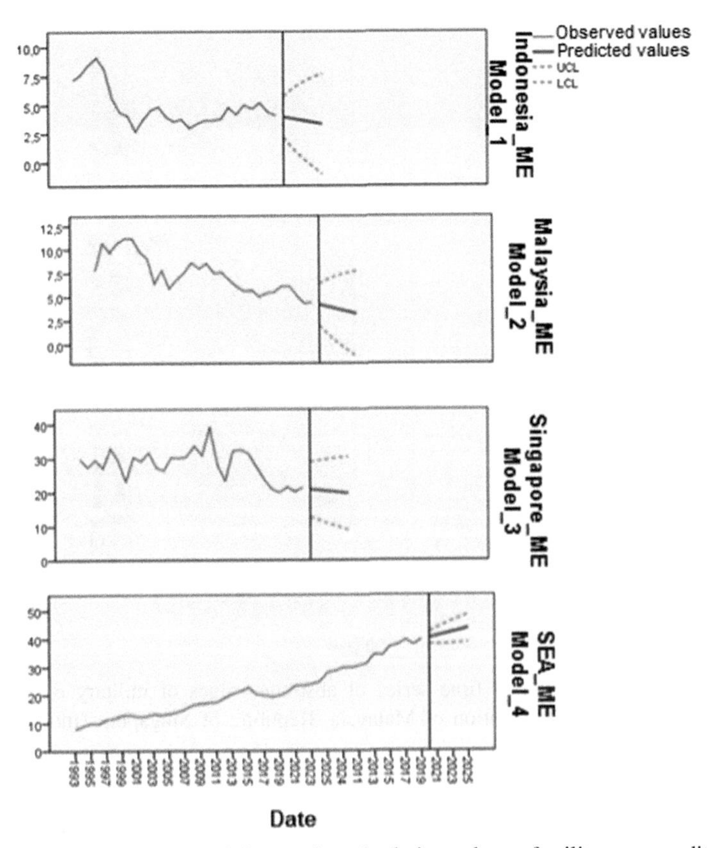

Fig. 2. Exponential smoothing of time series of relative values of military expenditures of the Republic of Indonesia, Federation of Malaysia and the countries of Southeast Asia (Indonesia_ME, Malaysia_ME, SEA-ME)

The research hypothesis is the development trend of the security architecture of the Strait of Malacca, within which the Republic of Singapore, the Republic of Indonesia and the Federation of Malaysia are not in line with the trend of militarization of Southeast Asia. Preliminary plots of the sequence diagram for the behavior of a series of GDP (in accordance with PPP), the relative values of the military expenditures of the Republic of Indonesia, the Federation of Malaysia, the Republic of Singapore (Indonesia_ME, Malaysia_ME, Singapore_ME) and the relative values of the military expenditures of the countries of Southeast Asia (SEA_ME) are presented below (Figs. 4 and 5).

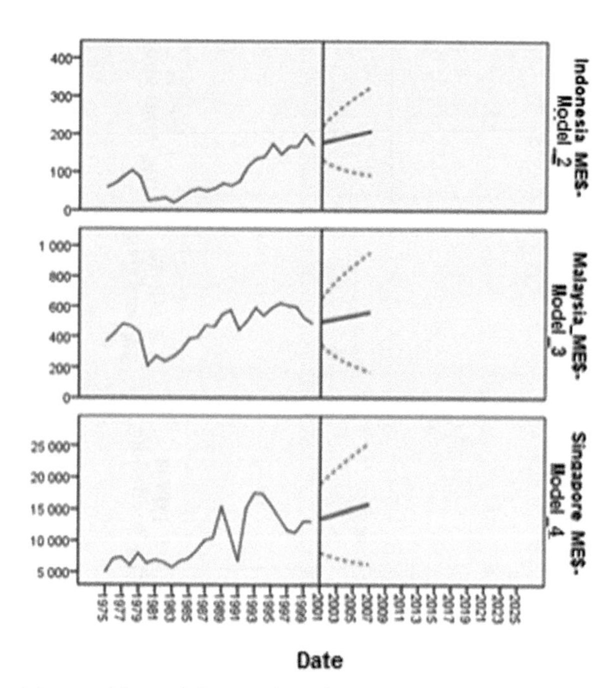

Fig. 3. Exponential smoothing of time series of absolute values of military expenditures of the Republic of Indonesia, Federation of Malaysia, Republic of Singapore (Indonesia_ME $, Malaysia_ME $, Singapore_ME $).

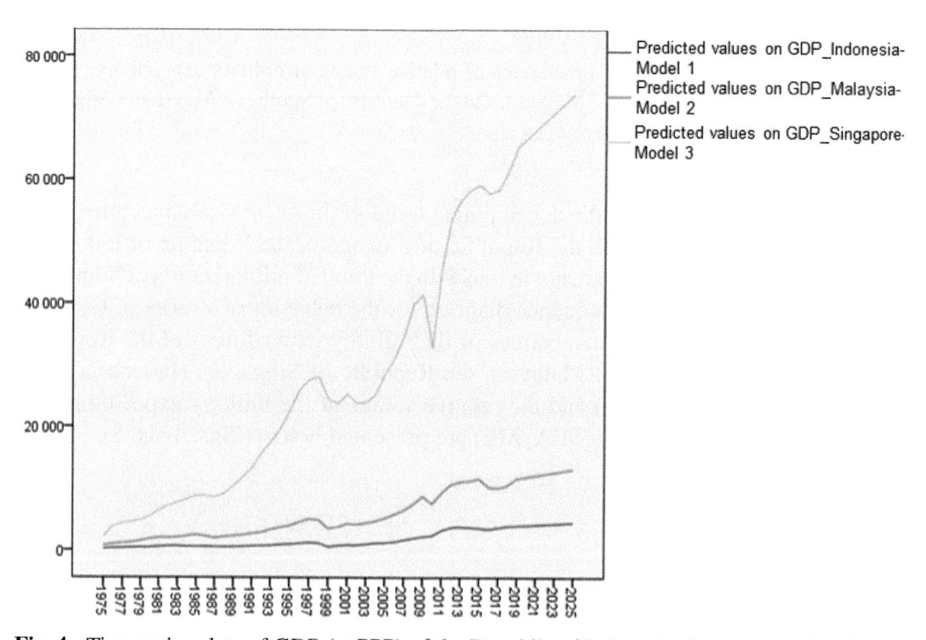

Fig. 4. Time series plots of GDP (at PPP) of the Republic of Indonesia, Federation of Malaysia, and the Republic of Singapore from 1975 to 2025.

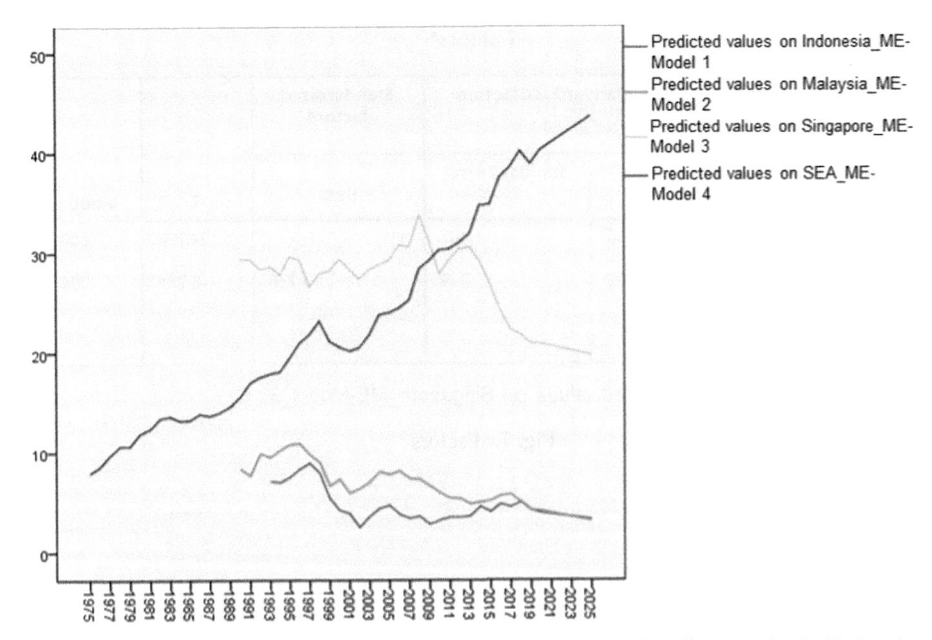

Fig. 5. Time series diagrams of military expenditures of the Republic of Indonesia, the Federation of Malaysia, the Republic of Singapore, and the countries of Southeast Asia from 1975 to 2025.

As part of constructing a linear regression, the following dependence models were built: Predicted values of military expenditures of Indonesia, Federation of Malaysia, Republic of Singapore depending on the Predicted values of military expenditures of Southeast Asian countries.

Summary for the model[b]

Model	R	R-squared	Adjusted R-squared	Standard error
1	,756[a]	,571	,559	2,60748

a.Predictors: (constant), Predicted values on SEA_ME-Model_4
b.Dependent variable: Predicted values on Singapore_ME-Model_3

Fig. 6. Summary for the model

Factors[a]

Model	Non-standardized factors		Standardized factors		
	B	Standard error	Beta	T	Value
1. (Constant)	36,380	1,467		24,796	,000
Predicted values on SEA_ME-Model 4	-,329	,049	-,756	-6,734	,000

a.Dependent variable: Predicted values on Singapore_ME-Model_3

Fig. 7. Factors

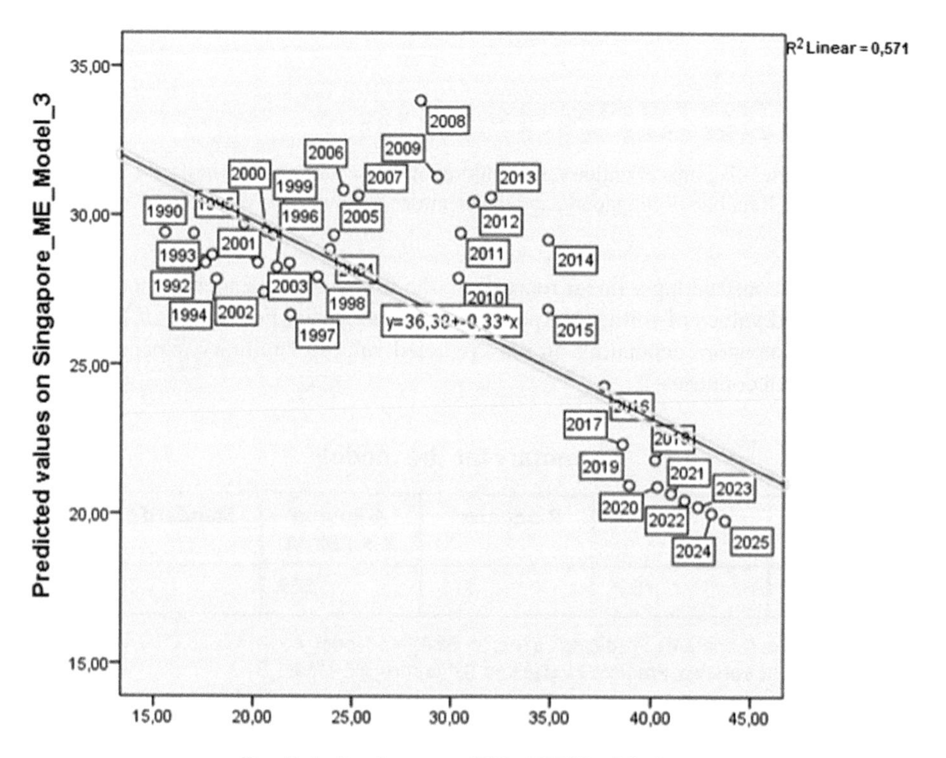

Fig. 8. Predicted values on SEA_ME-Model_4

Linear regression model of the predicted military expenditures of the Republic of Singapore to the military expenditures of Southeast Asian countries by 2025 (Figs. 6, 7 and 8).

Summary for the model[b]

Model	R	R-squared	Adjusted R-squared	Standard error
1	,467[a]	,218	,192	1,49995

a.Predictors: (constant), Predicted values on SEA_ME-Model_4
b.Dependent variable: Predicted values on Indonesia_ME-Model_1

Fig. 9. Summary for the model

Factors[a]

	Non-standardized factors		Standardized factors		
Model	B	Standard error	Beta	T	Value
1. (Constant)	7,357	,952		7,726	,000
Predicted values on SEA_ME-Model 4	-,090	,031	-,467	-2,937	,006

a.Dependent variable: Predicted values on Indonesia_ME-Model_1

Fig. 10. Factors

Linear regression model of the predicted military spending of the Republic of Indonesia to the military spending of Southeast Asia by 2025 (Figs. 9, 10 and 11).

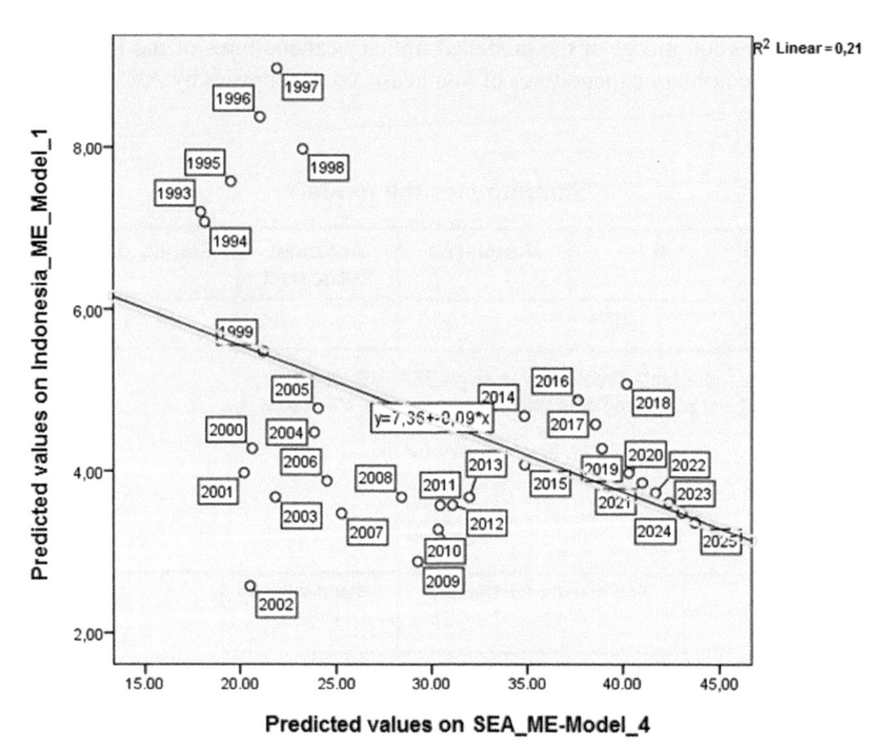

Fig. 11. Predicted values on SEA_ME-Model_4

6 Conclusion

Linear regression plots, the R-squared indicator of the model summary and coefficients show the inverse relationship between the trend in the development of military spending in the Malacca Strait countries in relation to the trend of military spending in Southeast Asia. Within the framework of seasonal fluctuations, despite the implicit expression, it seems possible to note the influence of the Asian financial crisis of 1997–1998 and the World financial crisis of 2008.

At the same time, in accordance with the analysis of the cases of multilateral cooperation in the fight against non-traditional challenges to national security on the example of the activities of the Strait of Malacca countries in the countering piracy in the 2000s and 2010s, it can be seen that the periodically manifested trends in the military spending of the Republic Singapore to an increase is in correlation with the security situation in the territorial waters of the Republic of Indonesia, the deterioration of the situation occurs during periods of both economic crises and the instability of the political situation associated with bursts of activity of extremist organizations and the role of the religious factor in the political process of the Republic of Indonesia, such as the situation of the Rohingya refugees in 2015. The exclusion of the Strait of Malacca countries from the general development trend of the countries of Southeast Asia is proof of the construction of a separate sub-regional model of security architecture. The analysis carried out by the authors makes it possible to designate the emergence of a fundamentally new model of regionalism in the political space of the Strait of Malacca, the conclusions of the study confirm the development trend of the current concept of the Indo-Pacific region with the dominant role of the Republic of Indonesia in the political space of the Strait of Malacca countries.

References

1. SIPRI Military Expenditure Database. https://www.sipri.org/databases/milex
2. World Bank Open Data. https://data.worldbank.org
3. Tarling, N.: Anglo-Dutch rivalry in the Malay world 1780–1824. University of Queensland Press, Brisbane (1962)
4. Tarling, N.: British policy in the Malay Peninsula and Archipelago, 1824–1871. Oxford University Press, Oxford (1969)
5. Bassett, D.K.: The British in Southeast Asia during the seventeenth and eighteenth centuries. University of Hull, Hull (1990)
6. Wright, H.R.C.: The anglo-Dutch dispute in the east, 1814–1824. Econ. His. Rev. **3**(2), 229–239 (1950)
7. Sandhu, K.S., Wheatley, P.: Melaka, The Transformation of a Malay Capital C. 1400–1980. Oxford University Press, Oxford (1983)
8. Khoo, K.K.: Malacca: First and foremost. Association of Malaysian Museums, Kuala Lumpur (1982)
9. Shuhaimi, N.H., Rahman, N.A.: Pre-Islamic Archaeological Coastal Strait of Malacca: Evolution or Migration. Penerbit Universiti Kebangsaan Malaysia, Selangor (2004)
10. Majid, Z.: Malaysia's Pre-History: Has Dark Days Been Bright. Universiti Sains Malaysia, George Town (1991)
11. Hussin, N.: Trade and Society in The Straits of Melaka. NUS Press, Singapore (2007)
12. Zainuddin, R.: Malaysia's Nationality. Oxford Fajar, Shah Alam (2010)
13. Abdul, W.Z.A.: Malaysia: Heritage and development. Fajar Bakti, Kuala Lumpur (1996)
14. Baharuddin, S.A.: Model of Ethnic Relations. Pusat Penerbitan MARA University of Technology, Shah Alam (2007)
15. Ramli, A.H.: Unity: The History and Foundation of Unity. Pusat Penerbitan MARA University of Technology, Shah Alam (2009)
16. INTAN: Our Country: History, Administration And Public Policies. National Institute of Public Administration, Kuala Lumpur (1990)
17. Ba, A.D.: Governing the safety and security of the Malacca strait: the nippon foundation between states and industry. J. Contemp. Asia **48**(2), 252–277 (2018)
18. Story, I.: Maritime Security in Southeast Asia: Two Cheers for Regional Cooperation in Southeast Asian Affairs. Institute of Southeast Asian Studies, Singapore (2009)
19. Möller, K.: Maritime Sicherheit und die Suchenachpolitischem Einfluss in Südostasien (Studie 35/2006). Stiftung Wissenschaft und Politik, Berlin (2006)
20. Hayashi, R.: Cooperation between coastal state and User States. The Malacca and Singapore Straits, pp. 20–22 (2002)
21. Top-20 2018. World Shipping Council
22. Dávid, A., Piala, P.: The strategic maritime canals and straits. 2 XI (2016)
23. Panda, A.: How a Canal Could Transform Southeast Asia (2013)
24. Acharya, A.: Constructing A Security Community in Southeast Asia: ASEAN and The Problem of Regional Order. Routledge, Abingdon (2009)
25. Mulgan, G.: The Art of Public Strategy: Mobilizing Power and Knowledge for the Common Good. Oxford

26. 1982 UN Convention on the Law of the Sea. https://www.un.org/depts/los/convention_agreements/texts/unclos/unclos_r.pdf
27. Mair, S.: Piracy and Maritime Security (SWP Research Paper) (2011)
28. Frécon, E.: Beyond the Sea: Fighting Piracy in Southeast Asia, RSIS Commentaries 2/2010. S Rajaratnam School of International Studies, Singapore (2010)
29. Loewen, H., Bodenmüller, A.: Die Straße von Malakka. SWP, Berlin (2011)
30. Hammick, D.: Turning the Tide. Jane's Defence Weekly, 28 November 2007
31. Story, I.: Securing southeast Asia's sea lanes. Asia Policy 95–127 (2008)
32. Valencia, M.J.: Cooperation in the Malacca and Singapore Straits: A Glass Half-Full (Policy Forum Online 06–103A 12/12/2006) (2006)
33. Ho, J. H.: The security of sea lanes in southeast Asia. Asian Surv. 46 (2006)
34. Bradford, J.: Shifting the tides against piracy in southeast Asian waters. Asian Surv. 48 (2008)
35. International Institute for Strategic Studies 2005 Ministerial Lunches Shangri-la Dialogue 2005. (www.iiss.org/conferences/the-shangri-la-dialogue/shangri-la-dialogue-archive/shangri-la-dialogue-2005/2005-ministerial-lunches)
36. Valencia, M.J.: The politics of anti-piracy and anti-terrorism responses. In: Graham G., (ed.) Piracy, Maritime Terrorism and Securing the Malacca Straits, pp. 84–102 (2006)
37. Ho, D.: The Malacca Strait: Will It Continue To Be Safe? RSIS Commentaries (2010)

Fluctuations in the Global Food Market: Politics Versus Economics?

Yury Gladkiy[1](✉), Viacheslav Sukhorukov[1], and Svetlana Kornekova[2]

[1] The Herzen State Pedagogical University of Russia, Saint Petersburg, Russia
gladky43@rambler.ru, suhor@herzen.spb.ru

[2] Saint Petersburg State University of Economics, Saint Petersburg, Russia

Abstract. The current state of the global food system is characterized by a rapid increase in the concentration of capital at each stage of the production and distribution chain of the agricultural sector of the economy. Gradually, the hegemony of agrarian multinational corporations in the food sector is being established, and their influence on the dynamics and structure of production and consumption of food is increasing. In parallel, the role of the political factor is being strengthened. Territorial expansion of multinational companies often leads to the degradation of agriculture in underdeveloped and middle-developed countries, provoking food crises. Some countries (for example, Russia) raise food security to the rank of national security and accelerate the development of not always competitive sectors of the agricultural sector. The political subtext can also be seen in the campaign against genetically modified products launched between the US and EU countries, which insist on conducting new research on the harmlessness of "mutant plants". In defending their commercial interests, both sides use political levers to put pressure on the World Trade Organization, which is designed to regulate trade relations in the world. When arguing their positions, the authors refer to specific facts of food consumption in key countries and regions of the world. The article focuses on Western Europe, Russia, and the United States.

Keywords: Food consumption · Globalization · Agricultural sector · Food independence · Policy of economic sanctions · Genomodified products

1 Introduction

The impact of globalization, as the leading trend in the modern world economy in recent decades, on the state of world agriculture and food consumption, can hardly be a serious subject of discussion. But it is hardly necessary to link success in solving the food problem with the globalization of the economy. The loss of competitiveness of the agricultural sector and its degradation in some countries is just a consequence of the increased dependence of the national market on imports of cheaper imported food [1].

It is not difficult to give examples of how the territorial expansion of multinational companies and financial markets leads to food crises in the economies of underdeveloped and middle-developed countries. So, Mexico - the birthplace of maize, which has huge

© The Author(s), under exclusive license to Springer Nature Switzerland AG 2021
R. Bolgov et al. (Eds.): *Proceedings of Topical Issues in International Political Geography*, SPRINGERGEOGR, pp. 197–210, 2021.
https://doi.org/10.1007/978-3-030-78690-8_18

tracts of arable land, at the beginning of the XXI century suddenly faced a sharp (60%) increase in corn prices. According to many analysts, this increase was made possible by the beginning of widespread use of corn for ethanol (biofuel) in the United States, where local farmers, having received considerable subsidies from the government, dramatically increased the acreage for this crop. A thorough analysis of the reasons that led to the almost doubling of prices for traditional corn tortillas showed that a significant role in the price increase was played by speculation by TNCs on the US needs for ethanol, which is a consequence of the "free trade" policy. The implementation of the North American free trade agreement finally plunged the Mexican corn sector into a chronic crisis due to the "influx" of cheaper American corn.

Similar reasons can be explained by the "rice" crisis in the Philippines, when the country paying a huge external debt had to eliminate quotas for almost all agricultural products and deprive the services that supported the agricultural sector of the economy of funding. The restructuring of national agriculture in the spirit of free trade and trade liberalization resulted in the transformation of the Philippines, a self – sufficient rice power, into a rice – importing country.

It is necessary to have a clear understanding of the negative changes taking place in the world food sector without the direct impact of those processes that are somehow associated with economic globalization. A close look reveals today many facts and trends that are not a direct consequence of economic globalization. Let's mention the main ones:

- traditional chronic malnutrition and spikes in acute hunger, as well as resulting diseases and premature deaths (especially in the Sahel and South Asia);
- involvement in the economic turnover of almost all arable land;
- in these conditions, the main means of increasing agricultural production is its intensification, which includes the mass use of mineral additives, herbicides and pesticides, which are sometimes poorly tested in environmental terms and pose a real threat to both agriculture (falling fertility) and livestock, as well as to humans. (According to the odious Nobel laureate, author of the doctrine of antibodies as factors of humoral immunity, Paul Ehrlich, "trying to feed a growing number of our own kind, we endanger the very ability of the earth to support any life at all" [2]);
- uncontrolled growth in the scale of fish and other marine organisms production and depletion of their stocks in a number of marine areas, which, at least, leads to a violation of the traditional food balance of the world's population;
- "failure" of the biosphere's regenerative mechanisms and the decline of biodiversity in nature and specific sectors of agriculture (including as a result of the triumph of the ideas of the "green revolution").

Of course, there are facts and trends with a "plus" sign, but they are extremely small. So, we are encouraged by the accelerated development of aquaculture and especially mariculture, associated with the cultivation of man-made marine plantations not only fish and round-mouthed, but also mussels, oysters, kelp. But the progressive pollution of coastal waters, including river discharges of toxic industrial and agricultural waste, threatens to stop the large-scale development of mariculture.

The more or less obvious negative consequences of agricultural globalization include, in particular, the loss of competitiveness and degradation of agriculture in dozens of countries; the increase in unemployment and poverty in underdeveloped countries due to the growth in the world production of ready-made food and agricultural raw materials; difficulties in adapting "global food products" to local needs and conditions, which, according to available research, often negatively affects the health of new consumers, etc.

It is difficult to make an econometric assessment of the impact of globalization on the food sector. This assessment (similar to what happens when evaluating the quality of services) is subjective, which is explained by the lack of reliable data and the truthfulness of the dynamics of institutional costs, the subjectivity of personal perception, and so on. Objective assessment of the quality of the food space may be given only its value, but that, unlike specific goods, which due to its consumer cost, not subject to cost measurement. But the subjective nature of the assessment does not mean that it is useless, if only because the food space of the country (region) has such an important property of self-organization as sustainability, the importance of taking into account which is undeniable.

2 Discourses in the Scientific Literature

The scientific literature that specifies the problems of global food consumption is an impressive list [3–9], etc. It mainly reflects the state of the world food market, as well as the impact of economic globalization on the food security of countries and on shifts in food consumption by the population of both individual states and macro-regions of the world. Less attention is paid to the analysis of the role of the political component, which is important for achieving certain regulatory goals (for example, ensuring food security of countries, eliminating hunger or malnutrition, etc.). Of Course, economic and political levers are often in the same hands, those who manage global food systems, that is, in the hands of multinational food companies with concentrated market power [10].

There is little coverage of issues related to the role of frequent sanctions "wars" between certain key countries in the deformation of the global food market [11–15], etc. The political subtext embedded in the global campaign against genetically modified food, which is most pronounced between Western European countries and the United States, reveals a latent connection with the economy. The call for more research on the harmlessness of such products is not related to the dispute about the food value of "mutant crops", but rather to the sphere of economic competition. In defending their commercial interests, both sides use political leverage to pressure the World Trade Organization, which is designed to regulate trade relations in the world.

Individual authors [16] analyze the links between the processes of globalization and conflicts arising from food insecurity. This refers to the displacement of traditional crops in developing countries by export crops (for example, coffee and cotton). It is a fair conclusion that export crops contribute to fueling conflicts during unstable market conditions, when prices fall, destabilizing households. Such conflicts become especially dangerous when the authorities ignore food security issues in their countries.

We will especially mention the collective monograph of the Institute of Africa of the Russian Academy of Sciences, which provides a thorough analysis of the problem of food security of the continent [17]. The collective nature of the research (with the participation of historians, economists, geographers and sociologists) provided an interdisciplinary spectrum of text based on the factual basis of the results of world conferences, global summits and regional assemblies.

Valuable analysis of certain aspects of the problem is contained in regular reports and documents of the FAO, using updated estimates of food balances, new demographic data and forecasts of the UN demographic program, land resource estimates, etc. [18–22].

3 Internet Discussion

Such discussions organically complement the polemics conducted in academic publications, monographs, and articles by authoritative scientists on the issues under consideration. In the specific conditions of Russia, the largest and most popular search indexes are "Yandex" (www.yandex.ru), "Rambler" (www.rambler.ru), "Google" (www.google.ru), "Aport2000" (www.aport.ru). Certain information about the state of the world food market and the characteristics of regional food consumption can also be found using various search engines, such as: "AltaVista" (http://www.altavista.tella.com), and "Yahoo!" (http://www.yahoo.com), "Lycos" (http://www.lycos.com), "InfoSeek" (http://www.infoseek.com); "Magellan" (http://www.mckinley.com); "Excite" (http://www.excite.com).

Of course, the content of many tweets from these systems is more suitable for training and reference products, rather than analytical research. Nevertheless, the use of the Google Scholar search engine, which indexes the full texts of scientific publications in all formats and disciplines and includes articles published in journals or stored in repositories, has brought real benefits to the preparation of the article. In the Russian language Internet the information site was also very useful ru.globalvoices.org, containing valuable socio-political analytics, and Inomics search engine that only searches for economic information.

4 Purpose and Methods of Research

The ideal in the trade - food security dichotomy is to achieve an optimal balance between national priorities and the common good. However, policy goals in the area of trade, as in many other sectors of the economy (primarily agriculture), are usually linked to various aspects of food security. Naturally, they vary markedly from country to country and may change over time depending on the country's economic potential, geopolitical threats, etc.

The main goal of this work is to clarify the specifics of shifts in the global food market due to the influence of economic globalization and political contradictions between different countries. Emphasis is placed on the connection of such contradictions with economic sanctions that force countries to adopt special programs for self-sufficiency in food to avoid a food crisis; on the implementation of countries' food self-sufficiency policies; and on the political aspects of expanding the market for GM products. The key country illustrating the authors' approaches is Russia.

The paper applies traditional approaches of economics to the analysis of social systems management: system, complex, functional, situational and process, which are used in combination.

5 Research Result

5.1 "Sinful" Union of Economics and Politics as a Key to Success

The problem of the conflict between economics and politics, as an expression of the structural contradiction of modern society, is not new in science. It is believed that hierarchical political structures of power have come into conflict with the modern network form of economic organization. Some modern researchers consider a self-governing creative association (first of all, a successful TNC) to be the highest form of production activity and emphasize that this form of organizing creative activity will become the basis of the next decades [23]. But the coexistence of two forms of power in the information society, they warn, threatens to seriously destabilize social development in the coming decades [24]. Although the latter conclusion is made in relation to Russian society, it can be interpreted more broadly.

Scientific publications of well-known economists describe many theories of international trade that have their own advantages and disadvantages. We do not refer to these theories, since the goal is a slightly different analysis. We are interested in the growth of corporate solidarity in the food sector, which increasingly appears in the essence of the "sinful" union of economics and politics.

Today, it is unlikely that anyone can be surprised by the sensational conclusions of an authoritative expert on nutrition and public health - Marion Nestle (USA) that the current American food companies use all the means at their disposal - legal, regulatory and public - to create an environment conducive to the sale of their products in a competitive market. It turns out that they have long lobbied Congress and the White house in favor of favorable legislation and trade agreements. But what is even more dangerous is that they do not selflessly advertise their dietary recommendations received from biased experts in accordance with a long-established corporate strategy. The instructions for such strategies explicitly explain that this particular tactic "is most effectively implemented by identifying leading experts… and hiring them as consultants or providing them with research grants" [25, p. 117].

Of course, Nestle's conclusions are based on the specific materials of the United States. But we should not forget that this country is a "showcase" of the Western world, where the ten largest food and alcohol corporations control more than half of food sales, and their share may increase. In addition, they control a significant share of the global food market. This suggests that food corporations linked by a network of "sinful" relationships do not always determine their strategy by science and the interests of human health. It is often based on the interests of a "selfish" economy, as well as policies that include issues and events in public life related to the functioning of states.

5.2 The "Echo" of the Sanctions Measures

The consequences of the economic sanctions policy of recent years are particularly pronounced in the Russian food market. At the same time, the Russian food embargo affected EU countries, Canada, Australia and even the United States, which lost a large market for food products. (By the way, in the Western press, Russia's response to Western sanctions was called "a tooth for a tooth").

Before the embargo, the EU member States were the undisputed leaders in the Russian food market and each other's main competitors.

Before the embargo, the EU member states were the undisputed leaders in the Russian food market, and they were each other's main competitors. In Russia food products from the EU has a competitive advantage compared with other suppliers' products:

- first, highly valued European standards of quality (prohibition on transgenic products, stabilizers, artificial preservatives and dyes);
- second, it ensured stable Russia must supply volumes in the short term, as well as qualitative logistic transportation;
- third, there was a desire to constantly update the food assortment and create new products in order to interest and satisfy the constantly evolving demand of customers.

Although European products are usually not unique (for example, there are hundreds of different manufacturers on the market of dairy and fermented milk products), they have won the hearts of Russian consumers not only by their quality and stability, but also by the history of demand. (For example, the Finnish company "Valio" has been operating since 1905, and since 1908 was a supplier of the Court of his Imperial Majesty). Despite the fact that European companies were not always able to provide low prices for their products (the cost of European yoghurts exceeded the cost of Russian analogues by 30–50%), the demand for it sometimes exceeded the supply.

Traditionally, the main competitors in the fight for the Russian food market are the EU member states -Finland, Lithuania, Latvia, Greece, Spain, Italy, the Netherlands, Poland, Estonia, Denmark, France, and Belgium. Other competitors include Belarus, Serbia, Ukraine, Norway, Iceland, Argentina, Chile, Brazil, Mexico, Israel, Morocco, South Africa, Egypt, China, and Thailand. The rigidity of competition is largely due to the aggravation of the struggle in the food market of the European Union between the leading leaders in the export of food products to the EU single market - Germany, the Netherlands, France, Belgium, Italy, great Britain, and Spain. According to official data, in the middle of the second decade of the XXI century, they accounted for more than 70% [26].

European producers before the sanctions supplied almost a third of all fresh fruits and vegetables to the Russian market, and the total volume of agricultural exports was estimated at €11 billion. Strikes by food producers have become a natural consequence of the Russian food embargo. In Europe, a number of cases were recorded when farmers destroyed their products in protest against Russian sanctions and dissatisfaction with the assistance provided to them.

After the introduction of the food embargo, only Belarus and Serbia were able to quickly increase the necessary amounts of food. These countries continue to supply Russia with milk and dairy products, meat and offal. A sharp increase in the volume of deliveries to Russia in a short period of time allowed these countries to significantly strengthen their positions in several segments of the Russian food market at once. One of the reasons for the growth of food supplies from Belarus and Serbia was not only the introduction of sanctions, but also the effect of the deflationary factor. The fact that the currencies of these countries are not pegged to either the US dollar or the Euro has had an impact. The preservation of prices in the currency of the producer country (in Belarusian rubles and Serbian dinars, respectively) led to an actual decrease in prices for goods imported from these countries in terms of dollars or euro. The same effect also affected the increase in food supplies from other countries whose currencies are not linked to world reserve currencies, primarily from Latin America and South-East Asia. (In the case of Belarus, the situation for Russia is complicated by the fact that its Federal service for veterinary and phytosanitary inspection records large-scale deliveries of fruits and vegetables from EU countries with a fake Belarusian addressee).

The frequent use of legal "loopholes" to circumvent bans and restrictions imposed by the Russian Government on agricultural imports, as well as non-price factors that increase the competitiveness of European food products in the Russian market, once again demonstrates the viability of the "sinful" union of economics and politics. Numerous publications attest to the contradictory nature of the policy of economic sanctions [27–31], etc.

5.3 Import Substitution Ideas Against Competitiveness of the Agricultural Sector

Real life shows that no country can completely refuse to import food products - otherwise, a deficit market is formed and prices for certain types of food products jump. However, in order to achieve food security (especially in the context of sanctions policy) Russia has to overcome the "imperfection of the market" by developing a special strategy and policy of import substitution. Its implementation is not a panacea, if only because the historical experience of many states indicates a decrease in the efficiency of national producers in this case, which in turn leads to a deepening of the crisis. This is what happened in several Latin American countries in the 80s of the last century. (For this reason, the term "import substitution", which is now full of Russian media, is perceived in Argentina as… a curse word).

Therefore, the policy of import substitution is a very responsible instrument of national economic policy. It is hardly possible to challenge the thesis that an uncompetitive economy is preferable to its absence. This thesis is confirmed by the fact that many countries - the current "showcases" of the market economy at the dawn of industrialization actively used the idea of import substitution. During the war of independence, even the United States had to resort to banning English goods ("Boston tea party").

Recall that the term "import substitution", which emerged in the middle of the 20th century, is due to the search for a new strategy for developing countries. It was designed to boost their economies with protectionist measures that could protect local markets from cheaper imported products. The founder of the concept of import substitution is recognized as a Nobel laureate R. Prebisch (an Argentine economist), who really made the most significant contribution to popularizing her ideas. However, similar ideas about the relationship between economic growth and the creation of a national production base to meet consumer demand have been formulated before. But Prebish was able to integrate various interpretations of the concept of "import substitution" and create a more or less complete concept of industrialization of underdeveloped countries (import substitution industrialization–ISI). The concept of "self-reliance" (or "collective self-reliance") has also spread in the China, Tanzania, and some other countries, along with the use of this concept.

Later, thanks to the efforts of well-known representatives of the Neo-keynesian school of economics (H. Chenery and M. Bruno, P. Lindert, N. Carter, et al.), the concept of import substitution was interpreted as one of the models of accelerated development of the domestic market, focused on the integration of the state into the global market. Unfortunately, it is often underestimated that such integration is possible only in conditions of fair and open economic competition and strict compliance with the rules of the World Trade Organization.

In implementing the policy of import substitution in the food sector, the Russian government takes into account the destructive trends and "failures" that are increasingly evident in the operation of the functional mechanism of the global economy. The obvious non-compliance with the rules of fair economic competition by all the main WTO countries and the leading "players" of the world market (anti-dumping arbitrariness, economic sanctions, tax preferences for their exporters, etc.) forces the government to make adjustments to both import and export policies in the agricultural sector.

In fairness, we note that no one in Russia among the authoritative economists associates the implementation of the import substitution policy, including in the agricultural sector, with a "historical "breakthrough." It is unrealistic for Russia to establish production of many types of high-tech science-intensive products, especially in the context of increasing global competition, as well as to achieve the level of profitability in agriculture in Western countries with developed market economies. The calculation is made on the "nationalization" of its own market in order to further accelerate economic growth and try to ensure the competitiveness of products.

You can not discount the specific climatic conditions of the country, making it the coldest Northern state in the world. Sharp continental climate, low winter temperatures, short transition seasons, frequent frosts in the warm season - these factors greatly limit the growth of labor productivity in the agricultural sector and do not contribute to the growth of its competitiveness in the global market. This circumstance once again indicates the need for state support for the agricultural sector.

5.4 GMO: The Background of Commercial Success

Avalanche spreading victorious stories about genetically modified organisms and products sometimes resemble the beginning of the famous fairy tale of Andersen: "…in China, all the inhabitants are Chinese and the Emperor is Chinese." But, among them there are works containing very serious and intriguing arguments. Supporters of GM products assure us that:

- since the introduction of transgenic organisms into practice (starting in 1994, with the transgenic tomato Peug Baug), no authoritative expert center in the world has found and proven damage to the health of both humans and animals when using genetically modified food;
- it is known that domesticated animals and agricultural crops have already been crossed and artificially selected over the centuries and therefore genetically modified;

- the inevitability of global expansion of GM crops is justified by the difficult food situation in underdeveloped countries;
- higher yield of GM crops automatically leads not only to a higher level of profitability of agricultural business, but also to improved environmental conditions;
- "plus" is a sharp reduction in the use of herbicides, insecticides, acaricides, nematocides, zoocides and other pesticides;
- GM crops do not necessarily have to become part of the nutritional diet of the population, but they will be useful for processing into biofuel, simultaneously contributing to the reduction of CO2 emissions into the atmosphere, etc. [32–35].

Those who do not accept GM foods with the same confidence convince us that the latter are definitely harmful and call them "Frankenstein food". The main critical arrows are as follows:

- genetic engineering has dangerous side effects, which consists in the fact that the "transevolutionary" and "interspecies" mixing of genes is fraught with many dangerous consequences for humans, which have not yet been fully clarified;
- GMOs should not be used as a food product, as they represent unhealthy food;
- genetically engineered vaccines (primarily for the prevention of hepatitis) - preventative agents with many unknown ones. They are used without proper laboratory expertise, without proper clinical trials in exporting countries, and are identified as "dangerous experiments on live children";
- uncontrolled production and distribution of GM products will negatively affect the regeneration mechanisms of the biosphere and lead to a decrease in biodiversity;
- independent research on the harmfulness of GMOs is silenced and deprived of funding, etc. [37–41] etc.

In any case, the rapid expansion of transgenic crops in the world indicates that the po-sitions of supporters of GM products are winning. "International Service for the Acqui-sition of Agri biotech Applications" (ISAAA) reports that from 1996 to 2018 the sown area of these crops has increased by 113 times, and their total area was 2.5 billion hectares. This confirms the idea that biotechnology is the fastest-growing crop technology in the world. At the same time, in the United States, Brazil, Argentina, Canada and India implementation of GM crops rates are close to 100%, and crops together occupy 91% of the world's biotechnological crop areas [42].

The true role of the "transgenic revolution" has yet to be evaluated by our descendants. Genetic engineering has many noble goals, including fighting world hunger, creating effective medicines, drastically reducing the proportion of herbicides and pesticides used, and so on. But the implacable opponents of GM products still have a lot of serious counterarguments. Among the less pedaled, but, from our point of view, very important - the threat to insects, especially bees. After all, the "bee – plant" symbiosis with its nectar is unique and indissoluble. It is bees that are involved in the reproduction of most plants. (In a few years after the death of the last bee, people will also die, - warned at the time A. Einstein, who was aware that without insects, the planet Earth will turn into a desert). The effects of genetically engineered vaccines-primarily in preventing hepatitis - have been poorly studied so far. Such vaccines are preventative agents with many unknowns.

They are imported without proper laboratory expertise, without proper clinical trials in exporting countries, and are identified by the expert society as “dangerous experiments on live children”, etc.

If we can ignore the frivolous arguments of rejection of the “transgenic revolution” (such as the inadmissibility of invading the sphere regulated by God), we can not dissociate ourselves from issues related to the less transparent side of the commercial success of GM products. The following points are most frequently mentioned in this regard:

- implementation of TNK strategy of total monopolization of the food sphere on the whole planet and concentration of full power in several corporations;
- lack of a reliable system regulating the development of GMO plants, livestock and medicines based on genetically modified materials of biological origin;
- the bad reputation of the giant chemical company “Monsanto Corporation” from St. Louis (USA, Missouri), one of the main GMO companies, which had a long experience of fraud, underhand struggle and bribery (it developed the deadly herbicide “Agent orange” to destroy the jungle in Vietnam in the 1960s);
- the opening of the GMO Pandora’s box in 1992 by George Bush-father due to a government decree that genetically modified plants are “substantially equivalent” to ordinary plants of the same species, which inspired agribusiness companies and became the “axis” of the entire GMO revolution, etc.

Many facts indicate that commercial “discord” is being transferred to the regional level. Let’s note, for example, the frequent confrontation of farmers related to the jurisdiction of GMO moratoriums on certain land plots. So, Australia has more certified organic land than any other country, but there are moratoriums in some states, and there are no moratoriums in others. A high-profile lawsuit is known due to the “contamination” of a site that had organic certification with GM rapeseed grown on another site. The trial, which gained great resonance, was considered in the Supreme Court of Western Australia [43]. Such conflicts, reflecting the confrontation between producers of organic and genetically modified crops, are increasingly appearing in court cases in various countries.

5.5 On Food and “Culinary” Sovereignty

The events of recent decades in the world have clearly confirmed the thesis that the independence of the state is determined by its real sovereignty, the essence of which is the ability of the state to independently, without external pressure, build foreign and domestic policies, join various international unions, etc. Without focusing on this issue, exhaustively analyzed by many authors, let’s focus on the interpretation of the concept of “food sovereignty” and the derived concept of “culinary sovereignty”. The relevance of this step is due, in particular, to the fact that the national security doctrines of a truly sovereign state, in addition to financial, economic and military-political provisions, usually include the food factor (and even civilizational, taking into account the cultural and genetic code).

The Doctrine of food security of the Russian Federation points to the organic relationship of food security of the country with the preservation of its statehood and sovereignty:

"Food security of the Russian Federation is one of the main directions of ensuring the country's national security in the medium term, a factor in preserving its statehood and sovereignty, an important component of demographic policy, and a necessary condition for implementing a strategic national priority – improving the quality of life of Russian citizens by guaranteeing high standards of life support" [44].

One of the poorly studied aspects of food sovereignty and food security is related to the nature of implementation of the provisions of such documents of state significance in the Federal state. It is not enough to consider this issue differentially by population groups with different income levels and provide targeted assistance to citizens who are in the zone of greatest social risk. It is important in the process of goal setting to formulate the goal of achieving at least relative equality in food consumption between the subjects of the Federation.

Rakhmanov A. develops the idea of the so-called "culinary sovereignty" of states. In his opinion, the essence of such sovereignty consists in the degree of orientation of its inhabitants to national cuisine. This indicator is proposed to be determined based on the ratio of national cuisine restaurants to the total number of restaurants in the leading cities of the corresponding country [45].

However, we are very critical of this method of determining the "culinary sovereignty" of countries, if only for the reason that in many countries (especially middle and low income, with a significant share of the rural population), the number of citizens visiting restaurants is not very large. On the other hand, if additional statistical information is available to make representative conclusions, this indicator may be useful.

As for the concepts of "culinary nationalism" and "culinary cosmopolitanism", their content depends not only on the degree of perception of other people's culinary recipes and dishes, but also on the nature of the relevant policies in this area carried out by the governments of countries.

It is difficult to dispute the opinion that the development of culinary art is largely connected with human culture, with the "national spirit", with the desire of the people to increase their competitiveness in comparison with their neighbors, and so on. In this regard, the author's use of the expression "culinary powers of different ranks" seems correct.

6 Points for Discussion

The study of the global food problem lies on the border of several economic and sociological theories. Thus, we can recall numerous theories of international trade; Gossen's theory of marginal utility (which determines the existing patterns of consumer behavior in the market, the formation of demand itself and the type of demand curve on the graph that reflects demand and supply); the theory of sustainable development, etc. At the same time, the food problem is inextricably linked to the consequences of other global problems - climate fluctuations, limited arable land, lack of fresh water, uncontrolled growth of the Earth's population, etc.

In this regard, the stated problem of conflict between economics and politics, as an expression of the structural contradiction of modern society, may seem much less significant. The authors proceed from the fact that this problem (especially the sanctions "wars" in the food sector) can lead to a tragic destabilization of the political situation in the world with difficult to predict consequences.

The role of food TNCs, GM crops, and GM organisms in shaping the global food market, and the methods for implementing the widespread idea of "food security" remain controversial.

7 Conclusion

The ongoing changes in the global food market require a comprehensive, "synergistic" study, taking into account the totality of factors that affect the production and demand processes, as well as the quality of food. They have different impacts on the food situation in countries, with unintended consequences for their food security. A special role among the factors in modern conditions is played by the policies of large food companies and the methods (including commercial manipulation) used by them in the sphere of trade and related industries.

There is no reason to disagree with the aforementioned authoritative us expert in the field of nutrition science, M. Nestle, who believes that leading food companies are doing everything possible to increase their profits. They not only actively fund food science research, but also influence government regulations, advise politicians, and potentially compromise objectionable researchers, and even lobby Congress and the White house for favorable legislation and trade agreements.

Significant changes in regional food markets occurred as a result of the sanctions policy. They are particularly evident in the food markets of the EU and Russia. In turn, the sanctions policy is also associated with shifts generated by the policy of import substitution, which most often contributes to reducing the competitiveness of the agricultural sector, but to strengthening food security. Quality fluctuations in the global food market are largely associated with increased sales of GM products.

References

1. Kovalev, E.V.: Potential of the world agro-industrial sphere: pluralism of assessments. World Econ. Int. Relat. **8**, 3–14 (2011)
2. Ehrlich, P.: The Population Bomb. NY (1968)
3. Nestle, M.: Food Politics: How the Food Industry Influences Nutrition and Health. University of California Press, Berkeley (2002)
4. Patel, R.: Stuffed and Starved: The Hidden Battle for The World Food System. Melville House (2008)
5. Kornekova, S.Y.: Geography of Food Consumption: A Systematic Analysis. Saint Petersburg (2017)
6. Corporate Power in Global Agrifood Governance. Cambridge, Massachusetts and London/England (2009)
7. Qiu, D., Chaoqun, Z., Xing, W.: An analysis of the China–US trade war through the lens of the trade literature. Econ. Polit. Stud. 148–168 (2019)
8. Fogel, R.W.: The Escape from Hunger and Premature Death, 1700–2100. Europe, America, and the Third World. NY (2004)

9. Pollack, A.: After loss, fight to label modified foods continues. New York Times (11/7/2012). Article in China Econ. J. **12**(4), 1–20 (2012)
10. Lyson, T., Raymer, A.L.: Stalking the wily multinational: power and control in the US food system. Agric. Hum. Values **17**, 199–208 (2000)
11. Boucher, J.C., Cameron, G.: I am a tariff man: the power of populist foreign policy rhetoric under president trump. J. Polit. **81**(2), 712–722 (2019)
12. Qiu, D., Chaoqun, Z., Xing, W.: An analysis of the China–US trade war through the lens of the trade literature. Econ. Polit. Stud. **7**(2), 148–168 (2019)
13. Qiu, L. D., Xing, W.: China–US trade: implications on conflicts. China Economic Journal, pp. 1–20 (2019)
14. Gladkiy, Y.N., Kornekova, S.Y.: Import substitution and competitiveness of the Russian economy: towards a dialectic of relationships. Society. Environment. Development **3**, 92–97 (2015)
15. Swanson, A.: Tramp Trade War with China is Officially Underway. The New York Times, July 5, 2018.
16. Messer, E., Cohen, M.: Conflict, food insecurity, and globalization: food, culture and society. Int. J. Multidiscipl. Res. **10**(2), 297–315 (2007)
17. Africa in the context of global food security. Moscow (2015)
18. Alexandratos, N., Bruinsma, J.: World agriculture towards 2030/2050: the 2012 revision: FSA Working paper, pp. 12–03. Rome (2012)
19. IFADI Fao WFP: The Linkages between Migration, Agriculture Food Security and Rural Development Rome (2018)
20. FAO, IFAD, UNICEF, WFP & WHO: The State of Food Security and Nutrition in the World. Building resilience for peace and food security. Rome (2017)
21. AO: he State of Food and Agriculture. Leveraging Food Systems for Inclusive Rural Transformation. Rome (2017)
22. FAO: Rome declaration on world food security and World food summit plan of action: World food summit, 13–17 November 1996. Rome (1996)
23. Drucker, P.F.: The Age of Discontinuity. NY (1996)
24. Inozemtsev, V.L.: Modern post-industrial society: nature, contradictions, perspectives. Moscow (2008)
25. Nestle, M.: Unsavory Truth: How Food Companies Skew the Science of What We Eat. Berkeley (2018)
26. Russia and the World. Moscow (2016)
27. Carter, B.: International economic sanctions: improving the haphazard US legal regime. Calif. Law Rev. **75**, 1159–1278 (1987)
28. Gevorgyan, K.: "Unilateral sanctions" and international law. Int. Life **8**, 93–104 (2012)
29. Hufbauer, G., Schott, J., Elliott, K., Oegg, B.: Economic Sanctions Reconsidered, pp. 17–18. Washington, DC (2007)
30. LaRae-Perez, C.: Economic sanctions as a use of force: re-evaluating the legality of sanctions from an effects-based perspective. Boston Univ. Int. Law J. **20**, 188 (2002)
31. US Sanctions on Russia. Congressional Research Service (2020)
32. Russo, E.: Special Report: The birth of biotechnology. Nature **421**, 456–457 (2003)
33. Zhou, Y., Lu, Z., Wang, X., Selvaraj, J.N., Zhang, G.: Genetic engineering modification and fermentation optimization for extracellular production of recombinant proteins using Escherichia coli. Appl. Microbiol. Biotechnol. **102**(4), 1545–1556 (2018)
34. Amarger, N.: Genetically modified bacteria in agriculture. Biochimie **84**(11), 1061–1072 (2002)
35. Sheridan, C.: Gene therapy finds its niche. Nat. Biotechnol. **29**(2), 121–128 (2011)
36. Oliver, M.J.: Why we need GMO crops in agriculture? Mo. Med. **111**(6), 492–507 (2014)

37. Gladkiy, Y.N.: Food security in Russia: opportunities for genetic engineering and agro-hightech. Soc. Environ. Develop. **4**, 199–203 (2011)
38. Gladkiy, Y.N., Kornekova, S.Y.: The world in captivity of genetically modified products: by geography production and consumption. Geogr. school **5**, 52–57 (2014)
39. Berg, P., Mertz, J.E.: Personal reflections on the origins and emergence of recombinant DNA technology. Genetics **184**(1), 9–17 (2010)
40. Head, G., Hull, R.H., Tzotzos, G.T.: Genetically Modified Plants: Assessing Safety and Managing Risk. London (2009)
41. Zhang, C., Wohlhueter, R., Zhang, H.: Genetically modified foods: a critical review of their promise and problems. Food Sci. Human Wellness **5**(3), 116–123 (2016)
42. Brief 54: Global Status of Commercialized Biotech/GM Crops (2018). https://www.isaaa.org/resources/publications/briefs/54/default.asp. (date of request 01.07.2020)
43. Dabrock, P.: Playing god? Synthetic biology as a theological and ethical challenge. Syst. Synth. Biol. **3**(1–4), 47–54 (2009)
44. The Doctrine of Food Security of the Russian Federation. 30.01 2010 (President's decree No. 120)
45. Rakhmanov, A.B.: Global culinary space and gastronomic strategies of cities Russia. ECO **3**, 91–103 (2017)

Sustainable Development and Environment Protection

Geo-Eco-Ideology: Modern World in Search of Common Values

Vladimir Belous[1(✉)], Vitaly Volkov[1], and Leonid Baltovskij[2]

[1] Saint Petersburg State University, Saint Petersburg, Russia
{v.belous,v.a.volkov}@spbu.ru
[2] Saint Petersburg State University of Architecture and Civil Engineering, Saint Petersburg, Russia
leonid.baltovsky@gmail.com

Abstract. Politics in the modern world is not driven by shared values and goals. The modern "globalist" and "anti-globalist" movement is not correlated with either a special theoretical concept or a new value system. This article is devoted to the formation of new common political values for humanity, the combination of which authors call geo-eco-ideology. The authors suggest a new look at the term "ideology". The old idea of ideological discourse was mainly based on the practical impulse of the Enlightenment. This kind of sum of ideas focused on political praxis is more appropriately called political "doctrines". In the modern era, ideas have lost their doctrinal character, but their moral and intellectual significance has not changed. The basis of geo-eco-ideology should be founded on a system of global values. The work is directly related to Geo-ecology, which is a emerging field at the intersection of geography and ecology. The focus of Geo-ecology is on the problems of the natural environment, taking into account the natural and anthropogenic processes occurring in them. The formation of geo-eco-ideology is undergoing a new stage, associated with the understanding of the value system relevant to existing threats. The effect of environmental glocalization unites the local places of possible environmental disasters and the entire space of global consequences in a single system. The main attention is paid to the problems of human relations with the environment, and the political factors of such interaction. We are primarily interested in the history of this process, or what factors are responsible for the emergence of new universal values. Answers to the challenges of the future arise based on new methodological principles and syntheses. Geo-eco-ideology as an interdisciplinary result of modern political knowledge and public consciousness, points to such systemic values as an ecological state, ecological political sovereignty, justice, and responsibility.

Keywords: Political ecology · Value system · Sustainable development · Full world · New enlightenment · Ecological political sovereignty · Justice · Responsibility

© The Author(s), under exclusive license to Springer Nature Switzerland AG 2021
R. Bolgov et al. (Eds.): *Proceedings of Topical Issues in International Political Geography*, SPRINGERGEOGR, pp. 213–225, 2021.
https://doi.org/10.1007/978-3-030-78690-8_19

1 Introduction

1.1 Purposes

Humanity's response to the challenges of our time must be based on a collective political strategy. The coronavirus makes this requirement even more relevant. In relatively recent times, such strategies were called "ideologies". Modern post-ideologies have a conflicting genesis. Unlike classical ideologies, the stages of their development may differ somewhat in their sequence. Classical ideologies began their evolution as new systems of views and values in the writings of individual thinkers, who were able, beyond the variegation of historical events and changes, to see the key contradictions of the modern era and certain ways out. Such a traditional system of values had a decisive importance for public life in case of its ability to influence the political and spiritual life of society.

When the influence of the system of ideas on social practice becomes historically significant, we are dealing with the views of a certain layer of society. Subsequently, the struggle for hegemony in the civil society space could result in the recognition of a particular ideology by the ruling elite. Thus, ideology reached the climax of its development, turning into a state one. In the modern political discourse, with all the diversity of opinions in the scientific community, ideology comes, first of all, as certain system of ideas, a system of value judgments of a worldview nature. This is such a picture of the world that can become a sustainable motivation for social practices aimed at transforming existing social relations, or at their legitimation. Social practice shows to what extent ideology proves to be historically successful in the development of specific states and peoples.

Our article is devoted to the formation of new common values for humanity, the combination of which we call geo-eco-ideology. The work is directly related to Geo-ecology, which is a young emerging field at the intersection of ecology and geography. The focus of Geo-ecology is on the problems of the natural environment, taking into account the natural and anthropogenic processes occurring in them. The main attention is paid to the problems of human relations with the environment, and the political factors of such interaction. We are primarily interested in the history of this process, or what factors are responsible for the emergence of new universal values. The methodological foundation of our study is a critical approach based on a political interpretation of the system values of an environmental worldview.

1.2 Methods

The authors' concept involves the consideration of the current principles of the new consciousness in the general context of the contemporary cultural crisis, which is discussed in a number of previous publications [1]. Ideas about the future are formed on the basis of the knowledge that people use in the present. In search of the meaning of existence, modern humanity focuses on the scientific (expert) community. Unfortunately, modern social science is focused on serving the multi-faced everyday life. Experts, at best, describe features of the surrounding reality. They are not ready to create new meanings, formulate ideals, and offer specific guidelines for the development of a person, society, and humanity as a whole. This is relevant to all intellectual practices – philosophical, political, sociological, and interdisciplinary. As an alternative to the prevailing empiricism should be the synthesis of three levels of knowledge – real, normative, and ideal. To break a vicious circle of the modern crisis of thinking, it is necessary to change epistemological principles. The authors consider the updated methodology oriented on the political reflection and scientific revision, with maximum separation from utopianism and postmodernism.

It is entirely applicable to the term that is key to this work. Historically, the concept of "ideology" appeared at the early 19th century, and after a century and a half, its content has been transformed to such an extent that concerning it the discourse of "death" has become generally used. Along with the "end of ideologies", ideologeme caused by political conflicts have gone into the past, having lost their original meaning; their place is taken by neutral lexeme. Meanwhile, neither the ideas that form the basis of any ideological constructions, nor the moral guidelines and values that determine people's behavior, disappear. It is the generally binding nature of ideas as political prescriptions that disappears. That is why we believe it is necessary to change the terminological labeling of known historical phenomena.

The totality of certain theoretical propositions that served as a basis for political action (praxis) should, in our opinion, be called "doctrine". As for the broader range of ideas – a person's view on his own place in the world – this is precisely an "ideology" in the proper sense of the word. While the doctrines' dominance (here we are talking only about political, not religious doctrines) embraces the period that approximately extends to the 19th and 20th centuries, the ideologies themselves accompany the entire history of mankind. Modernity, of course, is no exception. Ideology continues to perform a number of public functions on the present stage. Using the ideas that in different configurations form the actual ideology and ideologies, a person appropriates the world, turns it into a subject of research and understanding. Ideas allow a person to perceive the world as a task, an issue, a question. Due to the ideas, a person feels responsible for the world, society, and the environment. Ideology as a specific organization of the language space is designed to meet the fundamental human need for self-realization and social action.

It was interesting for us to consider new elements which arise in the modern scientific consciousness in assessing the prospects for human development. The authors consistently tried to combine logical and historical approaches in the general context of solving the stated task.

1.3 Highlights

a. The formation of geo-eco-ideology is conditioned by threats to the modern world;
b. The contemporary environmental consciousness needs a new system of values;
c. Today there is a trend when political interests orient people toward solidarity;
d. The future of each state is the responsibility for the whole world;
e. Today the concept of the lifeworld is becoming more and more relevant.

2 Findings

2.1 A Brief History of the Formation of World Ecological Consciousness

In the post-World War II period, the phenomenon of new ideologies arises, where the formation of geoecology (precisely as ideology, i.e. geo-eco-ideology) acquires special significance. It is born from an existential contradiction, namely, the possibility of self-destruction of mankind. A distinctive feature of the geo-eco-ideology formation is that the first two phases of its evolution, typical for classical ideologies, in fact, merge in one-time interval. Environmental alarmist movements arose in the mid-20th century not so much on the basis of theoretical conceptual studies of philosophers and scientists, but on the basis of the spread of statements about the real possibility of destroying life on the Earth in case of a nuclear conflict between states by the media. Some seemingly local issues have become existential. For example, such problems as environmental pollution, climate change, habitats, and the reduction of biodiversity in terms of their consequences are comparable to nuclear warfare. The only difference is the speed of the disaster approach.

In order for the very possibility of these problems involving into the public discourse, the accumulation of certain scientific facts critical mass was required. Since 1866, when E. Haeckel gave it a name, ecology, being at first one of many biological disciplines, it subsequently began to rapidly acquire an interdisciplinary nature.

In the 1960s, the awareness of the ability of civilization to self-destruction led to the creation of social movement, called environmentalism. The concept of ecology began to be filled with a new expanded content. The German philosopher G. Jonas has already considered ecology as a science, which subject was the relationship between man and nature, but one in which man was the cause of changes within this relationship [2].

Around the same time, a global rethinking of environmental problems began as the foundation of a new environmental consciousness in the struggle for hegemony in the civil society space. This process took place both in the European countries and in the Western Hemisphere. One of the most important centers of the theoretical understanding of environmental threats is the Club of Rome, the organization created in 1968. Club members focused on the processes unfolding in the global world along with their consequences, including environmental problems.

The theoretical base of geo-eco-ideology in the initial period was formed on the basis of classical ideologies as its certain variation. According to generalized research by J. Clark [3], several ideological trends of this kind have emerged, including: conservative environmentalism in libertarian [4] and the traditionalist version of environmentalism [5]; liberal environmentalism [6]; natural capitalism [7]; eco-socialism [8]. The ideology of eco-socialism is based on the principle that without a significant transformation of the capitalist system and its production relations, in particular, it is impossible to solve the environmental crisis. Within the framework of eco-socialism, the "ecology laws" elaborated by the ecologist and politician B. Commoner in "The Closing Circle" [9] have received a great recognition. The results of theoretical efforts to master environmental issues in the sphere of politics were reflected in solid compendiums [10–12]. However, despite a thorough understanding of environmental problems and their political consequences, the development of the ecological worldview took place within the framework of classical ideological concepts and did not lead to the formation of a geo-eco-ideology itself.

The main problem of old ideologies variations is their anthropocentrism. The idea of man's power over nature does not allow us to comprehend its active nature, and a man in spiritual unity with nature. The Anthropocene concept captures the enormous human geological influence on the existence and development of the Earth, but ignores the consequences of mankind activity. Modern society is like a child playing with matches. Awareness of the need to overcome anthropocentrism has led to the emergence of new theories that denied the special parameters and values of man. Based on this approach, Arne Naess formulated the concept of "deep ecology" as opposed to "surface ecology" [13].

2.2 Environmental Issues in the Reports to the Club of Rome

The first document on the environmental issues of the international organization, which got the name the Club of Rome, was the 1972 report "The Limits to Growth", prepared by a group of Donella and Dennis Meadows [14]. Based on mathematical modeling and analysis of five parameters (the use of unrenewable natural resources, environmental pollution, investment, population growth and food security), experts at the Massachusetts Institute of Technology predicted that at the beginning of the 21st century, mankind will reach the limits to growth in development. Recommendations were given to halt population growth through an active demographic policy, as well as to control production growth.

In 1992, the same authors published the report "Beyond the Limits" [15]. Ten years later, the work "The Limits to Growth: the 30-Year Update" was published [16]. These studies have shown that the development of civilization occurs in the context of the emerging environmental crisis. Under the auspices of the UN, a commission led by the Norwegian Minister of the Environment Gro Harlem Brundtland (Brundtland Commission) prepared a report "Our Common Future" [17]. As a result, the concept of "sustainable development" was introduced into scientific circulation. The search for reserves of economic growth in the framework of the concept of sustainable development in 1997 led to the publication of a new report, "Factor Four: Doubling Wealth, Halving Resource Use", where a scenario for a four-fold increase in labor productivity that does not go beyond growth was proposed [18].

The development of theoretical foundations that could form the basis of geo-eco-ideology was accompanied by self-criticism. The activity of the Club of Rome itself was criticized for pursuing the specific political goals of the transnational elite. Doubt was caused by the conclusions of reports and scientific research in the field of ecology. Some critics were very distrusted by the very fact of anthropogenic impact on climate change.

The opponents of ecological crisis researchers turned out to be the most diverse groups of the public, from individual scientists to representatives of business and politics. In the space of environmental problems discussion, different concepts encountered reflecting various worldview orientations, as well as the views of specialists in natural sciences and social disciplines. The problems of the ecological crisis, not being clearly theoretically justified, turned out to be quite deeply rooted in the public consciousness. Dividing society into opposing trends, it turned into a political problem itself.

2.3 The Main Theoretical Provisions of the Report to the Club of Rome (2018)

An analysis of current trends shows that geo-eco-ideology needs its own theoretical foundation, formed in the context of a critical attitude to the worldview values of the entire previous development of human civilization.

Representatives of the Club of Rome attempted to find such a ground in the framework of a new report published in honor of the fiftieth anniversary of the Club of Rome in 2018 [19]. The report was prepared by co-chairmen Ernst Ulrich von Weizsacker and Anders Wiikman, co-authored with thirty-four other members of the Club of Rome. The report claims to formulate the theoretical foundations of a new ecological worldview and a new geo-eco-ideology. It consists of three large parts. The first ("C'mon! Don't Tell Me the Current Trends Are Sustainable!") is devoted to criticizing the state of the modern world. The second part ("C'mon! Don't Stick to Outdated Philosophies!") is devoted to worldview shifts in values and attitudes.

The title of the third part ("Come On! Join Us on an Exciting Journey Towards a Sustainable World!") speaks for itself. This section of the report, devoted to applied issues on the widest range of problems from global government to agriculture, we will leave out of brackets. The most significant for us is an attempt to create a conceptual theoretical base of the emerging environmental consciousness.

What is the actual ecological picture of the world and environmental worldview? First of all, the authors of the report offer a new picture of the world: "full world" as the opposite of the "empty world". The concept of "full world" was originally proposed by the American economist Herman Daly [20]. According to his position, current trends in the economy cannot be preserved in the future. The reality of the modern world is not what we imagine it to be. Unless radical changes are made, the world will face a total loss of wealth and the possibility of environmental disaster.

Mankind has been forming for a long time in the space of the "empty world" with uncharted lands, seas, inexhaustible resources. In such a world, extensive development was prevailing. The worldview and value systems of the "empty world" were expressed by the ideals of anthropocentrism, progress, and the European Enlightenment of the 17th – 18th centuries. In the 20th century, mankind found itself in a "full world", encountering borders in its activity and existence. There are no more unknown lands and continents, the scarcity of resources becomes obvious, the economy overcomes all national borders and turns to be global. The economy must be transformed so that the world can become sustainable in the long run.

First, it is necessary to limit the use of all resources to the level at which the resulting waste can be absorbed by the ecosystem. Secondly, the exploitation of renewable resources is possible only to a level that does not exceed the ability of the ecosystem to re-create resources. Third, the depletion of unrenewable resources should not exceed the pace of development of renewable substitutes.

The report to the Club of Rome "The Limits to Growth" (1972) already showed the systemic nature of the crisis in the modern world. The growth of the Earth's population from one billion to almost eight billion causes a problem with food security, which entails the depletion of land, water and energy consumption. These changes entail climate change and increased pollution of the Earth, oceans, and air by industrial waste. Natural and man-made capital are more often complements than substitutes and that natural capital should be maintained on its own, because it has become the limiting factor. That goal is called strong sustainability. For example, the annual fish catch is now limited by the natural capital of fish populations in the sea and no longer by the man-made capital of fishing boats. Weak sustainability would suggest that the lack of fish can be dealt with by building more fishing boats. Strong sustainability recognizes that more fishing boats are useless if there are too few fish in the ocean and insists that catches must be limited to ensure maintenance of adequate fish populations for tomorrow's fishers [20].

From the point of view of E. von Weizsacker and A. Wijkman, the very concept and practice of capitalism were formed in the "empty world", and the ongoing financial crises in the 21st century are of a different nature than those described in the classic textbooks of political economy. Capitalism with its focus on short-term profit maximization, is moving us in the wrong direction – towards an increasingly destabilized climate and degraded ecosystems. In spite of all the knowledge we have today, we seem unable to change course, literally driving planet Earth to destruction [19].

The inconsistency of the current situation can be illustrated by the example of such an important macroeconomic indicator as the gross domestic product, which determines the policy of all states in the world, but in fact cannot adequately reflect sustainable development in the "full world", since GDP is aimed at a positive measurement of costs, even if they only compensate for the effects of destruction and disaster. For example, an oil spill increases GDP because of the associated cost of clean-up and remediation, while it obviously detracts from overall well-being. Examples of other activities that increase GDP include natural disasters, most illnesses, crimes, accidents and divorce [19].

The concept of a "full world" requires a change in a person's views on his own place in the world. In the ecological discourse, the concepts of environment and nature often distort the real state of things. The idea of a "full world" allows us to see modern problems in a new way.

It should be noted that ideas about the world can have several dimensions. There is an objective world as a result of knowledge of the natural sciences. It is opposed by the subjective world as a complex of ideas, values, experiences of an individual person. The lifeworld is thematized by the original meanings and defined by the circle of human problems. This is the world of everyday life, and going beyond it. The concept of a "full world" should clarify the relationship between different ideas, enriched with new content. At the same time, it already sets a certain metaphysics of geo-eco-ideology.

E. Weizsacker and A. Wijkman propose to consider the attitude of religious denominations to the ecological problem using the example of the second encyclical Laudato si (Praise be to you) of Pope Francis, (June 18, 2015). A message from this essential encyclical claims that humanity is on the path of self-destruction if rules are not adopted that seriously limit the utilitarian tendencies of the current economy. Anthropocentrism is erroneous even in relation to all living beings possessing the same perfection as a person created in a divine image and likeness. The anthropocentrism of the Enlightenment stems from the phenomenon of unhappy consciousness, which puts man in the place of God. Francis calls for a decisive cultural revolution.

The distorted perception of the modern world is based on many sources and their incorrect interpretations. It was the installations of the "empty world" that led to the dominance in the previous era worldview of the principles of the "invisible hand of the market" preference over state legislation; free trade based on comparative advantages, mutually beneficial for all market participants; competition, which is, supposedly, the evolution and progress driving force.

In the modern world, these principles are turned to be taken out of their context. For quite understandable reasons, the classics of political economy did not have the opportunity to apply the theory of comparative advantages to the transnational capital, the IMF with capitals of national proportions, and also to globalization processes. In the latter case, transnational capital has an absolute advantage over nation-states. C. Darwin's natural selection and competition neither in nature, nor in the economy, exhausts the mechanisms of evolution and development and is supplemented by the need to preserve diversity, solidarity and protection of small and medium-sized producers [19].

The core of the philosophy of Enlightenment and its epistemology are logical induction and empiricism. This approach, however, has its limits; it does not work in relation to biology, psychology, social reality. The criticism of reductionism is multifaceted. The authors suggest refer to other traditions and authors, such as F. Capra [21] and G. Bateson [22].

Changes in the worldview should be so fundamental and systemic that the authors of the report discuss the necessity of a new Enlightenment. It should be based on the experience of various civilizations. The ancient traditions of eastern culture are based on the idea of balance. First of all, a balance between reason and feelings should be achieved, that would result in a holistic worldview. The Chinese symbols Yin and Yang are an example of a balance of opposites. Western and Islamic traditions tend to distinguish and choose between good and evil, although the dialectic philosophy of George Hegel, or the integral psychology of Ken Wilber show that the tradition of balance is inherent in European philosophy likewise [19].

In the new Enlightenment, the principle of synergy should be the basis of the balance: between man and nature; between short- and long-term perspective; between speed and stability; between private and public; between women and men; between justice and reward for achievements; between state and religion. The balance between humanity and nature is designed to become the core of the new Enlightenment. If in the "empty world" the relationship between man and nature was natural, then in the "full world" the balance between them is a huge problem. Using the remaining oases of nature as resources to meet the needs of a growing population leads the planet to death, not balance. The New Enlightenment implies a rejection of anthropocentrism, but retains the ideal of humanism. The concept of "New Enlightenment" is the central ideological point of the report "Come On! Capitalism, Short-termism, Population and the Destruction of the Planet", both in meaning and in location.

2.4 On the Question of the Future Geo-Eco-ideology: Reflections and Statements

The ideas of the Future turn very quickly into a History of ideas. Despite the profound impression that the report makes, it is still dominated by the old enlightening intention of relying on the mind, depoliticizing the essential contradictions in the modern world. The conceptual position of the authors of this work on the issues discussed is presented below.

Political ecology creates a need for an in-depth understanding of the modern life world. Globalization largely removes the contrast between subject and object, human and environment. Modernity presupposes a person acting in the world, in intersubjective interaction with other people. The world defines a person just as much as a person defines the world. The content of the life world is coevolution. Nature enters the horizon of obvious human values.

Humanity cannot go beyond its own life world. Humanity can destroy the foundations of its life world by rejecting values that were in the circle of obvious values. Humanity can divide and break life ties, constructing a scientific world, a social world. Such a fate befell the concept of nature, which had been transformed from the highest authority among the ancient Greeks into the material for meeting the scientific needs in the New Ages. This is how the objective world of positive science is constructed, which is devoid of any subjectivity, and human subjectivity loses all value. The social world concentrates all possible meanings within itself, leaving the physical world without any meaning. An environmental crisis is a life world crisis, not an issue of environmental protection. Nature is the cradle of humanity and at the same time it is its Noah's ark. In the life world, nature hides itself behind things. It takes intellectual effort to see the nature of things and people.

The basis of geo-eco-ideology should be founded on a system of values developed within the framework of political ecology. An original attempt to justify the political ecology is made by B. Latour. According to his position, there is no single picture of nature in front of us. There are quite active actors who claim their own vision, and therefore political ecology does not promise peace. It's only beginning to understand in which wars it should take part and whom it should consider an enemy [23].

From our point of view, political ecology is a field of knowledge that explores the environmental crisis as the basis on which the modern world is divided into opposing political unities [24]. What is the essence of the environmental crisis? The environmental crisis is understood as such human activity, with the continuation of which in a limited foreseeable time, the conditions for the reproduction of this activity would be destroyed. The ecological crisis is rooted in human economic activity. Nature is part of human economic activity as a resource. Having exhausted the resource, economic activity denies itself as a fundamental form of human life.

Natural resources are no longer a commodity. Trade cannot provide guaranteed access to scarce resources that are becoming the subject of political strife. The state's eco-political sovereignty is ensured by resource security. Such security depends on whether the state can guarantee access to scarce resources by military and political means.

Modern states either export resources or import them. The scarcest resources are oil, water, and rare earth metals. The limited and irreplaceable natural resources lead to the fact that states are divided according to the principle of opposing interests and goals. Sovereignty, being a political and ideological structure, acts as an unconditional and highest value. This position gives us reason to talk about eco-political sovereignty. The meaning of this form of sovereignty is a guaranteed access to limited resources.

The concept of "eco-political sovereignty" is the first epistemological basis for the development of political ecology. In search of new values in the field of ecology, the resource-consuming countries form the concept of "Global Commons". We are talking here about natural resources as the common heritage of mankind, access to which must be ensured by transnational structures. Exporting countries, on the contrary, consider resources to be the most important property of their own states, which for greater protection may be the subject for nationalization. The prospects for the emergence of a world government in the form of transnational structures are not attractive even for the leading states.

On the other hand, sovereignty as an attribute of real statehood is becoming increasingly problematic in the global world, few states can defend their political sovereignty. To preserve their political identity, a significant number of states are forced to adhere to a bloc policy. Eco-political sovereignty is even more problematic. It requires significant multi-vectoral foreign policy efforts in different geopolitical regions. Competition for resources is becoming the leading content of world politics. Hydrocarbons, water, and rare earth metals form the basis of economic power, and hence, a state sovereignty. An alternative to this trend is the policy of not using non-renewable resources such as coal or nuclear fuel in the energy sector. The idea of eco-political sovereignty becomes an ideological impulse for social movements related to the rational use of resources and the use of renewable energy sources.

Even successful control over raw materials does not remove the global threats of the ecological crisis. Awareness of this fact by the subjects of global politics is the second basis for the political and environmental division of the world. The threat of anthropogenic climate change, the expansion of ozone holes requires coordinated decisions by the absolute majority of states. Collective decisions can be made on the basis of fairness recognized by all participants. However, certain groups of countries interpret justice in significantly different ways. The main division is indicated within the confrontation between the countries of the North and the South. Thus, the "northerners" interpret justice as an equal participation of all countries in environmental protection measures. The "Southerners" insist that the "North" should adopt the commitments for the socio-economic development of the countries of the South, financing the environmental activities, the transfer of environmental technologies, and elimination of imbalances in trade. Representatives of the South associate the principle of justice with the development right. From their point of view, the world in an era of environmental crisis should strive for unity both ideologically and materially. They are aimed at solving the problem through solidarity and cooperation with the countries of the North.A common point of view has not yet been developed.

The topic of political justice needs additional collective reflections and agreements. We believe that the ambiguous concept of "eco-political justice" should get its rightful place in the renewed system of global values. Only general discussions around the topic of "justice" will allow policy makers to identify themselves adequately with respect to the human-nature system. This concept should enter the system of human values and get its certainty. In fair relations, subjects are free and responsible. Eco-political justice involves the certainty of law, state and economy. Reaching agreement on issues of eco-political justice is possible only if the authority of international law and international institutions that implement its validity are revived. The uniqueness of the environmental challenge is that it makes previous political contradictions insignificant. Similarly, viral pandemics push countries to respond to global challenges in a unified way.

The consequence of the institutionalization of eco-political justice is that politically divided countries must inevitably become allies. Political pressure and violence are becoming unproductive in the world. The effect of environmental glocalization unites the local places of possible environmental disasters and the entire space of global consequences in a single system.

International political organizations are not exclusive subjects of environmental policy. States remain as the most important political actors linking various trends in world development. At the same time, modern statehood is undergoing significant transformations. Along with the fact that states strive to be democratic, legal and social, they still have to become ecological. The ecological state is a source of political decisions that mediate global and local interests. For an ecological state, the external environment is its life world. Caring for others in this life world is an essential manifestation of the principle of responsibility. Care as a state duty extends not only to its civil society, but to the entire planet. Environmental security of the entire planet is a condition for the state well-being.

The state is called upon to ensure the freedom and rights of every citizen. At the same time, the preservation of the world of life as a General condition for a free life must also be strictly observed. the policy of the ecological state is formed in the space of fundamental opposites, starting with the opposite of anthropocentrism and eco-centrism, ending with the opposites of the interests of civil society and business, the interests of present and future generations. The rights of future generations and the rights of nature are formed in the light of the formation of a new ecological legal consciousness. The ecological state shares and protects the values of an eco-centered world. A person cannot be politically free if he is not free to ensure the ecological basis of his existence. The values of a liberal state, such as property and individualism, become relative in an ecocentric state. The ecological state recognizes the conditions for the use of natural goods on the right to borrow them from future generations. Responsibility as a principle of the ecological state manifests the freedom and dignity of man as his nature.

3 Conclusion

The formation of geo-eco-ideology is undergoing a new (third) stage, associated with the understanding of the value system relevant to existing threats. In the formation of modern environmental consciousness, the alarmist stage can be distinguished, when attention to the self-preservation of life on the Earth was the main motive of social movements. The second stage can be described as the search for sustainable development. It was associated with the production of projects of a social and technological nature, aimed at containing the expansion of human civilization into nature, preserving the environment as an unconditional value. The third stage is associated with the development of political and environmental concepts, with its inherent potential to create a value system that can become a conceptual basis for the development of geo-eco-ideology.

The new ecological ideology rejects the old picture of the world based on the ideals of progress, growth and unconditional freedom of the individual. The new picture of the world is not yet fully presented, but it is already clear that in it a person is interwoven into the inseparable unity of his life world, filled with intersubjective social, technological and natural content. The horizon of values being built is such that any movement forward is a return back to the basics in search of harmony and balance of ecocentrism. The desire for eco-political sovereignty as a value of the state leads to the division of the world into weak and strong, which will lead to the birth of various forms of eco-imperialism. Justice, as a value of the international order, unites the poor on one side and the rich on the other. But it can't bring them together. However, there are values that can overcome these differences.

Today, perhaps for the first time, there is a trend when political interests orient people toward solidarity, rather than political dissociation in resolving crises. Political responsibility becomes relevant. This value is global in nature, although its carriers are turned to be individual political institutions. The state must take on another characteristic, becoming, along with legal and social, also an environmental one. A feature of the ecological state is the responsibility policy within the framework of the principle according to which, bearing responsibility for the planet, the state protects itself, and not vice versa. The development of a value system in the space of political ecology should lead to such concepts as eco-political sovereignty, eco-political justice, political responsibility. Along with the themes of "full world", "new enlightenment" (discussed in a report to the Club of Rome), they can constitute an ensemble of values that will form the basis of a new environmental awareness.

References

1. Baltovskij, L., Belous, V., Kurochkin, A.: Society and authorities: new mechanisms of communication in conditions of the network world. J. Appl. Sci. **15**(3), 538–544 (2015). https://doi.org/10.3923/jas.2015.538.544
2. Jonas, H.: Das Prinzip Verantwortung: Versuch einer Ethik für die technologische Zivilisation. Suhrkamp, Frankfurt am Main (1984)
3. Clark, J.P.: Political Ecology. Encyclopedia of Applied Ethics, vol. 3. Academic Press, San Diego (2012)
4. Anderson, T., Leal, D.: Free Market Environmentalism. Palgrave Macmillan, New York (2001). https://doi.org/10.1057/9780312299736
5. Bliese, J.: The Greening of Conservative America. Routledge, New York (2002). https://doi.org/10.4324/9780429496486
6. De-Shalit, A.: The Environment: Between Theory and Practice. Oxford University Press, Oxford (2000)
7. Hawken, P., Lovins, A., Hunter Lovins, L.: Natural Capitalism: Creating the Next Industrial Revolution. Little, Brown and Co, Boston (1999)
8. Dobson, A., Eckersley, R.: Political Theory and the Ecological Challenge. Cambridge University Press, Cambridge (2006). https://doi.org/10.1017/CBO9780511617805.009
9. Commoner, B.: The Closing Circle: Nature, Man, and Technology. Knopf, New York (1971)
10. Peet, R., Robbins, P., Watts, M. (eds.): Global Political Ecology. Routledge, London (2011)
11. Bryant, R.L. (ed.): The International Handbook of Political Ecology. Edward Elgar, Cheltenham (2015)
12. Perreault, T., Bridge, G., McCarthy, J. (eds.): The Routledge Handbook of Political Ecology. Routledge, London (2015)
13. Naess, A.: The shallow and the deep, long-range ecology movement: a summary. Inquiry **16**(1), 95–100 (1973). https://doi.org/10.1080/00201747308601682

14. Meadows, D., Meadows, D., Randers, J., Behrens III, W.: The Limits to Growth. A Report for the Club of Rome's Project on the Predicament of Mankind. Universe Books, New York (1972)
15. Meadows, D., Meadows, D., Randers, J.: Beyond the Limits: Confronting Global Collapse, Envisioning a Sustainable Future. Chelsea Green Publishing, Hartford (1992)
16. Meadows, D., Randers, J., Meadows, D.: Limits to Growth: The 30-Year Update. Chelsea Green Publishing, Hartford (2004)
17. Our Common Future (The Brundtland Report) (1987) Oxford University Press, New Delhi (2004)
18. Weizsaecker, E., von Lovins, E.B., Hanter, L.: Factor Four: Doubling Wealth, Halving Resource Use: A Report to the Club of Rome (1997)
19. Weizsaecker, E., von Wijkman, A.: Come On! Capitalism, Short-termism, Population and the Destruction of the Planet. Springer, Heidelberg (2018). https://doi.org/10.1007/978-1-4939-7419-1
20. Daly, H.E.: Economics in a full world. Sci. Am. **9**, 100–107 (2005). https://doi.org/10.1038/scientificamerican0905-100
21. Capra, F.: The Hidden Connections: Integrating the Biological, Cognitive, and Social Dimensions of Life into a Science of Substainability. Harper Collins, London (2002)
22. Bateson, G.: Mind and Nature: A Necessary Unity (Advances in Systems Theory, Complexity, and the Human Sciences). Hampton Press, New York (1979)
23. Latour, B.: Politics of Nature. Harvard University Press, Cambridge, MA (2004)
24. Borisov, N.A., Volkov, V.A.: In Search of a New Paradigm: An Essay on Political Ecology. Saint Petersburg (2014). (in Russian)

Climate Change and the UN 2030 Agenda for Sustainable Development

Marina Ermolina(✉), Anna Matveevskaya, and Mikhail Baranuk

Saint Petersburg State University, Saint Petersburg, Russia
m.ermolina@spbu.ru

Abstract. Throughout the second half of the 20th century, many developed countries started to raise concerns about environmental protection. Mostly it resulted from such crucial phenomena as scientific and technological progress and fast industrial development. Since then, states have been becoming more involved in environmental issues, and climate change now is one of the most important among them. Considering that the United Nations organization is one of the largest international bodies and it has a crucial role in the area of sustainable development and climate change. The goal of this paper is to analyze the issue of climate change in the scope of the UN activity and a scenario of possible ways of solving the problem.

Keywords: Environmental action · Sustainable development goals · Climate change · UN · Paris Agreement

1 Introduction

Given the complex nature of the issue of climate change several reasons define the relevance of the research topic. First of all, recently, the world has been going through a period of challenged globalization that can be characterized by the decreased level of international cooperation, and global attention shift away from environmental problems. All of it enormously complicates the achievement of the Sustainable Development Goals (SDGs), more importantly, Goal 13 and aims set by the Paris Agreement on global warming mitigation. Second, as the ongoing COVID-19 pandemic has shown, people are not prepared to deal efficiently with large scale catastrophes [1]. It means that in case of a similar global collapse caused by climate change it will be nearly impossible to respond to it after it happens appropriately, and the only way to fight climate change consequences is to carry out preventive measures to keep global warming under the average 2 °C increase. Third, economic crises, including the one caused by COVID-19, make policymakers give global environmental issues the lowest priority. However, such periods of time might be the best time to take environmental action because of their possible efficiency, lower costs, and economic potential.

One of the most critical strategies for climate change mitigation and the further development of humankind, which combines economic growth and the preservation of

© The Author(s), under exclusive license to Springer Nature Switzerland AG 2021
R. Bolgov et al. (Eds.): *Proceedings of Topical Issues in International Political Geography*, SPRINGERGEOGR, pp. 226–237, 2021.
https://doi.org/10.1007/978-3-030-78690-8_20

the natural diversity of the planet, is sustainable development. Consumption, industrial production, scientific and technological progress of the new millennium has a significant impact on climate, hence on our planet, humans, nature, and the entire world.

At the time the UN was created, which was more than half a century ago, environmental degradation was not considered a global threat. Moreover, it was not a common idea that it could cause or provoke international conflict and undermine human health, economic well-being, and social stability. At that time, the UN Charter did not even mention the word "nature" [2]. Today, the topic of sustainable development is extremely relevant.

It can be argued that over the past twenty-five years, ecological protection has become a fundamentally important area of UN activity. Currently, there are more than 200 international environmental conventions, as well as bodies such as the United Nations Environment Program, United Nations Framework Convention on Climate Change, or International Panel on Climate change [3]. But despite significant success in the development and ratification of treaties to prevent environmental disasters, the planet continues to deteriorate globally. No matter how inspiring are the growing trends for the fight against environmental problems, these efforts are not enough.

In the decades since the establishment of the United Nations, the condition of the Earth's natural systems has been deteriorating, and the pace has not slowed much. Global warming continues to enhance, ecosystems are disturbed throughout the world, fishing is irrational, poachers are operating, and the forests continue to decline, especially at the times of the ongoing COVID-19 pandemic when illegal activities are thriving as a result of the weakened control. All these trends have serious consequences for the well-being of people, including reduction of food for the starving, lack of drugs to cure diseases, air pollution in the cities, the destruction of traditional fishing for food and the disappearance of numerous island states that risk literally drowning from global warming [4].

2 Materials and Methods

It is widely known that humanity has been causing long-term irreversible damage to the planet. Numerous scientific researchers from such groups as the Netherlands Environmental Assessment Agency and the Royal Society confirm that the climate system and many of the world's vital ecosystems are at risk [5]. There are also serious concerns about a shortage of water, food, and energy. It is completely unclear how the world will satisfy the needs of the population, which is expected to reach the number of nine billion by 2050. Ensuring sustainable development should be the main goal of the international environmental movement. However, there are a small number of countries that are committed to sustainability as a goal or, at least, can agree on the costs or right policies to achieve it.

The need to protect the Earth's natural resources and, especially, the climate is a priority in the list of global concerns that the world community should pay more serious and precise attention to. The resolution of such problems requires effective work from national states and special guidance by international institutions, the main of which is, undoubtedly, the UN. In this regard, the study of UN activities in the field of climate preservation is extremely important. It determines the relevance of this work. In the 21st century, a more significant international organization has not yet emerged, nor has the world civil society gained enough leverage that can significantly and positively affect the achievement of sustainable development in the world more than the United Nations. Yet international environmental cooperation in the contemporary world is developing and, first of all, it is happening in the framework of the UN. Being the most universal organization in the world in many ways: the number of members, diversity of areas and activities, and this organization meets the democratic requirements of a progressive and developing world.

The goal of this work is to assess the existing practice, future strategy and effectiveness of the UN and its members in achieving the goal of keeping the global warming well under 2 °C comparing to the preindustrial levels and in completing the Goal 13: Climate action (SDG 13) in the conditions of the modern world.

In the course of this work, such tasks will be accomplished as analysis and summary of the main arguments of climate change supporters and critics of the UN activities in this field, the institutional features of the reforms to keep the global temperature increase below the sound limit, as well as the most important problems of the green economy. In order to achieve the goal of the research, it is also crucial to analyze climate change indicators, define possible pathways of development for global warming, and define whether the forecast is threatening. We will look into the climate change in the framework of the SDGs and Paris Agreement. As a part of assessing the effectiveness of the current climate change mitigation mechanism, it will be defined whether sustainable development is the right concept for it. Also the evolution of the UN, its conference system, and related organizations' environmental activity will be analyzed. As a part of it, in the final section, we will look into the political background of the climate change problem as well as define the existing and discuss the potential mechanisms to solve the climate change problem.

As we can see, the main subject matter of the study is climate change and environmental protection, while the scope of the study is the UN strategy and cooperation with its members to combat climate change problem under the Paris Agreement and on the backdrop of the UN 2030 Agenda for Sustainable Development. The scientific novelty of the research, thus, is defined by the latest data it analyzes, given the recent trends and events in the international sphere.

The research is mainly based on general scientific methods combined with experimental methods. The methodological basis for this study consists of the data analysis that helped interpret the evidence of the climate change, historical method, thanks to which, an analysis of the UN activities in the field of environmental protection was carried out to follow the key points in its evolution while the comparative analysis allows comparing the primary outcomes of the major UN environmental and climate change conferences. The analytical approach helped in analyzing the evidence of climate change. Finally, scenario development and modeling allow creating pathway development based on all the gathered information.

The key sources for this work are, first of all, primary sources such as the declarations of the main environmental conferences, climate change agreements and treaties, and various climate change reports of the UN-related organizations. Secondary sources include a vast volume of the many pieces of research looking into the various aspects of the development of climate change issue all around the world.

While the most importance is presented by the primary sources, a significant share of the research is based on analysis of the sustainable development as it represents the most efficient strategy to mitigate climate change. The concept of sustainable development is at the essence of the environment and development as well as conquering global warming. It allows us to realize how governments, businesses, and environmental groups should respond to it. Bill Adams' research paper, Green Development [6], is a clear and consistent analysis of sustainable development, both in theory and in practice, and is an important theoretical basis for the research. Green Development discusses the origins of thinking about sustainable development and its evolution, as well as ideas that dominate the concept of sustainable development (environmental modernization, environmental or green economy), and also form a clear perspective of the problems of environmental sustainability and socio-economic development. This scientific work is a synthesis of theoretical ideas in the field of sustainable development. It provides a theoretical basis, which is essential for an objective judgment and assessment of UN activities in the field of environmental protection and climate change mitigation. Green Development addresses topics such as the dilemma of sustainability, the origins, and development of the concept of sustainable development, conservation policies, as well as the most important issues of our time.

Another equally important source is Tim Jackson's book, Prosperity Without Growth [7]. As you know, the transition to a policy of sustainable development is one of the main goals of environmentalism. However, its achievement is impossible without joint efforts with the economic sector. As it was already stated above, cooperation is what the current stage of climate change mitigation lacks the most. And this work is a clear example of the need for cooperation between the environmental and economic sectors.

3 Results

Global warming, which in recent years has touched environmental activists all around the world is one of the most urgent global problems for humanity in the 21st century. Overpopulation, irrational use of natural resources, advanced scientific and technological progress, overconsumption – all of them have worsened ecological problems and, most importantly, helped enhance the irreversible process of climate change. The results started to become visible quite soon. The first time the international community started talking about global warming and the greenhouse gas effect was when the UN developed the issue and first voiced it in the 1970s.

Interest in this topic is related with many years of scientific research, which year by year has been registering a gradual increase in the average annual temperature of the atmosphere of Earth and the World Ocean. This rise is mainly connected with an increase in the concentration of greenhouse gases in our planet‘s atmosphere, increased solar activity, melting glaciers, and many other reasons. Nevertheless, climate change is not as obvious for all the researchers in the world. It is caused by the uncertainty of climate change skeptics in the factors that cause climatic changes. Is anthropogenic activity the reason, or is it just a consequence of natural cyclic climatic factors? This is the question that many scientists ask themselves. Many climate change dissidents use natural processes as the main explanation for global warming. There is a particularly high trend for the pessimistic approach to the anthropogenic cause of the climate change problem in Russian environmental science. However, in this chapter, we will analyze the main evidence of the climate change existence and try to challenge the opponents' views.

According to the Global Warming of 1.5 °C report, anthropogenic activity caused about 1.0 °C climate increase compared to the preindustrial levels. It reached the trend when the temperature increase each decade has become about 0.2 °C. Moreover, the temperature will not stop rising by itself. The IPCC projects the climate to get to the point of 1.5 °C in just 10 to 40 years [8]. According to the WMO Statement on the State of the Global Climate in 2019, the last five years have become the hottest for the whole history of climate observations [9]. Moreover, the consequences will only continue to deteriorate if humankind will not implement efficient countermeasures. However, it is not highly-likely in the circumstances when there are people who deny climate change. That is why the UN-family environmental activity in sharing scientific knowledge is that important.

Shifts in ecosystems would not be as threatening if they did not cause the changes that might be impossible to reverse. Among such changes, there can be a variety of new potential global problems. And as we can see from the recent COVID-19 outbreak, humanity is not at all close to being ready to adapt to the dynamic changes that could happen to our planet. The emergence of new diseases, a steep deterioration in water quality [10], the appearance of dead zones in coastal areas, shifts in the regional climate – this is just a shortlist of all the possible catastrophes that the uncontrolled human influence might cause for the planet.

More importantly, the degraded ecosystem and climate change increase the gap between social groups. Inequality becomes extremely noticeable at the times when the poorest are deprived of the essential basic resources like fresh air, water, food. All of these resources are heavily influenced by climate change. From 2006 to 2016, about two-thirds of the damage to crops was caused by floods, and almost 90% of the livestock loss was caused by drought [11]. The absence of the basic resources creates poverty and social tension, which only continue to decrease climate situation. According to the IPCC Climate Change and Land report, the most vulnerable to global warming are the groups with high demand for basic resources like food, feed, water that also lack technological advantage [12].

The growing demand for resources and services requires both the consumption of large volumes of available products and the increase in the agricultural industry. These measures have helped reduce the number of starving people and improve human health, but since 1961 consumption of food, timber, and energy has caused extreme and unprecedented rates of land and water use [13]. Agriculture, fishing, and forestry have been the three main components of a development strategy for countries for centuries. But all economic success is usually achieved due to the degradation of many ecosystems, as well as exacerbation of poverty in some social groups.

For a certain period of time, aggressive ecosystem use helped to improve the general standard of living of certain groups of people and avoid hunger. It led to a steady increase in the CO2 emissions caused by fossil fuels over the past two centuries. This increase was only interrupted by large-scale economic downturns such as recessions or oil crises [14]. However, this is not the way to develop in the 21st century. The social development issues in poor societies require a more complicated approach and decisions. While the intense natural resources extraction has its seeming economic advantages in the short term, such measures at the same time negatively affect other territories and groups of people, especially in the long run, because consequences of the ecosystem changes do not happen right away. A good example is the global temperature increase.

4 Discussion

The paper uses the works of many Russian and international scientists. However, the contribution of the latter is a significantly larger share. This might be due to the fact that, with all the achievements, Russian environmental scientific thought can be characterized by a more pessimistic approach towards the climate change issue partially because Russia belongs to the group of countries, which ensure their growth, mainly due to the raw materials industry. Thus, the topic of sustainable development in Russia is quite far from becoming the strategic goal number one for the development of the state which also explains the relevance of the present article. But, even though the Government of the Russian Federation at this stage is not the most interested in promoting the topic of solving environmental problems, this does not mean that the issue should not be discussed.

Climate change mitigation is a compound matter which is complicated by the existence of numerous constituents. It requires a deep analysis in order to define the right path for combating global warming, especially given that in order to successfully adapt or combat it; an immediate response has to take place.

People in all countries of the world should have access to vital goods, including healthy food, clean water, medical care, and clothing. The eradication of poverty is the UN Sustainable Development Goal №1 [15]. More generally, any progress made in eradicating poverty and hunger, improving health, will be in vain, and the environment is unlikely to be sustainable if the ecosystems on which humanity relies continue to degrade. Moreover, a situation is possible in which, due to the depletion of the planet's natural resources, the problem of providing residents with basic goods will not only remain but even get worse, especially in developing countries. Social, economic, and political problems that may arise due to overcrowding and lack of resources will have a negative impact on developed countries. For this reason, the identified problems and the responsibility for their solution are the responsibility of the entire world community. On the contrary, in this way, the sound management of ecosystems provides cost-effective opportunities to solve most environmental problems and achieve sustainable development.

The essence of the global environmental problem lies in the growing contradiction between the level of productivity of mankind and the state of stability of the natural environment. One of the main threats in case we maintain the current tendency to use the environment is the termination of the natural cycles of reproduction and self-purification of water, soil and atmosphere. The growing pressure of anthropogenic factors can lead to a rapid deterioration of the ecological condition of the Earth.

Already today, we can safely conclude that the environmental situation in the world has become critical. At both the national and global levels, evidence of environmental changes can be traced, such as:

1. An increase in the volume of carbon dioxide emissions into the atmosphere by 10–14% compared with the indicators of the late XX century (this substance is the main reason for the increase in global warming).
2. The release of tens of millions of tons of fuel combustion waste, as well as ash which often leads to the formation of acid rain that kills living organisms.
3. Pollution of the biosphere with various pesticides, radioactive waste, heavy metals, and, as a result, exceeding the allowable limits of air pollution by ten or more times.
4. Forest loss at a rate of 17 million hectares per year.
5. Soil erosion, resulting in the loss of one-fifth of the world's arable land.
6. Loss of diversity of flora and fauna.

Nowadays, even deforestation on the other side of the world carries a devastating blow, comparable to the attack of warships [9].

Thus, there is no simple solution to environmental problems, as they arise from the interaction of many others, including climate change.

Climate change from point of view of its impact on the global economy is not a simple large-scale natural hazard, but a powerful catalyst for multidirectional changes in many segments of the economy. Therefore, bound with the lack of freshwater, the food insufficiencies, natural disasters, migration, as well as the prospects for the development of a number of primary sectors such as energy, transport, construction, agriculture, climate change is closely twisted in a number of global economic processes [16].

Because the economic component of global climate change represents the growing harm to the global economy, as well as the increased costs of adaptation, it is necessary to create and improve economic policies related to the mitigation of climate change effects. These economic policies include the development of economic strategies that encourages the lessening of greenhouse gas emissions. While developing economic policy, it is necessary to take into consideration the features of the modern system of international relations [17]. According to the N. Stern's report "The Economics of Climate Change," published in 2016, when the temperature rises by 5–6 °C, the global GDP might descend by 14–15% [18].

While SDG 13 is supported by existing international law, its successful implementation also faces a number of obstacles. The success of SDG 13 as a whole, as well as that of the other SDGs, hinges to a great extent on the implementation of the Paris Agreement. While the Paris Agreement is applicable to all the State Parties that ratify or accept it, it is also a bottom-up instrument that relies on States establishing their own substantive targets in the form of Nationally Determined Contributions (NDCs), and its effectiveness relies on each Party's self-determined efforts. Its success thus requires the design of effective and coordinated policy, governance, and other legal efforts for its implementation at all levels, as well as on the individual ambitions of each Party.

Resource constraints affecting the State Parties may constitute another obstacle to the implementation of the SDG 13. Consequently, adequate and predictable climate financing, as recognized under the UNFCCC, the Paris Agreement, and SDG 13, will be critical for ensuring that the most vulnerable Parties can successfully adapt to the effects of climate change.

The global "carbon budget" may also exacerbate the issue of resource constraints, as governments and businesses will need to leave valuable fossil fuels in the ground in order not to exceed the maximum amount of carbon that may be released into the atmosphere while having a likely chance of limiting global temperature rise to 2 °C above pre-industrial levels. Efforts to remain within this budget raise questions about equity, including how to fairly allocate the rights to exploit remaining fossil fuel reserves. Fossil fuels are politically sensitive because many economic models rely heavily on fossil fuel revenues and recent discoveries of oil and gas in developing countries are expected to bolster budgets for socioeconomic development.

However, the effects of climate change are already being felt around the world, and urgent action is needed to adapt to and mitigate even greater impacts. Indeed, "the global nature of climate change calls for the widest possible cooperation by all countries and their participation in an effective and appropriate international response, with a view to accelerating the reduction of global greenhouse gas emissions." In spite of the international community's support for SDG 13, as well as other obligations under the UNFCCC and other international legal instruments, much depends on the effectiveness of the recent Paris Agreement and on the Parties' own efforts, innovations, and willingness to work together towards this goal.

Many scientists raised environmental concerns way before those issues have become established in current international political agendas. Many researchers around the world predicted and anticipated the change that was coming to life when this line of thought had not become common yet. It was in the middle of the 20th century when a famous Russian scientist Vladimir Vernadskiy has concluded that human influence on the environment can be compared to the geological epochs happening in shorter periods of time [19]. There are so many factors contributing to ecological problems in the world today, but a considerable number of global issues have anthropological origins, i.e., they are caused by human activities.

With all the scientific information collected and analyzed one would wonder where to start and how to settle all the issues since all the global challenges are so interlinked. The climate change problem requires a complex and all-inclusive solution. That is why the concept of sustainable development serves today as the main environmental strategy of the 21st century. According to it, the social, environmental, and economic areas have to be developed simultaneously and in a very particular way.

In such way of development the current economic and technological development of a state is consistent with the environmental situation and provides future growth at the same time [9]. The term itself came to use in 1980 when the International Union for the Conservation of Nature and Natural Resources introduced the World Conservation Strategy with the "common goal of achieving sustainable development through the conservation of living resources". The United Nations Environment Program (UNEP) later cultivated the sustainable development concept. The organization itself was created thanks to the efforts of participants on the Stockholm Conference in 1972. The Program's mandate focuses on a variety of areas and spheres, including the atmosphere, marine, and terrestrial ecosystems, green economic development, and environmental governance, and it popularized the concept of sustainable development into one of the biggest environmental strategies. According to the UNEP, sustainable development included key ideas such as help for the poor; the ideas of independent development in conditions of limited natural resources and of cost-effective development using non-traditional economic criteria; health monitoring, as well as equal access to technology, food, clean water and shelter for all; and finally, the need to implement initiatives aimed at improving people's lives [20, 21, 22, 23].

And for the benefit of future generations the purpose of environmental sustainability is to promote global life support systems permanently. In the most familiar interpretation of this conception, environmental sustainability is a desirable attribute of any model or method of human activity that is essential part of the development process.

5 Conclusion

According to scientific data, with a global warming of 1.5–2 °C, there is a threat of risks to natural, as well as socio-economic systems. Such risks are expressed in different units, and this makes it difficult to objectively assess the prevention of damage while maintaining a low level of warming. Limiting warming to 1.5 °C is possible, but this will require an immediate reduction of at least half of global anthropogenic $CO2$ emissions by 2030 and bringing them to zero by mid-century. At the same time, anthropogenic emissions of other greenhouse gases will also require a significant reduction.

In recent years, the concept of sustainable development has become the basic ideology for humanity in determining the future of the world. In fact, it is the official installation of world development in the 21st century. Along with numerous theoretical and scientific studies, the concept of sustainable development has become the basis for practical decisions by international organizations, many countries in the development of long-term national programs.

The reason why the concept of sustainable development seems the most suitable for solving global problems and, in particular, mitigating climate change is because it acts as a complex and multiarea measure that gradually eliminates highly interlinked problems that influence each other.

The UN system for organizing major conferences on environment and development has been criticized based on the question of what is reasonable to expect from global conferences. In best-case scenario, UN forums can help identify and manage economic, social, and environmental programs and establish the foundation for enhanced collaboration. As the countries committed to the Paris Agreement, they took responsibility to make changes on national levels and make their Nationally Determined Contributions. This is one of the most crucial elements of climate change mitigation, as it depends on international cooperation.

Complicated nature of climate change includes the high level of influence between the global problems and requires a seriously all-inclusive approach. Modern researchers find the concept of sustainable development suitable because, according to it, the social, environmental, and economic spheres have to be developed simultaneously and in a precise way when the economic and technological development of a country is consistent with the environmental situation and also ensures growth. Survival and continuous progress of humans are dependent on the health of the Earth's life support systems.

However uneven economic development, diversification of the problem of sustainable development, geopolitical diversity, as well as different management potential complicate its worldwide implementation. For developing countries, economic growth is impossible without external assistance and a high level of international cooperation.

UN has been the most active element of the evolution of environmental protection, and the crucial aspect of its activity is the conference system. Stockholm Conference in 1972 and Earth Summit in 1992 became groundbreaking, laying down the foundation for international cooperation and initiating the institutionalization of environmental protection in national legislations. These summits pushed the idea of a developing world consisting of states with shared environmental issues. However, they also raised fears among developing countries that saw potential restrictions and limitations to their industrialization. Not many further conferences achieved such progress; however, it is crucial to understand that at best, UN conferences can help identify and manage economic, social, and environmental programs and lay the foundation for collaboration, but the actual change should be happening on national levels.

United Nations Summit on Sustainable Development in New York and Climate Change Conference in Paris in 2015 pushed the issue and resulted in the adoption of the UN 2030 Agenda and the all-new 17 SDGs with the Goal 13 for Climate Action and of the Paris Agreement that established a system for informing, recording, and reporting greenhouse gas emissions of participating countries and ensuring fulfillment of their obligations. After that, four Climate Change conferences were held but did not show great results. Moreover, the last Climate Change conference in Madrid became famous for its distinct political inactivity in that it applied Rule 16 of the provisional UNFCCC rules of procedure 15 times, allowing any unresolved issues to be transferred to the next COP, thus, not motivating the countries to solve them.

Nevertheless, the Paris Climate Agreement launched the newly modified process of achieving a sustainable future. States parties agreed to try and keep the global temperature increase at 1.5C or lower compared to the preindustrial levels providing a "nationally determined contribution" (NDC) in limiting their emissions. The main goal is to set the changes on the levels of national legislatures; however, the national interests of the main international actors do not allow the economies to be modified easily. For example, many Arctic states like Russia or Norway see melting Arctic glaciers as an opportunity for gold, diamonds, rare earth metals, petroleum, natural gas, fish, access to potentially cost-saving new shipping lanes.

There is no doubt that in the conditions of economic crisis and ongoing COVID-19 pandemic, governments do not prioritize climate change fight. However, it is important not to overlook the problem and act when it might be even more economically reasonable to start rebuilding national systems according to the terms of the green economy.

Another important factor is the national interests. As the US exited Paris Agreement, there have been doubts about the potential success of the treaty with a loss of one of the biggest carbon dioxide polluters. However, if the other major international actors will keep up with their promises, there will not be any other way for the US except for the acceptance of the green economy. Moreover, with the US Presidential election this year, the balance of power can change in favor of environmentalists [24].

Also, the withdrawal of the US from the treaty does not deny its states of the right to follow the main guidelines of the Paris Agreement, and that will depend on the public opinion. As we had already observed the power of an individual in setting the agenda when Greta Thunberg made a speech at the UN, a similar trend can develop among the younger generation in the US.

A lot of existing potential has been realized in the process of combating climate change. There has appeared the first precedent of the successful court case against the government of the Netherlands about the right of citizens for the environment. This is also a highly remarkable stage of environmental evolution, but what the world needs to realize is that the scale of the climate change problem is huge. Even with a practically complete stop of human lives during the recent COVID-19 quarantine, self-isolation, and pause for almost every important sphere of human activity, the carbon dioxide drop during this period was not more than about 10% from the target goal in order to keep global warming under 2 °C.

References

1. Sakwa, R.: The pandemic, Russia and the west. Vestnik Saint Petersburg Univ. Int. Relat. **14**(1) (2021). https://doi.org/10.21638/spbu06.2021.101
2. Ermolina, M.A., Matveevskaya, A.S., Pogodin, S.N.: Efficiency of the UN action in the area of transport, environmental protection, health. MATEC Web Conf. **239** (2018). https://doi.org/10.1051/matecconf/201823904016
3. Kichigin, N.V., Khludeneva, N.I.: Focus on climate's law of the earth. Focus na klimaticheskiy zakon zemli (In Russian), pp. 153–161 (2016)
4. Ermolina, M.A., Matveevskaya, A.S., Matyashova, D.O., Kovalevskaya, N.V.: Priorities for sustainable urban development as exemplified by individual Asian countries. E3S Web Conf. **157** (2020). https://doi.org/10.1051/e3sconf/202015703013
5. Kram, T., Stehfest, E.: The IMAGE model suite used for the OECD Environmental Outlook to 2050. PBL Netherlands Environmental Assessment Agency (2012)
6. Adams, W.M.: Green Development: Environment and Sustainability in a Developing World. Routledge, London (2009)
7. Jackson, T.: Prosperity Without Growth. Routledge, London (2009)
8. Goodland, R.: The concept of environmental sustainability. Annu. Rev. Ecol. Syst. **26**, 1–24 (1995)
9. Jaiswal, R.K., Lohani, A.K., Tiwari, H.L.: Development of framework for assessment of impact of climate change in a command of water resource project. J. Earth Syst. Sci. **129**(1), 1–20 (2020). https://doi.org/10.1007/s12040-019-1328-x
10. Huang, M., Ding, L., Wang, J., Ding, C., Tao, J.: The impacts of climate change on fish growth: a summary of conducted studies and current knowledge. Ecol. Ind. **121** (2020). https://doi.org/10.1016/j.ecolind
11. Ma, J., Wang, P.: Effects of rising atmospheric CO2 levels on physiological response of cyanobacteria and cyanobacterial bloom development: a review. Sci. Total Environ. **754** (2020). https://doi.org/10.1016/j.scitotenv
12. Koch, K., Ysebaert, T., Denys, S., Samson, R.: Urban heat stress mitigation potential of green walls: a review. Urban For. Urban Green. **55** (2020). https://doi.org/10.1016/j.ufug
13. Biermann, F.: Curtain down and nothing settled: global sustainability governance after the 'Rio+20' Earth Summit. Environ. Plann. C **31**, 1099–1114 (2013). https://doi.org/10.1068/c12298j

14. Hickmen, T.: Voluntary global business initiatives and the international climate negotiations: a case study of the greenhouse gas protocol. J. Clean. Prod. **169**, 94–104 (2017)
15. Jung, J., Petkanic, P., Dongyan, N., Kim, J.H.: When a girl awakened the world: a user and social message analysis of Greta Thunberg. Sustainability **12**, 1–17 (2020)
16. Makarov, I., Stepanov, I.: The Paris Climate Agreement: Impact on the World Energy Sector and Challenges for Russia. Actual Problems of Europe, 1 (2018).
17. Fukuda-Parr, S., Hulme, D.: International norm dynamics and the 'end of poverty': understanding the Millennium Development Goals. Glob. Governance Rev. Multilater. Int. Organ. **17**, 17–36 (2011)
18. Kokorin, A.O., Kuraev, S.N., Yulkin, M.A.: Review of Nicholas Stern's Report "The Economics of Climate Change" (2009)
19. Vernadskij, V.I.: Scientific life as a planetary phenomenon. Nauchnaiya zhizn' kak planetarniy phenomen (In Russian) (1991)
20. Wood, S.: Sustainability in International Law. EOLSS. International Sustainable Development Law, I (2001)
21. Moss, N.: The Politics of Global Warming Environmental Sciences (1991)
22. Avdeeva, T.G.: Prospects for international negotiations on climate change. Perspectivy mezhdunarodnih svyazey po izmeneniyu klimata (In Russian) (2010)
23. Bolgov, R.V., Ermolina, M.A., Vasilyeva, N.A.: Open budget effects for urban development: Russia's cases. In: ACM International Conference Proceeding Series. 3rd International Conference on Electronic Governance and Open Society: Challenges in Eurasia, EGOSE 2016, pp. 184–188 (2016). https://doi.org/10.1145/3014087.3014116
24. Fominykh, A.E., Yarygin, G.O.: Greening of the US public diplomacy. Vestnik Saint Petersburg Univ. Polit. Sci. Int. Relat. **1**, 110–120 (2016)

Sustainable Development in BRICS Countries: From Concept to Practice

Jingcheng Li(✉)

Peter the Great St. Petersburg Polytechnic University, St. Petersburg, Russia

Abstract. As a prospective international association, BRICS (Brazil, Russia, India, China, South Africa) plays an important role in the economic development of the world and in the achievement of UN sustainable development goals. The active pursuit of sustainable development leads to fundamental changes in lifestyles and economic relations not only in the BRICS countries, but throughout the world. This article provides an analysis of the main stages in the formation of the sustainable development concept within the UN and its impact on the formation of the BRICS sustainable development discourse. The study also identifies approaches to the implementation of several essential fields of Sustainable Development Goals (SDGs) in the BRICS countries, namely poverty alleviation, education and health, environmental protection and green economy, which are the core activities of these countries at the current stage of development. Faced with the global challenges of the financial, environmental and climate crises, as well as major qualitative changes in the scientific and technological revolutions, the BRICS is increasingly seeking to transition to sustainable development, the aim of which is to eliminate inequalities between social strata, strengthen internal stability and expand international cooperation.

Keywords: BRICS · United Nations · Sustainable development · SDGs · Poverty alleviation · Environmental protection · International cooperation

1 Introduction

The unbalancing of the interaction of the "two worlds" – natural and artificial – is one of the essential problems of all mankind, and therefore the search for a solution should be a unifying beginning for all states [22]. Back in the XX century, the United Nations proposed the sustainable development concept, which has become a key strategy for development in the XXI century.

Under the auspices of the United Nations, the UN Conference on Environment and Development (UNCED) was held in Rio de Janeiro in 1992, as known as the Earth Summit. The UNCED adopted several important documents: Rio Declaration on Environment and Development, Agenda 21, Convention on Biological Diversity, UN Framework Convention on Climate Change (UNFCCC). It was at the Earth Summit that the goals and action plans for sustainable development were formulated for the first time.

© The Author(s), under exclusive license to Springer Nature Switzerland AG 2021
R. Bolgov et al. (Eds.): *Proceedings of Topical Issues in International Political Geography*, SPRINGERGEOGR, pp. 238–248, 2021.
https://doi.org/10.1007/978-3-030-78690-8_21

Rio Declaration has received a positive response worldwide, and most countries and international organizations have embarked on intensive discussions on sustainable development. The Chinese literature interprets sustainable development as a vital issue of modern society and harmony between man and nature. Sustainable development stresses that social and economic development should be coordinated with nature conservation and strive for harmony between man and nature. Development should not be carried out through the destruction of the environment, but should not only adapt to existing natural conditions, but also consider the interests of future generations and follow the path of sustainable development [11].

The leading principles of the Rio Declaration were the main challenges of the modern community, identified in the *Our Common Future* (*Brundtland Report*), prepared by the World Commission on Environment and Development in 1987: population and human resources, food and energy security, protection of ecosystems, industrial transformation, urbanization development and etc. [18]. The concept of sustainable development implies revolutionary changes in human society. As defined in *Our Common Future*, sustainable development means "development that meets the needs of the present without compromising the ability of future generations to meet their own needs [17]."

The World Summit on Sustainable Development (WSSD) in Johannesburg (2002) and the UN Conference on Sustainable Development – or Rio+20 (2012) reaffirmed the continuity of the transition to sustainable development. However, this course was fraught with several challenges. On the one hand, environmental scientists point to the harmfulness of the current pace of economic development to the natural world, and on the other hand, economists, especially from developing countries, argue that without intensified social and economic development, most of humanity remains in poverty. As Russian researcher V. Gorbacheva noted, sustainable development should reflect economic efficiency and social justice in the formation of a unified strategy of "socio-natural" development in order to reduce the "anthropogenic burden on the biosphere [10].

It should be noted that humankind has made great progress in various aspects of sustainable development, but this progress does not fully meet the requirements set by relevant international standards. Therefore, the concept of sustainable development should represent a balanced human development plan in the context of harmonizing the relationship between the artificial and natural worlds.

Sustainable development is necessary for all countries of the world and has its own priorities in each country. The most successful results in achieving the Millennium Development Goals (MDGs) set by the United Nations in 2000 were achieved with the strong support of social policies in China and India, which were able to significantly reduce poverty. It is important to note the decisions taken at the UN Sustainable Development Summit 2015, in particular the endorsement of *2030 Agenda for Sustainable Development*. It defines 17 Sustainable Development Goals (SDGs) covering three aspects: social, economic and environmental. The 2030 Agenda was the first policy document historically adopted by all UN member states in the field of sustainable development aimed at the development of all countries and international cooperation until 2030.

The concept of sustainable development has continued to evolve over the past 30 years, during which many international documents have been concluded. For instance, in late 1997 the Kyoto Protocol was concluded under the UNFCCC, which aims to limit greenhouse gas emissions in industrialized countries. However, the subsequent Copenhagen Accord and the Paris Agreement on measures to combat global warming were not fully implemented due to the withdrawal of the parties. It seems that the formation of unified global economic policy is unlikely to be possible in the near future. The contradictions in the development strategies of developed and developing countries clearly illustrate the unfavourable practical results.

In this world political environment, an important role in achieving Sustainable Development Goals can be played by the BRICS, as together these countries are home to more than half of the world's population, and thus the responsibility for the future of the planet to a large extent lies with the people of these countries. President of Russia V.V. Putin at the end of Xiamen BRICS summit in 2017, said: "BRICS is a very promising association <...>, and above all, it is associated with the structure of the economy and with our common desire to make the world economy more fair and noble [16]."

The participation of such major players as the BRICS countries – Brazil, Russia, India, China, South Africa in world global governance is one of the significant political and economic factors of modern world politics. Among the political principles, BRICS put the importance of organization in global governance to address global issues, namely sustainable development and ensure the implementation of its goals. These principles are reflected in BRICS agenda, national foreign policies and economic strategies of these five countries.

An important feature of the BRICS rapprochement is the common strategic desire to implement the concept of sustainable development. At the same time, each of the countries seeks to improve economic performance without predatory exploitation of natural resources and damage to the environment and ecology. The BRICS has established close economic, trade and financial cooperation, contributing to the stable development of both the five countries and many developing countries in general.

The aim of the article is to determine the implementation of the sustainable development concept in the BRICS. The study examined the national strategies for sustainable socio-economic development of the BRICS countries, while analyzed the results in achieving vital sustainable development fields: poverty alleviation, quality of education, good health and well-being, ecological protection, green economy. Discussion solutions were proposed to improve the effectiveness of cooperation between the BRICS countries in the direction of sustainable development and prospective forms of strengthening the internal sustainability of the BRICS.

Literature review on the topic under study. The role of BRICS in world politics and economics is widely studied by scientists and researchers from different countries. Theoretical provisions on the problems of constructing the BRICS discourse on sustainable development and evaluation of the BRICS activities in this field are contained in studies by Q. Zhao [27], W. Bai [3, 4], H. Tian [20], S. Bobylev and L. Grigoriev [5], N. Khmelevskaya [13], V. Gorbacheva [10], S. Ali et al. [1]. The conceptual researches and philosophies of sustainable development are contained in scientific works of G. Toloraya [21], C. Hu [11], N. Vasilyeva [22] and others.

2 Method of Research

Methodological basis of research is the application of combined scientific methods. Understanding and application of the sustainable development concept in the BRICS has become the theoretical basis for this article, which is determined by comprehensive sustainability in the socio-economic context of the BRICS countries. The historical method used to study the conceptual evolution of sustainable development is methodologically valuable. It is based on the analysis of UN global conferences and several international agreements. In building a discourse on sustainable development of the BRICS, a systematic analysis has been applied, which has made it possible to characterize the national specificity in the field of sustainable development. Comparison was used to identify the specific features of the BRICS countries in achieving some key objectives of sustainable development (SDGs) based on statistical data and measures countries on certain indicators.

3 Results of Research

3.1 BRICS Sustainable Development Discourse

The BRICS Political Parties, Think-Tanks and Civil Society Organizations Forum 2017 published a document entitled *"Think-Tanks + Finance" working for BRICS Development: BRICS Sustainable Development Report 2017*, compiled by the Centre for Financial Research & Development at China Development Bank and the China Council for BRICS Think-Tank Cooperation. This document proposed a BRICS Cooperation Development Index in terms of economic, trade and investment lines. The Report provides a systematic analysis of the BRICS contribution to the world economy, their economic development and the major challenges they have faced over the past decade and presents the concepts and experiences of China's development finance institutions [8]. And in January 2020, the *BRICS Sustainable Development Report 2018* was released. The main feature of the newest Report is that, in the context of South-South cooperation, the BRICS innovative investment and financing mechanism is considered a starting point for expanding cooperation, fully recognizing the importance of the investment and financial strategies of the BRICS countries.

BRICS countries have developed national policies in accordance with UN Sustainable Development Goals. The fight against poverty and the improvement of living standards have come to the top of national sustainable development priorities.

In Brazil. As noted in the previous part, Brazil played an important role in building international consensus on the sustainable development concept: Brazil hosted the 1992 Earth Summit and Rio+20 in 2012. These conferences have made significant contributions to sustainable development around the world. In its vision of achieving sustainable development, Brazil believes that the eradication of absolute poverty and hunger, access to health care and gender equity are at the core of its activities. Brazil also places high emphasis on sustainable infrastructure and financial sector of the national economy, which is as important as poverty alleviation and public health. As an energy-rich country, Brazil is working to develop clean and renewable energy sources and to preserve the environment. These provisions, which aim at achieving sustainable development, represent Brazil's commitment to achieving the goals set out in UN 2030 Agenda.

In Russia. At present, Russia is focusing on a series of official documents related to the implementation of the UN 2030 Agenda or sustainable development as a whole in the long term: the Federal Law *On Strategic Planning in the Russian Federation*, which defined the strategic goals and priorities of public policy in relation to social and economic development. In addition, Presidential Decree No. 204 *On National Goals and Strategic Objectives of Development of the Russian Federation until 2024, Strategy of Socio-Economic Development of the Russian Federation, Strategy of Scientific and Technological Development of the Russian Federation, Economic Security Strategy until 2030, The Forecast of Socio-Economic Development of the Russian Federation until 2036* and other fundamental documents have been developed. According to the *Human Development Report in Russian Federation 2016,* prepared by Analytical Centre under the Russian Government, "Sustainable development strategy for Russia with regard to Sustainable Development Goals until 2030" needs to be formulated, which "can be developed and built into Russia's strategic planning system, corresponding both to the need for balanced social and environmental development and to the country's international obligations [5, p. 12]."

In India. India is actively engaged in the implementation of the UN Sustainable Development Goals. Many SDGs are particularly important for India's economic and overall human development in the long term. India's main vision for sustainable development is to fight poverty and support low income population, which has become the focus of India's efforts to achieve the balance between industrial economic development and ecological protection. Both India and Brazil see the following SDGs as the main principles to be achieved by 2030: Goal 1: End poverty in all its forms everywhere, Goal 1: End hunger, achieve food security and improved nutrition and promote sustainable agriculture, Goal 3: Ensure healthy lives and promote wellbeing for all at all ages, Goal 5: Achieve gender equality and empower all women and girls, Goal 9: Build resilient infrastructure, promote inclusive and sustainable industrialization and foster innovation, Goal 14: Conserve and sustainably use the oceans, seas and marine resources and Goal 17: Revitalize the global partnership for sustainable development achievement [23, 24]. The two countries have already made significant progress in these activities.

In China. In recent years, China has paid great attention to build the "ecological civilization" and "Beautiful China". As President of China Xi Jinping stated at the National Conference on Environmental Protection (2018), "A civilization may thrive if its natural surroundings thrive and will suffer if its natural surroundings suffer. The natural environment is the basis of human survival and development, and changes to it directly impact the rise and fall of civilizations [26]." In 2016, *China's National Plan on Implementation of the 2030 Agenda for Sustainable Development* had been released, setting a timeline for achieving the most crucial SDGs for China. Firstly, the alleviation of absolute poverty (by existing standards) by the year 2020. The main indicators in this area are the eradication of poverty and hunger, access to health infrastructure for women and children, and access to housing. Secondly, by focusing on sustainable development in areas such as agriculture, health, public education and economic growth, these goals will be largely achieved by 2030 [9].

In South Africa. In 2012, South Africa released the *National Development Plan 2030: Vision 2030 – Our future, make it work* in order to accelerate economic and social transformation and to ensure that by 2030, South Africa will have eliminated poverty, reduced inequality, increased economic power and strengthened national capacity. South Africa will be able to comprehensively promote solidarity and cooperation in society. The Plan aims at creating a larger labour force, increasing investment in public infrastructure, improving national health system, enhancing the level of education. Furthermore, it seeks to improve labour relations, promote industrial diversification, etc. In 2013 South Africa promoted the adoption of *Agenda 2063 – "The Africa We Want"* within the African Union, which seeks to build a prosperous, inclusive and sustainable Africa. Mainly in order to achieve the SDGs, South Africa has focused on job creation, poverty alleviation, reduction of inequality and growth of inclusive economies by 2030.

3.2 BRICS Implementation of the SDGs

The BRICS countries, like other developing and emerging countries, face the challenge of making sustainable development strategies the basis for addressing a wide range of issues. While developed and developing countries have different social and economic conditions, therefore BRICS attaches high priority to the qualitative design and implementation of the SDGs. First of all, this is a set of measures to control anthropogenic impacts on climate change. Much has been said about this during the BRICS Summits, which resulted in the development of green economy based on the promotion of high-tech energy efficient and resource-saving solutions. This part is devoted to analysing the results of certain development goals in the areas of poverty reduction, education and health, environment and green growth.

Goal 1: No Poverty. Poverty as a social phenomenon is being defined differently at the global level. It is usually measured by the Human Development Index (HDI), Gini coefficient and the GDP per capita. Equality of society is an essential value of the SDGs. As noted by Chinese expert Bai Weijun, "To establish long-term cooperation in the fight against poverty, BRICS countries need to ensure stability and economic development first [4]."

According to the *Human Development Report 2019* of United Nations Development Programme [12], Russia remains the only country among the BRICS countries listed in the category of "very high human development", its HDI was 0.824 (49th place in the world). "High human development": Brazil – 0.761 (79th place), China – 0.758 (85th place) and South Africa – 0.705 (113th place). India was ranked as the "medium human development" with an index of 0.647 (129th place). On the one hand, these figures reveal that the BRICS countries have much chance to improve in terms of human development, and on the other hand, it shows the realization of national Sustainable Development Goals.

Goal 10: Reduced Inequalities. The Gini coefficient is mainly used as analytical indicator for measuring the difference in income distribution among the population. According to *BRICS Joint Statistical Publication 2018*, the Gini coefficient is 0.549 (2017) in Brazil, 0.412 (2016) in Russia, 0.367 in urban and 0.280 in rural of India (2011–2012), 0.467 (2017) in China, 0.639 (2015) in South Africa [7, p. 7]. The figures indicate that the overall gap between the nominally "rich" and "poor" is widely spread in BRICS countries. Further work is necessary to find a solution to this serious social problem and to monitor the reasonable gap between the strata of society.

Goal 8: Decent Work and Economic Growth. For BRICS countries it is important to develop comprehensive sustainable development cooperation, primarily in the economic, social and ecological spheres. According to the World Bank data for 2018 [25], GDP per capita in five BRICS countries was in average $8,008. The annual GDP per capita growth in Brazil was 0.5%, Russia – 2.3%, India – 5.7%, China – 6.1%, South Africa – -0.6%. By comparison, the world average GDP per capita growth in 2018 was 1.9%, OECD – 1.7%. Russia, India and China have maintained relatively robust and sustainable growth among the BRICS countries.

In the opinion of Russian expert N.G. Khmelevskaya, due to the redistribution of GDP the economic increase aimed at poverty alleviation in the BRICS countries may not reflect on the real incomes of most people [13]. However, Bai Weijun approaches this issue in a different way: in the fight against poverty China and India are playing the role of "suppliers of goods and services", while Brazil, Russia and South Africa are acting as "suppliers of raw materials and resources". The BRICS countries have mutually beneficial advantages in terms of supply and demand, thereby increasing the number of jobs and boosting incomes [3].

Goal 3: Good Health and Well-Being. The BRICS countries are constantly increasing the health budget. According to recent data [7, p. 6]: Brazil spends 4.9% of GDP on health (2013), Russia – 3.1% (2017), India – 1.2% (2016), China – 6.2% (2016) and South Africa – 4.2% (2016). The increase in the budget financing of public health has impact on the scientific-technical and infrastructural development of medicines, efficiency of infection control, improvement of health care systems, as well as on the improvement of the quantity and quality of medical staff. China, Brazil and South Africa have devoted relatively large resources to medicine and health care. However, India needs substantial public support to address the lack of resources in health care.

Goal 4: Quality of Education. In recent years, most of BRICS countries have experienced dynamic economic development and have increased their budgets in key areas of society such as health and education. Government support in these areas contributes to improving literacy and living standards (which is particularly relevant for China and India, home to more than billion people), and to increasing life expectancy and internal stability in long-term. BRICS public expenditure on education (percentage of GDP): Brazil – 5.7% (2014), Russia – 3.5% (2017), India – 2.7% (2017), China – 4.2% (2016) and South Africa – 6.9% (2016) [7, p. 6]. By comparison, OECD members spend in average 4.5% of GDP on education (2016) [15]. Spending on education in Brazil, China and South Africa generally corresponds to the developed country average.

Goal 7: Affordable and Clean Energy, Goal 13: Climate Action. As the Executive Director of the Russian National Committee on BRICS Research G. Toloraya noted, "An unnatural pattern of consumption has emerged that does not correspond to the real needs of the people at all [21]." Over the past decade, BRICS countries have emphasized sustainable development and have made strong efforts in such areas as the search for new sources of clean energy and environmental protection and restoration.

However, as the world's largest population and developing countries, China and India, their needs for energy and natural resources are huge, yet historically, pollution and ecological destruction have been serious. According to *BP Energy Outlook 2030*: by 2030, net growth in global coal demand will be provided by China and India, these two countries account for 63% and 29% of global coal growth respectively, while the Chinese market alone already accounts for 25% of global gas demand growth. China remains the largest consumer of coal, and India will overtake the United States in 2024 in terms of energy consumption [6]. BP experts noted that the current structural transformation in China's economy will lead to lower energy demand, especially after 2020. Energy demand is growing rapidly in South Africa and Brazil.

The question of how to balance resource consumption and environmental protection is therefore becoming increasingly important, especially, for developing and emerging countries. Chinese expert Zhao Qingsi believes that the present-day world climate change and low-carbon economy are the main challenges to global development. New and renewable energy sources will determine the division of labour in the new cycle of the industrial revolution [27].

Goal 9: Industry, Innovation, and Infrastructure. The BRICS countries have adopted strategies for clean technology and industry to ensure economic growth based on sustainable natural resources. It is noteworthy that the New Development Bank (NDB) as an international financial institution supports investments in prospective sustainable projects within the BRICS countries and externally. The priorities of NDB activities are development of renewable, clean and new energy, investment in urbanization and infrastructure, environmental protection, social welfare, health and education, etc. Thus, the BRICS will be able to implement Sustainable Development Goals also through the NDB.

4 Discussions of Research

The BRICS countries are strong supporters of the UN sustainable development initiative. As stated in the *BRICS-Summit - Fortaleza Declaration* (2014): "We stand for strengthening international cooperation to promote renewable and clean energy and to universalize energy access, which is of great importance to improving the standard of living of our peoples [19]." Furthermore, at the BRICS Xiamen Summit in 2017, one of the central themes was sustainable development.

For BRICS, Sustainable Development Goals are key strategic development concerns. Each country has specific approach to sustainable development based on its historical development path and socio-economic characteristics [14]. As noted by Russian expert

N.G. Khmelevskaya, the scale of resource consumption, the main directions of technical development and structural changes in the BRICS economies have their own values and impact on the implementation of SDGs by 2030 [13].

While actively participating in global governance and international financial system, BRICS countries face serious challenges to sustainable development – economic growth, structural change, social welfare, pollution, climate change, etc. Many of them will remain for a long time [20]. Within the framework of various UN programmes, BRICS has made significant progress in implementing Sustainable Development Goals in such areas as poverty alleviation, education and healthcare, environmental protection and green development.

However, without the efforts to implement the sustainable development concept, there could be no constructive changes in the harmonization of artificial and natural environments. As pointed out by Russian researcher L. Grigoriev, "the BRICS countries have significant differences in production, consumption and environmental situation, therefore, in the development and implementation of national strategies, each state should be guided by its own priority goals in terms of the current level of development [2]." In this process the BRICS countries are not only achieving national SDGs, but also formulating the comprehensive discourse and vision of sustainable development. The model of sustainable development can serve as an example for many other developing and emerging countries. In this regard, BRICS needs pay attention to further cooperation on sustainable development issues to achieve the common implementation of the SDGs [1].

5 Conclusion

In conclusion, the implementation of 2030 Agenda for Sustainable Development is relevant to national development realities and conditions. Each country has different measures to eliminate poverty and inequality in society, strengthen stability and foster market development. As one of the largest associations of the emerging countries in the world, BRICS is actively pursuing Sustainable Development Goals and has made major efforts in such areas as poverty alleviation, improvement of education and health, emission reduction and environmental restoration, innovative industrialisation.

BRICS has also consistently committed to moving towards sustainable development internationally by jointly working with the United Nations. During BRICS Summits and meetings at various levels, five countries are engaged in international cooperation based on sustainable development principles. Participation of BRICS in the implementation of 2030 Agenda presents expectations, as BRICS concentrates the Earth's major resources and accounts for about half of the world population. In the face of the double challenges associated with the international financial and climate crisis as well as the forthcoming scientific and technological revolutions, the BRICS countries are called upon not only to ensure sustainable development, but also to contribute to the economic transformation, attaching importance to the harmonization of artificial and natural world.

References

1. Ali, S., Hussain, T., Zhang, G., Nurunnabi, M., Li, B.: The implementation of sustainable development goals in "BRICS" countries. Sustainability **10**, 2513 (2018)
2. Analytical Centre under the Government of the Russian Federation: Cooperation between BRICS countries will help implement the SDGs (in Russian). https://ac.gov.ru/news/page/sotrudnicestvo-stran-briks-pomozet-vypolneniu-cur-21810. Accessed 1 Jan 2020
3. Bai, W.: A comparative study of anti-poverty policies in the BRICS. Mod. Econ. Res. **12**, 88–92 (2012). (in Chinese)
4. Bai, W.: Study on the BRICS anti-poverty cooperation mechanism. Reform Econ. Syst. **1**, 149–153 (2013). (in Chinese)
5. Bobylev, S.N., Grigoriev, L.M. (edi.): Human Development Report for 2016 in Russian Federation, p. 298. Analytical Centre under the Government of the Russian Federation, Moscow (2016). (in Russian)
6. BP: Energy Outlook 2030. https://www.bp.com/content/dam/bp/business-sites/en/global/corporate/pdfs/energy-economics/energy-outlook/bp-energy-outlook-2013.pdf. Accessed 1 Jan 2020
7. BRICS Joint Statistical Publication 2018/Statistics South Africa. Statistics South Africa, Pretoria, p. 290 (2018)
8. China Development Bank Project Team et al.: "Think-Tanks + Finance" working for BRICS Development: BRICS Sustainable Development Report. Dev. Finan. Res. **5**, 73–87 (2017). (in Chinese)
9. China's National Plan on Implementation of the 2030 Agenda for Sustainable Development. www.chinadaily.com.cn/specials/China'sNationalPlan(EN)(1).pdf. Accessed 6 Nov 2019
10. Gorbacheva, V.: BRICS: strengthening of partnership for the cloudless future. Rossiyskaya Gazeta, Spec. Iss. **246**, 7412 (2017). (in Russian)
11. Hu, C.: Sustainable development - new developments in materialist social-historical synergy theory. Forward Position **11**, 34–36 (2010). (in Chinese)
12. Human Development Report 2019. http://hdr.undp.org/sites/default/files/hdr2019.pdf. Accessed 10 June 2020
13. Khmelevskaya, N.G.: Contours of the BRICS dialogue on sustainable development in the realities of foreign trade relations. Bull. Int. Organ. **4**, 74–95 (2018). (in Russian)
14. Li, J., Pogodin, S.: "Made in China 2025": China experience in Industry 4.0. In: IOP Conference Series: Materials Science and Engineering, vol. 497, p. 012079 (2019)
15. OECD Family Database: Public spending on education (Public policies for families and children). https://www.oecd.org/els/soc/PF1_2_Public_expenditure_education.pdf. Accessed 1 Jan 2020
16. Putin called BRICS a promising association (in Russian). https://ria.ru/20170905/1501767809.html. Accessed 1 Jan 2020
17. Report of the World Commission on Environment and Development: Our Common Future. https://sustainabledevelopment.un.org/content/documents/5987our-common-future.pdf. Accessed 10 June 2020
18. Rio Declaration on Environment and Development. https://www.cbd.int/doc/ref/rio-declaration.shtml. Accessed 1 Jan 2020
19. Sixth BRICS Summit – Fortaleza Declaration. https://www.mea.gov.in/bilateral-documents.htm?dtl/23635/Sixth+BRICS+Summit++Fortaleza+Declaration. Accessed 10 June 2020
20. Tian, H.: Priority areas and policy options for BRICS cooperation on sustainable development. Int. Econ. Cooperation **8**, 19–23 (2017). (in Chinese)
21. Toloraya, G.: Sustainable Development and BRICS. Problems of strategy development for Russia (in Russian). http://nkibrics.ru/posts/show/53fce0c36272697ee4050000. Accessed 1 Jan 2020

22. Vasilyeva, N.A.: UN: necessity of ecological management. Bull. Tula State Univ. Hum. **1**, 139–149 (2010). (in Russian)
23. Voluntary National Review Report on Implementation of Sustainable Development Goals. India 2017. https://niti.gov.in/writereaddata/files/India%20VNR_Final.pdf. Accessed 10 June 2020
24. Voluntary National Review Report on Implementation of Sustainable Development Goals. Brazil 2017. https://sustainabledevelopment.un.org/content/documents/15806Brazil_English.pdf. Accessed 10 June 2020
25. World Bank. GDP per capita (current US$). https://data.worldbank.org/indicator/NY.GDP.PCAP.CD. Accessed 10 June 2020
26. Xi, J.: Pushing China's development of an ecological civilization to a new stage. QiuShi J. (Engl. Ed.) **2**, 3–23 (2019)
27. Zhao, Q.: BRICS and global energy governance: roles, responsibilities and pathways. Contemp. World Soc. **1**, 147–152 (2014). (in Chinese)

The Development of Cooperation Between Member States of the Arctic Council 1996–2019: Environmental Partnership or Struggle for Resources

Anastasiia V. Sboychakova(✉)

Saint-Petersburg State University, Saint-Petersburg, Russia

Abstract. The formation of cooperation policy in the field of environmental protection and sustainable development of the Arctic is connected with international institutions. At the beginning of the 1990s, transformations in the international arena took place that were also affecting the Arctic region. The collapse of the USSR changed the geopolitical situation, while globalization raised public awareness of global issues. The fact that the Arctic States were parts of different political blocs during the Cold War hindered their political and economic interaction in the early 1990s. The then-existing international organizations present in the Arctic did not cover issues related to soft security and international cooperation. It was necessary to ease tensions between the Arctic States and thus prepare the ground for further enhancement of cooperation.

At the same time, environmentalism within the world community peaked at the turn of the decade, affecting the work of international organizations at the political level. After the 1987 report by the Brundtland Commission, the Arctic strategy for sustainable development was only a matter of time. The proposals made in Mikhail Gorbachev's Murmansk speech in 1987, as well as Finland's initiative in the Rovaniemi Process, came to pass due to the Arctic States' readiness to cooperate on non-military matters. The choice fell on environmental protection and sustainable development. The article examines the activities of the Arctic Council as a unique organization that was created to ensure environmental cooperation in the Arctic. The main research question is to study the place and role of environmental cooperation for the member states of the Arctic Council and assess the possibility of a joint environmental policy in the Arctic.

Keywords: Arctic · Arctic Council · Cooperation in environmental protection · Sustainable development

1 Introduction. From the Rovaniemi Process to the Arctic Council

At the preparatory stage of the Rovaniemi Process, Finland advocated the creation of a long-term joint strategy and policy to eliminate pollution in the Arctic, leveling the risks of further pollution, and the need for structured joint legislative and practical policy

© The Author(s), under exclusive license to Springer Nature Switzerland AG 2021
R. Bolgov et al. (Eds.): *Proceedings of Topical Issues in International Political Geography*, SPRINGERGEOGR, pp. 249–256, 2021.
https://doi.org/10.1007/978-3-030-78690-8_22

measures on these issues. A clause on the possibility of a legally binding agreement was included in the draft declaration, and its acceptance was considered during further negotiations [2].

The Russian Federation, which became the legal successor of the USSR, was in a technological and financial crisis; it was more interested in exploring the Arctic resources and navigation opportunities in the region. Although Russia did not emphasized on the Arctic environmental security, the necessity to collaborate with other Arctic states turns the country to participate in the negotiations.

Due to the increasing number and degree of environmental risks, Canada favored the conclusion of an agreement with subsequent implementation of specific environmental measures into the national legislations, as well as expressed reluctance to include non-Arctic participants into the Arctic cooperation [3]. Norway was critical of any regional legal obligations and argued that the security issues of the region should be part of the global agenda, while in the Arctic it was necessary to explore the issues that were not covered by the then-existing international agreements [2].

Sweden, having no access to the Arctic waters, but eager to participate in Arctic policy, insisted on forming a common Arctic strategy, creating the rules and supervision for its implementation.

In aim to avoid conflicts between member states [4] the US insisted on not engaging in military matters and on specifying the scope of interaction to the issues of sustainable development and environmental policy.

In 1991, the foreign ministers of Finland, Norway, Sweden, Denmark, Iceland, Canada, the United States and the USSR met to sign the Rovaniemi Declaration on the Protection of the Arctic Environment [5]. The Arctic Environment Protection Strategy (AEPS) [6] and Arctic Monitoring Assessment Program were adopted as the common action plan. The main objectives of the cooperation, according to the declaration, were the elimination and prevention of pollution in the Arctic [7, 8].

The Arctic Council as a forum for promoting interstate cooperation "involving the Arctic indigenous population and other Arctic residents" [9] was established on September 19, 1996. The Arctic Monitoring Assessment Program became part of it, later converting into one of the first working groups of the Arctic Council.

Denmark, Iceland, Canada, Norway, Russia, USA, Finland, Sweden became members of the AC [10]. At the same time, there are three types of participation in the Arctic Council: member states of the Council, permanent participants with an consultative role and no rights to vote during decision-making process [11], and observer states (states from outside the Arctic, NGOs). Ministerial meetings, in which Arctic states represents by foreign ministers are took place twice a year with two-years chairmanship of each country. At the beginning, Arctic Council was funded ad hoc without any donations from any Arctic State.

2 Methods of Research

The research studies the features of environmental cooperation among the eight Arctic states under the auspices of the Arctic Council. Using the historical method, the study reveals the parameters of the emergence and development of the Arctic Council, identifies the features of cooperation between the eight Arctic states in the forum, key factors contributing to the transformation of relations between the AU member states, periodization of the stages of the AC activity.

The study of the institution of chairmanship within the AU allows to understand the degree of influence of the presidency on the development of the AC. A content analysis of the AC's founding documents and declarations complements the findings, allowing us to trace the accompanying policy changes within the Arctic Council.

The study proves that, to a greater extent, the environmental policy of states within the AC is declarative, while the national interests of each participant extend to the natural resources of the Arctic, which is consonant with the basic postulates of the realistic paradigm of international relations. The role of the AC is minimized and the organization of the forum, to mind of researchers requires reforms [1]. The article divides the work of AC into 3 periods: 1 – from the Rovaniemi Process to the emergence of AC as a high-level forum, 2 - period 1996–2000: forming of cooperation between member states, the setting period during which the forum was being formed and no significant events occurred, 3 – 2000–2013, great activation of AC, which usually connects with active discussion on climate change in the Arctic, 4 - 2013- 2019 – start of the second circle of chairmanships, Kiruna process and the three legally-binding agreements under the auspice of the AC. This period is not completed, the study ends with the year of the beginning of the Icelandic chairmanship in the AC.

3 The Arctic Council 1996–2000: Forming of Cooperation Between Member States

In the first meeting of the Council (1998) the Iqaluit Declaration of the Arctic Council and Sustainable Development Program and Working Group were adopted. The Council reaffirmed its commitment to offer measures to promote sustainable development, including opportunities to protect the environment and preserve the well-being of Arctic residents [12].

The Barrow Declaration of the Arctic Council (2nd meeting) of 2000 [13] outlined further scientifically, informational and advisory partnership in the region, and emphasized the headship of the Arctic Council in region sustainable development issues. The Arctic Council's Arctic Contaminant Action Program was developed, and the ACAP working group was established under the chairmanship of Norway, and reports on the Arctic Council were planned for the World Summit on Sustainable Development 2002, in Johannesburg.

4 AC from 2000–2013: The Impact of Climate Change

At the Ministerial gathering held in 2002 in Inari, specific accentuation was laid on the job of the Arctic Council as a worldwide accomplice in worldwide and territorial endeavors to secure nature. The Arctic Monitoring and Assessment Program reports affected a few regional peaceful accords – the Aarhus Protocol on Persistent Organic Pollutants and the Stockholm Convention on Persistent Organic Pollutants. It was decided to continue the efforts to eliminate pollution and adopt a regional action program to protect the Arctic marine environment from land-based activities [14].

At the Ministerial gathering, a discussion stimulated on the consequences of two significant logical activities of the Arctic Council: the Arctic Climate Impact Assessment and the Arctic Human Development Report. The way toward setting up the Project Support Instrument (PSI) has started, to turn into the lasting subsidizing system for funding the Arctic Council working groups.

At the Salekhard Ministerial meeting (2006) discussion concentrated on the accompanying issues: adaptation to the climate change, contamination avoidance and decrease of harmful emissions, insurance of the marine condition and biodiversity preservation [15].

After 2006, the Arctic Council Ministerial gatherings were incidentally surceased, which could be connected to a few occasions: the Russian logical undertaking toward the North Pole (2007) and the Ilullisat declaration of Russia, Norway, Canada, Denmark and USA (2008), with no Iceland, Finland and Sweden.

One of the points of the Russian campaign toward the North pole in 2007 was to take tests from the base of the Arctic Ocean to demonstrate that the Lomonosov and Mendeleev Ridges are coasting of the territory of Russia. The reaction of the US, Canada, Norway and Denmark was mixed, as this expedition directly affected their resource interests [16]. Considering these occasions, the Declaration of the meeting revering the "unique position to address these possibilities and challenges" of the Arctic Ocean [17], was most likely planned to lessen misunderstanding between them. Although biodiversity conservation efforts and sustainable development promotion policies in the region continued to be a priority, the absence of the observer-states and NGO's and indigenous peoples associations from the Ilulissat, as well as the absence of three of the members, hardly make it possible to call the Ilulissat Declaration a step to Arctic collaboration. Nevertheless, this step could be caused by the intention of the Arctic states to resist the pressure of non-regional international actors [18].

The resumption of the Arctic Council meetings at the high level was set apart in April 2009 with Tromsö Declaration [19], which again pointed on the leadership of the Arctic Council in Arctic environmental issues and, subsequently, of every one of its participants in ecological, economic and social cooperation in the Arctic. States that did not have access to the Arctic Ocean became more active and sought to emphasize the need to ensure well-being of the indigenous population of the Arctic. Having recently taken non-formal interest in the Arctic Council, Iceland turned into the pioneer of the Arctic Ocean Review (AOR) in a joint effort with USA, Canada, Norway and Russia. The objective of the venture is to intermittently audit the status and significance of global and territorial understandings and the guidelines identified with the Arctic marine condition, just as to examine the applicability of the treaty to the Arctic Ocean.

In 2011, Sweden adopted the Arctic Strategy, which focused on climate change issues and the situation with indigenous people. As a non-coastal Arctic state with no access to Arctic deposits, Sweden during its Arctic Council Chairmanship (2011–2013) focused working with permanent participants of AC.

The Agreement on Cooperation on Aeronautical and Maritime Search and Rescue in the Arctic (2011) [20], actively developed by Russia, the US and Norway became 1st non-declarative treaty under the auspices of AC. The principal common exercises in the Arctic occurred in October 2011 on the territory of the depositary of this agreement.

Reforming and strengthening of the institutional format of the AC was also underway. There was no talk of expanding the Council agenda, so it worked on less-contentious matters (economical and financing issues, technological infrastructure, flow of communication with non-AC actors).

In Nuuk ministerial meeting the demands and procedure for new observers was developed "as a measure to strengthen the Arctic Council" [21].

5 Arctic Council 2013–2019: The Second Circle of Chairmanships

The main outcome of the Kiruna process (2013) was adoption of the Agreement on Cooperation on Marine Oil Pollution, Preparedness and Response in the Arctic and launch of Scientific Cooperation Task Force in aim to prepare one more treaty.

According to the Ottawa Declaration, the Arctic Council is a discussion forum, and agreements are concluded between States, so we cannot talk about the de facto transformation of the Arctic Council into an international organization.

As Kao S-M. states, the appropriation of the legally binding agreement is a proof of new level in regional cooperation, however It doesn't imply that regional treaties are relevant to every Arctic issue [22]. In 2013 Russian Federation provided the third appointment of the AC delegates in Salekhard and the third International Arctic Forum "The Arctic: Territory of Dialog" [23].

The next chair state of AC - Canada took this part secondly during 2013–2015. Back in 2012, the expanded oil extraction in the Arctic caused controversy concerning case of "The Prirazlomnaya platform", which in fact was the high-profile Greenpeace protest [25], ended in lengthy legal proceedings with participants in the next year. The Ukrainian crisis became the reason of sharpness between Russia and other Arctic states [26], although later they were eased largely owing to cooperation in the AC.

Political events of the Globe (the struggle with terrorism in the Midle East actions of ISIL, the Syrian war and EU sanctions against Russia, the Arctic nations militarization in the Arctic) caused disquietudes concerning Arctic stability as well [26]. As parties of EU severe Arctic states participated in sanctions against Russia. The Canadian boycott of meeting of AC task-force in Russia in 2014 could be caused not only by the situation in the Ukraine, but in addition by the pending issue over the North Pole between mention states. In 2015 Russian foreign minister skip the meeting in Canadian Iqaluit. Nevertheless, despite Russian focus on domestic problems [27], work on a scientific cooperation treaty was launched by the creation of corresponding mandate [28].

The key findings of the US Chairmanship (2015–2017) of the Arctic Council were the signing of an agreement on scientific cooperation [29], the assessment from AMAP "Snow, Water, Ice and Permafrost in the Arctic", a common initiative of the AC working groups "Circumpolar Local Environmental Observer Network" (CLEO).

In Fairbanks in May 2017, it was decided to continue Arctic Council projects concerning conservation of flora and fauna, study the results of ice dissolving and extend Arctic collaboration to sustainable use of Arctic seas [30], which also noted as a key point of the US work as a chair of the AC [31]. The final declaration of the meeting settled down a task force on development new energy infrastructure.

During Finland's Chairmanship (2017–2019), AC states came to consensus on the first Long-term strategic plan of the Arctic Council, which enshrines the Council's common goals, to focus on the issues of decreasing pollution in the Arctic, and to intensify cross-border cooperation in pollutants disposal, but until now this task has not been completed [32].

In 2019, Iceland became the Chair of the AC. The results of the chairmanship have not yet been summed up, but it is already clear that Iceland has adopted the most successful features (as recognized) of the previous chairman's policy and supports projects to preserve biodiversity and prevent pollution of the Arctic waters [33].

6 Conclusions

At the beginning, the foundation of the Arctic Council showed the commitment of the Arctic States to joint interstate collaboration in the field of sustainable development and saving nature. However, the cooperation mechanism does not have any supranational functions or functions of an international organization. The structure of the Arctic Council is so far consistent with the Ottawa Declaration.

After the publication "Arctic Climate Impact Assessment", the Arctic attracted the attention of various NGOs and states from outside the Arctic using the Arctic Council as a tool to get involved in regional events. From the other side, deliberate attention to the Arctic has enhanced the collaboration inside the AC. While the global events of 2012–2013 that were not directly related to the Arctic affected the political situation, they did not lead to complete cessation of cooperation in the Arctic.

Since 2016 the intensification of the AC activities is observed: in addition to the increase in the number of published scientific and analytical papers, the Rules of Procedure and the requirements for observers to the Arctic Council have been developed and reformed, the approach to the chairmanship has been changed, the Secretariat along with a subsidiary financial instrument has been established.

Three legitimately treaties were effectively marked with the direct involvement of the AC's working groups. The signing of such agreements underscores the willingness of the resource-developing countries to take responsibility for possible environmental consequences, which can be seen as a positive trend. Owing to the work of the Arctic Council a new stage of cooperation in the Arctic is provided.

However, research shows that Arctic Council members build their environmental policy and the cooperation from the position of their national interests, which include the development of subsoil resources and the water area. While the Arctic states with no direct access to the Arctic waters also hoped to legitimize their rights to be present in the region through the creation of a regional international organization. It is becoming increasingly clear that environmental cooperation (the main declared goal of the AC) is secondary to the interest of the states in managing the natural resources. This is especially demonstrates by Russia, Canada, USA, Norway and Denmark, who has broaden access to the Arctic richness. The need to meet the demands of the changing world has also touched the AC. Iceland's chairmanship will end in 2020, summing up the results of which it will be possible to talk about qualitative changes in the work of the AC.

References

1. Smieszek, M.: Do the cures match the problem? Reforming the Arctic council. Polar Rec. **55**, 121–131 (2019). https://doi.org/10.1017/S0032247419000263
2. Tennberg, M.: Arctic Environmental Cooperation: A Study in Governmentality. Ashgate, Aldershot (2000)
3. Canada's Northern Strategy: Our North our Heritage (2009). The Government of Canada website. http://www.northernstrategy.gc.ca/index-eng.asp. Accessed 08 Nov 2017
4. Bergman, R.: Perspectives on Security in the Arctic Area/Report for Danish Institute for International Studies (2011). https://www.files.ethz.ch/isn/133473/RP2011-09-Arctic-security_web.pdf. Accessed 25 Nov 2017
5. The Rovaniemi Declaration on the Protection of the Arctic Environment (1991). Arctic Circle website archive. http://arcticcircle.uconn.edu/NatResources/Policy/rovaniemi.html. Accessed 08 Nov 2017
6. Arctic Environmental Protection Strategy (1991). The Arctic Portal Library. http://library.arcticportal.org/1542/1/artic_environment.pdf. Accessed 08 Nov 2017
7. Young, O., Osherenko, G.: Polar Politics: Creating International Environmental Regimes, 1st edn., pp. 186–187. Cornell University Press, Ithaca (1993)
8. Vidas, D.: Protecting the Polar Marine Environment: Law and Policy for Pollution Prevention. Cambridge University Press, Cambridge (2000)
9. Ottawa Declaration on Establishment of the Arctic Council (1996). Official website of the Arctic Council. http://hdl.handle.net/11374/85. Accessed 23 Nov 2017
10. Lagutina, M., Kharlampieva, N.: Transnational model of the Arctic governance in the XXI century. Arctic North **3**, 64–84 (2011)
11. Oldberg, I.: Soft Security in the Arctic-The Role of Russia in the BEAC and Arctic Council. UI Occasional Papers, no. 4. Swedish Institute of International Affairs (2011)
12. Iqaluit Declaration of the Arctic Council (1998). Official website of the Arctic Council. http://hdl.handle.net/11374/86. Accessed 23 Nov 2017
13. Barrow Declaration of the Arctic Council (2000). Official website of the Arctic Council. http://hdl.handle.net/11374/87. Accessed 23 Nov 2017
14. The Inari Declaration of the Arctic Council (2002). Official website of the Arctic Council. Accessed 23 Nov 2017
15. Salekhard declaration of the Arctic council, 2006. In: Ivanov, I. (eds.) The Arctic Region: Problems of International Cooperation, Reader, vol. 3, pp. 201–206. Aspect Press, Moscow (2013)
16. Russian expedition to the North pole.Comments and evaluations, official website of the RIA-Novosti news agency. https://ria.ru/society/20070803/70307088.html. Accessed 23 Nov 2017
17. The Ilulissat Declaration of the five states, coastal to the Arctic ocean. In: Ivanov, I. (eds.) The Arctic Region: Problems of International Cooperation, Reader, vol. 3, pp. 206–208. Aspect Press, Moscow (2013)
18. Heininen, L.: Security of the global Arctic in transformation – potential for changes in problem definition. In: Heininen, L. (eds.) Future Security of the Global Arctic: State Policy, Economic Security and Climate. Palgrave Pivot, London (2016)
19. Tromso Declaration. On the occasion of the Sixth Ministerial Meeting of the Arctic Council (2009). Official website of the Arctic Council. http://hdl.handle.net/11374/780. Accessed 23 Nov 2017
20. Agreement on Cooperation on Aeronautical and Maritime Search and Rescue in the Arctic. Nuuk. Greenland (2011). Official website of the Arctic Council. http://hdl.handle.net/11374/531. Accessed 23 Nov 2017

21. Ramos, B.: Strengthening the Capacity of the Arctic Council: Is the Permanent Secretariat a First Step? http://www.arcticyearbook.com/images/Articles_2013/SANCHEZ_AY13_FINAL.pdf. Accessed 23 Nov 2017
22. Kao, S.-M., Pearre, N., Firestone, J.: Adoption of the Arctic search and rescue agreement: a shift of the Arctic regime toward a hard law basis? Mar. Policy **36**(3), 832–838 (2012)
23. Report on the results and the main activities and medium-term priorities of the Ministry of Foreign Affairs of the Russian Federation in 2013. Moscow. 2014, Website of the Ministry of Foreign Affairs of the Russian Federation. http://www.mid.ru/BDOMP/Brp_4.nsf/arh/E543DA81B862140144257CC9003E2CF9?OpenDocument. Accessed 23 Nov 2017
24. Attack on Prirazlomnaya and Arctic development projects (Federal media Monitoring: 13–26 August 2012), Materials of the Arctic Info news agency. http://www.arctic-info.ru/analytics/monitoring-federal-mass-media/27-08-2012/ataka-na-prirazlomnyu-i-proekti-po-osvoeniu-arktiki-monitoring-federal_nih-smi-13-26-avgysta-2012-goda-/. Accessed 23 Nov 2017
25. Konyshev, V., Sergunin, A., Subbotin, S.: Russia's Arctic strategies in the context of the Ukrainian crisis making in the Arctic. Polar Rec. **42** (221) (2006)
26. Heininen, L.: High Arctic stability as an asset for storms of international politics – an introduction. In: Heininen, L. (eds.) Future Security of the Global Arctic: State Policy, Economic Security and Climate, p. 2. Palgrave Pivot, London (2016)
27. Bolgov, R., Chernov, I., Ivannikov, I., Katsy, D.: Battle in Twitter: comparative analysis of online political discourse (Cases of Macron, Trump, Putin, and Medvedev). In: Chugunov, A., Misnikov, Y., Roshchin, E., Trutnev, D. (eds.) EGOSE 2018. CCIS, vol. 947, pp. 374–383. Springer, Cham (2019). https://doi.org/10.1007/978-3-030-13283-5_28
28. Scientific cooperation: Making a Good Thing Even Better. https://www.arctic-council.org/index.php/ru/our-work/news-and-events-ru/192-scientific-cooperation-making-a-good-thing-even-better. Accessed 23 Nov 2017
29. Ministers of the Arctic States met to sign an agreement on scientific cooperation and transfer the Finnish presidency, official website of the Arctic Council, news section. http://www.arctic-council.org/index.php/ru/our-work/news-and-events-ru/451-fairbanks-04. Accessed 23 Nov 2017
30. Fairbanks Declaration (2017). Official website of the Arctic Council http://hdl.handle.net/11374/1910. Accessed 23 Nov 2017
31. Zhuravel, V.: The Arctic Council: transition of the presidency from the USA to Finland, further strengthening of Russian-Finnish cooperation. The Arctic and the North 28 (2017)
32. Barrya, T., Daviðsdóttira, B., Einarsson, N., Young, O.R.: The Arctic council: an agent of change? Glob. Environ. Change **63**, 102099 (2020). https://doi.org/10.1016/j.gloenvcha.2020.102099
33. Zhuravel, V.P., Kudryavtseva, R.-E.A., Chistalyova, T., Eidemiller, K.Yu.: The first results of Icelandic chairmanship over the Arctic Council. In: IOP Conference Series: Earth and Environmental Science, vol. 539, no. 1 (2020). https://doi.org/10.1088/1755-1315/539/1/012050

Evolution of the Environmental Policy of the European Union: Stages, Actors and Trends

Yuri Kovalev, Alexander Burnasov(✉), Anatoly Stepanov, and Maria Ilyushkina

Ural Federal University, Yekaterinburg, Russia

Abstract. The purpose of this article is to analyze the features of the formation and evolution of the EU sustainable development policy, to assess the influence of internal and external actors on its formation. The European Union is one of the main actors in the global sustainability policy. The ecological crisis gave an impetus to the formation of an environmental protection policy, the development of the discourse of the "quality of life" of the population, which gradually transformed into a modern strategy for the sustainable development of the region based on green technologies. The current EU sustainability policy comes down to a policy of environmental modernization, which is a transition to a new level of technological development of the social system without a radical transformation of existing social institutions. The main stages of the establishment and development of the EU sustainable development policy are reflected in the EU Environmental Action Programs. An analysis of the Environmental Action Programs reveals three stages in the development of the EU's sustainable development policy. Environmental economic growth is becoming the dominant idea of sustainable development. Since 2012, sustainable development policies have been heavily influenced by the concepts of a green economy and environmental transformation. The European Commission plays a key role in implementing sustainable development policies. The European Parliament and the Council of Ministers of the EU carry out deliberative and legislative functions. Their activities are greatly influenced by diverse groups of actors, which pursue their own goals.

Keywords: EU · Environmental protection policy · Sustainable development · Green economy · Environmental modernization · Post growth

1 Introduction

In 1967 The European Community was formed as a result of the merger of three organizations - the European Coal and Steel Community (ECSC, 1951), the European Economic Community (EEC, 1957) and the European Atomic Energy Community (Euratom). It did not pay any attention to environmental issues at the beginning of its existence. The main goal of the political association was to accelerate the economic growth of the member states, poverty alleviation and improvement of the population well-being. The solution to these problems was seen as the basis of political stability for the entire region. Only

© The Author(s), under exclusive license to Springer Nature Switzerland AG 2021
R. Bolgov et al. (Eds.): *Proceedings of Topical Issues in International Political Geography*, SPRINGERGEOGR, pp. 257–269, 2021.
https://doi.org/10.1007/978-3-030-78690-8_23

the permanent growth of the general population well-being of the Community countries (under the ideological tutelage and material support of the United States) proved the superiority of the Western economic system on irrefutable facts, eroded left and right political extremism, "cooled" the zeal of reformers, which was of paramount importance during the years of confrontation between the East and the West. The environmental costs of economic development were considered unavoidable in achieving these goals. In this policy, the environment was assigned the role of a supplier of resources and the sphere of waste disposal (solid, liquid, gaseous) production. Only in the early 1970s heated discussions about the decline in the life quality in Western countries (E. Mishan 1967), environmental pollution (R. Carson 1962, B. Commoner 1966, 1971), as well as the growing political activity of social environmental movements gave impetus to the formation of *the environmental policy of the Community.*

2 Methodology and Data

The study of peculiarities of formation and evolution of the sustainable development policy of the EU, assessment of the impact of internal and external actors on the formation has been intensively conducted since the late 1990s. Among the Russian researchers who have made significant contributions to the development and study of sustainability, it is necessary to mention Danilov-Danilyana V.I., Moiseeva N.N., Loseva K.S. The EU Policy in the field of sustainable development is reflected in the works of Pakhomova N.V., Sherbak, I.N., Belova Y.A., Shelamova N.A. Foreign researchers such as Saks V., Gel E., Tio L., Marhold H. made a significant contribution to understanding the policy of sustainable development. Lenshow A., Burns C. Zito A. assess the trajectory of the European Union environmental policy and integration in the light of the rising pressure for policy dismantling and disintegration in their research [20]. Brunix K., Ovaere M. And Delarue E. proved the impact of the market stability reserve on the EU emission trading system [4]. The possibility of implementation of energy policy of EU in post-soviet states is analyzed by Dudin M. [11]. The issues of forming the sustainability policy and implementing its goals in the field of environmental protection (Guseev A.A., Redinkova T.V., Averina), agriculture (Kvochkin A. I., Pokrovskaya S. F., Chaika V.), economy (Savinsky A.V., Klavdienko V.P, Krass M.S.), energy sector (Salygin V.I., Guliyev I.A.) have been studied profoundly. Governing sustainability in the EU is examined in the book by Domorenok E. [10]. But, less attention is paid to the study of the peculiarities of the formation and evolution of the EU's sustainable development policy (Mayer J.H, Huenemerder K.F.). Moreover, a comprehensive study of the formation and evolution of the EU's sustainable development policy has not been developed.

The methodology of the current research is based on a retrospective structural and analytical approach. A discursive analysis of the main documents in this area, adopted by the EU in the period 1970–2020, makes it possible to determine the trajectory of the organization's sustainable development policy and assess the impact of various factors in the direction of its evolution. The main sources of this research are the EU Environmental Action Programs (1973–2013), as well as scientific works of experts in this field.

In the late 1960s and early 1970s, some European politicians sounded the alarm about the critical state of water and forest objects on the territory of the EEC countries (for example, G. Grohl's speech in the Bundestag on the extinction of forests). In 1970, a number of members of the European Parliament from different countries demanded that the EEC Commission introduce immediate measures to protect the river Rhine from the industrial pollution [21: 58]. The Rhine is the central waterway of Europe, flowing through the territory of 6 countries, the founders of the EEC. Even today, the water supply of the population of neighboring countries and the existence of river and coastal ecosystems depend on the water quality of the river. By the beginning of the 1970s, the situation was disastrous due to the emissions of toxic effluents from industrial enterprises (mainly German ones) into the river. The mass extinction of fish in the river and the detection of chemicals in drinking water required comprehensive cross-border measures at the supranational level. The EEC States were obliged to take legislative measures to save the Rhine. The topic of air pollution in Community cities was discussed by the European Parliament in 1968. The response to these demands was European Commission's presentation "First communication of the Commission about the Community's policy on the environment" in 1971. In 1972, the second report was published. They formed together the basis for the creation of the first EEC Environmental Action Programme (EAP).

On November 22, 1973, this programme was proclaimed [Declaration of the Council of the European Communities, 1973]. This happened a year after the publication of the report of the Club of Rome "The Limits to Growth" and the UN Conference on the Environment in Stockholm. The goals and guidelines adopted at the Stockholm conference in the field of environmental protection are fully reflected in the first action programme in this area of the EEC: 1. Limitation, reduction and eliminating environmental pollution; 2. Maintaining ecological balance and protection of the biosphere; 3. Rational use and management of resources; 4. Quality orientation of economic development; 5. Improvement of working and living conditions; 6. Greater attention to environmental aspects in urban planning and planning activities; 7. International cooperation and searching for common solutions, especially in international organizations [Ibid.: 5]. In the preamble of the programme, the main goal of these events was mentioned. It was the desire to improve not only the standard of living of the Community's population (which was reflected in the Treaty of Rome), but also the conditions and quality of their life [Ibid.: 5]. Here we see a departure from the "economic growth at any cost" paradigm that was dominant in the West at that time. Stimulating the development of "quality of life" indicators became the primary task of the community [Ibid: 252]. The focus on "quality growth" laid down in the first program of the EEC became a practical foundation for further development of the concept of environmental modernization (green economy) and sustainable development. Indeed, as J. H. Meyer pointed out in his research, the concept of "sustainable development" introduced into the political lexicon by the UN Commission on development and the environment (Brundtland Commission) in 1987 was based on the ideas already formed in the 1970s by individual international actors in the field of environmental protection from European countries in the 1970s [21: 56].

However, considering in detail the formation of a new environmental policy in the structures of the European community in the early 1970s, it can be noted that the EEC itself did not have a common view on the further development of the organization in the field of environmental protection at that time. Both in the European Parliament and in the EEC Commission there was a struggle between two ideological directions: technogenic environmental modernization and integrated environmental transformation of the EEC. The position of two Commissars was decisive (the position of Italian commissar of industry A. Spinelli and the agricultural commissar S. Mansholt). Both commissioners saw the greening of the EEC policy as the "key" to improving the living conditions of Europeans. However, they had different views on how to solve environmental problems. A. Spinelli was a technocrat and did not see any contradictions between technological (economic) development and ecology. In his opinion, improving the quality of life and working conditions of the European population is possible only with further economic growth and technological progress. On the contrary, the Dutch politician S. Mansholt pursued radical concepts of environmental protection and was opposed to the development of the nuclear energy. The club of Rome report published in 1972 had a profound impact on his worldview. On his proposal made in 1972, Europe needed a new economic plan that would not aim at economic growth, but would be aimed at preserving the ecological balance of the EEC countries [16: 229]. Such a radical shift from the principles of market economy required strict planning of economic activity in the form of democratic socialism, which was what environmental social movements were trying to achieve at that time [Ibid.: 231]. However, S. Mansholt's plan was not supported in the structures of the EEC. The development of the first action programme for environmental protection of the EEC was carried out under the patronage of A. Spinelli's department.

The second programme of action was adopted by the European community in 1977. It paid special attention to reducing the environmental burden, protecting and managing natural resources, fauna and flora, recycling garbage and waste, and cooperating with the EEC States in these matters [Resolution of the Council of the European Communities, 1977]. Prior to 1992, two more environmental protection action programmes were approved, which were mainly aimed at continuing the priority areas of environmental protection laid down in the first programmes. The fourth action programme adopted by the EEC in 1987 after the announcement of the creation of the single common market of the EEC countries emphasized the importance of expanding the Treaty of Rome to include the chapter on the protection of the environment (part VII, article 130) for the Community's environmental policy. The programme also consolidated the dominant position of supranational environmental protection priorities over national and local ones. Certain legal lines and norms in the field of environmental protection became mandatory for the Countries of the community. However, the economic component was also crucial here as the common market required common environmental standards for economic participants in order to avoid distortions of competition. Different environmental standards of the community countries would create unequal conditions of competition. Therefore, unification was inevitable.

The 5th Action Programme of the European Union was formed in February 1992 and adopted on May 17, 1993. It was of historical significance. Its content was primarily influenced by the UN conference on sustainable development held in Rio de Janeiro in 1992. In fact, the 5th environmental action programme of EU was the first sustainable development programme of the organization. Its very name "For long-term development" reflects the main principles of the UN concept of sustainable development. It also places more emphasis on the growing role of the EU as a global actor in the new sustainability policy than previous programmes [Resolution of the Council and the Representatives, 1993: 12]. The most important topics of the EU international policy are the following: global climate change, the disappearance of the ozone layer and the reduction of the planet's biological diversity [Ibid: 12]. To achieve the set goals of sustainable development, the programme specifies ways (measures) of environmental transformation of the EU economy sectors – industry, energy, transport, agriculture and tourism. The document emphasizes that these measures do not have strict legal force, they are not binding [Ibid: 13]. In addition, the document did not provide specific quantitative indicators of the desired development of both the European Union as a whole and its individual constituent states. This was the weakness of this program. Many of the changes outlined in the programme were not achieved by 2000.

In July 2002 the EU published the 6th environmental action programme until 2012. Article 1 of the programme states that its main goal is to promote the integration of the environmental component into all EU political fields and contribute to the sustainable development of the community before and after its expansion [Beschluss Nr. 1600/2002/EG des Europäischen Parlaments und des Rates, 2002: 3]. Priorities mentioned in the programme were as follows: 1. Climate change; 2. Natural and biological diversity; 3. Environment, health and quality of life; 4. Natural resources and waste. Each of these priorities included a number of specific goals. Thus, the fight against climate change set the goals of ratification and implementation of the Kyoto Protocol by the EU, reduction of greenhouse gas emissions in industry, energy, transport and agriculture [Ibid: 7]. Article 5.5 pointed out the importance of the EU's international engagement in the fight to stabilize the earth climate: the fight against climate change has become an integral part of the EU's foreign policy and the top priority of its sustainable development policy [Ibid: 8]. The document outlined EU regulations (which are mandatory) in the field of environmental protection, which, unlike other programmes, found a positive response in non-governmental environmental organizations.

The current 7th action plan for the environmental protection entitled "Living well, within the limits of our planet" was adopted by the EU Parliament and Council on 20, November, 2013. The leitmotif of the programme was again the fight against climate change at the global, regional and local levels. The programme considers the reduction of the natural capital of the European Union countries due to the expansion of the economic space and anti-environmental technologies as the main obstacle to achieving environmental and climate goals [Decision No. 1386/2013/EU of the European Parliament and of the Council, 2012: 176]. The EU Commission is increasingly concerned about climate change and its negative impact on the natural, social, humanitarian and economic capital of the territories. Thus, the programme notes that the costs of natural disasters partly caused by global climate change, amounted to more than 300 billion euros in 2011 [Ibid: 177]. According to the authors of the programme, effective counteraction to these trends is possible only with the transition to a new type of management, which will be based on resource-efficient, nature-saving and low-carbon technologies. The term "green economy", which gained its political life at the conference in Rio de Janeiro in 2012 (Rio + 20), has become the main semantic element of the program [Ibid: 182]. The establishment of a "green economy" in the EU requires investment in environmental innovations, design, optimization of products and processes, minimization of emissions and production waste, and transition to a cyclical closed economy [Ibid: 184]. The transition to an eco-friendly "green economy" is understood as the basis for further sustainable economic growth and increasing the welfare of EU residents, creating new jobs, investments, etc. [Ibid: 182]. By 2020 The European Union plans through "green" activities to reduce greenhouse gas emissions by 20% and at the same time improve the overall energy efficiency of EU countries by 20% as well [Ibid: 182].

The 7th Environment Action Programme directly points out the ways to overcome environmental costs at the regional, national and global levels. The main mechanism for achieving sustainability is environmental modernization of the economy. Only on the basis of new technologies, environmental innovation processes, and re-equipment of the main sectors of the economy, it can be possible to stabilize the global ecosystem. Environmental modernization is becoming a leitmotif of the concept of sustainable development.

3 Results and Discussion. Institutions and Actors of Sustainable Development Policy in the EU

The EU sustainable development policy is formed by complex interaction of various structures of the organization that are both in hierarchical and horizontally networking relations. Whereas its modus operandi is contingent on internal (national, regional, private economic, social, political, party and other interests) and external (treaty regime, emergency situations) factors. The Multi-Level-Governance mechanisms, created at the very beginning of the organization's establishment in the 1950s, turn the process of making and implementing political decisions into an extremely complicated one, that requires continuous communication between structures of the local, regional, national and supranational political level. Interest groups formed within the structure of the European Commission (by Directorates-General), the EU Parliament, the EU Council of Ministers (Coreper) often have informal relations and work together to achieve common goals. In addition, transnational actors (TNC, NGO), which have their own lobbying bodies in key EU structures, play an important role in EU policy formation. Kaveshnikov N. [18] states "EU policy in any area of its competence and, accordingly, the method of management in this area is the product of numerous consecutive decisions taken at different levels of the multi-level system. Although many actors are capable of operating at different levels, each of the three listed has its own key actors and its own rationalization of the decision-making process. Nonetheless, it is important to understand that the borderline between management levels is not clear, and the correspondence between them and the types of decisions is not absolute." Within a sustainable development policy, various political fields such as energy, transport, agriculture, industry, trade, ecology, labor and finance, constantly intersect. The search for a political compromise that satisfies all negotiating parties is an extremely labor-intensive. A multi-level system involves the exchange of information of different content and quality between different levels and horizontal structures of the EU political system. Political decisions taken at the upper (supranational) level in the field of sustainable development (directives, decisions, recommendations) directly (through their implementation in national legislation) or indirectly (soft power policy, "naming and shaming") affect regional and communal policies. European Commission, European Court of Justice, European Council, European Parliament Committees, European Environment Agency carry out regular monitoring of the implementation of legal acts by national, regional and local authorities. There is also a counter-motion in this system, when changes at the lower (regional, communal) levels affect generating initiatives at the upper level. Thus, in December 2019, the decision of the Supreme Court of the Netherlands to force the current Government of the country to take additional measures to reduce greenhouse gas emissions is likely to affect the climate objectives of the next EU Environmental Action Program. In its turn, as one of the central actors in the global policy of sustainable development, the EU has an enormous influence on the formation of the main development directions of such international organizations as the UN, OECD, World Bank, WTO. The European Union, for instance, is assigned the role of the vanguard in international climate policy [20]. Without its initiatives, many of the modern advances in combating global climate change would have been impossible.

Among the main EU institutions in sustainable development policy formation are the European Commission, the European Parliament, the European Council, the European Court of Justice, and the European Environment Agency. Their interaction determines the perspective of the adopted regulations and measures in sustainable development. European Commission is the linchpin of the policy formation. It acts as an initiator, controller and event executive. Without the incoming initiatives of the European Commission, discussion and decision-making by the Council of EU is impossible [19]. However, both the Council and the European Parliament can oblige the Commission to draft new proposals or improve existing ones. This results in a constraint on the Commission's monopoly as an innovator.

Since the early 1970s, the European Commission has been pursuing an active environmental protection policy of its member states. During this period almost all initiatives in this field emanated from the Commission. The decisions on the initiative formation are made by the College of Commissioners (currently consisting of 27 members) by a simple majority vote. Prior to this, the Directorates-General (DGs, correspond to ministries in national states) prepare appropriate proposals (Fig. 1).

During the period from 1970 to 2020, there has been a strengthening of the environmental pillar in the activities and structure of the European Commission. This is reflected in the formation of new Directorates-General related to sustainable development and environmental protection issues, as well as the establishment of new divisions of competence in these fields in other Directorates-General. Thus, until 1973, only one department in the Directorate-General for Trade was engaged in environmental problems. In 1973, within the structure of the European Commission an independent Directorate-General for Environment was created. Its competence covered the development of initiatives in the field of environmental protection, monitoring the fulfillment of instructions and directives of the European Commission in national states. To date, one of the main goals of the Directorate is to stimulate the development of The Green Economy in the EU countries as an essential component of sustainable development. Since the 1980s the number of the employed in this Directorate has grown tenfold and now stands at 500 people [Umwelt Direcktoria 2020]. In addition, virtually every Directorate-General of the EU Commission has its own commissioners for environment. The Directorate-General for

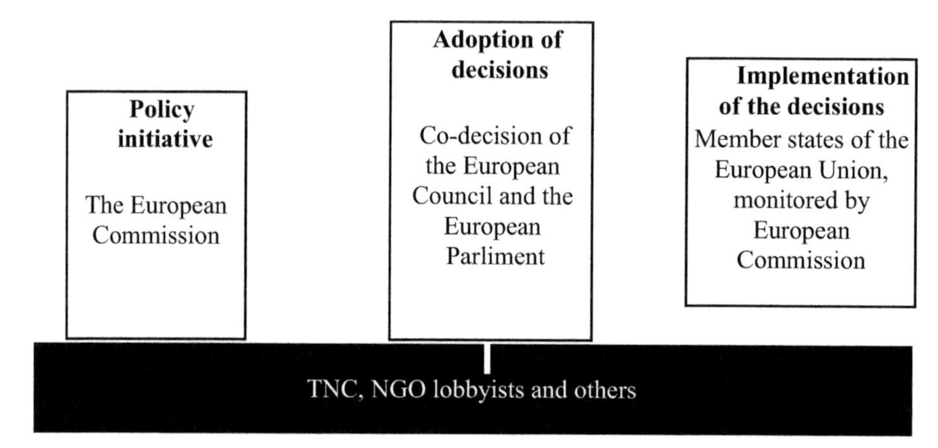

Fig. 1. The main EU institutions for initiating, adopting and implementing political development strategies. Compiled by the author based on [Simonis 2017].

Agriculture and Rural Development and the Directorate-General for Trade even have their own environmental departments [19]. The activities of the autonomous Directorate-General for Climate Action, created in 2010, are directly related to the greening of EU structures and the implementation of the concept of sustainable development. One of the activities of this Directorate has become the development of EU climate strategies, providing for the creation of a low-carbon, resource-efficient, environmentally friendly economy [EU climate action, 2010]. Since November 2019, the Directorate-General has been under the direction of the European Commissioner for Climate Action Frans Timmerman. Council of the EU and the European Parliament are the most important political decision-making institutions of the organization. Decisions on practically all legislative acts, directives, legal lines and decrees are made by these bodies. At the same time, the main function belongs to the Council of Ministers. The Council of the EU does not represent a single entity, but consists of 10 ministerial groups dealing with issues in their own areas of responsibility. Thus, there is the Justice and Home Affairs Council, the Agriculture and Fisheries Council, the Employment, Social Policy, Health and Consumer Affairs Council, the Economic and Financial Affairs Council, the Environment Council, etc. They are composed of national ministers who lead the respective ministries in their own countries. Decisions on a particular legislative act are also made by a simple majority vote. Each meeting is accompanied by intense preparatory work carried out in the Committee of Permanent Representatives and Working Groups. The position of the presiding country of the Council has a huge impact on its final decision.

The European Parliament is the second most important legislative body in the EU. Since its inception, it has played an active role in environmental policy of the European Union. Its requirements in this area were more radical than the proposals of the European Commission and decisions of the Council of Ministers. Unlike national parliaments, the European Parliament has only limited legislative power and is forced to share it with the Council of Ministers. In the legislative process, the work of the European Parliament is reduced to three main functions: hearings on draft laws, the work of parliamentarians in Commissions and voting on legal decrees, directives, lines in plenary sessions [15]. The Commissions of the European Parliament play an essential role in the preparation of draft laws. Their specialized competence coincides with competence of the European Commission's Directorates-General. The environmental issues of the European Union are addressed by the Committee on the Environment, Public Health and Food Safety (ENVI), which was established in 1973. The Committee on Industry, Research and Energy (ITRE), the Committee on Transport and Tourism (TRAN), the Committee on Agriculture and Rural Development (AGRI) and the Committee on Economic and Monetary Affairs (ECON) are responsible for the issues of ecological transformation of the economy. The European Parliament Committees are composed of members from various factions. Thus, in the Committee for Environmental Protection, out of 162 members, 26 belonged to the European Green Party or the Group of the European United Left – Nordic Green Left (calculated by the author from: [15]). The party proportions of the political composition of the Committee directly reflect the party composition of the European Parliament. Nowadays, the European Parliament is dominated by conservative and social democratic parties (182 and 154 seats, respectively). The continual growth of the European Green Party since the early 1980s is also worth mentioning. In the period from 1984 to 2019, their proportion in the European Parliament increased from 4.6 to 10%. In addition to standing committees, temporary committees are also formed at certain times. Between 2004 and 2009, the Committee on Climate Change (CLIM) functioned and, between 2009 and 2014, – the Committee on the Financial, Economic and Social Crisis (CRIS). Since 2013, the Committee on Budget (BUDG) for sustainable development has been operating under the EU Parliament. The European Court of Justice and the European Environment Agency also play an important role in the sustainable development policy. Their responsibilities include monitoring and holding accountable the leadership of the EU countries (or its regions) for failure to comply with the EU legal regulations. Along with the central EU institutions for initiating, adopting and implementing decisions in the field of sustainable development, lobbying organizations play an important role, especially TNCs and large non-governmental organizations.

Transnational corporations have exerted and continue to exert a tremendous influence on the legislative process in the EU. In the 1990s, their tough position of ignoring environmental problems gave way to a strategy of environmental adaptation, in which concerns could, on the one hand, position themselves as pioneers of environmental transformation, and, on the other, discover new sources of profit. Combining environmental goals with economic goals has become a new development strategy for many companies. Created in 1995, the World Business Council for Sustainable Development (WBCSD), which includes 200 TNCs with a capital turnover of 8.5 trillion dollars, now operates in more than 70 countries around the world [26]. In terms of its political influence on environmental transformation processes, this organization has no analogues in the business world. According to the organization, about half of the TNCs (47%) that make up the WBCSD are located in Europe. Most of them have direct representatives in Brussels. Economic lobbying structures in the EU are the most powerful in terms of the number of employees and financial support. The interaction between politics and business often takes place on an informal, closed level. As Knill notes, the influence of lobbying organizations is especially significant on the formation of political initiatives of the

European Commission. The European Commission often uses proposals and evaluations of experts from the ranks of business lobbyists [19]. Also, the Council of Ministers is an advantageous platform for protecting economic interests. However, national business structures have a great influence here. European lobbying organizations often act as a coordinator in distributing pressure to national political bodies of EU countries. Purely ecological lobbying organizations are less representative than economic ones. The reasons for this lie in the fact that the EU from the very beginning of its existence was seen as an economic organization. Therefore, economic lobbying was established in the EU structures earlier than environmental. Added to this is a lower level of cohesion among environmental organizations for the protection of their interests. One of the most significant environmental lobbying associations that have a direct impact on the formation of environmental policy in the EU and individual states is the Climate Action Network (CAN) Europe. On the territory of the EU, it unites about 1,700 NGOs. The goal of the organization is to influence the EU structures to develop effective climate change policy [Climate Action Network Europe, 2019]. The association receives more than 250 thousand euros annually from the EU budget for preparing reports and communications for the European Commission and for organizing expert meetings. Long-term work of these two independent organizations practically united their development priorities.

4 Conclusions

The main stages of formation and elaboration of sustainable development of the EU are reflected in the Environment Action Programs. During the period from 1973 to 2020, the European Parliament and the Council of Ministers adopted seven such programs. Analysis of the Environment Action Programs reveals three stages in the elaboration of the EU sustainable development policy. During the initial period (1970s), the formation of a discourse of sustainability was inextricably linked with the formation of a supranational environmental policy. The transnational problems of the EEC member states (pollution of the Rhine River) required the adoption of common political solutions. In the first phase of the development (1972–1992), political actions were aimed at minimizing the negative consequences of the economic growth. The main idea of this period was to improve not only conditions, but also the quality of the population's life. Since 1987, environmental policy has become an independent EU political field. The market consolidation of the member countries required common environmental standards. During the second phase (1992–2012), the EU environmental policy was strongly influenced by the documents on sustainable development adopted at the UN conference in Rio de Janeiro. Ecological economic growth is becoming the dominant idea of sustainable development. The goals of 'sustainability' are stated in the political documents of the EU and nation states. Since 1993, the EU has become a central actor in the global sustainable development policy. Since 2012, the sustainable development policy has been strongly influenced by the concepts of a green economy and ecological transformation of the economy. The transition to a new type of economic management is seen as the foundation for further long-term socio-economic growth.

The European Commission plays a key role in the formation and implementing sustainable development policy. The European Parliament and the Council of Ministers perform deliberative and legislative functions. Their activities are greatly influenced by a variety of groups of actors who pursue their goals. In the last 20 years, there has been a strong tendency towards global business greening. Global business is becoming one of the main actors in the policy of environmental modernization and sustainable development. Most of the innovations in this area come from the structures of TNCs. Thus, it can be assumed that the world economic system is entering a new period of its development – the period of 'green capitalism'. The ecological paradigm of growth forms a new accumulation regime.

Along with the approval of the new EU development strategy by TNCs, NGOs, political parties, its criticism from a number of social movements, science, supporters of alternative development does not cease. A British researcher Tim Jackson, a major critic of the green economy, emphasizes that the strategies of environmental modernization introduced in the past have not brought the expected results as the consumption of natural resources, energy, waste production, greenhouse gas emissions did not decrease, but increased many times over [17]. In developed countries, the introduction of certain green technologies has created a 'rebound' effect, expressed in an increase in the absolute and relative terms of resource and energy consumption. Externalization of the environmental costs of the Western countries has improved their environmental situation, but worsened it on a global scale. Therefore, as some researchers note, the dominance of the environmental modernization narrative in the global political discourse of sustainable development makes it difficult to achieve the comprehensive goals of sustainability. This strategy is narrowly focused and does not call into question the transformation of international political institutions, industrial relations, consumption patterns, international order in the North-South and West-East systems, as well as property, social norms, people's way of life. The transition to sustainable development requires radical changes in individual and collective forms of life and relations in all structures of society, at various levels of its social and territorial organization [3]. With or without a green economy, there could be no sustainable development without fundamental changes in these areas.

In addition, the development of a green economy, "ecological" re-equipment of the energy economy, production, transport, digitalization of living space require enormous expenditures of non-renewable resources. Recent studies show that the ecological footprint of an electric vehicle, when calculated together with the environmental cost of producing it, is on par with the footprint of a conventional vehicle. The green economy might prove to be a new political myth that is certainly profitable for concerns, but harmful to the foundation of planetary life. According to Jackson [17], the only way of complex greening of the society is a complete rejection of the doctrine of economic growth and the creation of a post-growth economy. However, today no political system in the world dares to implement its principles. The green economy remains the only path into the future, and the European Union is its main global apologist.

References

1. Bemmann, M., Metzger, B.: Ökologische Modernisierung. Zur Geschichte und Gegenwart eines Konzepts in Umweltpolitik und Sozialwissenschaften. Campus Verlag, Frankfurt/New York (2014)

2. Nr, B.: 1600/2002/EG des Europäischen Parlaments und des Rates vom 22. Juli2002 über das sechste Umweltaktionsprogramm der Europäischen Gemeinschaft, https://eur-lex.europa.eu/legal-content/DE/TXT/PDF. Accessed 05 June 2020
3. Bohn, C., Fuchs, D., Kerkhoff, A., Müller, C.J.: Gegenwart und Zukunft sozial-ökologischer Transformation. Nomos, Baden-Baden (2019)
4. Brunix, K., Ovaere, M., Delarue, E.: The long-term impact of the market stability reserve on the EU emission trading system. Energy Econ. **89**, 104746 (2020)
5. Climate Action Network Europe: European NGO coalition on climate and energy. http://www.caneurope.org/about-us. Accessed 15 Sept 2020
6. Communication on the European Green Deal. https://ec.europa.eu/environment/basics/green-economy/index_en.htm. Accessed 20 Sept 2020
7. Decision No. 1386/2013/EU of the European Parliament and of the Council of 20 November 2013 on a General Union Environment Action Programme to 2020 'Living well, within the limits of our planet'. https://eur-lex.europa.eu/legal-content/EN/TXT/HTML. Accessed 27 Oct 2020
8. Declaration of the Council of the European Communities and of the Representatives of the Governments of the Member States Meeting in the Council of 22 November 1973 on the Programme of Action of the European Communities on the Environment. https://eur-lex.europa.eu/legal-content. Accessed 12 Sept 2020
9. Development and international economic cooperation: environmental issues. Report of the World Commission on Environment and Development. https://www.un.org/ru/ga/pdf/brundtland.pdf. Accessed 11 Oct 2020
10. Domorenok, E.: Governing sustainability in the EU: From political discourse to policy practices. Governing Sustainability in the EU: From Political Discourse to Policy Practices, 1 January, pp. 1–190 (2018)
11. Dudin, M.N., Zasko, V.N., Dontsova, O.I., Osokin, I.V.: The energy politics of the European union and the possibility to implement it in post-soviet states. Int. J. Energy Econ' Policy **10**(2), 409–416 (2020)
12. Entschließung des Rates der Europäischen Gemeinschaften und der in Rat Vereinigten Vertreter der Regierungen der Mitgliedstaaten vom 19. Oktober 1987 zur Fortschreibung und Durchführung einer Umweltpolitik und eines Aktionsprogramms der Europäischen Gemeinschaften für den Umweltschutz (1987-1992). https://eur-lex.europa.eu/legal-content/DE/TXT/PDF/. Accessed 05 June 2020
13. EU climate action. https://ec.europa.eu/clima/policies/eu-climate-action_en. Accessed 10 Oct 2020
14. European Environment Commission. https://ec.europa.eu/dgs/environment/index_en.htm. Accessed 20 Sept 2020
15. European Parliament. https://europa.eu/european-union/about-eu/institutions-bodies/european-parliament_de. Accessed 05 June 2020
16. Hünemörder, K.F.: Die Frühgeschichte der globalen Umweltkrise und die Formierung der deutschen Umweltpolitik (1950–1973). Franz Steiner Verlag, Stuttgart (2004)
17. Jackson T.: Prosperity Without Growth. Economy for a Planet with Limited Resources. M.: AST-Press (2013)
18. Kaveshnikov, N.: Management methods in the European union. World Econ. Int. Relat. **8**, 49–60 (2015)
19. Knill, C.: Europäische Umweltpolitik. Steuerungsprobleme und Regulierungsmuster in Mehrebenensystem. VS Verlag für Sozialwissenschaften, Wiesbaden (2008)

20. Lenshow, A., Burns, C., Zito, A.: Dismantling, disintegration or continuing stealthy integration in European Union environmental policy. https://www.scopus.com/record/display.uri?eid=2-s2.0-85082683790&origin=resultslist&sort=plf-f&src=s&st1=Evolution+of+the+Environmental+Policy+of+the+European+Union&st2=&sid=ccf111958f19fbe05f9980e2ac9818e0&sot=b&sdt=b&sl=74&s=TITLE-ABS-KEY%28Evolution+of+the+Environmental+Policy+of+the+European+Union%29&relpos=1&citeCnt=0&searchTerm=. Accessed 15 Oct 2020
21. Meyer, J.-H.: Bürgerschaftliches Engagement über Grenzen? Europäische Umweltpolitik und transnationale Vernetzung von Umweltgruppen in den 1970er Jahren. In: Nachhaltige Stadtentwicklung. Infrastrukturen, Akteure, Diskurse. Jens Ivo Engels, Nina Janich, Jochen Monstadt, Dieter Schott (Hg.) Campus Verlag, Frankfurt/New York (2017)
22. Resolution of the Council of the European Communities and of the Representatives of the Governments of the Member States Meeting The Council of 17 May 1977 on the continuation and implementation of a European Community policy and action programme on the environment. https://op.europa.eu/en/publication-detail/-/publication/634c0a49-1819-47d4-9950-2908f0da613d/language-en. Accessed 15 Oct 2020
23. Resolution of the Council and the Representatives of the Governments of the Member States, meeting within the Council of 1 February 1993 on a Community programme of policy and action in relation to the environment and sustainable development. https://eur-lex.europa.eu/legal-content/EN/TXT/PDF/. Accessed 11 Oct 2020
24. Rifkin, J.: Der Globale Green New Deal. Campus Verlag, Frankfurt/New York (2019)
25. Simonis, G.: Handbuch Globale Klimapolitik. Utb. Paderborn (2017)
26. World Business Council for Sustainable Development (WBCSD). https://www.wbcsd.org/Overview/About-us. Accessed 12 Oct 2020
27. 1,2 Milliarden Euro für null Transparenz. https://www.spiegel.de/wirtschaft/eu-kommission-finanziert-ihre-eigenen-kritiker-a-1141392.html. Accessed 17 Sept 2020

The Development of International Sustainable Industrial Tourism on the Example of Germany

Yulia Kozina(✉) and Nadezhda Bogdanova

Peter the Great St. Petersburg Polytechnic University, St. Petersburg, Russia

Abstract. This article is aimed at addressing issues of sustainable tourism development within current international system. The paper analyzes the formation of sustainable industrial tourism as a leisure activity. Using the example of the Federal Republic of Germany, the article examines the results that have been achieved due to the concept of sustainable development in the field of industrial tourism. The article contains recommendations of strengthening international cooperation in this area based on a methodological approach to collecting and interpreting tourism development data, which combines indicators of three dimensions of sustainable development (economic, socio-cultural and environmental). Benchmarking is used as a strategic management tool, which is based on a combination of primary market analysis using statistical data in order to create a theoretical basis for sustainable tourism in the international arena.

Keywords: Sustainable development · Industrial tourism · International tourism · Ecology · The Federal Republic of Germany

1 Introduction

International relations is a combination of economic, political, ideological, legal and other ties between states and non-state actors.

Nowadays, tourism, being an area that makes a significant contribution to the state's economy due to the intensification of globalization, plays a fundamental role in the economic, cultural and political development of not only the recipient state. International tourism contributes also to the development of international relations, just as the sphere of international relations has a reciprocal effect on the tourism industry. In the modern world, according to the degree of development of the tourism sector in the state, we can give an indication of the relationship between various political units and the level of international security.

The *relevance* of the topic depends on the fact that it is important for modern researchers to consider such multidimensional concept as tourism, without interruption from pressing global problems, namely environmental and ecological problems. Consequently, the concept of sustainable development comes to the fore in the field of international tourism.

© The Author(s), under exclusive license to Springer Nature Switzerland AG 2021
R. Bolgov et al. (Eds.): *Proceedings of Topical Issues in International Political Geography*, SPRINGERGEOGR, pp. 270–280, 2021.
https://doi.org/10.1007/978-3-030-78690-8_24

The term "Sustainable Development" was first used in 1987 in the report "Our Common Future" by the World Commission on Environment and Development. The doctrine of sustainable development is based on the idea that economic development should be conducted without the threat of depletion of natural resources, so that future generations would be able to satisfy their needs in order to maintain a high quality of life [1]. Therefore, sustainable tourism represents all forms of tourism development and management that do not contradict the natural well-being of established societies for an unlimited period.

The *innovativeness* underlies the identification of how international organizations and other political units influence the development of modern tourism in the framework of sustainable development.

The *aim* of the study is to develop recommendations on improvement of the implementation of the sustainable tourism development concept by the combined efforts of the international community.

In compliance with the designated aim, the following *purposes* were put forward:

- to find out how the sphere of international relations affects the development of international tourism, in particular, by the example of transition from standard mass tourism to the concept of sustainable development;
- to determine what results have been achieved in the field of sustainable industrial tourism development using the example of the Federal Republic of Germany;
- to develop suggestions and recommendations on how to improve the sphere of international tourism in the context of international relations.

2 Materials and Methods

The theoretical and methodological framework of this study is based on general scientific methods of cognition, namely:

- on a review of general scientific, economic, and sociological literature;
- on using the classification method as part of a study on the development of industrial tourism in Germany.

In addition, the method of content analysis was applied when working with text massive and processing received data.

The comparative method has proven to be no less important in the research process, since it allowed us to compare the results that were achieved in various federal states of Germany as part of the modification of former industrial facilities in accordance with the principles of sustainable tourism, without harm to the environment and the local population.

The official websites of federal ministries and departments of the Federal Republic of Germany were the most valuable for studying the topic. In particular, the German National Tourist Board [2], the German Tourism Association [3], the Federal Environment Agency [4] and the Federal Minister for the Environment, Nature Conservation, and Nuclear Safety [5].

From a theoretical point of view, the following works were useful: the work of Nawar Al-Saadi "The role of international relations in the development tourism sector", where the author considers the general issues of the tourism sector functioning in the context of modern international relations [6], the paper of Shiva Jalalpour and Jamshid Shojaeifar "The tourism industry and the international relations", in which they analyze how international tourism affects international relations [7], and the book of V.S. Novikov "Innovation in Tourism" [8], where the author sets out the principles of modern sustainable tourism and distinguishes standard mass tourism and the concept of sustainable development.

3 Problem Statement

3.1 The Establishment of Sustainable Tourism

At the present stage, tourism is a rapidly developing field of activity and is one of the largest economic sectors. So, according to the latest statistics published by the World Travel and Tourism Council (WTTC) in 2019, the contribution of the tourism industry to world GDP was 10.4% [9].

The impact of tourism, both domestic and international, on the host country is enormous. However, no one argues that this influence is exclusively positive.

Tourism brings definitely benefits not only in the socio-economic sector. It also helps in the development of political relations between countries and peoples by creating a positive image of the state or region. Moreover, international tourism contributes to a deeper understanding of the cultural aspects of other people's lives, which in turn creates favorable conditions for the flourishing of tolerance and respect for foreign customs and traditions.

Nevertheless, a developed network of tourist attractions and a favorable image and the so-called "popularity" of the region suggest a large tourist flow, which is a serious negative burden on the local ecosystem. A good example of this is the destruction of coral reefs in Thailand. The concept of sustainable tourism is called upon to solve or at least mitigate the problem of causing damage through the tourist use of the area.

For a more complete disclosure of the topic, we must first start with the history of the formation and development of the concept of sustainable tourism.

A change in outlook on tourism and a turn towards environmental security and protection was made at the end of the 20th century. Therefore, the concept of sustainable tourism is not as new as it might seem at first glance. The term and fundamental principles of this concept were distinguished by the World Tourism Organization (UNWTO) in the late 1980s. They were protecting the ecological systems of the planet and conserving natural resources by using them wisely so as not to deprive future generations of the opportunity to satisfy their needs [10]. Sustainable tourism has come to mean tourism activities that use tourism resources rationally and ensure their safety.

Sustainable tourism should not be confused with other various types of tourism, such as recreational or business tourism. Sustainable tourism is just a "set of rules", which can help this area develop not only its economic sector, but also protect the environment.

The principles of protecting the biosphere were consolidated at the United Nations Conference on Environment and Development in Rio de Janeiro in 1992, where a program plan for the 21st century (Agenda 21) was elaborated [11].

Thus, since 1992, the concept of sustainable tourism has been officially enshrined in the documents of the World Tourism Organization, which was created to promote tourism as a factor in "ensuring international peace and understanding".

In 1995, at the platforms of such organizations as the World Tourism Organization and the World Travel and Tourism Council, an action plan in the field of sustainable tourism was outlined for the governments of states and sectoral divisions:

- assessment of both economic and environmental activities of organizations operating at the national level, in terms of the concept of sustainable tourism;
- development of programs for the implementation of the concept and public education in the relevant field;
- ensuring the participation of all governmental and non-governmental initiative groups in the work of sustainable tourism development programs;
- development of new tourist products in accordance with the basic principles of sustainable tourism [8].

A good example of the implementation of these recommendations is the modification of industrial tourism facilities in Germany.

3.2 The Basic Provisions of the Concept of Sustainable Tourism Development

In its modern form, the presented format of tourism development and management meets the following principles, which were finally consolidated in 2004 in the documents of the World Tourism Organization:

- ensuring tourism activities in harmony with nature and respect and enforcement of laws aimed at protecting the environment;
- contributing to the protection of natural ecosystems, for example, by regulating and reducing tourist flows;
- ensuring cooperation in the field of an open economic system and involving citizens in the development of local tourism and solving problems in this area;
- combating trends in protectionism in the tourism environment;
- international exchange of experience in introducing effective means of tourism development and management [10].

Since the main rule of sustainable tourism is the design of local tourist locations, taking into account the peculiarities of the area (both landscape and cultural-historical) by involving the forces of the local population, it is possible to achieve an increase in the standard of living of residents in the respective regions without any special expenses.

In greater detail the aspect reflecting the relationship between the tourism development of peripheral areas and the concept of sustainable development will be demonstrated later on using the example of industrial tourism in the Federal Republic of Germany.

4 Research Questions

4.1 The Concept of Sustainable Tourism Development in Germany

There are two departments that are involved in the development and promotion of both the country as a tourist destination and individual tourism products in the Federal Republic of Germany: the German National Tourist Board (Deutsche Zentrale für Tourismus) [2] and the German Tourism Association (Deutscher Tourismusverband) [3]. It is noteworthy that despite the fact that each federal state of Germany is a separate tourist destination, and they often provide competing products to the country's tourism market, the represented agencies are engaged in the general promotion of Germany at the external level, creating a positive image and increasing the tourist flow to the country. In addition to these organizations, each municipality in Germany develops a tourism strategy at the local level.

Since the Federal Republic of Germany is one of the favorite European tourist destinations (Germany takes the third place according to the Statistical Office of the European Union after Spain and France), as part of the concept of sustainable tourism, the Federal Ministry for the Environment, Nature Conservation, and Nuclear Safety (BMU) could not but be interested [5].

During the 2002 this Ministry, together with the Federal Environment Agency, conducted an in-depth analysis of the impact of tourism on the local ecosystem. Qualitative assessments were made of the total load of the tourist flows to the amount of primary energy consumption, harmful emissions into the environment (into the atmosphere, soil and water resources) and the noise level was also measured. These factors negatively affect land use and management and biodiversity. Among the main problems created by tourists, the following were highlighted:

- Contribution of tourist flows to the climate change. It's a question of carbon dioxide (CO2) emissions from traffic (mainly due to travelling by car, bus, plane and ship). It is also worth emphasizing that, according to these criteria, the most harmful mode of transportation is travelling by airplanes, since their emissions into the atmosphere at a certain altitude have great consequences for climate change.
- Violated air circulation with the accumulation of polluting particles at the surface of the earth. The ecologists are particularly concerned about the use of fireworks as part of tourism events.
- Water shortage and water pollution. In this case, there is competition in the distribution of water resources between the needs of tourists (for example, water for hotels) and the needs of the local community and agriculture.
- Pollution and alteration of soil structure due to overuse, for example, for the construction of hotels or parking places. This leads to the loss of soil natural functions and can even become a risk for seasonal flooding, etc.

Every two to three years, the Federal Ministry for the Environment, Nature Conservation, and Nuclear Safety holds a "Federal Competition on Sustainable Tourism Destinations in Germany" to encourage the most active federal states in this field and stimulate further development. The most striking year was 2017 as it was the time that the United Nations declared the year of international sustainable tourism. To this day it was the last year of this competition. The main evaluation method is benchmarking, which implies comparing products and services of one company with similar products and services of another in order to improve aspects of comparison [5].

The assessment of the sustainability of a tourist site was carried out according to the following criteria:

- Management - The tourism destination has developed strategies and concepts based on sustainability principles. For this, there are control mechanisms to ensure the best possible conditions for a stable orientation of the object. Management actively influences the development of proposals for sustainable tourism.
- Economy - Tourism contributes to the economic well-being of the population and helps to create regional economic structures. Tourism providers have a wide range of high-quality events.
- Ecology - Protection of the environment, nature, biodiversity and landscape is carefully maintained. Tourism organizations use natural resources especially efficiently. The sponsors of the tourism organization are developing sustainable mobility in the region and are actively advertising public transport options. This supports climate protection, energy conservation and resource efficiency measures.
- Sociocultural trends - Tourist offers make the region unique, thereby strengthening traditions and cultural heritage, the general well-being and quality of life of the host community. Local residents, for example, have the opportunity to participate in tourism development processes.

As a winner was announced the Swabian Alb Biosphere Reserve, becoming the best by the following criteria:

- protection of nature and landscape;
- high quality of life in the region;
- regional prosperity.

The jury paid special tribute to:

- close and partnership-based cooperation between the Biosphere Reserve Office and tourism decision-makers, what created synergies in developing further ideas and recommendations;
- development of using the practical guide “Sustainable Tourism Development in Germany”;
- the reserve actively participates in such initiatives as “Green destinations”, “Green tourist map of Germany”, TourCert certificate “Sustainable tourist destination”;
- sustainable mobility offers, such as SaarlandCard or Destination Nature membership, provide impetus for the environmentally friendly transportation of guests (Jury recommendation: ride on a 501 biosphere bus accompanied by a certified guide) [12].

4.2 The Current State of Industrial Tourism in Germany

German sustainable tourism development projects focus not only on areas that are currently suffering from climate change, such as the alpine region. Attention is also paid to territories that, as a result of historical and industrial development, have encountered environmental problems and have been turned into lands "forbidden" for tourists. Speaking of Germany, we immediately recall such federal states as North Rhine-Westphalia with its Ruhr region and the Saarland.

Here, with some former industrial enterprises, which have now been modified into places of sports and cultural leisure activities, the problem was solved in a certain way. Vivid examples of this are:

- The Zollverein Coal Mine Industrial Complex (German: Zeche Zollverein) [13] in the city of Essen and Landschaftspark Duisburg-Nord [14] in Duisburg, North Rhine-Westphalia, Ruhr district;
- The Völklingen Ironworks (German: Völklinger Hütte) [15], Saarland.

The already inoperative coal mine, Zeche Zollverein, which completely ceased operations in 1993, boasts a rich cultural life in the twenty-first century. Firstly, it is one of the stations of the regional project "The Way of Industrial Culture" of the Ruhr region. Secondly, there is a big variety of museums which are opened for visitors, for example, the Coal Road Museum, the Ruhr Museum, the North Rhine-Westphalia Design Center and the Ceramics Museum. Exhibitions of contemporary art are periodically held in the pavilions here, and a permanent exhibition of the Soviet artist I. Kabakov is located in the former coke plant [13].

Landschaftspark Duisburg-Nord was developed at the site of a former metallurgy combine in 1991. This place is interesting not only in terms of landscape, but also of leisure centers located on its territory. For example, now in the former gas tank - a special tank for storing various gaseous substances - there is a diving center where scuba divers can enjoy not only artificial coral reefs, but also sunken ships and cars. The place of the old power station was reserved for concert and festive events for lovers of open-air cinema, rock climbing and parkour [14].

A good example of the coexistence of industry and the usual place for leisure and recreation is the creation of a landscape park, one of the largest in the world, on the territory of the former metallurgical plant. The Völklingen Ironworks, which worked from 1873 to 1986 and was the largest iron smelter in Germany, was the first industrial monument to receive international status in the world. Most often, the territory is used for rock concerts and chamber music concerts in the open air [15].

It is worth noting that the Zollverein Coal Mine Industrial Complex and the Völklingen Ironworks were included in the UNESCO World Heritage List in 2001 and 1994, respectively, which emphasizes their special cultural and historical significance. Coal mines in the Ruhr region (Bergwerke des Ruhrgebiets), which were once used for open pit mining, have now become a place of attraction for athletes from all over the world, as these careers provide an excellent opportunity for training and preparation for mountain cycling competitions.

Such metamorphoses on the site of former industrial enterprises made it possible to diversify the tourism and economic sectors of the represented federal states. According to the official data of the North Rhine-Westphalia Travel Agency published in 2019, the North Rhine-Westphalia accepted 51.9 million tourists officially registered in hotels or hostels, forty-three percent of which is of the internal tourist flow. The total number of vacationers increased in federal state from 2009 to 2019 by almost 11.9 percent. The number of tourists in the Ruhr region, where the majority of mines, metallurgical plants and coal mills are located, in the corresponding period increased by 42.6 percent from 781 thousand vacationers in 2009 to one million one hundred forty-one thousand in 2018. In addition, the most popular types of leisure activities and recreation in 2019 in North Rhine-Westphalia were cultural and concert events (32.4% of the total number - the most popular), sports activities (8.6%) and relaxing outdoors (7%) [16].

The tourism industry of the Saarland Federal State is developing according to a similar scenario, however, at a lower rate, since historically the Saarland was not a tourist place at all. Despite the successful development of industrial tourism, the external tourist flow is still at a rather low level. The official Tourism Agency of Saar in 2020 set a goal to increase the attractiveness of the region by using its tourism potential in typical forms for the land. In addition to developing a cultural complex on the territory of the former metallurgical plant in Völklingen, the Saar government has developed a number of multi-day cycling routes for the development of outdoor activities that meet the principles of sustainable tourism, as well as making most of the attractions of industrial tourism easily accessible for people with disabilities. An illustration of this unique modification of industrial tourism facilities is the Saar testing area (Bergehalde Ensdorf), a monument to the coal industry, awarded the Prize for Architecture and Urbanism in Saar in 2017 [12].

International Relations and International Tourism. It is obvious that there exists a close relationship between international relations and international tourism. First of all, this is reflected in the fact that even the presence of tourist flows from one country to another is explained by a certain level of relations that the states managed to achieve. The situation in the international arena (the presence of diplomatic relations, political conflicts, general socio-economic, environmental and other projects) directly affects the tourism sector of the country. Notable example in this case is situation that occurred with Russia not so long ago. It is about a plane crash that happened on October 31, 2015 with an Airbus 320 aircraft over the Sinai Peninsula, which resulted in a long termination of air traffic between the Russian Federation and Egypt. This in turn entailed significant losses from the host state.

In general, tourism can even serve as a mean of diversification for those countries that occupy narrow positions in the global economy and it can help them reach a qualitatively new level in the international arena. For example, commodities export countries. In this case, the state needs to pay more attention to the tourism sector [6, 7].

The tourism industry operates also in a special legal field, which is also determined by international treaties and agreements. An example of this is the Schengen Agreement (signed in 1985) on the simplification of passport and visa control at the borders of European states that agreed to the conditions. The Federal Republic of Germany is also a part of this agreement since 1990.

The Global Sustainable Tourism Council [17] considered the possibility of creating a unified certification of tourism organizations. Thus, the widespread introduction of this certification allows not only travel companies to achieve individual business benefits, such as reducing costs, increasing employment and improving their image, but also provide environmental and social guarantees to the local communities.

Nevertheless, the international legal field may impose some restrictions on tourist flows, as in the case of the current difficult epidemiological situation due to COVID-19, when the Italian bilateral initiative to create a tourist passageway was rejected by the German side in May 2020.

As a whole, the international environment, the absence of strict regional and systemic constraints, as well as global challenges, has helped Germany to become an attractive tourist destination. The country has a rich cultural, historical and architectural heritage. Germany today has become one of the global scientific and economic centers; this is especially manifested on a European scale. In addition, the Federal Republic of Germany is not involved in acute political conflicts and the country is not in danger of international isolation, which could negatively affect the tourism business.

According to the World Tourism Organization, Germany in 2019 entered the top ten most visited countries in the world (39.4 million tourists) [10]. Due to such popularity and a positive image of the country on the international scene, Germany can take advantage of such a position to attract other interested countries to support the concepts of sustainable tourism development in Germany and to implement them within national borders. As a key player in European integration, Germany is given the opportunity to promote these ideas at the political level within the institutional system of the European Union.

5 Conclusion

Thus, on the basis of the study, it was found out that international tourism is an integral part of the system of international relations as well as the political and economic spheres. We have revealed the relationship between these two phenomena. On the one hand, a stable international tourist flow is ensured by a favorable and peaceful foreign policy situation, the level of political, economic and other types of cooperation between states.

On the other hand, international tourism affects the perception of the country by the actors of international relations by creating a certain image in the framework of the provision of tourist services.

We also indicated that in the world where global problems reign, the tourism industry is being restructured so that if the danger is not eliminated, then at least it should not be exacerbated on a large scale. This primarily concerns global environmental issues, such as climate change or water pollution. In order to preserve and restore ecological systems, the tourism sector has joined the popular concept of sustainable development, which aims to keep the environment and biodiversity for future generations.

In the Federal Republic of Germany, in particular, the concept of sustainable tourism is supported by specialized ministries and departments. Thus, the German Federal Environment Agency and the Federal Ministry for the Environment, Nature Conservation and Nuclear Safety also carry out educational work with the population by constantly publishing information on sustainable tourism and special tourist guides on official websites. For example, the Ministry recommends choosing nearby places for leisure and recreation. This is connected with the use of various vehicles by vacationers that have a negative impact on the environment, and it is also proposed to abandon air travel completely. Citizens of Germany are asked to choose suppliers of tourism services that meet environmental standards. In this case, the enterprises, engaged in such activities, are interested in meeting quality standards.

Based on its own experience, the Federal Republic of Germany could promote similar projects at the European and global levels.

In conclusion, we offer some recommendations that would help to increase the development level of sustainable tourism in the context of modern international relations:

- it is necessary to use leading countries as an example for their neighbours and the rest of the world community;
- it would be necessary to intensify the work of foreign ministers at the platforms of specialized international organizations, such as the World Tourism Organization, as well as involve non-governmental organizations and initiative groups in these issues;
- it is important to organize international events within the framework of a certain type of tourism in order to attract the attention of the international community and diplomatic circles. During such events, a good opportunity for the discussion and implementation of joint projects on sustainable development appears, the search for the right investors, as well as the exchange of experience;
- the idea of intensifying the integration of the global tourism sector and planning an international tourism free zone, which would provide preferential tax and customs conditions for entrepreneurs from the tourism business industry and equalize foreign units with national rights, is being promoted.

References

1. Development and international economic cooperation. Report of Brundtland Commission. UN General Assembly (1987). https://www.un.org/ru/ga/pdf/brundtland.pdf. Accessed 10 June 2020
2. Deutsche Zentrale für Tourismus. https://www.germany.travel/de/index.html. Accessed 10 June 2020
3. Deutscher Tourismusverband. https://www.deutschertourismusverband.de/. Accessed 10 June 2020/10
4. Nachhaltiger Tourismus. Umwelt Bundesamt. https://www.umweltbundesamt.de/themen/wirtschaft-konsum/nachhaltiger-tourismus#bedeutung-des-tourismus. Accessed 10 June 2020
5. Nachhaltiger Tourismus. Projekte und Initiativen. Bundesministerium für Umwelt, Naturschutz und nukleare Sicherheit. https://www.bmu.de/themen/wirtschaft-produkte-ressourcen-tourismus/tourismus-sport/nachhaltiger-tourismus/projekte-und-initiativen/. Accessed 10 June 2020
6. Al-Saadi, N.: The Role of International Relations in the Development Tourism Sector, 1st edn. LAMBERT Academic Publishing, Saarbrücken (2014)
7. Jalalpour, S., Shojaeifar, J.: The tourism industry and the international relations. World J. Environ. Bioscie. **6**, 68–72 (2014)
8. Novikov, V.S.: Innovation in Tourism, 1st edn. Academia (2007)

9. World Travel & Tourism Council. https://wttc.org/. Accessed 10 June 2020
10. The World Tourism Organization (UNWTO). https://www.unwto.org/. Accessed 10 June 2020
11. Agenda XXI. Rio de Janeiro (1992). https://www.un.org/ru/documents/decl_conv/conventions/agenda21_intro.shtml. Accessed 10 June 2020
12. Tourismus Zentrale: Angebote (2020). https://www.saarland.de/dokumente/thema_tourismus/Brosch_Angebote_2020.pdf. Accessed 10 June 2020
13. Zeche Zollverein. https://www.zollverein.de. Accessed 10 June 2020
14. Landschaftspark Duisburg-Nord. https://www.landschaftspark.de. Accessed 10 June 2020
15. Weltkulturerbe Völklinger Hütte. https://www.voelklinger-huette.org/willkommen. Accessed 10 June 2020
16. Tourismus in Nordreihn-Westfalen. Daten und Fakten. https://www.touristiker-nrw.de/wp-content/uploads/2019/06/TOU082318_Marktforschungsbroschuere_2018_online.pdf. Accessed 10 June 2020
17. Global Sustainable Tourism Coincil. https://www.gstcouncil.org/sustainable-tourism-training/certificate-in-sustainable-tourism/. Accessed 10 June 2020

Migration and Socio-Demographic Processes

Muslim Community in France: Problems of Adaptation

Anastasia Umnova[1], Anna Matveevskaya[1], and Sergey Pogodin[2](✉)

[1] Saint Petersburg State University, Saint Petersburg, Russia
[2] Peter the Great Saint Petersburg Polytechnic University, Saint Petersburg, Russia

Abstract. An elevated level of ethnic migration is currently observed. This is due to such reasons as: 1. local armed conflicts; 2. the policy of oppression and suppression of the right to ethnic self-determination; 3. the desire to migrate to the more developed countries of Europe and North America from Latin America, South Asia and Maghreb; 4. an increasing need for natural security. All this requires attention and a serious approach at the global level. For the first time since the existence of migration processes, the world community has faced such a serious problem as the migration crisis in 2014. The Muslim world has become the main source of labor and humanitarian migration to France. As a result, communities of a different cultural and civilizational orientation have formed in the host country. The answer is a natural tendency to preserve ethnic identity. Muslim communities in France are culturally differentiated. An analysis of the cultural component of Muslim society shows that immigrants are distinguished by an extremely depressed cultural and psychological attitude. The growth of migration flows leads to augmentation of anti-emigration sentiments among the population within the state due to the formation of ethnic stereotypes and propaganda nationalism within French society. In this regard, France must create a special cultural space for possible cultural dialogue and social interaction between two civilizations: Western and Islamic - in order to design a public foothold for building a strong, pragmatic and effective political discourse later.

Keywords: Muslim communities · France · Migration processes · Problems of adaptation

1 Introduction

The decision to hold a High-level Dialogue on International Migration and Development by the General Assembly in 2003 drew attention to international migration issues at the global, regional and national levels [1]. The reason lies in the unprecedented growth in labor migration, which, according to the UN, amounted to 154.2 million migrants in the 1990s in the world [2].

Over the last twenty years, migration has become global in social nature due to intense economic transformation. So, the labor force of donor countries, in this case we are talking about the third world, is aimed at compensating for the deficit in the

© The Author(s), under exclusive license to Springer Nature Switzerland AG 2021
R. Bolgov et al. (Eds.): *Proceedings of Topical Issues in International Political Geography*, SPRINGERGEOGR, pp. 283–297, 2021.
https://doi.org/10.1007/978-3-030-78690-8_25

labor force of recipient countries. The interdependence and mutual influence of states is a consequence of these processes. Migrants look toward the most developed and important centers of Europe, in particular France. The result of the massive movement to France is the formation of a community of migrants. The desire to preserve their language, culture and traditions, the way of life for some of the migrants is an important aspect. This complicates the processes of adaptation and integration of newcomers into the host society.

Moreover, in the twenty-first century, the European community is faced with an unprecedented flow of refugees and migrants due to the migration crisis in 2015. This time is also characterized by a high level of interest in the life of Muslim communities and the Islamic religion in general on the part of the public and political leaders. It is important to note, however, that this interest has two sides of the same coin. On the one hand, this is curiosity and a desire to learn another culture. On the other hand, it is the fear of terrorism and Islamist ideology, which is fueled by the significant media.

The priority for the French Government in the early 2000s is the issue of reducing family migration and increasing the role of labor migration in the context of the global crisis [4]. In this instance, the country's parliament even considered the possibility of introducing a DNA test for migrants arriving for family reunification.

However, as we can see, the policies of N. Sarkozy and F. Hollande only served as a catalyst for internal tension among the Muslim community. This is due to the fact that the reforms in the period from 2000 to 2014 touched on for the most part basic Muslim values and were based only on the restriction of religious self-identification and self-expression, which ultimately caused a wave of indignation.

At the present stage, the policy of the representative of the centrist movement E. Macron is aimed at updating the institutional French values, introducing new trends into the already existing realities, which are associated with unprecedented Muslim influence and power in a globalizing society. The French government is doing its best to minimize public dissatisfaction with migrants. Each year, the government allocates over 5 billion euros for the development and promotion of programs for the integration of newly arrived migrants into the social life of the country [5].

Today, it seems an impossible step to fully limit and neutralize the huge Islamic diasporas, which are constantly replenishing due to legal and illegal immigration. The development of sustainable and pragmatic policies can influence a softer and smoother process of integration or "political involvement" of Muslim communities in French society.

Muslim communities in France are culturally differentiated. The integration of the Maghreb, Turkish, and Berber populations is proceeding extremely slowly and, in some cases, "heterogeneous". This is due to the fact that the second and third generation of migrants strive to carry out internal reforms within their communities regarding the perception of democratic values, as well as adapt family norms and Islamic traditions to existing realities.

The problem is that the "identity dilemma" that characterized the second wave of migrants is being imposed on the new generation. That is why, in order to overcome the existing differences in the sphere of culture, education and employment, the child should be allowed to develop from a family different from the official "culture" - freely and independently. In this case, the goal of effective and fruitful implementation of the integration process will be achieved at the social and political level.

Futhermore, the solution to the problem of integrating the Muslim community into an autochthonous society seems impossible if the policy pursued is carried out through the "prism of French republican values" [6]. The reason lies in the fact that this political attitude is aimed at the complete or partial omission of ethno-national differences in society, in other words, the illusion of the absence of real ethno-confessional conflicts is created. Otherwise, communities of Muslim representatives can take the solution into their own hands. In particular, the younger generation can defend their ideas from the standpoint of "Islamist radicalism", which later becomes the cause of mistrust and fear of Islamic culture in general and gives rise to a range of significant and intractable social problems.

2 Materials and Methods

Despite the fact that the sense of national identity of France, formed after the Great French Revolution of 1789, turn a blind eye to ethnic discourse as a tool for developing domestic policy, this state remains the most attractive for migrants from completely different regions throughout its history. That is why the migration measures of the French state are aimed at implementing the integration policy through the use of national mechanisms. These mechanisms are: mandatory study of language, culture, history, as well as the legislative framework of the Republic by migrants.

Migration to France is 7.85 million people, or 11.9% of the total population (66.2 million), according to Eurostat statistics from 2000 to 2015 [7]. In 2018, Insee (Institut national de la statistique et des études economiques) states that immigrants make up 6.5 million or 9.7% of the total population (66.9 million), of which 4.3 million (6, 4%) were born outside Europe. Also, in 2018, more than 120 thousand applications for obtaining official status, including 20 thousand children, were officially submitted in France [8].

INED (Institut national d'études démographiques) reports that according to the 1999 census, among the population living in France, about 13.5 million people have full or partial foreign origin for three generations [9]. Gerard Noiriel explains that in 2002, about a third of French people are of foreign origin. In 2011, according to an article by Michel Tribalat, there are 19.2% of people of foreign origin in France within two generations and almost 30% for three generations [10]. Pascal Blanchard reports that in 2015, 12 to 14 million French people, that is, 18 to 22% of the total population, have at least one close relative who was born in non-European territory.

Muslims in the countries of the European Union are the largest religious community, which is extremely diverse in its ethnic essence [11]. Moreover, France, being one of the European states, is the leader in the number of Maghrebians living in the country. However, the ban on conducting an ethnic census and creating official religious statistics is still in effect in the state, despite the current trend towards a change in the ethnic makeup in favor of the representatives of the Arab world. However, this phenomenon does not apply to surveys that can be conducted by the French national institutions that advise. These institutions act in accordance with the measure required for the purposes of processing and after the approval of the National Commission for Informatics and Freedom (CNIL

- Commission nationale de l'informatique et des libertés) and the National Council for Statistical Information (CNIS - Le Conseil national de l'information statistique).

Thus, in 2009, the percentage of the Muslim population, according to the French Institute of Public Opinion (Ifop - Institut français d'opinion publique), was 5.8%, which is equivalent to 3.5 million people. A study published in October 2010 by the National Institute for Demographic Research (Ined - Institut national d'études démographiques -) and the National Institute for Statistics and Economic Research (Insee - Institut national de la statistique et des études économiques) showed that the Muslim population in France has decreased up to 2.1 million people [12].

Already in 2015, about 6 million people (9% of the population) were immigrants from the Middle East and the Maghreb and, according to Insee [13]. It is important to note that 11% of Muslims in France, in spite of the fact that they are French citizens, were born into an immigrant family.

As of March 2020, the population of France is about 67 million people, 12% of whom are migrants of the Islamic faith [14]. In this regard, it should be emphasized that approximately 325 thousand immigrants stay legally on the average in the territory of the state [15]. Nevertheless, experts note that most (about 94% of migrants) remain in the position of "illegals" [16].

Age statistics, according to Ifop, indicate that Muslims living in France, along with migrants, make up 5.6% of people over 15 years old, and only 10% under 25 years old, and 84% under 50 years old. It follows from this that is a fairly "young" social group [17].

The ethnic component of migration flows to France remained practically unchanged. Despite the existing European migration crisis, requests for citizenship or asylum in France came from the Maghreb countries between 2014 and 2018. That is why 43.6% of migrants are Maghreb. Thus, Algeria accounts for 25% of migrants, Morocco 11.5% and Tunisia 7.1%, respectively [18].

Geographic statistics in the context of migration flows and resettlement of the Muslim population of France is uneven, as well as increased concentration in only a few regions of the country. Thus, about 62% of visitors and their descendants live in three regions - Ile-de-France, Auvergne - Rhone - Alpes and Provence - Alpes - Cote d'Azur. At the same time, 40% of immigrants are in the urban part of Paris, while the French make up only 20% of the inhabitants of these quarters. Moreover, the capital of France has twenty administrative entities, of which five are recognized as Muslim [19].

Boris Dolgov, an expert at the Center for Arab Studies at the Institute of Oriental Studies of the Russian Academy of Sciences, states that three generations of migrants from the Maghreb and the Middle East are in France by the beginning of the XXI century. For example, the first generation are migrants, on average 60 to 70 years old, who arrived in France after North Africa gained colonial independence, as well as after World War II. The second generation is represented by the children of the "first" migrants. It is noteworthy, that the older demographic cohort treats with respect and tolerance to the established republican traditions of the recipient country, also express gratitude to a greater extent than their descendants. They, in turn, express certain levels of dissatisfaction with the inability to have and exercise their rights on an equal basis with local residents. Finally, the children and grandchildren of the second wave of migrants together make up the third generation of migrants. But there is a paradox. Representatives of the third group have a stronger connection with the secular culture of France, they are brought up and developed through European educational standards, liberal culture and the media to a greater degree than their ancestors. However, the reality reflects the fact that this generation tends to radical branches of Islam, as well as an absolute consciousness of their belonging to an "other" culture, Muslim [20].

3 Results

The importance of analyzing the cultural component of Muslim life lies in the ability to identify the special aspects of the civic position of representatives of the Islamic faith, communicative behavior in relation to existing political and social discourses, as well as a choice in favor of religious or republican values.

The authors begin their analysis of the cultural peculiar properties of the Muslim community in France by looking at the traditional way of life in the Muslim family, which directly affects the development of social skills and integration in general. As you know, the patriarchal system of relations within the Maghreb and Middle Eastern communities both in the ethnic homeland and on the territory of France is a classic model of genealogical relationships up to the present day. Children should recognize the authority of their fathers and the older generation, follow not only Sharia norms, but also traditional methods of education. A woman is perceived only in two modalities - a wife and a mother, from this we can deduce that the position of Muslim women is fundamentally different from the status of French women in one social community. The impossibility of close contact with the opposite sex, and in some families even simple communication with representatives of the European community of any genders, is perceived extremely ambiguously by the young generation of immigrants. On the one hand, they are forced to face the democratic, liberal and secular lifestyle of their peers every day in state institutions. Peers are able to influence the formation of the outlook of newcomers and, in the future, identification. On the other hand, family conditioning acts as a barrier, borders, through which it is difficult, and sometimes even impossible, to pass without risking to face misunderstanding on the part of both the Muslim community and the European one. In other words, many children studying in secondary schools for the first-time face directly, independently with a new and completely different culture, which in most cases attracts them. As a result, they adopt some social practices and refuse to accept the value system imposed on them by their parents. Therefore, many representatives of the third generation do not know and are not willing to learn Arabic. They do not see the need for this, being in a country with different linguistic realities, as well as the standards of morality and character training.

Moreover, as a part of the investigation., we were able to find out one interesting practice of Maghreb families in the scope of the percipience of the classic French greeting (a mandatory gesture in the form of two to four kisses on the cheek for both women and men) by representatives of the older generation. They believe that according to the Muslim religion, a woman is obliged to maintain her piety, therewith contributing to the protection and the maintaining of the honor of the whole family, until the wedding. Until that moment, she is prohibited from any "intimate", "close" contact with men, including the observance of ethical French rules. That is why, in order to warrant the implementation of this dogma, many children from a Muslim family are sent "for re-education" to their homeland (Algeria, Tunisia, Morocco). This mainly concerns girls. Parents believe that it is indispensably to keep in memory and follow religious traditions with the Eastern world.

But along with that one cannot ignore the fact that, along with immigrants with a more liberal perception of Islamic values, there are extremely interested and engaged young people. They are unable to give up family religious practices. They are confident in the strict observance of all dogmatic norms of Islam, and, most importantly, in the involvement of elements of private life in the sphere of public interests. The current situation regarding the "wearing of religious attributes" in France since the 2000 is a confirmation of this.

An important event in the cultural life of Muslims is the adoption in 2004, the "Law on the Observance of the Principle of Secularity in Public Institutions" and the "Law on the Prohibition of Hiding the Face in France" in 2010, which prohibited the wearing of clothing that hides the face, including niqabs [21]. However, this contributed to an escalation of tensions both within and outside Muslim communities. The problem was related to the fact that considered by many as an attempt to discriminate on the basis of religious and ethnic principles against representatives of Islamic culture [22]. If a woman's veil is a "visible" or "noticeable" sign, then a breast cross can be hidden under clothing, and therefore only an individual believing in Islam is discriminated against. But at the same time, there is a need to ensure the safety of French citizens through the adoption of such measures. It is caused by an increased threat from radical groups, often preaching the Muslim religion. So, even after 10–15 years of the implementation of this regulation disputes over the need to wear religious attributes does not subside.

Statistics show that approximately 62% of Muslim respondents and 37% of non-Muslims believe that women should have the right to wear hijab and niqab in public. Significantly women are more supportive of wearing face-covering headgear. 18% of women reject it, 26% of men disagree. It is noteworthy that, at the same time, 45% of women claim that they have never worn the distinctive elements of Islam, 7% only stopped wearing the niqab over time [23]. It follows from this that the majority of Muslim representatives are still closely associated with traditional cultural values imposed by religious dogmas, which only a small part of the third generation of immigrants intend to abandon.

The practice of using a double name is another interesting feature of the Muslim population of France. One is traditionally a Muslim name and the other is Christian. However, this desire for duality and mental mediation between two cultures - French and Muslim - is only a social construct that cannot so much affect the self-identification of the Muslim himself, but improve his perception in society, thereby facilitating social integration. It can be deduced an inference that each individual ultimately only develops a unique attitude towards such cultural hybridity, and does not become a hybrid in the cultural sphere [24].

The education of immigrants in France is one of the stumbling blocks to the political integration of the Muslim population. The problem is related to the desire to preserve identity, which is expressed through history, religion and language, on the part of the representatives of the Maghreb and the Middle East. Maghrebians are more quickly integrated into the French educational system due to the influence of historical discourse. According Insee, 21% of migrants from the Maghreb countries consider French as their native tongue [25].

However, the problem of finding new employment is associated with an insufficient level of education among migrants, opening the way only for low-paid jobs with an increased risk of unemployment. So, according to research by the statistical office of the European Union in France in 2016, there was a situation where 43% of migrants in the age range from 14 to 70 years did not have a diploma of complete secondary education [26].

Modern realities dictate their own demands that destroy the secular nature of the French public education system. They are based on the religious beliefs of Muslim students and parents. The French publication Le Figaro reports that teachers in educational institutions are forced to face misunderstanding and condemnation from parents from Muslim families, who, in turn, oppose courses on the Holocaust, Crusades and completely refuse to accept history, culture, and values of the French Republic [27].

Childhood illiteracy among the Muslim population is also at a high level. For example, many children at the time of the beginning of elementary school know only 4 hundred words, which is significantly less than the children of the French, whose vocabulary is 1500 words. Moreover, representatives of Islamic culture deliberately use a version of French adapted for the Arabic language for convenience and the manifestation of their ethnic identity. In this vein, migrants have poor communication skills, which become a stumbling block when they are looking for a job.

The education system does not have the urgent levers of social impact to overcome social inequality, even in spite of the measures that the government is taking, annually allocating 1.4 billion euros for the main educational zones, in which 21% of students study. Representatives of the third generation of migrants more often than the local population stay for the second year; 18% of migrant children do not finish school to the end, which is 6% higher than the children of the autochthonous population [28].

The lowest rates of educational achievement can be seen among the male part of the representatives of North Africa, as well as Turkey, as consequence of which they are less competitive in the labor market.

As for the sphere of employment, the Muslim population faces an equally inflow of problems, which are expressed through discrimination on racial, ethnic and religious grounds [29]. The fight against this phenomenon is a significant sector of the policy of the French authorities.

Research data from the Montaigne Institute states that in the period from 2013 to 2014, 45% of representatives of the second wave of immigrants once in their lives became victims of professional discrimination. Thus, Catholics are a priority among employers more often than Jewish, and twice as often as Muslims. Muslims receive invitations for interviews four times more questionnaires than Catholics [30].

Interviews with immigrants from the Maghreb countries studying in France at the Higher School of Social Sciences confirm this fact. 14 out of 23 respondents personally faced the problem of employment based on ethnicity. They were rejected by many organizations, possessing the necessary list of skills to obtain both low-skilled and prestigious jobs. If the representatives of the Maghreb culture replaced the Arab surname with the European one, then the majority of the companies answered positively and were ready to hire them.

In 2014, the problem of unemployment among residents of "sensitive zones" came to the fore. Experts report that unemployment was 27.1% for the autochthonous part of the French and 28% and a half for new arrivals [31]. Nevertheless, highly qualified specialists of the Muslim diaspora were most vulnerable to this phenomenon, since each entrepreneur focused his attention on the ethnic characteristics of the selected personnel. Although the law "On Equality of Opportunities" in 2006 gave employers the right to select new staff through a system of anonymous questionnaires, employers' attitudes remain highly biased and "xenophobic". So, it should be added that about 19% of employers consider the attachment of a photo to be a mandatory condition for drawing up a resume [32].

Economist Gerard Pincet wrote an essay in 2013 based on Insee's numbers, pointing out that foreigners and their direct descendants contributed to 65% of France's public finance deficit in 2011 [33]. It can be caused by their low tax payments and income. Indeed, government figures state that only 40% of people of working age will have a job.

France is not the most affected country by the 2015 migration crisis, like Germany or Italy, but according to the European Statistical Office, the number of forced migrants arriving there also increased: from 64 thousand in 2014 to 84 thousand in 2016 [34]. Difficulties are associated with the fact that the influx of migrants is not dictated by the need to increase the labor force in labor markets, which is why their arrival can create problems of unemployment or parasitism within the country. According to the data, 24% of migrants over 18 years of age are considered poor because their incomes were below 60% of the national average, 10% suffered from severe material deprivation [35].

It goes without saying that the socio-economic adaptation and integration of migrants is complicated by the impossibility of their self-realization in the labor market. As compensation, the upper echelons offer allowances and payments. However, the effect is diametrically opposite, as migrants begin to realize that looking for work is less profitable and effective than living through state support on benefits, which ultimately contributes to the spread of social dependency [36].

Thus, the situation on the labor market for representatives of the Muslim population really presents certain difficulties, which are associated both with social organization in the remit of cultural perception, and the desire of each individual of a foreign cultural community to identify himself by religious, ethnic or political characteristics.

4 Discussion

The beginning of the new century marked the emergence of a new complexity for the Muslim diaspora: problem of determining their status. Muslims cannot fully practice their religion, being a religious minority in European society. The "identity dilemma" arises. This process, which is associated with the problem of self-identification, gives immigrants a choice: to define themselves according to limited religious norms or generally accepted republican values.

There is an alternative. It concerns the possibility of "acculturation of Islam", that is, a Muslim has the opportunity to reveal his identity from the standpoint of Islamic culture, and not from the standpoint of religion. However, French society itself is ambiguous in its perception of the Muslim diaspora. For example, in 2015, former head of the Cabinet of Ministers, Manuel Valls, announced the existence of "territorial, social and ethnic apartheid" in the country. He recognized discrimination on the basis of race and ethnicity as a valid problem [37]. More than 76% of the French share this opinion, believing that there is discrimination based on ethnicity in the country [38].

Surveys by the French Human Rights Commission report that 77% of French people perceive French Muslim as equal citizens. 45% of French people believe that Muslims want to adapt to the traditions and lifestyle of the country that has accepted them. The authors suggest that public opinion changed with the emergence of a third generation of Muslims, who perceive the democratic values of France with greater respect than their ancestors. However, not all Muslims still have a positive attitude towards the marriage of their children to non-Muslims. 15% of Muslims are against the marriage of a son with a non-Muslim and 32% of a daughter with a non-Muslim [39].

The religious issue of the country's territory exists. In 2010, 55% of French people considered the influence of Islam to be pervasive and destructive; by 2019, this figure had risen to 63%. In addition, 43% of French people currently believe that the presence of the Muslim community in France poses a threat to the country's identity, and 17%, on the contrary, consider it a factor of cultural enrichment and deception by experience.

At the same time, 33% of French people believe that Muslims are well integrated into French society. And 47% regret that Muslims are forced to live separately in some areas of the state and that economic difficulties, such as lack of work, hinder their integration.

It is essentially to add that anti-Islamic sentiments were recorded in 2014 in France. They got their expression in protests and demonstrations directed against the Muslim population. The report of the Collective Against Islamophobia emphasizes that the Muslim diaspora is forced to put up with the constant manifestation of Islamophobia and elements of racism. Thus, about 4 hundred actions against French citizens of the Islamic confession were registered, as well as 40 aggressive expressions against mosques. The survey showed that about 70% of the respondents perceive Islam as a religion that contradicts the basic values of French republicanism. Every second Frenchman believes that the Muslim community is striving to "impose its way of life" on the autochthonous population [40].

The authors conclude that the French public expresses loyalty and toleranc'e towards Muslim culture and their position in the country. However, the existing fear of loss of identity, the problem of identifying the French themselves (as well as immigrants in general), the increased threat from Islamic radical groups, which is reflected in the general perception of the Muslim religion by Europeans, cannot but influence the escalation of the negative reception of the local population towards immigrants.

The phenomenon of integration did not appear immediately in French society. Until 1980 the French government actively used assimilation as a tool to develop and support the process of integrating immigrants into the European community. However, the migration flow only intensified every year, xenophobia and discriminatory regards on the majority of the French grew, which caused to the establishment of ethno-confessional societies within the state, refusing to accept the cultural component of the host country. For this reason the integration policy of France was considered the most politically correct and effective for newly arrived Muslims.

In France, the main emphasis is on the individual, who is considered outside of his ethnocultural areola. Consequently, the authors state that the success of integration into French society depends only on it [41]. A similar process take's place through the acceptance of the democratic values of the Republic, legislative norms, respect and knowledge of the state language, coupled with the desire to get secondary and higher education in France. It is important to note that the latter method is not so much a part of the culture as an important social element that contributes to high, effective and successful integration.

One of the important and significant phenomena in the system of adaptation of migrants in any country is the "culture shock" developed by the Canadian anthropologist Kalervo Oberg. It represents an element of disorientation experienced by a person who is faced with an unfamiliar conduct of life in the wake of visiting a foreign country, immigration, changes in the social environment or simply a way of life [42]. The most distinctive forms or signs of "shock" are: information overload (most often lack of culture of differentiation), language barrier, generational gap, technological gap, longing (for home, property, friends, etc.), and also in lots of cases feeling inferiority in the face of rejection and xenophobia from the local population [43].

The authors emphasize that the Muslim population of France has in many ways all the characteristics inherent in "culture shock". This is the problem of finding and defining identity; communication in many cases in different languages or dialects; generation gap (the first, second and third generations of migrants who enter into internal conflicts in the family and external conflicts with society in an attempt to impose or distance themselves from Islamic culture); certainly, the homesickness inherent in the older generation.

The next obstacle to successful integration is the rise in the level of "radical Islamism" in France, which contributes to the escalation of "xenophobic sentiments" within civil society. The local French population often does not see the difference between practicing "peaceful" Muslims and representatives of radical movements. This leads to a lack of understanding, rejection of another's culture and fear.

The Internet has become an active and large-scale platform for propagating extremism among the younger generation. Moreover, according to data from French intelligence, about 15,000 adherents of radical Islam are currently registered, of which 18% are minors. Half of the representatives of this branch in Islam pose a serious threat to national security.

Public negative attitudes towards the Muslim population are fueled by the appalling rise in terrorist activity. In this vein, it should be emphasized that the threat of terrorism ranks first in the top of the negative consequences of the migration of the Muslim population in French society. Concerned citizens express extreme distrust towards the representatives of the Islamic confession in connection with such events as: the attack in the editorial office of Charlie Hebdo on January 7, 2015, a series of attacks in Paris and its suburbs on November 13, 2015, the attack in Nice on July 14, 2016.

Of course, the threat to national security is the main priority of the French government in the current realities. The authors believe that the government needs to control the development of radical movements on the territory of the state, to conduct special courses to inform the French population. This will weaken the desire of young people to engage in self-identification in the ranks of radical Islamist forces, as well as change the attitude of the local population towards the "moderate" representatives of the Muslim world.

However, "xenophobia" and "racism" on the part of the French towards the Muslim minority is expressed not only in internal relationships, but also at the political level. For example, the influence of the public and its support for right-wing radical forces suggests that French society is indeed experiencing difficulties in accepting Islamic culture, which is especially acutely felt after such an international phenomenon as the migration crisis of 2015.

In 2017, migration took one of the most substantial places in French political life. This confirms the fact that the leader of the National Union Marine Le Pen has entered the second round of the presidential elections. It is important to note that the control of immigration flows from the very beginning was proclaimed the central direction of the party's policy. Party members have repeatedly received accusations of xenophobia and anti-Semitism [44]. The National Front offers its own steps for passing the migration crisis. The suggestions are:

- reduction of legal immigration from 200 thousand to 10 thousand people, while giving preference to highly qualified specialists;
- a sharp decline in the admission of refugees seeking asylum in France;
- compulsory measures to punish any cruel manifestation of racism against French citizens;
- application of the policy of national priority: (all other things being equal, employers should give preference to representatives of the French nation).

The party hints at the "purity" of the French nation, in which there is no place for foreigners [45].

This is surely why such populist slogans are becoming even more popular in the mass consciousness in connection of the increase in the presence of terrorism in the country.

In France, a number of laws exist despite similar statements from the political elite. They help immigrants to socialize and integrate into the life of French society. Among the main ones:

- conclusion of a binding contract of integration, according to which a migrant or a new citizen is obliged to take French courses, lectures on civil rights and obligations, consultations about employment;
- the adoption of the "Law on the deprivation of French citizenship" in relation to persons guilty of terrorism as a result of terrorist attacks in 2015 [46];
- The French government pays benefits to unemployed migrants. It is 281 euros for an adult and 184 euros for a child [47];
- migrants receive preferential health insurance;
- children of migrants receive citizenship by birthright, and parents are paid a childbirth allowance [48].

Moreover, a special project "Intercultural city" exists at the level of the Council of Europe and the European Commission. At the present stage, three French cities are participating. These are Lyon, Strasbourg and Paris. The main measure is the creation of the Equality Council, which develops and brings together city service activities and local action initiatives foster to contribute the development of a variety of approaches to local cultural policy-making. So, an interesting example is the experience of one recreational facility in Lyon. In this place, immigrant women share their experiences of living in Muslim countries, talk about the cultural characteristics and traditions of their families and etc. The population is increasingly interested in Islam, despite the rise of Islamophobic sentiments observed in France. This was evident in the increase in sales of Islamic literature in 2015 [49].

The importance of the migration of the Muslim population lies in the fact that it has a positive effect on the country's demography. Thus, according to the French census, every fifth child in 2017 (19%) was from an immigrant from the Maghreb countries. In 2009, this figure was only 16%. Thus, the contribution of immigrant women to the fertility rate in France really deserves attention [50].

5 Conclusion

The problem of integrating people belonging to a different civilizational culture and traditions into French society is the most important and key thread of conversation in the international arena in the XXI century. The slogan: "Freedom is freedom of movement" is now overshadowed by the problem of defense against the countless influx of migrants with different cultural, religious and ideological characteristics.

Uncontrolled migration flows affect the national composition of host countries. Moreover, at the present stage, migration has an ethnic connotation. The main source of migration processes was the Muslim world. A new kind of culture, "civilizational orientation" is being formed in immigrating countries. A regulated lifestyle, depending on specific religious forms, unusual patterns of behavior and an unfamiliar worldview become the cause of the development of misunderstanding and alienation between the majority and the other - the Muslim population.

In modern realities, the springboard for rapprochement of cultures is the process of integration, which allows the new population to merge into the social realities of a cultural field different from it, with the possibility of preserving and developing their own characteristics of way of life and traditions, subject to their agreement and reconciliation with the official political discourse of the state. Different ideological and cultural worlds represented in France can not only coexist, but also complement each other's traditional components without displacing the earlier ones.

Political measures aimed at overcoming the problems of adaptation of the Muslim population in France, basically, consist of the desire to limit the influence of the Middle Eastern states on existing Muslim institutions and associations (we are talking about material support), as well as to provide both new arrivals and already citizens of the Republic the possibility of realizing a decent standard of living in accordance with the country's democratic values.

Thus, the existing problems in the field of Muslim adaptation in France are not insoluble and exclusive. Undoubtedly, the natural gap between cultures, educated and developing at the end of the XX century, requires a sufficient amount of time to effectively and efficiently overcome them. However, for this, the French government must not only revise and modernize the institutional mechanisms of the existing system of power, but also change the vectors of political and ideological propaganda, in particular in the media, in order to begin to build a favorable and "safe" image of Muslim culture in France.

References

1. Report of the Secretary General of the UN General Assembly "International migration and development" [In Russian] (2010)
2. Dumont, G.F.: Les migrations internationales au XXI siècle. Des facteurs recurrents ou nouveaux? Actuelles de L'IFRI, 4 September 2015. https://www.ifri.org/fr/publications. Accessed 13 Jan 2020
3. Kulik, S.V., Eidemiller, K. Yu., Biktimirova, R.R., Vanicheva, M.N.: Islamic diffusion in the countries of the Arctic region: the processes of Muslim migration to Scandinavian countries in the context of the transforming Islamic world. In: IOP Conference Series: Earth and Environmental Science, vol. 302 (2019)
4. Nielsen, J.S., Otterbeck, J.: Muslims in Western Europe. 4th edn., pp. 1–232 (2015)
5. Kuropyatnik, A.I.: Immigration and national society, France, pp. 147–156 (2005). [In Russian]
6. Salt, J.: Migration in Post-War Europe: Geographic Essays, pp. 83–85 (1978)
7. Foreign-born population by country of birth. Eurostat. Eurostat. https://ec.europa.eu/eurostat/statistics-explained. Accessed 11 Dec 2019
8. France, portrait social, édition 2018. www.insee.fr. Accessed 25 Sept 2019
9. Tribalat, M.: Une estimation des populations d'origine étrangère en France en 1999, pp. 51–81 (2004)
10. Tribalat, M.: Statistiques ethniques: le décryptage de Michèle Tribalat. https://www.lefigaro.fr. Accessed 05 Dec 2019
11. Goerzig, C., Al-Hashimi, K.: Radicalization in Western Europe: Integration, Public Discourse and Loss of Identity Among Muslim Communities, pp. 1–174 (2014) https://doi.org/10.4324/9781315817118
12. France: comment est évalué le nombre de musulmans. https://www.lefigaro.fr. Accessed 25 Apr 2020
13. Number of Muslims in Western Europe Pew Research Center Religion & Public Life. https://www.pewforum.org. Accessed 02 Apr 2020
14. Population change – Demographic balance and crude rates at national level. Eurostat. http://appsso.eurostat.ec.europa.eu. Accessed 05 Dec 2019
15. Total number of long-term immigrants arriving into the reporting country during the reference year. Eurostat. http://ec.europa.eu/eurostat. Accessed 27 Dec 2019
16. Gusev, D.: Asylum in France: the chamber of Accounts on the high cost and lack of control of the system. [In Russian]. http://inosmi.ru/world. Accessed 04 Jan 2020
17. Tabet, M.: Religion, famille, société: qui sont vraiment les musulmans de France. https://www.lejdd.fr/Societe. Accessed 03 Mar 2020
18. Bouet, P.: Les flux migratoires et trajectoires des médecins situation en 2014. https://www.conseil-national. Accessed 03 Mar 2020
19. Generals of the sand quarries – the suburbs of French cities turned into ghettos for migrants. [In Russian]. https://www.1tv.ru/news. Accessed 18 Nov 2019
20. Dolgov, B.V.: Islam in France: Muslims and Secular Republic Vostok-Oriens, pp. 123–133 (2015)

21. Deminceva, E.: Being an Arab in France, pp. 120–123 (2008)
22. Pauly Jr., R.J.: Islam in Europe: Integration or marginalization? pp. 1–191 (2013)
23. Tabet, M.C.: Religion, famille, société: qui sont vraiment les musulmans de France. https://www.lejdd.fr. Accessed 02 Apr 2020
24. Rosello, M.: France and the Maghreb: Performative Encounters. http://proxy.library.spbu.ru. Accessed 04 Jan 2020
25. Monso, O., Gleizes, F.: Langue, diplômes: des enjeux pour l'accès des immigrés au marché du travail
26. Population by educational attainment level, sex, age and country of birth and etc. Eurostat. http://ec.europa.eu/eurostat/data/database. Accessed 23 Dec 2019
27. Relever Les Defis De L'Integration A Lécole. http://www.lefigaro.fr/assets/pdf/rapport-hci.pdf. Accessed 04 Apr 2020
28. Tassel, F.: Descendants d'immigrés au pays des inégalités. http://www.liberation.fr. Accessed 23 Mar 2020
29. Matveevskaya, A.S., Pogodin, S.N.: Integration of migrants as a way to diminish proneness to conflict in multinational communities. Vest. Sankt-Peterburgskogo Univ. Filosofiia Konfliktol. **34**, 17–23 (2018). https://doi.org/10.21638/11701/spbu17.2018.110
30. Institut Montaigne Discriminations religieuses à l'embauche: une réalité. http://www.institutmontaigne.org. Accessed 23 Apr 2020
31. L'Observatoire national de la politique de la ville. http://www.onpv.fr/uploads/media_items/rapport-onpv-2015.original.pdf. Accessed 07 Apr 2020
32. Amadieu, J.-F., Rodier, A.: La crise a favorisé le discriminations au travail. http://www.lemonde.fr. Accessed 08 Sept 2019
33. Pince, G.: Les Français ruinés par l'immigration. https://www.polemia.com/les-francais-ruines-par-limmigration-de-gerard-pince/. Accessed 02 Feb 2020
34. Asylum and first-time asylum applicants by citizenship, age and sex. Annual aggregated data. Eurostat. http://ec.europa.eu/eurostat. Accessed 05 Mar 2020
35. Population by educational attainment level, sex, age and country of birth and etc. Eurostat. http://ec.europa.eu/eurostat. Accessed 17 Apr 2020
36. Institut national de la statistique et des études économiques. Être né en France d'un parent immigré. https://www.insee.fr/fr/statistiques/2575541. Accessed 21 Apr 2020
37. Valls évoque "un apartheid territorial, social et ethnique" en France. www.lesechos.fr. Accessed 22 Apr 2020
38. Dix ans de politiques de diversité: quel bilan? http://www.institutmontaigne.org. Accessed 22 Apr 2020
39. Goodwin, R., Krzysztof, K., Sun, S., Ben-Ezra, M.: Psychological distress and prejudice following terror attack s in France. J. Psychiatric Res. **91**, 111–115 (2017)
40. Occupy le mosque: France's new far-right nativism. http://www.sbs.com.au/news/article/2012/11/06/occupy-le-mosque-frances-new-far-right-nativism. Accessed 15 Apr 2020
41. Matveevskaya, A.S., Pogodin, S.N.: The essence of cross-cultural conflict (Presentation of a problem). Vest. Sankt-Peterburgskogo Univ. Filosofiia Konfliktol. **33**, 115–118 (2017). https://doi.org/10.21638/11701/spbu17.2017.112
42. Macionis, J., Gerber, L.: Sociology, p. 54 (2010)
43. Oberg, K.: Cultural shock: adjustment to new cultural environments. Pract. Antropol. 177–185 (1960)
44. Notre Projet: Programme Politique du Front National. http://www.frontnational.com. Accessed 15 Apr 2020
45. Mikä Islam Todella On. https://aikapommi.wordpress.com. Accessed 03 Apr 2020
46. Fyodorov, S.M.: France. Sovremennaya Evropa, pp. 126–129 (2016). [In Russian]
47. Semenova, O.Yu.: Daily life in modern Paris [In Russian]. E-reading. https://www.e-reading.club/book.php?book=1016398. Accessed 20 Nov 2019

48. Zholudeva, N.R.: Muslims in France: life in Muslim neighborhoods, pp. 132–137 (2019). [In Russian]
49. Sales of Islamic literature in France tripled. Islam News. [In Russian]. https://islamnews.ru/news-456435.html. Accessed 28 Apr 2020
50. Volant, S., Pison, G., Héran, F.: La France a la plus forte fécondité d'Europe. Est-ce dû aux immigrées? Popul. Soc. 1–4 (2019)

The Sino-Russian Relations in the Field of International Tourism

Irina Plastinina[1], Victory Pogodina[2], Vladimir Evseev[3], and Tamara Tarakanova[3](✉)

[1] Saint Petersburg State University, Saint Petersburg, Russia
[2] Saint-Petersburg State University of Industrial Technologies and Design, Saint Petersburg, Russia
[3] Peter the Great Saint Petersburg Polytechnic University, Saint Petersburg, Russia

Abstract. Russian-Chinese relations are currently developing quite successfully in many areas. Tourism is one of the most significant and promising areas of cooperation between Russia and China. Cooperation in this area should help to increase cultural ties and increase mutual understanding between Russian and Chinese citizens. The study presents the main types of tourism that are most popular among tourists, as well as the most popular tourist destinations in Russia and China. The most popular types of tourism for Russian and Chinese citizens are cultural, educational, environmental, business and educational. The authors analyzed the main laws that govern relations between Russia and China in the field of tourism and bilateral agreements, that are aimed at strengthening and facilitating interaction in the process of tourist exchange. A large number of measures to enhance the safety of Russian and Chinese tourists are being taken at the legislative level and at the level of intergovernmental agreements. Bilateral agreements are mainly aimed at facilitating the visa regime. The article examines the activities of two states in the framework of UNWTO, as well as provides an overview of other measures taken by the parties to increase the quantity and quality of tourist exchange. All projects supported by Russia and China within the organization are aimed at increasing the attractiveness of their tourist destinations for citizens of the partner state, as well as facilitating the interaction of staff and foreign tourists. The article analyzes current problems that require timely solutions, but so far remain ignored by governments.

Keywords: International tourism · Russia and China · International relations · Tourist destinations

1 Introduction

Russian-Chinese tourism is important in the socio-economic sphere of both states. Firstly, it is able to provide a social and communicative role, allowing travelers to communicate with each other in an informal setting, to get to know each other better. Thanks to this, the citizens of the Russian Federation and China have predominantly good opinions about each other. Secondly, tourism contributes to the recreation and recovery of travelers,

© The Author(s), under exclusive license to Springer Nature Switzerland AG 2021
R. Bolgov et al. (Eds.): *Proceedings of Topical Issues in International Political Geography*, SPRINGERGEOGR, pp. 298–309, 2021.
https://doi.org/10.1007/978-3-030-78690-8_26

which leads to improved health and productivity of citizens of the two states. Third, we note the importance of tourism in the social and economic spheres. The development of tourism should help reduce unemployment in Russia and China, because more and more people are becoming employed in this area. Tourism also contributes to the development of transport infrastructure, communications, construction, agriculture [1], which is a kind of catalyst for the economic development of the two countries.

However, Russian-Chinese tourism does not always have a positive effect on the development of the two states, and there are a number of problems. These are problems associated with the negative image of a Chinese tourist in the eyes of local residents. This image was formed as a result of the uncivilized behavior of many of the Chinese visitors. Another problem is the lack of tourism revenue for Russia. All Chinese groups coming to the territory of the Russian Federation are served by Chinese companies that own transport, restaurants, hotels, shops, as well as guides. All profits from Chinese tourism are returned back to China, and the costs associated with the arrival of an increasing number of Chinese tourists have to be paid from the Russian budget.

Russian-Chinese took place during the heyday of overland trade between Europe and Asia along the Great Silk Road. Attempts to establish friendly contacts were made even under Ivan the Terrible and Vasily Shuisky. The Russian embassies never reached China. Relations between the two powers were rather poorly developed until the XVII century, when the barrier between Russia and China was first overcome. The poor development of relations during this period, due to the lack of roads and the necessary infrastructure [2]. In the XVII - XVIII centuries. the territories of Russia were already wide due to the annexation of Siberia. Countries had to resort to forming a common border. The states paid visits to each other for the purpose of negotiations on border issues. Treaties were created (Treaty of Nerchinsk in 1689, Treaty of Kyakhta in 1727), which later became important steps on the way to further active development of relations. In the XIX century. residents of Russia and China began to show interest in studying each other's culture and life. Awareness of the Chinese population is growing in Russian society. The Russian Orthodox Spiritual Mission has made a special contribution to the development of relations by studying the history, culture and spiritual life of China. During this historical period, a scientific center - Asian Museum was established. This center laid the foundations for scientific research in China and staff training began at St. Petersburg State University at the faculty of oriental studies [3]. In the XIX century. scientists begin to travel in order to study each other's culture and life. Thanks to such trips, many beliefs about Chinese society were dispelled, and many traditions of the Chinese people even attracted educated Russians and they were eager to visit this country.

At the beginning of the XX century. a sharp decline in bilateral relations between Russia and China has occurred. The main reason was that Russia took part on the side of a number of Western powers in the military intervention against China. In 1898, a popular uprising began in China, and in 1900, Russia entered the territory of Northern China and took part in suppression of the uprising along with the European powers. These events caused strong anti-Russian sentiments in China, because the de facto Union state acted as the interventionist [4]. All this contributed to the interruption of established cultural ties. From the 1920 to the late 1950s, there was warming in bilateral relations. This is due to the establishment of the Communist Party's power in China. The states carried out joint actions to counter Japan's plans to seize China and the Far Eastern part of the USSR. But despite the general improvement in relations, tourism was rather poorly developed during this period. In China, the civil war was underway at that time, and Soviet citizens could not leave the country for travel.

Other problems have occurred since the 1960s until the end of the 1980s, which led to the aggravation of relations over the border issue. This led to an almost complete split in relations between the two states. The Chinese side believed that all the treaties signed during the tsarist Russia were inequitable and unfair to China. The Soviet side was convinced that there was no need to revise issues related to the border, since it was finally formed in the XIX century. The Mongol question contributed to the growth of contradictions, in addition to sharp disagreements over border issues. Mongolia at that time was an ally of the Soviet Union and enjoyed its protection. The country allowed the USSR to deploy its troops on its territory. Naturally, this led to a sharply negative reaction from the People's Republic of China.

Certain steps were taken when M.S. Gorbachev came in power in the USSR. They contributed to the normalization of relations with the PRC [5]. Thus, the withdrawal of troops from Afghanistan, the significant reduction of Soviet troops in Mongolia, and the readiness to develop an intergovernmental agreement on the border issue on the basis of equality favored the building of mutual strategic trust between the USSR and the PRC. This trust became a key foundation for improving Soviet-Chinese relations [6]. Relations have stabilized, creating the basis for active interaction between Russia and China subsequently, including in the XXI century. The expansion of interstate relations laid the foundation for business and then mass tourism. September 24, 1988 is the official date of the beginning of the tourist exchange, when the tourist groups were exchanged. The most various types of tourism began to develop later. These are such types of tourism as medical, sightseeing, vacation, research, etc. [7]. The agreement on visa-free group trips was concluded in March 1991 between the Chinese Foreign Ministry and the Soviet Embassy in the China. However, this agreement only applied to the border regions (in Russia—the Primorsky and Khabarovsk territories, the Amur and Chita regions, and in China—Heilongjiang, Liaoning, Jilin). The agreement provided that groups of tourists, according to agreed lists, can travel through the territory of another country without visas.

2 Materials and Methods

The methodological basis of this research uses a number of scientific methods that allowed us to comprehensively investigate the problem. The chronological method of research made it possible to find out how the interaction between Russia and China has changed with the course of history and how the general development of the two states has led to the growth of bilateral tourism. The method of document analysis allowed us to determine the basis of the legal regulation of Russian-Chinese tourism. The comparison method made it possible to see the differences and similarities in the activities of the two states in order to develop tourism. The bibliographic base of the research includes scientific works of Russian and foreign political scientists, researchers and specialists in the field of international relations, as well as the works of politicians and diplomats. Research materials can be grouped into four large groups of sources:

- Historical research. The work of Blagoder Yu. and Mints S. [8] describe the contribution of Russian sinologists of the XIX century. The articles demonstrate changes in Russian people's perception of China with the increase in the amount of knowledge about it, the growth of interest of Russian society in the study of this country, consider the stages of the formation of the image of China in Russian society from the XVII to the XX centuries. The author traces the idea of how the the Russian society's perception of China changed with the increase in the amount of knowledge, and how educated Russian people tried to learn more about this "country-mystery" and establish strong ties with it.
- Works that consider the economic factors of tourism development in this region. The works of Efremova M., Chkalova O. and Bi R. [9], Kudryashova S. [10] presents the economic reasons for the development of Russian-Chinese tourism, clearly outlines the main directions of cooperation between two states in the field of tourism. In the work of Chirkov D. and Chirkova Yu. [11] the tourism sector is shown as a complex economic system, the reasons for the increase in the influx of Chinese tourists, which did not bring the expected revenues to the country's budget. Special attention is paid to the implementation of "gray" schemes in the tourism sector and how they undermine the economic security of Russia. The problem of transnational organized crime in the field of tourism is considered. Article by Machalkin S.and Moreva S. [12] also focuses on the fact that despite the growth in the number of Chinese tourists, the quality of the influx leaves much to be desired for the Russian economy. The study identifies the reasons why only relatively poor segments of the Chinese population visit Russia, and the "high-income" Chinese tourist flow simply ignores it, choosing other destinations. The article by Liu H. [13] examines the factors influencing the development of tourism in China. In particular, special attention is paid to the problems that prevent tourism from becoming an even more efficient and profitable industry. The authors recommend working to improve the activities of travel agencies, make prices more affordable for tourists, transport more convenient, and use the features of their culture to attract foreign tourists, including Russian.
- Research on the peculiarities of tourism development in the region. The research of Kulgachev I., Lepeshkin V. and Manteifel E. [14] presents the main trends in tourism in Russia and East Asia in general, and China in particular, based on statistics. It is important that the article describes the problems, that can be solved to increase the number of tourists from East Asian countries to Russia, and, in addition, to increase the income from tourist services. The article by Novozhilova I. and Tokusheva D. [15] examines and compares the features of receiving Chinese and Japanese tourists in Russia. The conclusions suggest that the methods of serving Asian tourists in Russia need to be improved, since they are very different from European ones. It is noted that a completely different organizational approach to the reception of Asian tourists is needed. This is necessary for Russia to become more attractive for them as a tourist destination. The research of Polozhevich V. [16] considers several areas of humanitarian relations between Russia and China. The author notes that over the past few years, not only the flow of Chinese tourists has increased to Russia, but also vice versa, China is becoming more and more popular with Russian tourists. It is also noted that Russia and China exchange information, experience and employees in the field of tourism. This contributes to improving the interaction between Russian and

Chinese travel agencies and the development of tourism infrastructure. The research of Ochkasova I. and Sheleg E. [17] pays special attention to the changing needs of the modern Chinese tourist, which are largely ignored by the host Russian side, hence many problems arise. The article reflects the factors that hinder the normal development of Russian-Chinese relations in the field of tourism.
- Works dedicated to certain types of tourism. The authors of the article "Development of red tourism in the perspective of the Russian' [18] consider the so-called "red tourism". This is a unique type of tourism that is characteristic only for Russia and China and involves visiting significant places associated with the history of communism. The authors of the article note that it is necessary to develop "red tourism", and therefore it is necessary to train specialists to serve tourists on the red routes. The study showed that it is necessary to jointly search for sources of financing for this tourism industry; and it is proposed to use the possibilities of public-private partnerships. The article by Liu J., Pan H., Zheng S. [19] examines the impact of environmental quality, in particular air quality, on tourism in China. Environmental pollution and the rapid urbanization in China, according to the authors, reduce the attractiveness of China for foreign tourists. The article gives an idea of China's internal problems that hinder the development of tourism A study by Zhao M., Dong S., Xia B., Cheng H., Li Y., Li Z., Zheng J. [20] examines the creation of a new economic path and its role in the development of tourism in three countries. It is noted that this corridor will contribute to the expansion of cultural ties between states, including tourism. It is assumed that this will benefit the development of bilateral tourism between Russia and China. Research by Chow C., Tsui W. [21] identifies key factors that explain the rapid development of cross-border tourism from Russia to China. The main such factors, according to the authors, are the socio-economic and transport infrastructure. Major Chinese transport infrastructure projects are expected to attract more Russian tourists to China. The authors note the need for both governments to facilitate the free flow of tourists across the border.

3 Results

The most popular types of tourism are cultural, educational, business, environmental and educational. Cultural and educational tourism. Russia and China have a rich history, many architectural and art monuments that attract tourists. Popular cities for Chinese tourists in Russia are those that are included in the so-called "Red Route" (Moscow, St. Petersburg, Kazan and Ulyanovsk) [22]. Visiting these cities is attractive for Chinese tourists not only because they want to visit places associated with the activities of Lenin V.I., but also because the leading figures of China - Mao Zedong and Zhou Enlai once visited these places [23]. Also, the history of Russia and China is connected by the fact that at a certain period China largely followed the example of the Soviet Union. For this reason, many sectors of the economy in China developed on the the Soviet model, especially in industry.

For cultural and educational purposes, Russian tourists most often visit such Chinese cities as Beijing, Shanghai, Harbin, Henan, Jilin, and Shandong. The most famous sights of ancient times, architectural structures of different styles, magnificent masterpieces of art are located herex [24]. Russian tourists are interested in visiting tours of the Great Wall of China and the "Silk Road", historical sights of Beijing, as well as Hagia Sophia in Harbin, the Temple and Tomb of Confucius, the Five Great Mountains [21].

Ecological tourism is becoming more and more popular. This is due to the fact that residents of both Russia and China, especially those who live in cities, are increasingly willing to move away from cities, visit natural monuments, and get acquainted with the ecosystem of another state during their travels. The largest number of forest protected areas and forest parks in the world are located in China. These are 1,658 forest parks and 1,757 forest protected areas. Russian tourists tend to visit Lake Baiyangdian, Vudalyanchi Park, "Five Lakes", a geological park of world significance with a natural zone "Primeval Crater Forest", Sianshai nature reserve, and the reserve "Lake Khanka" [25]. The last reserve is most interesting because it is located on the border with Russia. One part of it belongs to China, the other to Russia. The open-air Geological Museum of Benxi National Park is especially popular with Russian tourists. The Chinese are interested in Lake Baikal and its surroundings, national parks and national reserves of Primorye. Increasingly, Chinese tourists visit the Altai mountains. In winter, Chinese tourists are very interested in seeing such a unique phenomenon as the northern lights in Russia. There is a growing interest in ecological tours in the Russian North [26].

Bilateral business tourism is also developing. According to the WTTC, in 2017 in China, only 22.8% of the total spending on the tourism sector was allocated to business tourism. This indicator is quite high compared to other developing countries and according to UNWTO forecasts, it will rise by 4.9% by 2025 [27]. Popular cities that are most frequently visited by Russian citizens for business purposes are Beijing, Shanghai and Guangzhou. These cities are the most important for business tourism. Significant commercial and industrial organizations are located in these cities. The main flow of foreign investment goes here, there are a huge number of opportunities for business development, and the infrastructure is the most developed [28]. It is important to note that the APEC business travel card issuance system is valid for citizens of the Russian Federation and China. The APEC Business and Business Travel Card implies that the holder has the right to visit the APEC member countries for business travel without a visa. An additional advantage is the ability to speed up customs formalities upon arrival at the airport of any of the APEC member countries [29].

The development of Russian-Chinese educational tourism is also important. The number of Chinese students choosing Russian universities for higher education is growing every year. Statistics of the FGANU "Sociocenter" reports that in 2017–2018 is about 30 thousand Chinese citizens studied at Russian universities. 277 thousand foreign citizens were trained in Russian universities during this period [30]. It is obvious, that among other foreign students, the number of Chinese citizens is very significant. First of all, Chinese students come to study in Moscow, St. Petersburg, as well as universities in cities in the Far East [31]. Russian citizens also often travel to China to study certain specialties and languages. In recent years, Russian students have come to study in the Northeastern provinces of China (Heilongjiang and Liaoning), as well as in the capital and Shanghai. According to the Ministry of Education of China, in 2018 more than 19,000 Russian students studied at Chinese universities [32].

Russia and China take an active part in UNWTO activities. Thus, the UNWTO General Assembly has already been held in both states, in China - in 2017 in the city of Chengdu, in Russia - in 2019 in St. Petersburg. Important decisions were made during the Assemblies in China and Russia. These decisions will affect the development of international tourism until 2030. Thus, at the 22nd session of the UNWTO General Assembly in China, "The Chengdu Tourism Cooperation Initiative within the Belt and Road" was published [33]. This initiative is based on the fact that the "Belt and Road" can be used not only for economic purposes, but also for the humanitarian ties of the countries of the region, especially tourism. The initiative provides that the number of contacts in the interested states of the "Belt and Road" will be resolved, as far as possible, improving the quality of tourist exchange, training qualified specialists in the industry, as well as working together to reduce risks and improve the safety travelers. It is important to note that during the 23rd session, the Chinese and Russian sides discussed the prospects for cooperation, drawing attention to the fact that the development of the medical and cruise tourism segments is gradually gaining momentum in our country. The possibility of opening a Russian Language Center on the territory of Hainan Island was discussed. However, due to the subsequent events related to the COVID-19 epidemic, all this remained at the discussion stage. During the 23rd session, the Moscow Government and the UNWTO signed a Memorandum of understanding, which is an important step, since Moscow has been an affiliated member of the UNWTO since 1999, but no such document has been signed yet. The Memorandum is aimed at ensuring that the organization provides support for Moscow's activities in the field of tourism, as well as to enable Moscow to to interact more actively with other members of the organization. Representatives of the Tourism Committee expressed hope that this Memorandum will become the basis for close cooperation between the parties in the future [34].

As part of UNWTO, China promotes China Friendly program, which is a set of recommendations for receiving Chinese tourists. Each such set of recommendations is developed for each specific country only after its request [35]. One of the last events was an Ice and Snow Tourism Training Workshop in Heilongjiang, which was initiated by the UNWTO in China. UNWTO experts worked with members of the Heilongjiang Provincial Department of Culture and Tourism to develop plan for the regional tourism sector. The plan aims to strengthen the province's status as a leading seasonal tourist destination and will remain in effect until 2030. A number of bilateral agreements exist under which tourist exchange between Russia and China is carried out. The most important of these are the Agreement between the Government of the Russian Federation and the Government of China on visa-free group tourist trips (as well as the Protocol on Amendments to the Agreement), the Memorandum of Cooperation between the "World Without Borders" non-profit partnership and the Chinese Association of Tourist Companies, the law on tourism of China and the federal law "On the fundamentals of tourist activities in the Russian Federation.

The basic document regulating tourism activities in China is the Tourism Law of the People's Republic of China of 2013 [36]. This law protects the rights of tourists and tour operators, is aimed at maintaining a reasonable use of tourist resources, as well as supporting a steadily developing tourism industry. The law pays special attention to the process of tourism development, which must comply with the principles of economic, social and environmental benefits. The main document of the Russian Federation that regulates any activity in the field of tourism is the Federal Law "On the Basics of Tourism Activities in the Russian Federation" [37]. It contains the rights and obligations of tourists, the powers of the state authorities of Russian Federation in the field of tourism. The law prescribes the goals of state regulation of tourist activities, which generally coincide with the goals of state regulation of tourism in China (ensuring citizens' rights to rest, freedom of movement and other rights when traveling, environmental protection, creating new jobs, increasing the income of the state and citizens, preservation of tourist sites, rational use of natural and cultural heritage, etc.).

4 Discussion

Various programs and events aimed at improving interaction in the field of tourism are held annually in order to improve the quality of bilateral tourism and increase the tourist flow. The most important steps in promoting mutual understanding in this area are the years of tourism in Russia and China: 2012 was the year of Russian tourism in China, and 2013 was the year of Chinese tourism in Russia. During these years, numerous events were held. They are aimed at increasing the cultural understanding between the citizens of the two countries, as well as increasing the awareness of potential tourists about local tourist destinations.

In March 2012, the Tourism Association "World Without Borders" became the organizer of the first Russian-Chinese tourism forum, which marked the opening of the year of Russian tourism in China. It was devoted to a general analysis of existing problems and discussion of ways to further development of mutual tourist flows. In the same year, the second summit of the Russian-Chinese Tourism Forum was held in Shanghai. It was devoted to the issues of investment cooperation. The memorandum was signed, according to which the parties will promote the strengthening of bilateral ties and investment cooperation. In addition, a Cooperation Agreement with the Chinese Academy of Tourism was signed. It secured the format of the Forum's summits as the main platform for the Russian-Chinese dialogue in the field of tourism. The parties signed a Strategic Partnership Agreement with the Chinese banking association Union Pay. The aim is to expand the use of the Union Pay payment system by Russian and Chinese tourists when traveling from one country to another [38]. The Beijing-Moscow motor rally passed through 4 regions of China and 18 regions of Russia. The holding of this event made it possible to draw attention to the problems of automobile tourism, to a certain extent, even contributed to the improvement of transport infrastructure, as well as the creation of interregional routes. In addition, the rally demonstrated the capabilities of the Russian tourism sector, which, in turn, contributed to the growth of the attractiveness of the tourist product for the Chinese tourist. The data of a poll conducted by Radio China International in certain regions of the Russian Federation in 2013 report that almost 100% of respondents would like to take part in the events of the Year, since their level of knowledge about the Middle Kingdom leaves much to be desired (half of the respondents indicated that they know about China "little", another 20% said that they know "very little") [39].

Representatives of Russia and China discussed the existing problems with the so-called "gray schemes" in the sphere of Russian-Chinese tourism at the Forum Summit in 2019. During the discussions, the need to combat illegal tour operators was identified. Experts believe that this is only possible with the joint efforts of both Russia and China, and this requires the participation of both from private business in the tourism sector and government agencies that regulate activities in this industry. In addition, the procedure for introducing electronic document management was discussed. This procedure greatly facilitates the preparation of all the necessary documents before traveling to another country and speeds up the process of obtaining a visa. According to tourism association "World Without Borders", the number of Chinese tourists who arrived in Russia as part of organized tourist groups in 2019 amounted to 1,200 thousand people (an increase of 12% compared to the previous year) [39]. The main reasons for such rapid progress in outbound tourism are, first of all, the improvement of the quality of life of Chinese citizens, the increase in wages, as well as the signing of an Agreement between the Government of the Russian Federation and the Government of China on visa-free group tourist trips.

However, a huge amount of possible profit from Chinese tourism goes past the Russian treasury and returns back to China. This is due to the current so-called "gray schemes" of tourism. In China, a large number of organizations exist that offer tourist groups trips at the lowest possible prices. Chinese tourists come to Russia and meet, are served by their own compatriots. The organizers bring groups of tourists to their own special shops, restaurants, which open "on a call from their people". As a result, Chinese tour operators may not even think about competition from legal Russian organizations [40]. Naturally, such participants in the "gray schemes" do not pay any taxes, and all profits from them are returned back to China. In addition, there are no professional guides for Chinese tourist groups. The guides are Chinese compatriots who either live and work in Russia, or students who have come to Russia to study at universities. And they often tell such facts that discredit Russia in the eyes of Chinese citizens. For example, the guides say that the Russian Far East has always belonged to the Chinese, and the Russians illegally hold it, or that Moscow was built by the Chinese, etc.

The harm for the Russian side from such tourism is enormous. First, Russia is losing a huge amount of profit, which it has every right to receive Chinese tourists on its territory. Secondly, groups of Chinese tourists served under such "gray schemes" receive an extremely low quality of service, which is issued as the norm in Russia, resulting in an extremely negative opinion about Russian tourist service as a whole. Many Chinese remain disappointed that they are not shown a lot of attractions that they would like to see, and instead offer to spend money on various souvenirs and other goods. Such tourists leave their negative reviews about their trip to Russia on the Internet. As a result, more affluent Chinese citizens, who can afford individual travel with a visa, do not even consider Russia as an attractive tourist destination, considering that Russia with its low level of service is suitable only for citizens with low income. All efforts of the Russian side to increase the attractiveness of Russia as a tourist destination are simply in vain for a huge number of Chinese tourists. Unqualified Chinese guides provide groups of tourists with false information about the culture and history of the country. All this does not contribute to the growth of mutual understanding between the citizens of the two countries, and enhances the negative image of Russians in the eyes of Chinese citizens.

Tourism is a form of soft power and should create a certain positive image of the country visited. However, in the case of China and Russia, quite often things do not turn out quite like that. A significant problem is the negative image of Chinese tourist in the eyes of Russian citizens. This is due to the behavior of Chinese tourists themselves when traveling in Russia. Cases of Chinese tourist guides being beaten up over minor disagreements are common. Fights with local residents for various reasons happen. Quite often, the Chinese desecrate monuments, thereby showing extreme disrespect for local culture and traditions, littering and violating the rules of conduct on airplanes. Conflicts occur between tourists and local Russians. For example, a crowd of Chinese tourists tried to storm the ferry and attacked the crew members in the summer of 2017 on Lake Baikal. This happened due to a large influx of tourists and, as a result, the lack of seats on the ferries for everyone. In 2019, on Lake Baikal, again, a clash between Russians and Chinese tourists happened over a ferry. This time the video was published. Chinese tourists tried to block the Russians access to the ferry, making way for their compatriots. It is also known that in the summer of 2019, in line in Tsarskoe Selo, two Chinese tourists beat a Russian woman, after which she was hospitalized [40]. Such conflict situations arise and in the vast majority of cases, Chinese tourists are not responsible for their actions, which strengthens their confidence in impunity. The problem is that Russian citizens in most cases prefer not to contact law enforcement agencies, but to solve everything themselves. The situation is aggravated by the fact that Chinese tourists are often robbed in Russia. The Association "Tourism Safety" reports that this happens most often in St. Petersburg [41].

5 Conclusion

Tourism and culture are an integral part of people's lives. International tourism contributes to the dissemination of knowledge about the cultural values of other countries, which in turn ensures the development of international relations between people. In addition, tourism contributes to the preservation of cultural achievements of different peoples [42]. International tourism provides an exchange of cultural values between people. This creates mutual understanding between citizens of different states, which leads to the development of friendly relations between countries. We can observe this in the example of Russia and China.

All this leads to an increase in mutual hostility between the citizens of the two states. It is clear that such cases are not able to influence the general policy of the two states, including in the field of tourism, but this seriously affects public opinion and leads to the formation of certain negative stereotypes about each other. For example, many Russians already have an extremely bad idea of the Chinese as cultureless brawlers who do not respect other people's traditions. On the other hand, many Chinese believe that only deception and misunderstanding can be expected from the Russians. Unfortunately, every year this representation of the citizens of the Russian Federation and China is only increasing, no matter what the authorities officially declare about the friendship of peoples and the growth of mutual understanding. The governments of the Russian Federation and China are doing a lot to make bilateral tourism as mutually beneficial, safe and attractive as possible for tourists from both states [43]. Governments have sought to reduce restrictions on the more active flow of tourists by signing the Agreements. However, in practice, not everything worked out so well. The authorities must see real problems and take effective steps to eliminate them, then bilateral tourism will become truly mutually beneficial.

References

1. Matveevskaya, A., Pogodina, V., Tarakanova, T., Evseev, V., Nesterova, I.: Technologies of tourism in the modern urban environment. Int. J. Civ. Eng. Technol. **9**(10), 1566–1574 (2018)
2. Russia and China: four centuries of interaction. History, current state and prospects for the development of Russian-Chinese relations (2013). (in Russian)
3. Kirmasov, B.: Development of Russian Sinology in the XIX th century: historical pages. Znanie. Ponimanie. Umenie (2013). (in Russian)
4. Chen, B., Sevastyanov, D.: Formation, contemporary state and the prospects for the development of international tourism in China and Russia. Vest. Sankt-Peterburgskogo Univ. Seriya Geol. Geogr. **1**, 100–108 (2010)
5. Tarakanova, T., Pogodin, S.: Russia and China in the Shanghai cooperation organization: problems of political and economic cooperation (2018). (in Russian)
6. Balakin, V., Syaoin, L.: China and USSR in the 1960s–1980s: from confrontation to equal cooperation. Vestnil Yuzhno-Ural'skogo universiteta (2016). (in Russian)
7. Matveevskaya, A., Pogodin S, Wang, J.: Russia and China in the field of international tourism. Vest. Saint Petersburg Univ. Philos. Conflict Stud. **35**, 384–393 (2020). https://doi.org/10.21638/spbu17.2020.214
8. Blagoder, Yu., Minc, S.: The image of China in the minds of the Russian educated society of the 17th- early 20th centuries. Vestnik Rossiyskogo universiteta druzhby narodov (2011). (in Russian)
9. Efremova, M., Chkalova, O., Bi, Zh.: Analysis of the development of international tourism between Russia and China. Econ. Analiz: Teoriya Praktika (2017). (in Russian)
10. Kudryashova, S.: Forms of humanitarian interaction between the Russian Feredation and the China in the XXI century. Gumanitarnaya diplomatiya: lichnost', socium I mir (2018). (in Russian)
11. Chirkov, D., Chirkov, Yu.: "Shadow" business in tourism and economic security of Russia. Servic plus (2018). (in Russian)
12. Machalkin, S., Moreva, S.: Analysis of the quality and condition of the tourist flow from the People's Republic of China. Soc.-Ekon. Yavleniya Process. (2019). (in Russian)
13. Liu, H. The analysis of the influence factors of China's outbound tourism market. In: MATEC Web of Conferences, vol. 228, p. 05004 (2018). https://doi.org/10.1051/matecconf/201822805004
14. Kul'gachev, I., Lepeshkin, V., Manteifel', E.: Tourist exchanges between Russia and East Asian countries: state and development prospects. Mezhdunarodnaya torgovlya I torgovaya politika (2017). (in Russian)
15. Novozhilova, I., Tokusheva, D.: Ways and methods of serving Chinese and Japanese tourists in Russia, considering their national characteristics. Simvol nauki (2020). (in Russian)
16. Polozhevich, R.: Humanitarioan cooperation of the Russian Federation with the People's Republic of China at the present stage. Gosudarstvennoe upravlenie (2017). (in Russian)
17. Ochkasova, I., Sheleg, E.: Analysis of the development of Russian-Chinese tourism in the Russian Federation. Vest. Asoc. vuzov turizma serv. (2019). (in Russian)
18. Dzhandzhugazova, E., Blinova, E., Orlova, L., Romanova, M.: Development of red tourism in the perspective of the Russian – Chinese economic cooperation. Espacios **1** (2018)

19. Liu, J., Pan, H., Zheng, S.: Tourism development, environment and policies: differences between domestic and international tourists. Sustainability **11** (2019). https://doi.org/10.3390/su11051390
20. Zhao, M., et al.: Development patterns and cooperation paths of tourism industry within the China–Mongolia–Russia Economic Corridor. In: IOP Conference Series: Earth and Environmental Science, vol. 190 (2018). https://doi.org/10.1088/1755-1315/190/1/012067
21. Chow, C., Tsui, W.: Cross-border tourism: case study of inbound Russian visitor arrivals to China. Int. J. Tour. Res. **21**, 693–711 (2019). https://doi.org/10.1002/jtr.2297
22. Widawski, K., Wyrzykowski, J. (eds.): The Geography of Tourism of Central and Eastern European Countries, pp. 375–435. Springer, Cham (2017). https://doi.org/10.1007/978-3-319-42205-3
23. Matveevskaya, A., Pogodina, V.: Russia's participation in international tourism projects. Geographicheskaya nauka, turism I obrazovanie: sovremennie problem I perspectivy razvitiya (2018). (in Russian)
24. Chistilina, A., Shalaev, D.: Analysis of tourist destinations in China for citizens of the Russian Federation. Vest. Asoc. Vuzov Turizma I serv. (2019). (in Russian)
25. Ermolina, M., Kapustina, M., Matveevskaya, A., Pogodina, V.: Legal regulation of ecological tourism in Arctic. In: IOP conference Series: Earth and Environmental Science, vol. 302, p. 012037 (2019). https://doi.org/10.1088/1755-1315/302/1/012037
26. Official website of UNWTO. https://www.unwto.org
27. Li, J., Pogodin, S.: "Made in China 2025": China experience in industry 4.0. In: IOP Conference Series: Materials Science and Engineering, vol. 497, no. 1, p. 012079 (2019). https://doi.org/10.1088/1757-899X/497/1/012079
28. Official website of Russian APEC research center. (in Russian). http://apec-center.ru
29. Official website of FGANU "Sociocenter". (in Russian). https://sociocenter.info/ru
30. Mayorova, O., Li, L.: Analyzing Sino-Russian border tourism cooperation. In: 14th International Conference on Services Systems and Services Management, ICSSSM, p. 7996212. (2017). https://doi.org/10.1109/ICSSSM.2017.7996212
31. Official website of Xinhua News Agency. http://www.xinhuanet.com
32. Official website China Internet Information Center. http://russian.china.org
33. Official website of Interfax-Tourism. (in Russian). https://tourism.interfax.ru
34. Official website of the Travel Association "World without Borders". (in Russian). http://www.visit-russia.ru
35. Tourism law of the People's Republic of China (2013) http://en.pkulaw.cn
36. Federal Law of the Russian Federation "On the basics of tourist activities in the Russian Federation" (2020). (in Russian). http://docs.cntd.ru
37. Official website of LLC Marketing, Exhibitions, Consulting. (in Russian). http://www.archive.expochina.pro
38. Official website of "Russian newspaper". (in Russian). https://rg.ru
39. Official website of "Komsomol'skaya pravda". (in Russian). https://www.spb.kp.ru
40. Official website of "Eurasia. Expert". (in Russian). https://eurasia.expert
41. Official website of Association "Tourism Safety". (in Russian). https://www.tourismsafety.ru
42. Matveevskaya, A.: Tourism as a political interstate dialogue. Nauka Krasnoyar'ya (2017). (in Russian)
43. Sidorenko, T.: The scope of economic cooperation between Russia and China and future prospects. Probl. Desarrollo **45** (2014). https://doi.org/10.1016/S0301-7036(14)70849-1

International On-Line Collaboration in the Context of Contemporary Higher Education

Anna Riabova(✉) and Olga Pavlova

Peter the Great Saint-Petersburg Polytechnic University, Saint Petersburg, Russia

Abstract. The purpose of the paper is to reveal and identify the role of collaborative on-line international learning projects in the context of contemporary pedagogical challengers and principles. The novelty of the research is emphasized by the necessity to consider the appearance of the cultural behavioral patterns which are based on the on-line ethnicity, and the necessity to explore such principles of Pedagogy as Socially Conscious Pedagogy, Accessibility, Open Pedagogy and Educational Resources. The Universities' mission to teach young people around the world respect and international understanding is highlighted. The role of collaborative international projects in promoting virtual mobility through COIL (Collaborative On-line International Learning) is viewed and emphasized. Con-ducted in 2019, the international project between Russian and American students provided a collaborative educational experience focused on the challenges of migration. The paper explores the implication that borders; migration; migrants; stereotypes and ethnocentrism have on international relations. Even though this is a virtual experience, it is highly beneficial to students' development in inter-cultural competence. The need for Universities to give their students the opportunities and environment to enhance their intercultural competence is discussed. The COIL model enhances skills in team – teaching, and the students 'ability to produce and present work collaboratively. The paper reveals the necessity for individual human potential realization and considers the content of the permeable structures that develops cross – disciplinary interaction, soft skills and critical thinking development.

Keywords: Higher education · Socially conscious pedagogy · Accessibility · International project · Collaborative on-line international learning · Intercultural competence

1 Introduction

The universities worldwide have incorporated the COIL approach as a part of internationalization. The goal of the paper is to identify the role of international projects for teaching distance collaboration in the context of contemporary higher education. The tasks of the paper are: to analyze the principles of contemporary higher education; to discuss the faculty experiences of participating in the Collaborative On-line International

© The Author(s), under exclusive license to Springer Nature Switzerland AG 2021
R. Bolgov et al. (Eds.): *Proceedings of Topical Issues in International Political Geography*, SPRINGERGEOGR, pp. 310–317, 2021.
https://doi.org/10.1007/978-3-030-78690-8_27

Project; to emphasize the need for the shared cultural vision and international collaborative project–based learning. It is crucial to achieve intercultural competence while participating in collaborative projects. The intercultural competence can be described as the ability to communicate effectively and appropriately in the situations when the experiences, culture, ethnicity and language are not shared by the individuals involved in the process of interaction.

Socially conscious pedagogy explores ways to make the classrooms transformative spaces of resistance to injustice. Institutions and systems often do not practice questions of justice, so the classrooms can feel like contradictory spaces for students. The second principle that is built from socially conscious pedagogy focuses on making the classroom accessible. It is important to think about accessibility throughout the process of developing and teaching the courses. The principle of Universal Design for Learning urges educators to integrate variety, choice and flexibility into course design, avoiding the same approaches to teaching in different classrooms. Writing pedagogy considers the role of writing in students' course by promoting critical thinking and clear expression of thought through writing. Reflective letters are thought to be a tool to consider dynamic movement towards better understanding of the course, and are considered as the essential component of the project. Better perception what kind of people the students are, how they got to our classrooms, and where they want to go next can greatly enrich the teaching and can be estimated with the help of reflective letters.

Education and intelligence are becoming a core value in the new XXI century. In the global age, all the advanced powers will rely on the acquisition of new knowledge and the creation of new technologies [12]. There should be a new level of personal commitment to experience a challenge of high-quality learning. The teaching staffs who are able to fit such an expanded professional awareness can find the teaching activity rewarding [5]. In order to benefit from COIL, the Universities must be in geographical regions with different cultural backgrounds. COIL should involve collaboration between two or more faculty members, who teach related courses; can create both a shared syllabus and course material, and assume shared responsibility in mentoring the students during the whole project. The length of the project can reach a semester. The main idea is to help the students to be more global thinkers and gain some beneficial development in intercultural competence as they work collaboratively with people who belong to different cultures. The teachers should seek ways to challenge students by engaging them in real-world issues that are authentic and collaborative. Collaborative projects can be challenging to implement, but with proper planning and reflection, the teachers can develop effective, manageable, and motivating learning experiences in the classroom.

2 Problem Statement

There is not enough investigation targeted at highlighting different perspectives of collaborative project–based learning and the usefulness of collaborative activities toward the intercultural competence development. Some researchers [3, 8] explain that the shared goals of interculturalism are based on the ability to cooperate and comply with various points of view, maintain and develop the relationships and effective communication with little distortion. Other researchers highlight the necessity to see people first, to learn other cultures' values [18, 21]. There is the special need for some practical suggestions to the teachers interested in pursuing international collaborative activity and developing a better understanding of what professional skills might be involved in conducting a collaborative international project. Critical reflection is used increasingly in the academic context as a tool of focusing the students' attention onto their own experience gained during the project activity [16]. Although the critical thinking activity is the requirement of professional practice and academic study there is a great challenge in doing this practice within a framework of international educational conventions. Within English–language–learning contexts, project–based learning has been shown to be a powerful method that enables the integration of academic, social and linguistic communication skills with the application of real world issues and contexts. The relevance of the research is emphasized by the necessity to consider the appearance of the cultural behavioral patterns, which are based on the on-line ethnicity [6, 17] and updating the pedagogical knowledge regarding cross–cultural competence while developing mutually beneficial relationships that can contribute to international problems solution [14]. Learning is seen in terms of variation, professional skills' development and increased confidence. Teaching is challenging for the most committed instructors and the evolving social realities create contexts that impact the lives of both faculty and students [11, 13]. The research questions are connected with the analysis of modern higher education in the context of new soft skills implementation and the development of new learning environment that is to shift in viewpoint from ethnocentric to a more multi–culturally accepting perspective. The new perspective of participating in the Global Exchange Project collaboration is set forth; the aspects and challenges of different stages of collaborative work are assessed. The role of collaborative international projects as the foundation of successful change and the basis of the students' research and analytical skills development is estimated.

3 Theoretical Background

There is a growing stream of practical books, aimed at guiding both new and experienced university teachers. Such researchers as Karasik [12], Komarova [14], Alkarzon [2], and Eaton [9] captured the major trends and perspectives of teachers' education and put forth a vision for the future that is grounded in current education and research. The role of technology is evident. The networks make it possible to get the access to information that is situated in domains of practice. The number of authors considered the topic of blended learning implementation and the challenges of digital teaching environment: Allen and Seaman [1], Brisk and Harrington [7], Gashkova, Berezovskaya and Shipunova [10] compared the blended learning with on-line classes and measured the positives outcomes for students in blended format. Simola described the Finnish model of higher education and the links between the Universities and local environment with the emphasis on political geography [19]. Becket and Slater [4], Brisk [6] highlighted the necessity of international collaborative projects' careful planning that integrates both intercultural and English language learning development. The networks allow the students to join different professional communities, video tools can help the teachers evaluate and improve their curriculum to fit the international requirements. It was found that the relationship development became more pragmatic once the initial connection had been formed although good communication was still vital for success.

4 Methodology

The research methods included the comparative method, and the pragmatic–communicative method. The object of the paper, the tasks and research methods are to be considered in relation to the impact of innovation on the educational strategy. Judgment by others and self, direct observation by others from the host culture, student interviews, and case studies are the assessment methodologies unanimously acceptable by academic administrators involved in institutional internationalization programs.

5 Results and Discussion

The impact of international collaborative activity on the learning outcomes is discussed; the efficacy of blended learning is emphasized; new approaches to teaching are outlined. The promotion of international learning opportunities by developing a COIL (Collaborative On-line International Learning) module is highlighted. The skills necessary for successful project co–development are identified. If international collaboration works it creates synergy and develops mutually beneficial relationships, and university partnership provides a great number of opportunities for students and staff alike. Blended learning tuition involves a combination of elements, activities and inputs; the success of this type of learning depends on the understanding and commitment of the trainers, the effectiveness of well–organized and reliable administration, and the wholehearted engagement of learners, prepared to participate in a course for innovation to continue. Cultural contacts nowadays occur in a much more globalized and faster way. More than ever, intercultural communication is to be practiced, since the scope can encompass the relations between remote Eastern and Western cultures [3, 20]. The most significant manifestation of identity politics in modern society is multiculturalism. Collaborative projects are aimed at conducting research and motivating students' creative skills development more than traditional lesson framework [17]. Thus multiculturalism recognizes cultural diversity together with the fact that the differences should be respected and the importance of beliefs and values, sense of self-worth and various ways of life should be acknowledged. Cultures can enrich each other by extending cultural understanding and transcending the benevolence [22]. Instructors can use a variety of learning practices aimed at providing feedback. This work ranges from asking students to produce their own knowledge through original research, community engagement projects, and creative expressions to integrating texts and materials that are outside of disciplinary canon into the syllabus. On the first day of class instructors may invite students to choose between the topics that will be studied throughout the semester, crowd source questions that can orient selection of material, and/or ask students to suggest course materials that are not part of the traditional canon. As the semester unfolds it is very important to encourage students to reflect on the processes (write reflection letters). The critical reflection structures that are focused on developing effective study and following the requirements of using insights to effect change are selected. The learning goals should be clearly communicated to the students. Instructors can use a variety of activities that promote students' direct engagement with course material. These include small groups with task-oriented in-class activities where students are invited to respond to a problem proposed by the teacher, to reach consensus within the group, to posit questions to each other, and / or to practice reading a text. When reading and working together to attend to a question, task, or problem, students can learn how to listen deeply, to build trust, and to communicate generously, with the teacher as a model for these practices.

The COIL Project: Borders, Migrants and Migration can serve as a good example of socially conscious collaboration in the classroom. University of Arizona, Tucson, LaGuardia Community College, NY and Peter the Great Polytechnic University, Saint-Petersburg participated in the international on-line project. This international project brought together students from three geographically remote urban campuses to explore multiple meaning of border, migrants, and migration. The participants of the international collaborative project were supposed to develop a more critical awareness of their own perspectives on borders and migration, including the role of campus spaces, established institutional practices, and socio-political contexts in shaping people's perspectives on these global issues. The purpose of the Russian-American collaborative project was to explore stories and discourses of borders and migration collected on respective campuses. The first activity that the teachers decided was an icebreaker activity through a Skype session where students met each other for the first time. They talked about their major, hobbies and why they had decided to take this course. Skype was selected because it is free of charge and is the interactive technological tool that can be used in the classroom [9]. For the second Zoom session, Russian and American students were asked to read the article about migrants and migration before coming to the class. The activity was meant to build confidence on both groups. The students talked about the attitude towards migrants and migration in their countries, shared the information they considered as ethical dilemma. The students got involved in a cross-cultural dialogue about liminality and movement – conditions that can be enriching and devastating, transformative and unsettling. At the end of the project, participants were expected to create digital group narratives about the impact of borders and migration on their own and their global partners' communities. The students were engaged in a deep, reflective activity through global discussion, use of multimedia, geo-location tools, and multimodal narratives (digital stories). The first stage was connected with the definition of borders, migration and migrants given by the students from three Universities. The Russian and American students described borders as physical or mental structures that prevented people from doing something or going somewhere. The students described not only visible but invisible borders as well paying special attention to privacy. The second stage was connected with discussions and debates. The students were divided into four groups and the moderators inspired them to give their opinions on the articles that had been chosen by the instructors and were targeted at the migration processes' assessment. During the whole project the reflective activity that the participants got used to sharing with their partners and teachers helped them to assess the challenges of that new experience and to deepen the critical writing skills. The students reported that it was difficult at the beginning, as they did not know what to expect but as the semester went on they were able to get to know each other. In the curriculum of four steps, students explored stories and discourses of borders and migrations collected on their campuses and reflected their own relationship to the topic. The participants created the Special Learning Journey that was divided into four steps. During the 1st step the students prepared a multimodal account of their walk for sharing with global partners via Project Mapping Platform and Collaborative Blogging Space. The students met with their global partners via Zoom to discuss the "borders" on their campuses. During the 2nd step the participants interviewed the "border crossers" informally, prepared ethnographic report about their activity and shared it with the global partners via Project Mapping Platform and Collaborative Blogging Space. The students discussed their ethnographic discoveries via Zoom. The 3rd step was connected with the stories of Migration: Where is home? What does it mean to become a local? The participants read the articles about migration and migrants, shared their thoughts and reflections with global partners via Collaborative Blogging Space. The 4th step was aimed at discussing the biggest discoveries, identifying the most significant personal stories and digital artifacts of border crossing the students collected over the weeks of project activities. Each student wrote a reflective essay about the participation in this global project and submitted the reflection to the course instructor. The ability to scrutinize and share experience is essential in academic accomplishment and international collaborative projects help to facilitate this. According to the reflection letters' assessment the participants were able to enhance the following skills: professional skills; research skills; critical thinking; social skills; communication skills; coping with challenging/unexpected situations; interpersonal skills; awareness of one's own abilities.

When the COVID 19 interfered with the development of COIL project the following questions were to be discussed by international student:

- How did COVID-19 pandemic transform borders and boundaries between people, countries, private and public spheres?
- How can a global, student-centered dialogue about the pandemic foster deep reflection, resilience, and healing across borders? What would we want to remember about the world and ourselves during the pandemic and why?

The synchronous discussion was organized where the students were asked to reveal how COVID 19 pandemic transformed borders and boundaries between people, countries, private and public spheres. The participants were involved in a global, student –centered dialogue; they mentioned that pandemic fostered deep reflection, resilience, and healing across borders. The stories that were connected with different recollections of peoples' behavior during the pandemic were the most impressive. The students led a small-group collaborative poetry activity, in which they wrote a collaborative poem with the international peers about their most vivid memory of the pandemic.

Integrating writing into the curriculum can increase student engagement, keep them on track with reading, and prepare them for the digest writing course. Now we are in the situation where on-line technologies, mobile devices, and electronic learning contexts can be combined to create a functional learning environment. At best, information technology, and related applications enhance interactivity and motivation for learning, as well as provide the students with genuine opportunities to develop their skills in the area they use to study and do the research in. Collaboration has become international in nature. However, as participants have become more experienced and sophisticated in blended learning, they are more critical. In a course covering a wide range of management areas, some activities are more appropriate or effective than others. For instance, reading articles, which have an angle on a topic, and can then be mapped onto the participants' own contexts seems to generate most discussions. The results of the students' questionnaire revealed that nearly all of the students considered that an on-line learning management system helped them improve their knowledge.

6 Conclusion

The survey results pointed out that the students generally perceived collaboration as positive experience, which is consistent with the literature. The students were grateful they had had real experience with people from different cultures pursuing a common goal. Critical thinking is a skill crucial to all students studying on academic courses. Critical thinking requires elements connected with the emotional side of learning and is associated with the freedom the teacher allows [20]. The major themes that we focused on included: time orientation, communication styles, team dynamics, trust building, socializing, and creating a reflective environment.

The students are encouraged to solve the problems independently, to convince their peers into their ideas and to analyze the results. International collaboration and project management is based on presentation of examples as well as discussions, debates and experience comparison. Participants have the opportunities to develop personal and professional skills for collaboration, to learn some unknown information about the life of their partners from abroad, master basic tools for virtual communication and evaluation through workshops. The collaborative process gives a clear and elaborative way to achieve the goal through joint work. Trust among the members overcomes initial hesitance [21]. It is not surprising that the integration of critical thinking and soft skills in the process of teaching is considered as essential to prepare learners for modern workplaces. Many opportunities (of publishing, creating professional connections) depend on the ability to explain the nature of the research. This was an enriching experience. Firstly, it is important that the professors from different universities were able to build a course with a common goal. Secondly, the successful experience was achieved because Russian and American students got to know each other and managed to push through a project despite their cultural differences, since they worked together pointing out common areas. Thirdly, according to the results, it can be seen that students can be classmates even though they have barriers to overcome such as distance, cultural differences, language.

It is important for the teachers to present their research as part of the big picture of the discipline, as teaching requires the participants to find a balance between the general and the specific as you consider students as your audience. The course design and learning goals should be clearly highlighted; the research and teaching activities should feed each other since they are at the core of professional practice. Teachers are supposed to collaborate in the implementation of project–based–learning and support their students and each other throughout the process. Collaborative on-line international learning can be a new method of learning and evaluation.

References

1. Allen, E., Seaman, J.: Online Report Card: Tracking Online Education in the United States Babson. Survey Research Group and Quahog Research Group LLC (2016)
2. Alkarzon, A.: The Influence of faculty exchange programs on faculty members' professional development. Res. High. Educ. J. **30**, 1–16 (2016)
3. Alvino, E.F.: A Central Concern: developing intercultural competence: School for International Training (SIT) Occasional Papers Series, Addressing Intercultural Education, Training and Service (2000)
4. Becket, G.H., Slater, T.: The project framework: a tool for language, content, and skills integration. ELT J. **59**(2), 108–116 (2005)
5. Bekisheva, T.: Blended Learning: Modern Tendencies in Higher Education. Modern Research of Social Problems 11.2 (67) (2016)
6. Brisk, M.E. (Ed.): Language, Culture, and Community in Teacher Education, pp. 97–102. Mahwah, N.J.: Lawrence Erlbaum, Mahwah (2017)
7. Brisk, M.E., Harrington, M.M.: Literacy for All Teachers. 2nd edn. Mahwah, N.J.: Lawrence Erlbaum Associates, Mahwah (2007)
8. Dobbins, M.: Higher Education Policies in Central and Eastern Europe: Convergence Towards a Common Model? MacMillan, Palgrave (2011)
9. Eaton, S.: How to use Skype in the ESL/EFL classroom. Internet TESL J. **18**(11), 1–14 (2010)
10. Gashkova, E., Berezovskaya, I., Shipunova, O.: Models of self-identification in digital communication environments. In: The European Proceedings of Social & Behavioural Sciences 35, pp. 374–382 (2017). https://doi.org/10.15405/epsbs.2018.02.44
11. Hativa, N.: Teaching for Effective Learning In Higher Education. Kluver Academic Publishers, Dordrecht, the Netherlands (2012)
12. Karasik, V.I.: Discorsive personalization. Language, communication and social environment. Voronezh, pp. 78–86 (2007). [In Russian]
13. Kiely, R.: Small answers to the big questions: learning from language programme evaluation. Lang. Teach. Res. **13**(1), 99–116 (2009)
14. Komarova, Z.: Methodology, method, methodics and the technology of scientific investigations in linguistics. Ekaterinburg (2012). [In Russian]
15. Love, A.G., Dietrich, A.J., Fitzgerald, J., Gordon, D.: Integrating collaborative learning inside and outside of the classroom. J. Excell. Coll. Teach. **25**(3–4), 177–196 (2014)
16. Martín, P., Pérez, I.K.L.: Convincing peers of the value of one's research: a genre analysis of rhetorical promotion in academic texts. English Spec. Purp. **34**, 23–36 (2014)
17. Miller, L., Hafner, C.A., Fun, C.N.K.: Project based learning in a technologically enhanced learning environment for second language learners: students' perceptions. E-Learn. Dig. Media **9**(2), 183–195 (2012). https://doi.org/10.2304/elea.2012.9.2.183
18. Rubtsova, A.: Socio-linguistic innovations in education: productive implementation of intercultural communication. In: IOP Conference Series Materials Science and Engineering, p. 497, no. 1 (2019). https://doi.org/10.1088/1757-899X/497/1/012059
19. Simola, H.: The Finnish miracle of PISA: historical and sociological remarks on teaching and teacher education. Comp. Educ. **41**, 455–470 (2005)
20. Ter-Minasova, S.: Language and intercultural communication. URAO Publishing House, Moscow (2000).[in Russian]
21. Torre, M.: The selection of personnel for international service. Int. Exec. **6**(2), 5–6 (1964). https://doi.org/10.1002/tie.5060060202
22. Wolff, F., Borzikowsky, C.: Intercultural Competence by International Experiences? An Investigation of the Impact of Educational Stays Abroad on Intercultural Competence and Its Facets. J. Cross C. Psychol. **49**(3), 488–514 (2018). https://doi.org/10.1177%2F0022022118754721

Sociocultural Problems of Adaptation of International Students in a Non-native Language Educational Environment and Techniques to Overcome Them (at Universities of Russia, Finland, Sweden and China)

Nikita Ivannikov[1], Dmitry Kolesnikov[1], Marina Sablina[1], and Alexey Tsyb[1,2(✉)]

[1] Peter the Great St. Petersburg Polytechnic University, St. Petersburg, Russia
[2] Sociological Institute of FCTAS RAS, St. Petersburg, Russia

Abstract. The article annotates the study of sociocultural issues and behavior patterns of international students in the educational environment; problems of students' adaptation in a non-native language environment in an unfamiliar culture; the study of various aspects of a surrounding society impact on a students' personality and algorithms of a student behavior, comprehensive interdisciplinary study of general theoretical and special topics related to sociology, psychology, cultural studies; studying the possible impact of the educational system of the university and academic institution of higher education on a student's personality and scale of priorities, studying the identification of sociocultural differences, developing communication skills of young people belonging to different sociocultural models on the example of Peter the Great Saint-Petersburg Polytechnic University, Saimaa University of Applied Sciences, Kymenlaakso University of Applied Sciences (Finland), Mälardalen University College (Sweden), Tsinghua University (PRC). As well as the elaboration of the revised concept study of "The psycho-pedagogical support of the adaptation of foreign students in the educational environment in the system of higher professional education" and the toolkit development for monitoring and processing the research results (preparing questionnaire forms and software development), briefing Peter the Great St. Petersburg Polytechnic University monitoring specialists.

Keywords: Education · University · Student adaptation · Culture shock

1 Introduction

The problem of adaptation to changing conditions is now one of the central ones in the social, historical, humanitarian and psychological sciences. This relevance is caused by a whole complex of processes taking place in the world today: globalization, digitalization,

© The Author(s), under exclusive license to Springer Nature Switzerland AG 2021
R. Bolgov et al. (Eds.): *Proceedings of Topical Issues in International Political Geography*, SPRINGERGEOGR, pp. 318–327, 2021.
https://doi.org/10.1007/978-3-030-78690-8_28

urbanization and other ones that change human and social life radically (influencing both anthropological and ethnographic aspects) [1]. Adaptation in the sphere of education is the most important direction of research, because the society of the XXI century should consist of highly qualified specialists capable of solving new emerging problems in both local and global scale, otherwise the prospects of this society are doubtful.

The two-level system of higher professional education adopted by the Russian Government has contributed to Russia's accession to the European Bologna Process movement, modernization of the Russian higher school and unification of interests of international and Russian higher educational institutions. The two-level system of education allowed creating the unified educational space in Europe. Besides, thanks to these innovations, new problems have arisen related to the education of Russian students abroad and the problems of international students studying in Russian higher educational institutions. The first level (bachelors' degree) prepares a student for the jobs involving executive functions in productive or socio-economic areas. When studying at the second level (master's degree) future specialists are focused on activities that require analytical and project-oriented skills as well as on research activities.

However, when considering the schemes of organizing the educational process in Russian universities and universities of China, Finland and Sweden, some fundamental differences for both bachelor's and master's degree levels can be observed. Note that the main objective of the Bologna process is to achieve comparability and, ultimately, harmonization of the national educational systems of higher education in Europe. How to achieve comparability and harmonization of Russian and Chinese educational systems of academic higher educational establishments and how to help a student in the process of adaptation while studying in a non-native language educational environment? To develop options for a possible transfer from one studying system to another.

The relevance of the study is determined by intensification and enhancing international (including educational) contacts. Cooperation in the field of education is the most essential and urgent task for many countries, as training and education of young people play an important role in the process of spiritual rapprochement of peoples and integration into the global community. Cooperation in the sphere of education is getting particularly intensive between countries with common geographic boundaries. Cooperation in higher education between Russia, China, Finland and Sweden is an illustrative example in this respect.

Integration of the domestic system of higher education into the European one, establishment of their interaction creates new challenges and also changes the existing order on the market of educational services, in particular their export. The success of exporting education directly affects credibility of the country and of the educational institution in the world. This increases the requirements for Russian universities, which must be met in order to be able to compete with other participants in the educational market.

Within regarded issue, an important role is given to the adaptation of international students to the new language, everyday life and academic environment. The possibility of such adaptation at a sociocultural level, which can be seen as "a process of establishing a certain correspondence between an individual and his or her new non-native language social environment. The process of socio-cultural adaptation includes several components that are closely related to each other and are designed to overcome cultural and linguistic barriers" [15].

At the same time, it must be understood that the success and timing of adaptation depends not only on objective reasons, but also on the individual characteristics of the student. In particular, it is possible to include the psychological aspect of adaptation, when personal qualities, perception abilities and readiness for new experience, value orientations of an individual come to the foreground to influence the success of the process. All of them play an important role. Another key perk in the structure of professional training is being stress resistant.

Thus, the fundamentality of higher education in Russia is a significant factor that attracts students from other countries and regions. However, the arrival of a student and admission to an educational program is not the final decision, it is much more important to retain students. Here the processes of adaptation come to the fore, because without its successful completion it is impossible to talk about the success of the international educational program. Studying the problems of adaptation of international students in the Russian Federation justifies the relevance of this work. On the other hand, having adapted the world experience in educational systems, Chinese universities annually get into World's Top 100 Universities lists according to various authoritative rankings (QS World University Rankings, ARWU). The given circumstances positively make China an appealing country for international students, so the relevance of this project is beyond doubt for both Russian and Chinese sides. For its part higher education in universities of Finland and Sweden is very prestigious for hundreds of Russian and Chinese students.

2 Problem Statement

The new phase of comprehensive partner ship and strategic cooperation between the Russian Federation and the People's Republic of China, as well as a continuous humanitarian bridge between Russia, Finland and Sweden, suggests 'intensifying humanitarian exchanges, especially through the implementation of the action plan for development of Russian-Chinese cooperation in the humanitarian sphere', as well as current Russian-Finnish and Russian-Swedish relations. At this phase the issue of studying the problems of sociocultural adaptation of Chinese, Finnish, Swedish and Russian students in a different sociocultural educational environment, as well as creating technological models to overcome these problems, becomes urgent. This involves:

1. the integrated multidisciplinary study of general theoretical and special topics related to sociology, psychology, cultural studies in the field of adaptation;
2. the study of sociocultural issues and behavior patterns of students in the educational field of a different social environment and culture, problems of student adaptation in a foreign environment in an unfamiliar culture;
3. to rethink the conditions of students adaptation to a different sociocultural environment on the basis of 'The Theory of Communicative Action' by Jürgen Habermas in accordance with the types of social action: strategic action, normative action, dramaturgical action, communicative action;
4. the study of various aspects of the related society influence on individuals and algorithms of student behavior;

5. the study of the educational system of the university and academic higher education institution influence student's personality (object of action) and his/her scale of priorities, as well as students' opportunities for active/inactive actions (subject of action) in the process of adaptation to a different educational environment;
6. the detection of binary relations of similarities and differences in sociocultural environments of the educational environment of Russia, China, Finland and Sweden, development of communication skills of young people belonging to different sociocultural models.

In addition, it is essential to develop the adaptation techniques for students on the basis of these theoretical studies.

Majority of all international students faces certain physiological and socio-psychological challenges. Among physiological difficulties, the main ones are getting used to the climate and cuisine, other amenities.

Difficulties of socio-psychological nature are emerging because of the attitude of surrounding people. A significant place is occupied by adaptation to other social living conditions, including a new system of education. Adaptation to educational activities is often quite difficult. Thus, during almost the whole period of study students have difficulties in learning a material in a foreign language. In this case, extralinguistic context also may provide unnecessary challenges to learning process [15].

3 Research Questions

The main objective to solve this problem is to study the anthropological and sociocultural aspects of the problems of adaptation in a foreign language environment and the human/environment interaction by applying the typology of the social action according to 'The Theory of Communicative Action' by Jürgen Habermas. which will help to tackle the problem innovatively (from the fundamentally new positions) and, as a result, to develop methods of general and specific nature, combining the educational process in four countries (Russia, China, Finland, Sweden), and to solve the following research tasks at the same time: to indicate the substance of sociocultural adaptation of international students; to explore the main directions of their sociocultural adaptation; to identify problems arising in the course of the sociocultural adaptation of international students to Russian, Chinese, Finnish and Swedish educational environment; conduct thematic surveys among international university students on the issues of their sociocultural adaptation; to conduct a comparative analysis of sociocultural adaptation problems of international students in Russian, Chinese, Finnish and Swedish educational environments on the basis of the survey data; to make conclusions on the problems of sociocultural adaptation in the educational environment of four countries on the basis of the obtained data; to develop practical recommendations aimed at optimizing the process of sociocultural adaptation conditions of international students at Russian, Chinese, Finnish and Swedish universities.

It should be noted that the process of adaptation to the educational environment is as complex as any other adaptive process, and is a composite phenomenon that occurs in the interaction of various factors that form the socio-cultural environment. Three stages of this process can be distinguished - entry into a new space, development of necessary personal qualities, their rooting in the personality, which leads to the formation of a new, adapted consciousness of the individual. Such socio-cultural adaptation is a complex process of personality collision and new environmental conditions. In its course, foreign students, who have their own special cultural, social, ethnic and psychological background, need to successfully solve new problems caused by the mismatch of the previous norms of morality, behavior, environment and everyday life. If a student copes with new challenges, he or she develops and assimilates new forms of activity and behavioral norms for full functioning in the environment.

The importance of successful adaptation cannot be diminished, as it is largely a determining factor in the success of the educational program in general. The problems that students need to solve first of all can be divided into the following groups:

1. Psychophysiological (climate change, psycho-emotional tension due to changing environment).
2. Educational (lack of language practice, differences in the implementation of educational programs in the host country and home country - requirements, ways of teaching).
3. Sociocultural (entrance to the new cultural space, language barrier in communication with both teachers and administrative services, as well as with fellow students, adaptation to regional and local cultural peculiarities).
4. Domestic (dissatisfaction with everyday life, provided amenities and financial situation).

The following measures can be taken to improve the situation of the student and accelerate his or her adaptation:

1. It should be taken into account that a student's adaptation is not only within the framework of mastering the disciplines. The organization of extracurricular activities of various kinds helps to improve the course of adaptive processes, as it introduces and/or deepens their understanding of the new reality and its values;
2. Improving the quality of education, providing clarity of competencies and skills, justifying their necessity, and providing practical reinforcement should all be part of educational policy.

Successful application of these measures, as well as the achievement of full adaptation by other means, will have a positive impact both on the image of the educational organization and the entire perception of the host country by the world community, and not only at the academic level [2]. The process of adaptation is multidimensional and complex, requiring students to interact with their environment at all levels - domestic, social, cultural and psychological. Its result is the mastering of new roles, the adoption of new values.

4 Purpose of the Study

The aim of the study is to analyze the main problems of international students' adaptation to the new socio-cultural environment. It is truly relevant because majority of contemporary research in this area is focused on sociocultural adaptation [8]. In line with the target, the main trajectories are:

1. definition of the main aspects of interpreting the sociocultural adaptation;
2. definition of conceptual and categorical framework of the sociocultural adaptation study; analysis of the main theoretical approaches to the sociocultural adaptation study;
3. identification of the problems which may arise or emerge within the process of the sociocultural adaptation of international students in both Russian and Chinese educational environment;
4. development of practical recommendations which objective is to optimize the course of the sociocultural adaptation of students in a non-native language environment.

It may also be interesting to note that in order to be successfully integrated into an academic community while living in a region with a non-native language, students do not need to intentionally contact with non-native host speakers, as they may as well understand the culture of the host country through communication with other students and who have similar cultural and linguistic backgrounds [4].

5 Research Methods

The theoretical and methodological basis of the study: basic provisions of the concepts of adaptation and socialization of the individual, ideas of axiological approach, including the value development of a person in the society and the role of values in person's socialization, adaptation to the environment, as well as definition of value identity and relevance; culturological approach revealing the category of culture, the concept of interaction between cultures and the principle of intercultural interaction, the leading ideas of the activity approach in determining the interaction of international students with representatives of a new society.

The experimental part of the study will comprise a number of techniques:

1. Express-questionnaire «Tolerance index» [13], aimed to determine aspects of tolerance: general level and/or personality trait (reveals a general attitude towards the environment; attitude towards other people; social attitudes in various areas of interaction where tolerance and intolerance of a person are manifested); types of tolerance: ethnic (determines attitude towards people of a different race, ethnic group, own ethnic group; attitudes in the field of intercultural interaction) and social (attitudes towards certain social groups (minorities, criminals, mentally disabled, beggars); communicative attitudes (respect for the opponents' opinion, readiness for constructive conflicts resolution, for productive cooperation).

2. E.Bogardus social distance scale [12, p.216–218], aimed to evaluate the degree of psychological closeness of people: maximum social dinstance; minimum social distance. Rates attitude to a particular nationality.
3. D.Phinney methodology, measuring the rate of ethnic identity [14], aimed to assess: the degree of cognitive component of ethnic identity; the degree of affective component of ethnic identity.
4. Methodological guidelines on the analysis of the level of culture shock expression [3], allowing estimating the level of culture shock among students in a new sociocultural environment.
5. I.A. Sholokhovs' Questionnaire Adaptations to life and study in a foreign cultural environment [10] aimed to determine communication difficulties in a foreign language, motivation to study, general adaptation to life in a new environment, loyalty to a foreign society and culture.

The research process does not exclude the use of additional techniques. The research tools are: questionnaires, surveys, texts, internet resources, data collection. The elaboration of the concept of 'Psycho-pedagogical support of adapting international students in the educational environment in the system of higher professional education', which includes the following aspects: conditions of emotional adaptation in a foreign educational environment; influence of a microgroup in a foreign educational environment on the personal plans of a student; cognitive conditions of student adaptation in a foreign educational environment; conditions of a student's behavior correction in a foreign educational environment; development and analysis of techniques of the psycho-pedagogical support of students adaptation in a foreign educational environment; sociopsychological portrait of a student in the process of adaptation in a foreign educational environment. The sample of the present part of the study will be formed from the pool research results of the students from St. Petersburg Polytechnic University and foreign universities according to the following project scale:

1. Psychological phenomena based on the diagnostics results; motivational characteristics: motivation for educational and professional activities; interests: inclination to performing/creative work, inclination to be professionally oriented.
2. Cognitive characteristics: memory, attention, abilities, thinking. Individual personality characteristics. Professional aptitude of an individual: general workability, temperament, accentuations, professional contraindications.
3. Individual educational road map. Educational environment. Training period.
4. The social contacts result.

6 Findings

Practical details of the concept will be developed, basic existing research areas will be considered; basic research areas will be identified, contacts with foreign educational institutions will be established; Psycho-pedagogical support of adaptation of international students in the educational environment in the system of higher professional education' research concept will be finalized; tools for monitoring and processing research results will be prepared (questionnaire forms prepared and software developed); monitoring specialists will be instructed. Besides, St. Petersburg Polytechnic University pilot sites are planned to create in three of the partner universities from the list:

1. Chengdu University, Shenzhen University, Changsha University of Science and Technology, Tianjin University, Harbin Institute of Technology, Shanghai International Studies University, South China Normal University, East-China Normal University, South China University of Technology in the People's Republic of China;
2. Mälardalen University College (Mälardalenshögskola) in Sweden;
3. Saimaa University of Applied Sciences (Lappeenranta & Imatra) in Finland.

It will be arranged for testing the model of psychological and pedagogical support of adapting international students in the educational environment. It is intended to draft a glossary containing the terms relevant to psycho-pedagogical guidance of the student adaptation in a foreign language educational environment as well as a package of variable programs to support the development of student's personality in the context of adaptation in a foreign educational environment; to test the results obtained in the courses conducted by the project members; to participate in Russian and international seminars on the subject; to organize seminars for students participating in the exchange program.

There are four main factors (host's communicative competence, host's social communication, the foundations of cross-cultural transformation, and advanced cross-cultural transformation) that are important for the development of cross-cultural adaptation of international students. The existence of a relationship between these paths is significant and positive due to synergetic effects. This relationship provides an informative theoretical basis for studying cross-cultural adaptation of international students, as well as strong empirical support for the structural model of cross-cultural adaptation [5].

Another important aspect is that cultural diversity and the participation of host country citizens in the system of social and emotional support for students are important factors that portend a successful cross-cultural adaptation. This demonstrates the importance of encouraging initiatives that facilitate interaction between the host country and its population and the diversity of higher education relationships that are designed to foster the process of cross-cultural adaptation [11]. Such relations can also be maintained through social networks; however, in this case the results of adaptation depend mostly on one's own language skills in the region of stay, as well as personal and professional characteristics of those with whom students communicate [7]. Another important aspect of working in such an environment is the quality of students, which may be called cultural intelligence (CQ), its role in confrontation or clash with a new culture and possible shock. In its turn, cultural shock is largely, but negatively related to both psychological and socio-cultural adaptation. The CQ may be regarded as a "moderator" which reduces the negative impact of culture shock on both psychological and socio-cultural adaptation of international students [6].

Also the set of beliefs in the value of one's own life, in the experiencing of success and gaining prestige and such personal qualities like motivation for creating new ideas, rejecting or limiting desire to entertain oneself, willingness to change live habits and lifestyle, lack of desire to self-develop have a major influence on the results of these kinds of adaptation [9].

7 Conclusions

The results of the study can contribute to intensification of scientific and practical activities on psychophysical adaptation of international and Russian students in the process of international education, training of international students, organization of international activities in Russian universities; can help to create a friendly environment for foreign citizens in Russian universities, to integrate bachelors' degree programs and basic training for foreign citizens, to implement a set of measures to achieve the effective positioning of Russian education.

References

1. Chesnokova, N.V.: Sociocultural adaptation of foreign students in Russia: the example of the Vladimir region. Dissertation. N.N. Miklukho-Maklai Institute of Ethnology and Anthropology, Moscow (2012). (in Russian)
2. Krivtsova, I.O.: Sociocultural adaptation of foreign students to the educational environment of the Russian university (on the example of the N.N. Burdenko Voronezh State Medical Academy). Fundamental Research, 8 (part 2), pp. 284–288 (2011)
3. Mnatsakanyan, I.A.: Students adaptation in new sociocultural conditions: PhD thesis in psychology. Yaroslavl State Pedagogical University named after K.D.Ushinsky, Yaroslavl (2004). (in Russian)
4. Myles, J., Cheng, L.: The social and cultural life of non-native English speaking international graduate students at a Canadian university. J. Engl. Acad. Purp. **2**(3), 247–263 (2003). https://doi.org/10.1016/S1475-1585(03)00028-6
5. Peng, R.Z., Wu, W.P.: Measuring communication patterns and intercultural transformation of international students in cross-cultural adaptation. Int. J. Intercult. Relat. **70**, 78–88 (2019). https://doi.org/10.1016/j.ijintrel.2019.03.004
6. Presbitero, A.: Culture shock and reverse culture shock: the moderating role of cultural intelligence in international students' adaptation. Int. J. Intercult. Relat. **53**, 28–38 (2016). https://doi.org/10.1016/j.ijintrel.2016.05.004
7. Raymond, J., Hua Wang, R.: Social network sites and international students' cross-cultural adaptation. Comput. Hum. Behav. **49**, 400–411 (2015). https://doi.org/10.1016/j.chb.2015.03.041
8. Sarmiento, A.V., Perez, M.V., Bustos, C., Hidalgo, J.P., del Solar, J.I.V.: Inclusion profile of theoretical frameworks on the study of sociocultural adaptation of international university students. Int. J. Intercult. Relat. **70**, 19–41 (2019). https://doi.org/10.1016/j.ijintrel.2019.02.004
9. Shamionov, R.M., Grigoryeva, M.V., Grigoryev, A.V.: Influence of Beliefs and Motivation on Social-psychological Adaptation among University Students. Procedia Soc. Behav. Sci. **112**(7), 323–332 (2014). https://doi.org/10.1016/j.sbspro.2014.01.1171
10. Sholokhov, I.A.: Psychological peculiarities of adaptation of international students: PhD thesis in psychology. Moscow (2002). (in Russian)
11. Shu, F., Shujaat, F.A., Pickett, M.L., Ayman, R., McAbee, S.T.: Social support perceptions, network characteristics, and international student adjustment. Int. J. Intercult. Relat. **74**, 136–148 (2019). https://doi.org/10.1016/j.ijintrel.2019.11.002

12. Sonin, V.A.: Psychodiagnostic knowledge of professional activity. Publishing house Rech, Saint Petersburg (2004). (in Russian)
13. Soldatova, G.U., Kravtsova, O.A., Huhlaev, O.E., Shaigerova, L.A.: Psychodiagnostics of individual tolerance. Publishing house Smysl, Moscow (2008). (in Russian)
14. Stefanenko, T.G.: Ethnopsychology: textbook for students, 3rd edn. Aspect Press, Moscow (2004). (in Russian)
15. Tikhonova, E.G.: Features of adaptation of foreign students in a higher education institution of the region. Regionolology, 2 (2010). (in Russian). http://regionsar.ru/ru/node/507. Accessed 21 June 2020

Role of Volunteer Organizations in Providing Food Security for Discriminated Groups

Nikita Ivannikov(✉), Olga Lofichenko, and Daria Peregudova

Peter the Great St. Petersburg Polytechnic University, Saint Petersburg, Russia

Abstract. This article is devoted to the study of the influence of the volunteer movement on solving the problem of food safety in the context of existing in society various types of discrimination. The main goal of the work is to reveal the link between discrimination and food security, as well as to identify the role of volunteering in solving the above problem. To achieve this aim, we have given different interpretations of the term "discrimination" by domestic and foreign experts, identified similarities and differences in its understanding. Based on the approaches described in the work, the authors' definition of the concept of "discrimination" was proposed. The authors also examined various types of discrimination, each of which has a negative impact on the citizens of a particular country, as well as on the global community, making it difficult to combat such global problem as hunger. The factors that aggravate the problem under consideration are identified. The article presents volunteer programs that are aimed at solving the problem of hunger, and their results. Moreover, the importance of working on this problem of volunteer movements on an equal basis with state and intergovernmental actors is indicated. In the course of the study, the following methods were used: structural and functional analysis.

Keywords: Discrimination · Food security · Volunteering · Human rights

1 Introduction

The beginning of the XXI century is rightly not only a time of global challenges to humanity, but also a period of concentration of society on the individual. The spread of ideas related to humane and ethical attitude towards the human being, establishment of equality in rights between different categories of citizens, eradication of hatred and intolerance of members of society towards each other has increased significantly. The so-called democratic values go beyond the framework of official documents and acquire true significance in the society of the XXI century, becoming an international trend of modernity. Growing democratization of the world community sets tasks for specialists from different countries, which are related to ensuring human rights and liberties.

It should be noted that anti-discrimination measures play an important role in the equality of citizens of any democratic state. Discrimination that excludes granting equal rights and freedoms to all people without exception [14], cannot be part of a democratic political regime, commitment to which is a relevant idea for modern society.

© The Author(s), under exclusive license to Springer Nature Switzerland AG 2021
R. Bolgov et al. (Eds.): *Proceedings of Topical Issues in International Political Geography*, SPRINGERGEOGR, pp. 328–334, 2021.
https://doi.org/10.1007/978-3-030-78690-8_29

Ensuring food security and eliminating hunger is unreservedly connected with the ensuring fundamental human and civil rights. In compliance with the ICESCR, which acknowledged the right to be free from hunger, participating states are obliged, using the variety of scientific and technical knowledge, to manage food resources rationally and fairly [11].

Therefore, in order to provide the entire population with quality food, it is necessary to solve an important socio-political task - to implement the provisions of this pact. However, the ability to implement some of the provisions of this pact and the implementation of human rights regarding the provision of food in the regions of Asia, Africa, the Caribbean, etc. is still in question, in particular due to low incomes, discrimination on the basis of gender and race, price discrimination.

Food insecurity is a consequence of the crisis in modern political, economic and environmental systems. Currently, there is no problem of insufficient food production on a global scale, so that often the problem of hunger is caused by unequal and inequitable distribution of benefits among the population of the Earth [10].

There is no doubt that the solution of these tasks remains an important priority of government activities. However, if a democratic society is taken into account, it is necessary to consider its activities in this direction. In particular, an important component of the fight against global problems of humanity is volunteer work. Today it is an important area of research for social and humanitarian specialists. This happened due to the popularization and actualization of volunteerism around the world. Since it is a modern tool and criteria of activity in the formation of public consciousness and behavior, the level of development of volunteerism is one of the main indicators of welfare of the social sphere.

The scope of actions taken in the framework of volunteer assistance is quite extensive. It does not stop at the local and national levels, but extends to the international community as a whole. The concept of "volunteerism" is used to denote volunteer work as an activity held by people on a gratis basis and aimed at achieving socially significant goals and solving community problems. In particular, to solve the problem of hunger and overcome discrimination in society.

2 Objectives

The main purpose of this article is to examine the importance of discriminatory processes in ensuring food security, as well as to identify volunteer activities as a necessary and important part of this process. This will require consideration of the subject matter of discrimination, its impact on food security, and volunteer work related to food security.

3 Materials and Methods

For further detailed consideration and analysis of the practical material chosen as a research topic, it is necessary to first identify the essence of discrimination and its types.

According to Article I of the Universal Declaration of Human Rights, "all human beings are born free and equal in dignity and rights" [22]. The denial of human rights on the basis of race, gender, age, religion, sexual orientation, disability and other similar grounds is discrimination. Legal scholars interpret this concept in a slightly different way. Researchers E.I. Unzhakova and O.G. Beldina in their work "Discrimination in labor relations" define discrimination as "negative attitude, bias, violence, injustice and deprivation of certain rights of people because of their belonging to a certain social group" [24: 2].

In A.B. Borisov's Law Dictionary, discrimination is understood as "a general legal term referring to infringement of the rights of the state, legal entities and individuals (as compared to other states, legal entities and individuals)" [2:171].

According to the American Psychological Association, discrimination is "unfair or biased treatment of individuals and groups based on characteristics such as race, gender, age, or sexual orientation" [5].

The official Russian version of the term "discrimination" can be considered the interpretation specified in Article 136 of the Criminal Code of the Russian Federation, which states that discrimination is "a violation of the rights, freedoms and lawful interests of a person and a citizen on the basis of his/her sex, race, nationality, language, origin, property and official status, place of residence, attitude to religion, beliefs, membership of public associations or any social groups" [1].

Since discrimination has been an acute social problem for many years, some of its types have become "classic". These include racial, sexual, age, and religious discrimination [7]. This concept has already become so commonplace that it does not require detailed decoding. There are, however, other areas of discrimination - less known, but therefore no less socially dangerous.

Discussing the relevance of volunteer work, it should be noted that in line with United Nations resolution A/RES/56/38 [19] and its specific recommendations, all participating countries are encouraged to support volunteering at the state level in order to achieve sustainable development goals. On this basis, it is necessary to consider volunteerism as a socially active participation in solving global problems.

According to L.E. Sikorskaya, volunteering is nothing but a "way to preserve and strengthen human values, such as kindness, gratuitous assistance to any person regardless of his position in society, cultural and ethnic peculiarities, religion, age, gender. Volunteerism is a creative social force that contributes to building a more humane and just society through universal cooperation" [21].

To achieve the required result, the article uses the method of structural and functional analysis.

4 Research and Discussion

Nowadays, global political permanency depends to a large extent on food security because it includes fundamental human rights to a healthy and fulfilling life. That is why developments based on the methodology of regional public opinion monitoring are so important [12], because they may indicate existing problems in the field of discrimination.

It should be emphasized that discrimination is a whole system or a line of behavior/relationships towards discriminated groups of people at various levels - from domestic to state. In this regard, there are other, deeper classifications of discrimination. Anja

Rassek, a German expert in this field, notes in her article "Discrimination: definition, types, assistance" [18] the following subcategories of discrimination:

1. **By orientation**: group and personal. Group discrimination is a negative attitude towards a whole group of people who supposedly do not have an individuality (Jews, Muslims, migrants); and personal - transfers individual negative features of one representative of the group to others;
2. **By source**: structural, institutional, linguistical. Structural discrimination is a lack of political/economic/social system of the state. It is institutionalized and associated with the infringement of large groups. Institutional discrimination is a shortcoming of individual, as the name suggests, social institutions. Linguistic discrimination lies in the language itself, where there are negative colored words that humiliate a certain group of people and invisibly tune native speakers against the discriminated group.;
3. **By terms of impact**: direct and indirect. Direct discrimination is quite obvious - its causes include appearance, skin color, age and other similar signs. Concealed or indirect discrimination, in turn, is not so visible that it is easy for the discriminated person to prove that his or her rights have been infringed. This may be due, for example, to a deliberate delay in an employee's promotion.

There are many studies that confirm the relationship between hunger and discrimination on one or another basis. For example, racism has been shown to directly hinder the achievement of the required level of effectiveness in food security [15]. This is confirmed by a 2018 study that found that young mothers of Hispanic and African-American origins did not receive enough medical care, including psychological care, which undermined both their health and that of their children [17].

Indeed, gender discrimination is a massive problem. Among other things, restrictions on women's rights to education ultimately affect both employment in agricultural production in rural areas and the ability to earn a decent wage and thus provide themselves with high-quality food in adequate quantities [16]. In principle, women's empowerment in some regions and other aspects of it are very important not only in terms of democratization of life, but also for the eradication of hunger [3]. These problems are especially urgent in rural areas [20], in particular with regard to the employment of women in food production [9].

Discrimination in education, employment, housing and health also affects the enjoyment of human rights to food - all of which directly hinder food security [8]. These areas also include income discrimination, as well as poverty caused by socially inequitable distribution of benefits [26].

Another important obstacle is the specifics of NGO work in this area. For example, situations when the state does not allow such organizations, whose activities are directed against the infringement of citizens' rights, to work in full measure [13]. Such counteraction has a negative impact on food security.

There are also other negative factors affecting the work of volunteer organizations. Since many NGOs strive to provide assistance to needy groups of population, the prevailing items are ordinary food, water, basic necessities, medicines, rent, housing for victims, etc. Since these things are necessary in this sphere constantly, there is no need for innovations and clearly visible development in this sphere, as there is no need to invent some new technologies or improve the technical component of such organizations for their activities.

This plays a dual role: on the one hand, the sphere is becoming unpopular for sponsors, as it may seem irrelevant, and on the other hand, in pursuit of attracting investors to volunteer services, the latter are trying to develop "modern" projects, which in fact in practice are not needed, but are being developed to attract capital, while their development and even partial implementation of a large number of such necessary funds in other areas.

At present, the work of volunteer associations for food security is quite actively developed, exists as a specialized FAO volunteer program [6], as well as common UN programmes, whose work also has a positive impact on the food problem [23]. It is worth mentioning the WFP's volunteer initiative, which is not only about providing food, but also about training, fighting for the realization of human rights [25]. Important here will be also volunteer campaigns and organizations that seek to develop ideas of diversity and create sustainable, fair and democratic food systems free from discrimination [4].

5 Results

Thus, based on the above, it should be concluded that discrimination, being a concept not new, unfortunately continues to develop in the modern world, acquiring additional variations. Summarizing the above-mentioned approaches, we will accept in our work for discrimination the intentional unjustified bias towards a person, related to his gender, age, nationality, religion, sexual orientation, health and other personal characteristics; worsening his position in society and violating his rights.

Collaborative decision-making, as well as the exchange of experiences beyond race, class, level of education, gender and age, are useful and necessary to develop meaningful, effective and lasting changes in food and nutrition security. However, working together in this direction will not be an easy task. Issues of racism and discrimination, as well as cultural differences, will continue to rise and may actively impede both the development and implementation of solutions. On the other hand, voluntary work that is provided in favor of those who need it is really necessary and should not be unaffordable or simply absent, as it is often a necessary element in working with discriminated groups.

References

1. Article 136. Violation of equality of human and civil rights and freedoms. The Criminal Code of the Russian Federation of 13.06.1996 N 63-FZ (as amended on 31.07.2020). Consultant Plus (2020). [In Russian]. https://www.consultant.ru/document/cons_doc_LAW_10699/67c198fece5202f893460246a15f884f72173c28. Accessed 20 September 2020
2. Borisov, A.: Great Law Dictionary (2011)
3. Clement, F., et al.: From women's empowerment to food security: Revisiting global discourses through a cross-country analysis. Global Food Security, 23, 160–172 (2019). https://www.sciencedirect.com/science/article/abs/pii/S2211912417301086. Accessed 20 September 2020
4. Community Food Security Coalition. Volunteer Match. https://www.volunteermatch.org/search/org447235.jsp. Accessed 20 September 2020

5. Discrimination: What it is, and how to cope. American Psychological Association (2019). https://www.apa.org/helpcenter/discrimination. Accessed 20 September 2020
6. FAO regular volunteer programme. Food and Agriculture Organization of the United Nations, 1–6 (2017). http://www.fao.org/3/I7351EN/i7351en.pdf. Accessed 20 September 2020
7. Forms of discrimination. Information point of the Council of Europe in Minsk. https://coe.bsu.by/index.php/ru/topics-ru/diskr-ru/54-formy-diskriminatsii. Accessed 20 September 2020
8. From disparities to discrimination: getting at the roots of food insecurity in America. Children's HealthWatch. https://childrenshealthwatch.org/from-disparities-to-discrimination-getting-at-the-roots-of-food-insecurity-in-america/. Accessed 20 September 2020
9. Galiè, A.: Leveraging Gender for Food and Nutrition Security Through Agriculture. Encyclopedia of Food Security and Sustainability, 3, 426–431 (2019). https://www.sciencedirect.com/science/article/pii/B9780081005965215680. Accessed 20 September 2020
10. Gallegos, D., Chilton, M.: Re-Evaluating Expertise: Principles for Food and Nutrition Security Research, Advocacy and Solutions in High-Income Countries. Int. J. Environ. Res. Public Health **16**(561), 1–16 (2019). https://www.mdpi.com/1660-4601/16/4/561/htm. Accessed 20 September 2020
11. International Covenant on Economic, Social and Cultural Rights (1966). http://www.un.org/ru/documents/decl_conv/conventions/pactecon.shtml. Accessed 20 September 2020
12. Ivannikov, N.: Regional peculiarities of united nations food security policy. Dissertation submitted for the degree of Candidate of Political Science (2019). https://disser.spbu.ru/files/2019/disser_ivannikov.pdf. Accessed 20 September 2020
13. Jenderedjian, A., Bellows, A.: Addressing food and nutrition security from a human rights-based perspective: A mixed-methods study of NGOs in post-Soviet Armenia and Georgia. Food Policy, 84, 46–56 (2019). https://www.sciencedirect.com/science/article/abs/pii/S0306919217302920. Accessed 20 September 2020
14. Longman, A.., Schmidt, A.: Dictionary of Human Rights [In Russian]. http://www.biometrica.tomsk.ru/ftp/dict/encyclo/5/discrim.htm. Accessed 20 September 2020
15. Odoms-Young, A.: Examining the Impact of Structural Racism on Food Insecurity: Implications for Addressing Racial/Ethnic Disparities (2019). https://doi.org/10.1097/FCH.0000000000000183
16. Pesek, A.: 10 Impactful Organizations Tackling Hunger and Food Insecurity Around The World. https://www.charitycharge.com/10-impactful-organizations-tackling-hunger-and-food-insecurity-around-the-world/. Accessed 20 September 2020
17. Phojanakong, P., Brown Weida, E., Grimaldi, G., Lê-Scherban, F., Chilton, M.: Experiences of Racial and Ethnic Discrimination Are Associated with Food Insecurity and Poor Health. Public Health, 16(22), 4369 (2019). https://www.mdpi.com/1660-4601/16/22/4369. Accessed 20 September 2020
18. Rassek, A.: Diskriminierung: Definition, Arten, Hilfe. Karrierebibel. https://karrierebibel.de/diskriminierung/. Accessed 20 September 2020
19. Resolution adopted by the General Assembly on 10 January 2002. Recommendations on support for volunteering. A/RES/56/38 (2002). https://www.unv.org/sites/default/files/A%20RES%2056%2038.pdf. Accessed 20 September 2020
20. Savari, M., Sheykhi, H., Shokati Amghani, M.: The role of educational channels in the motivating of rural women to improve household food security. One Health, 10 (2020). https://www.sciencedirect.com/science/article/pii/S2352771420300355. Accessed 20 September 2020
21. Sikorskaya, L.: Organization of volunteer activities in the urban environment (2008)
22. Universal Declaration of Human Rights (1948). https://www.un.org/en/universal-declaration-human-rights/. Accessed 20 September 2020
23. UNV partnering with FAO. UN Volunteers. https://www.unv.org/partners/unv-partnering-with-fao. Accessed 20 September 2020

24. Unzhakova, E., Beldina, O.: Discrimination in the sphere of labor relations, Hum. Prog. J. **2**(3), 1–8 (2016). [In Russian]. https://cyberleninka.ru/article/n/diskriminatsiya-v-sfere-trudovyh-otnosheniy-1/viewer. Accessed 20 September 2020
25. Volunteer with WFP and help us achieve Zero Hunger. United Nations World Food Programme. https://www.wfp.org/careers/volunteers. Accessed 20 September 2020
26. Vulnerable People. More than meets the eye. Let's fight racism UN Initiative. https://www.un.org/en/letsfightracism/poor.shtml. Accessed 20 September 2020

Cultural Dimension of International Relations

From Class to Culture: Ideological Landscapes of the Left Thought Collective in the West, 1950s–1980s

Andrei A. Znamenski(✉)

University of Memphis, Memphis, USA

Abstract. The paper explores the evolution of the political and ideological landscapes of the left thought collective in the West. Heavily influenced by the classical Marxian paradigm prior to the 1950s, it gradually shifted to the matters of culture and identity between the 1950s and the 1980s. In the left ideological paradigm, this transformation became known as the "cultural turn"; some early left authors also referred to this shift as "Cultural Marxism." The latter became a favorite word of choice for scholars and writers on the right. Social scholarship on both sides of the political spectrum have frequently stressed the important role of the so-called Frankfurt School in pioneering the abovementioned transformation. This paper argues that, as far as the mainstreaming of the cultural turn, there were more important intellectual sources that not only came to privilege culture and identity but they also expanded the geography of the left thought collective beyond Europe. These sources included racialized Marxism of C.L.R James, William Dubois, and Frantz Fanon (1940s-1960s) and British Cultural Studies, which gradually phased out economic determinism and the class-based approach of classical Marxism by shifting attention to the Third World, race and gender minorities within Western countries, and identity matters in general.

Keywords: Marxism · Cultural turn · British Cultural Studies

1 Introduction

In 2010, sociology professor Rick Fantasia [1], a member of the Democratic Socialists of America, struggled to explain the results of US Congress elections that were disastrous to democrats and that, at that time, brought a majority to the republicans. Fantasia was part of a Social Forum, a 15,000-strong army of left activists who gathered in Detroit, Michigan. Observing this convention, he noted that most people who arrived at this convention mostly represented various minority organizations that were either involved into identity politics or represented immigrant workers. Fantasia also noted a heavy presence of countercultural and environmentalist elements, including New Age seekers. At the same time, the activist scholar pointed out that one important element was missing: working-class people, especially white workers. With frustration, Fantasia noted that there were only a few white workers: "The whites were mostly educated members of

© The Author(s), under exclusive license to Springer Nature Switzerland AG 2021
R. Bolgov et al. (Eds.): *Proceedings of Topical Issues in International Political Geography*, SPRINGERGEOGR, pp. 337–354, 2021.
https://doi.org/10.1007/978-3-030-78690-8_30

the middle class, organizers, activists, representatives of philanthropic organizations and academics." The described gathering and the expressed concern were a microcosm that reflected the shift in the entire ideology and the social base of the current Western left for the past fifty years.

What worried Fantasia was not some aberration or a temporary flaw in the left strategy and tactics. In fact, this was the result of a natural evolution of the left mainstream. Since the 1960s, it drifted away from concerns about an economic growth and class-based politics, which were associated with the old left. Instead, the left began shifting toward culture, race, and identity issues as well as environmentalism. This metamorphosis is sometimes labeled by a loose umbrella expression the "cultural turn." On the level of ideas, this turn is usually associated with the emergence of the often-mentioned post-modernism, and it includes several intellectual trends and political practices that came to a forefront as traditional Marxism was eclipsing. The most important among these trends were post-colonial studies, critical theory, feminism, multiculturalism, and political correctness. Some authors on the right refer to all this by an umbrella term "Cultural Marxism" – a pejorative expression that serves to point to genetic links between the current cultural left and the old Marxian left.

2 How We Name It: Critical Theory, Cultural Marxism, and the Identitarian Left

Below I explore the sources of the cultural turn among the left and the development of their passion for identity matters, which resulted in the phenomenon pinpointed by Fantasia. Mainstream scholarship and literature on the cultural left is dominated by the following popular notions. On the right, there is a widespread conviction that "Cultural Marxism," primarily through the "malicious" activities of the Frankfurt School, set out to erode the Western civilization. In the meantime, the easily triggered left have been ascribing any critique of political correctness and the "sacred cows" of race and gender to the evil forces of fascism and racism. If we "deconstruct" the history of the actual left's evolution toward culture and identity, we might problematize both approaches and tone down the heated debates around the cultural turn. Moreover, the understanding of the gradual evolution of the modern left from economic determinism (along with their fixation on the proletariat) to the privileging of culture, identity, and lifestyle activism will help us understand better how and why literally every aspect of human life became politicized in the eyes of the current left. In other words, the history of the cultural turn will shed more light on the origin of the popular left notion that personal is political.

Exploring the emergence of the cultural left, one needs to paint a bigger picture by showing that, besides the often-mentioned Frankfurt School, there were other essential sources that fomented that identitarian turn. Among other things, we need to address the significance of the landmark year 1956 for the entire left collective. The picture will not be complete without exploring celebrity sociology W. Right Mills' crusade against "Victorian Marxism." Furthermore, we need to examine the writings of C.L.R. James, William Dubois, and Frantz Fanon who were pivotal in refurbishing popular Marxist memes (the proletariat, class domination and oppression, the new man, false consciousness, and center-periphery) along racial and non-Western lines. Most important, one needs to examine the activities of British communist historians, Birmingham Institute of Cultural Studies, and such important periodical as *New Left Review*. Without them, it will be hard to understand the historical role of the 1960s-1970s' New Left, which acted as an intellectual bridge between old economic- and class-based Marxism and current cultural left that are heavily steeped in identity politics.

3 Cultural Turn: Scholarship and Methodology

In existing debates about the cultural turn, the term "Cultural Marxism" has aroused most controversy. Current identity-oriented progressive writers and scholars do not like this expression. Their favorite term of choice is Critical Theory and the host of expressions derived from it: Critical Cultural Theory, Critical Racial Theory, Critical Legal Studies and so forth. However, earlier left authors did not see any problems with that expression. In fact, several of them pointed out that this very term captured well the essence of the socialist ideology that was undergoing an adjustment to the new times. For example, in the 1990s, Ioan Davis [2], Dennis Dworkin [3], and, as late as 2004, Douglas Kellner [4] did not find it controversial using expression "Cultural Marxism."

In contrast, current cultural left, especially non-academics [5–8], who did not take time to explore the history of Marxism and neo-Marxism, have been quick to label Cultural Marxism as a hate taboo term, claiming that it promotes fascist, Nazi, and anti-Semitic ideas. Using such smear metaphors, they want to intellectually link all critics of the identitarian left on the right to Hitler's propaganda workers who had talked about "Cultural Bolshevism." Furthermore, downplaying the historical links between pre-1960s "scientific socialism" and the current cultural left, some left authors [9], who are totally fixated on identity matters, stress that they moved beyond Marxism and that they hardly have anything to do with Marxism.

In fact, that is exactly what traditional Marxist authors [10, 11], who still privilege a class-based approach, have been arguing, stressing that the identarian left have nothing to do with Marxism. The Marxian "traditionalists" label their wayward cultural comrades as traitors to the cause and dismiss them as "bad Marxists". Moreover, several scholars (historian Paul Gottfried [12] and philosopher Helen Pluckrose [13]) who are critical of both traditional Marxism and the current identitarian left, too spoke against using expression "Cultural Marxism." While correctly stressing that the post-Marxist left stopped prioritizing economic determinism and class, they have simultaneously downplayed genetic links that connected the Old Left and the current identitarian left who share the same *gestalt* and many ideological memes (e.g. oppression/domination, production, false consciousness and so forth).

Several conservative authors (e.g. Kerry Bolton [14] and Jeffrey D. Breshears [15]) have come to view Cultural Marxism as a grand conspiracy on the part of the left. They have portrayed it to as a sinister plan masterminded by the so-called Frankfurt School that allegedly sought to uproot Western civilization and Christianity. The most grotesque version of the Cultural Marxism conspiracy theory links it exclusively to the activities of German-Jewish scholars (who had indeed dominated the Frankfurt School [16]). That theory goes as follows. A group of mostly Jewish intellectuals, who were part of radical socialist and communist forces in the 1920s' Germany, were upset about the failure of the 1917 Communist revolution in Europe and decided to modify the Marxist-Leninist project of world revolution by mixing Marx and Freud.

Their goal was to smash capitalism not through the cultivation of the working-class indignation but through undermining Western culture and civilization (traditional family, gender hierarchies, and sexual norms). In the 1930s, being kicked out by national socialists from Germany, the Frankfurt School cabal moved to the United States, where it became the "Trojan horse" of the radical left, setting out to undermine the culture and values of the United States – the economic and political hub of the Western civilization. One of the major proponents of this view has been writer William Lind [17], who in fact was instrumental in popularizing the very expression "Cultural Marxism" among the right. It is mostly by drawing on his writings that the left journalists came up with the argument that this term serves as an anti-Semitic dog whistle.

Although it is not directly related to the discussed theme, it is essentially to note that, while such conservative authors as Lind have singled out the Frankfurt School to be demonized as the major intellectual culprit, progressive authors [18, 19], who have been working within the paradigm of neoliberalism studies, became similarly fixated on searching for an intellectual "cabal" to be held responsible for spearheading neoliberalism. To some of them the chief culprit became the Mont Perelin Society, an intellectual collective that was shaped by F. A. Hayek and Milton Friedman and that advocated free market and individual liberty. The irony of the situation is that both pejorative memes "Cultural Marxism" and "neoliberalism" [20] do describe social trends that have been unfolding in society. Far from being the products of the grand conspiracies, they reflect what has been going on in the intellectual culture and on the ground among various segments of society.

Incidentally, several scholars and writers [21–24] have recently explored the content of Cultural Marxism, trying to separate the conspiracy elements from actual intellectual links between Marxism of old and the current cultural left. Although I believe that the term "Cultural Marxism" still can be useful especially when we need to stress the continuity between the old Marxian socialism and the present-day cultural left, it indeed might be too narrow. So, I personally prefer to use such broad definitions as the "cultural left" and "identitarian left."

Behind the rise of the cultural left, there stood a large thought collective that did reflect genuine concerns of various segments of the left and social movements. The writings of the Frankfurt scholars, who both analyzed Western society and did issue utopian suggestions about how to transform it, were marginal until the 1960s. Their scholarship, which helped to shift the left's priorities from class to identity and culture, would have remained marginal had it not been for wide and vocal audiences that for various reasons picked up and consumed them. To summarize, the "Frankfurters" were not a sinister alien cabal that was preying on Judeo-Christian civilization with the sole purpose to destroy it. One can describe their effect on society by an old saying: when a student is ready, a teacher comes. In the 1960s and the 1970s, their ideas, which had earlier been marginal, suddenly began to resonate with thousands of progressives in the West and beyond. Such "Frankfurters" as Herbert Marcuse (1898–1979), Theodore Adorno (1903–1969), and Erich Fromm (1900–1980) described the development of mass consumer society, patriarchic family, the effects of propaganda on masses, criticized industrial society, and

Western mass culture, frequently issuing sweeping condemnation of the entire "soulless" Western civilization.

The reason these ideas came in vogue was simply because, by the 1960s, the West and the rest of the world experienced profound socio-economic changes: the decline of traditional working class and the rise of intellectual professions, the massive involvement of women into all spheres of life and the end of the male-oriented societal ethic, which until the 1960s had been considered normal, the emergence of new technologies, an industrial pollution, and concerns about how to better handle an economic growth. Furthermore, the world saw the rise of Third World national liberation movements, the collapse of old colonial empires, and the emergence of minority movements in Western countries. Finally, the Soviet Union, which a large portion of the left had earlier considered the great new hope, lost its image as the ultimate socialist utopia. Facing those changes, the old left began to crumble. There was a need to refurbish the left ideology and identity. In the 1960s and the 1970s, during student anti-war movements, the rise of Third World national liberation movements, civil rights protests among the people of "color," the expansion of women and gays rights movements, the ideas disseminated by the above-mentioned intellectual "power centers" of the left resonated well with thousands of protesters. One cannot simply dismiss the culture and identity matrix that became popular in the left thought collective as something imposed from above on the "innocent populace."

4 Toward "Socialist Humanism" and Away from Traditional Marxism (1956 and Beyond)

1956 was a pivotal year for the socialist thought collective. This was the year when the Soviet nomenklatura elite partially exposed Stalinism, trying to polish the tainted image of socialism. The communist bureaucracy was tired of living in a constant fear, and, after the death of Stalin, it sought to secure its privileges and to somehow reform communism to make it more appealing. During the same year, taking advantage of the limited destalinization, people of Hungary openly rose in an anti-communist revolt against the Soviets. The suppression of the Budapest rebels by the Soviet tanks was a devastating blow at the moral of the millions of left idealists around the world who still believed that the Soviet Union was the hub of the bright future. There was a growing frustration with the Soviet model of socialism that was tied to a total nationalization and centralized planning. Moscow was rapidly losing its status as the blueprint for the socialist utopia. It was natural that the year of 1956 signaled the emergence of the so-called New Left who sought to disentangle themselves from the Soviet experience [25].

In the meantime, the working-class people in the West were dramatically improving their living conditions and were not expressing any desire to go to barricades to battle capitalism. Social democrats were shedding the last vestiges of Marxism, and communist parties were increasingly losing their membership. The left, especially their radical wing, were poised to turn into rebels without a cause. In fact, the major character from a 1956 John Osborn [26] play expressed it best when he uttered a phrase that became classic: "There aren't any good, brave causes left."

There was not much to gain for the left by sticking to the economic playing field, where "rotten" capitalism was improving people's living standards and securing an economic growth. Those who wanted to keep a radical left agenda alive had to rekindle the traditional left subculture. The Trotskyites, cosmopolitan Marxist-Leninist heretics, who were the victims of vicious political assaults from their Stalinist rivals, did arouse a sympathy among dissident communists seeking a socialist alternative beyond the Soviet experience. After all, the Trotskyites were the first to struggle to preserve the radical elan of the Marxist creed, while simultaneously attacking both capitalism and Stalinism. Yet, with their old and worn out mantra about the primacy of an economic basis, vanguard party, and false claims about an increasing misery of the industrial working class, the Trotskyites were out of touch with reality. They simply appeared as reenactors of the bygone era and could not generate any visible support among workers, quickly degenerating into an esoteric intellectual sect.

Cornelius Castoriadis [27], a prominent left theoretician, captured well the whole dilemma faced by the left who were frustrated about the proletariat that failed to fulfill its prophetic mission: "The proof of the truth of the Scriptures is Revelation; and the proof that there has been Revelation is that the Scriptures say so. This is a self-confirming system. In fact, it is true that Marx's work, in its spirit and its very intention, stands and falls along with the following assertion: The proletariat, as it manifests itself as the revolutionary class that is on the point of changing the world. If such is not the case – as it is not – Marx's work becomes again what in reality it always was, a (difficult, obscure, and deeply ambiguous) attempt to think society and history from the perspective of their revolutionary transformation – and we have to resume everything starting from our own situation, which certainly includes both Marx himself and the history of the proletariat as a component." Issues that became more relevant by the 1960s were the US war in Vietnam, the rise of Third World anticolonial movements, civil rights struggle, women liberation. Traditional working-class issues became less irrelevant, whereas the issues of race, gender, and culture that earlier had occupied a marginal place on the left's agenda, now were coming to the forefront. The mainstream radical left had to rethink their agenda and customize it to the changes.

Since 1956, to dissociate themselves from the Soviet brand of socialism, the Western left sought to humanize Marxism. Hence, a natural shift away from economic determinism and economic efficiency toward the issues of culture and identity. Later, this trend manifested itself in the emergence of such contemporary memes as "socialism with a human face," "democratic socialism," "socialist humanism," and "Marxism-Humanism."

5 "Sense of Classlessness" and British Cultural Studies, 1950s-1970s

One of the major trailblazers of that drift away from economic determinism and toward humanized Marxism and culture was a dissident group of British Marxist intellectuals who were later labelled as the New Left. Several of them came from so-called Communist Party Historians Group that emerged within the British Communist Party in 1946. Others were communist fellow travelers and independent Marxists. At the end of the 1950s, when the Moscow commanding heights began to question Stalin's infallibility, these historians, sociologists, and literary scholars either quit on the party or moved away from traditional Marxism-Leninism, challenging its Stalinist theory and practice. These dissident intellectuals included such prominent figures as E. P. Thompson (1924–1993), Herbert Hoggart (1918–2014), Christopher Hill (1912–1996), Raymond Williams (1921–1988), Christopher Hill (1912–2003) Stuart Hall (1932–2014), Raphael Samuel, (1934–1996), John Saville (1916–2009), Eric Hobsbawm (1917–2012), George Rudé (1910–1993) Rodney Hilton (1916–2002). Several of them (Hall, Hobsbawm, Hoggart, Thompson, and Williams) had a profound impact on Western social scholarship, especially in English-speaking countries. For example, Hall and Williams literally laid the foundations of current cultural studies. In their turn, Thompson and Hobsbawm heavily effected history scholarship, shifting its mainstream narrative toward writing about the past "from below."

These New Left dissidents began questioning the old Marxist notion that the end of capitalism was linked to the increasing immiseration and economic degradation of the proletariat. Gradually downplaying the immiseration thesis, which represented a wishful thinking, they stared to argue instead that the need for socialism was arising from the bourgeois affluence and consumerism. Furthermore, these ex-communists cracked the traditional Marxist conviction that economic class interests conditioned all politics, social life, people mindsets, and culture. Gradually shedding off economic determinism, these left scholars who had invested their whole careers into "scientific socialism," found a new outlet to continue their intellectual pursuits – retrieval of the popular culture of working-class people.

Their intellectual quest eventually gave rise to *New Left Review*. Launched in 1960, it became the major periodical of the Western New Left. In fact, the very expression "the New Left" originated from the collective that was congregating around this magazine and that was hanging in and around the Partisan Coffee House in Soho, a bohemian area of London, and the Birmingham Institute of Cultural Studies. Searching for a new identity, the New Left changed the very concept of political, moving away from the traditional left "sacred sites" such a factory and a trade union to the realm of labor culture, folklore, lifestyles, and individual behavior. Hall, who was part of that ideological collective, noted that he and his comrades were looking for a better place to anchor their radical elan. Incidentally, one of his speculative essays [28] carried a characteristic title "A Sense of Classlessness." Hall specified that the major way for him and his comrades to ground themselves was politicizing various issues surrounding college life, high schools, movie theaters, art and other walks of life and institutions. Jumping ahead, I want to stress that for the current cultural left politicizing the issues of lifestyle is one of the major ways of sustaining their identity. Hall defined this New Left ideological quest as "the proliferation of potential sites of social conflict and constituencies for change." The famous slogan of radical feminism "the personal is political" captured well the essence of that search. Overall, as Hall stressed, all kinds of issues, including personal troubles and complaints could be amplified and opened to politicization.

Far from ditching the "chosen people" of traditional Marxism, social scholar Williams, another member of the emerging New Left collective, came up with a neat alternative to the worn-out narrative of the industrial working class as an economically deprived group. He suggested embracing a cultural approach to the proletarian "noble savages" and learning about their "authentic" ways. Formally remaining a member of the British Communist Party, in the 1950s, Williams gradually drifted toward the Labour (social democratic) platform. To dramatize his opposition to the economic materialism and determinism of Marxism, the scholar labelled his suggested angle as "cultural materialism." To be fair, Marxist-oriented anthropologists, many of whom too were phasing out economic determinism in the 1960s, were simultaneously peddling a similar notion in their scholarship. Because of Williams' heavy media presence, his ideas about the working-class culture and group identity trickled down into Western humanities, where later they were used as a methodological blueprint for feminist, racial, gay, and queer theories.

To ideologically legitimize the cultural shift, the dissidents appealed to the authority of foundational Marxist texts and used relevant quotes from the founders. Just as their Soviet counterparts who, when partially cleansing the house of Stalinism, turned to Marx and Lenin, the Western New Left had their intellectual "Reformation" by invoking the early writings of Marx. Besides such writings as *The Eighteenth Brumaire of Louis Napoleon*, they particularly became drawn to so-called *Economic and Philosophic Manuscripts* (1844). The latter represent vague and abstract notes made by young Marx about humanism and alienation. Excavated and published by Bolshevik scholars in Moscow in the 1920s, those notes appeared to contemporary radical socialists as irrelevant: they did not yet contain the famous pillars of Marxist "science" such as surplus value, the primacy of economic basis, socio-economic formations, and the role of working class as the historical savior of the world from oppression. In the wake of the 1917 revolution, being busy with class battles and ready to harness the "laws of history" in order to usher in the radiant communist future, Bolsheviks and their radical left allies in other countries did not pay much attention to those manuscripts, considering them raw speculations of the great mind in its infancy. Yet, in the light of the unfolding New Left revisionism, which aimed to mute economic determinism and the Stalinist totalitarianism, those vague notes suddenly became relevant and "mature." What especially resonated with the British dissident Marxists and the New Left in general was Marx's generalizations about alienation of humans in modern Western society.

The goal was to revise the traditional Marxist canon, which preached that economic basis conditioned political and cultural "superstructure," and to place instead an emphasis on the "superstructure." In his *Culture and Society*, Williams [29] furnished relevant quotations from the writings of Marx and Engels to make a case that the cultural superstructure should not be reduced to the economic basis. Instead of old speculations about the economic conditions of the working class in England, the scholar was on the quest for the traditional working-class culture, which he romanticized as organic, wholesome, and authentic. Moreover, Williams sought to disentangle it from "artificial" bourgeois mass culture. A sympathetic contemporary aptly remarked that the intellectual quest of Williams and his New Left colleagues, who sought to pinpoint an "authentic" proletarian culture, was an attempt to merge "imaginative literature and socialist humanism."

Marxist sociologist Hoggart, who founded the Birmingham Center for Contemporary Cultural Studies in 1964, too portrayed the idealized working-class culture as organic and natural, casting it against the "non-authentic" bourgeois culture. According to Hoggart

[30, 31], mass bourgeois culture was undermining and phasing traditional and wholesome working-class ways. It was natural that Williams and Hoggart, who celebrated the bygone traditional labor culture, became drawn to Romantic poets and writers who celebrated Merry Ole England. In fact, intellectual speculations of these two scholars surprisingly resembled dismissive rants of conservative critics regarding modern British culture. Irving How [32], a walk away American Trotskyite and socialist sceptic, who was closely watching the cultural speculations of his English comrades, could not resist making a comment: "I suspect that in their stress upon the working-class neighborhood and its indigenous culture men like Williams and Hoggart are turning to something that is fast slipping away."

Historian Hobsbawm [33], another prominent member the same group of dissident Marxists historian whose books became must read in many history and anthropology courses, gives us an example of a true-believer who was literally tormented by the idea of how and where to find a "class-savior" at that age of "classlessness." Unlike his wayward comrades who quit on official communism, Hobsbawm, chose to remain in the British Communist party. Moreover, at the turn of the 1950s, still infested with the idealism about the proletariat as the ultimate victim-savior, the historian put his two cents in the famous debate about the effect of the Industrial Revolution on the living conditions of the working class in England. In the spirit of classical Marxism, Hobsbawm was trying to argue that by 1800 the life of the factory laborer had become miserable if compared with the preindustrial Britain. Yet, at the very end of the 1950s, with the debacle of the immiseration thesis as the familiar economic playground of classical Marxism was shrinking, Hobsbawm slowly began to drift toward new "pastures" in the Third World. At the turn of the 1960s, he took several trips to Latin America, exploring revolutionary movements in that part of the world, falling for Cuba, and engaging Peruvian peasants into talks about the level of their oppression. At some point, Hobsbawm became so excited about the revolutionary potential of Latin America that he defined it as the engine of the future socialist revolution.

In 1959, the scholar published *Primitive Rebels* [34]. Although the current identitarian left will find his title too patronizing and Eurocentric, *Primitive Rebels* did clear the ground for the cultural turn in the general shift away from the proletariat. The book represented a history account that romanticized people whom Hobsbawm defined as social and noble bandits, from English Robin Hood types and Sicilian mafia to peasant communism in Italy and Ukraine and to Spanish anarchists of the 1910s-1930s. The indirect message of the book was that all those segments fomented a spontaneous social justice by undermining oppressive systems. The book became a runaway bestseller in the English-speaking world, and it was also translated in all major European and Asian languages. In fact, the enthusiastic reception of the text demonstrated that Hobsbawm did stir the popular aspiration among the left to find new revolutionary "chosen ones" to letch on.

Those independent New Left, who were not constrained by ties to the communist movement like Hobsbawm, went further and began to completely debunk the role of workers as the "chosen people" destined to save the world from capitalism. In 1960, flamboyant American sociologist C. Wright Mills (1916–1962), an emerging intellectual guru of the New Left, openly challenged the "labor metaphysic" of the old comrades.

He scorned the romancing of the proletariat as "Victorian Marxism" [35] and dismissed it as a survival of the past. Trying to fill the old Marxist clichés with a new content, the sociologist insisted that in the new "post-industrial" conditions, the true catalyst of revolutionary changes would be Western, Soviet bloc, and Third World intellectuals. Those among the New Left (e.g. E. P. Thompson) who continued to believe in the proletarian class struggle were confused and upset about such arrogant attack on the sacred pillar of Marxism. Their confusion was typical for those true believers whose traditional identity was crumbling. On the one hand, the communist dissenters wanted to exorcise Stalinism and economic determinism from "scientific socialism." Yet, on the other hand, they were too attached to the old ideological meme of proletarians as the "chosen people" to simply cast aside this foundational stone of the Marxist theology.

6 C.L.R. James, William Dubois, Frantz Fanon, and the "Curse" of the Western Civilization

Despite the reticence to completely ditch the major pillar of traditional Marxism, heretical ideas of Mills, Hoggart, and Williams continued their intellectual journey, trickling into the left thought collective. Eventually, these ideas blended with "racialized Marxism" of the Third World, opening doors to the emergence of the identitarian left. Moving further away from the sacred memes and sites of classical Marxism (a factory, economic growth, the working class, and the Soviet Union), New Left intellectuals gradually began to take on the Western civilization in general, bourgeois life-styles and culture, embracing the Third World and non-Western cultures. The slowdown of class battles and sluggish radical socialist activism in the West contrasted with the great awakening in the Third World in the 1950s and the 1960s. In this part of the world, emerging national liberation movements directly challenged European colonialism. Cast against the "dormant" and "corrupted" Western working-class, the Third World appeared to the New Left as the potential hub of revolutionary activism.

Among influential early voices that had triggered the identitarian revision of Marxism was W. E. B. Dubois (1868–1963), an African American social scholar, nationalist, Soviet fellow-traveler, and a convert to communism at the end of his life. The other one was C.L. R. James (1901–1989), an independent Marxist bohemian, novelist, and theoretician from a British Caribbean colony of Trinidad. The third was Frantz Fanon (1925–1961), yet another Caribbean-born black intellectual from the French-owned island of Martinique. All three had been instrumental in merging Marxism, Pan-Africanism, and Third World nationalism. Since the 1960s, the New Left and their successors among the cultural left have been holding those personalities in a high esteem. In fact, in academia there grew the whole publication industry around those intellectual icons.

Early in his career, along with socialism, Dubois absorbed then popular contemporary race and "folk soul" ideas when he was studying in Germany between 1892–1894. After completing his studies in Europe, he applied those ideas to his budding Black nationalism. In his 1897 *The Conservation of Races,* Dubois [36] called for the cultural unity of the "Black race" to replicate the efforts of the Teutons, Slavs, Anglo-Saxons, Latins, Semites, Hindu, and Mongolians who were busy, as he explained, consolidating their own civilizations. Dubois envisioned such black racial solidarity as a counterweight to the contemporary domination of the "whiteness of the Teutonic" and their soulless civilization that was fixated on individualism and economic enterprise. Moreover, he assigned for the American blacks the role of the enlightened vanguard in that utopian racial commonwealth. Very much like his racially-conscious Germanic contemporaries, who lamented the degradation of the Aryan soul by corrupt forces of modern industry and commerce, Dubois generalized about the bourgeois civilization of the West corrupting Africa – the primal and vital center of the black race.

In his *Souls of Black Folk* (1903) and *Negro* (1915), Dubois [37] spoke in favor of segregating "black culture" from "white civilization" and speculated about an abstract black soul, race, and culture irrespective of any local and linguistic differences. In fact, later in 1934, Dubois severed his connections with the National Association for the Advancement of Colored People that was working to eliminate racial segregation in the United States because the organization's activities contradicted the racial utopia he contemplated. Interestingly, in his novel *Dark Princess* (1928), Dubois [38] portrayed the Atlas Shrugged-type society of non-white expatriates who formed the Great Council of the Darker Peoples. Represented by "dark" superheroes, that society was planning to take over and engineer a happy future on the planet after white institutions collapsed.

Dubois relied on European romantic memes of the noble savage (collectivist, generous, wholesome, happy, simple), which he applied to all blacks as a race. Also, drawing on his parochial experience as a black American, the writer singled out race as the central factor in the entire world history. In a similar manner, he depicted slavery as the experience that defined not only the history but also conditioned the future behavior of the black "tribe." Dubois welcomed the 1917 Bolshevik Revolution that he considered a scorching wind that was to cleanse the modern world, washing away bourgeois civilization. Since the Soviet Union crusaded against the West, he automatically viewed that country as an ally: the enemy of my enemy is my friend.

By 1935, Dubois came to conclusion [39] that the Soviets would destroy the "rotten" Western civilization and help construct a new non-Western cultural order. The writer praised Stalin as a great liberator, and the 1956 revelations of the Soviet crimes did not shatter this conviction. Reflecting contemporary political preferences of Third World anti-colonial leaders and spokespeople, Dubois called the Soviet Union and China the shining models of the future. Moreover, in 1961, he moved to Ghana where he became a senior advisor to Kwame Nkrumah, the head of that country who claimed building socialism. Simultaneously, Dubois converted to communism. Both acts were quite symbolic: on the African soil, his black nationalism, which was saturated with romantic memes of European primitivism, met Marxian socialism. There is no need to stress that Dubois writings has been a must read in many humanities and social science courses across Western academia since at least the 1970s.

The evolution of C.L.R James [40], who became another must reference for Western social scholarship, moved in a reverse direction, although the result was essentially the same. From early on, in the 1930s and the 1940s, he prophesized his loyalty to Marxism. Yet, gradually, James began to play down class exploitation, amplifying the significance of racial and colonial oppression. It was natural because these issues were personally more relevant to him than far-away class battles in distant and alien Europe. James at first embraced anti-Stalinist Trotskyite version of communism and its prophecy of the world revolution. Yet, later, driven by anti-colonial concerns and by a desire to identify a new reliable revolutionary force to act as a surrogate proletariat, he gradually shifted his attention to the Third World.

Essentially, both Dubois and James were refurbishing the classic Eurocentric prophecy of Marxism along the Third World lines. Marx [41] had welcomed colonialism as a progressive system that had been sucking underdeveloped areas into the global commodity economy, pushing the world closer to capitalism, and creating an economic basis for a revolution and for a subsequent leap into the radiant communist future. For Marx, slavery was an archaic mode of production that, at the dawn of human history, had boosted economic development and then disappeared, giving rise to more progressive stages of human evolution such as feudalism and then capitalism. According to the founder, slavery survived in some backward areas of the globe (US South, Latin America, Africa) that did not catch up yet with the industrial West. In contrast to this classical Marxist vision, James and Dubois argued that slavery was not a vestige of the bygone socio-economic formation and portrayed it as an essential component of modern capitalism. Moreover, both writers insisted that that resources the West extracted from slave labor and colonies were more important than the exploitation of European proletariat.

All that allowed them to suggest that the entire Western prosperity was accomplished at the expense of non-Western people. Again, in contrast to Marx who viewed capitalism as a progressive stage on the way to communism, James and Dubois began to argue that capitalism was not a progressive but a regressive system – a European cultural institution imposed on the rest of the world for the purposes of exploitation. It is notable that, when asked to single out two books that had affected his worldview, besides Leon Trotsky's *History of the Russian Revolution* (1932), James mentioned *The Decline of the West* (1918) by Oswald Spengler. The latter text, which was saturated with a deep pessimism, prophesized the decline of the Western civilization. Incidentally, Spengler too greatly affected Dubois who began taking for granted that the West was in the state of a perpetual decline. Both writers found in the German philosopher's text what they were looking for: a radical critique of the entire Western civilization.

The crucial role in shifting Marxism toward race and identity matters belonged to Fanon, a popular anticolonial writer whose landmark text *The Wretched of the Earth* (1961) became a book of choice [41, 42] for the entire generation of the Third World national liberation activists in the 1960s-1970s. Fanon's bashing of the West also won him numerous acolytes in the countercultural circles and among the New Left both in Europe and the United States. Since the 1980s, assimilated by the academic left (post-colonial studies and critical race theory) into the educational and media mainstream, his writings became an important intellectual fountainhead for identity studies and identity politics [43].

A psychotherapist by profession, Fanon was a French-speaking intellectual who took part in the Second World War and then in the Algerian War of independence (1954–1962). In his writings, he focused not on economic liberation but on the cultural and psychological decolonization of the Third World. Drawing on Marxist class clichés, Fanon revised them along racial lines: "You are rich because you are white, you are white because you are rich." Fanon [44] insisted that the colonial periphery became the mentally imperiled by Western values, which natives needed to shed off because these were alien "white values": "Come, comrades, the European game is finally over, we must look for something else. Let is not imitate Europe. Let us endeavor to invent a man in full, something which Europe has been incapable of achieving." In his view, Europe was deadly sick and the keys to the liberation of humanity were in the hands of the Third World that was destined to bring to life the so-called New Man; incidentally, the latter meme also originated from Marxism. Fanon's friend Jean-Paul Sartre, a famous French philosopher and Soviet apologist, felt happy that "the most ardent poets of negritude are at the same time militant Marxists." Yet, in his naivete, repeating the mantra of the old left, Sartre simultaneously stressed that the mingling race with that class was "not a conclusion" but a transitional stop on the way to a greater color-blind commonwealth. When Fanon [43] read these Sartre's words, he felt utterly offended as if he was robbed of his identity. Contrary to what his philosopher friend believed, for Fanon, "racialized Marxism" was the conclusion.

Traumatized by the brutalities of the French he witnessed during the Algerian liberation war, Fanon insisted that nothing connected the colonizers and the colonized except racist violence. Ignoring the multitude of social, economic, and cultural relations in the contemporary colonial and post-colonial periphery, he argued that "the colonial world is a Manichean world." In the irreconcilable "black and white" world portrayed by Fanon, oppressed victims held the ultimate truth because of their sheer status of being colonized people. To Fanon, morality and truth were relative. They depended on how well these two things served a liberation cause. This included lying and committing violence, provided these vices served a good cause. Stressing that truth was a matter of political expediency, he wrote, "Truth is that which hurries on the break-up of the colonialist regime; it is that which promotes the emergence of the nation; it is all that protects natives, and ruins foreigners. In this colonialist context there is not truthful behavior: and the good is quite simply that which is evil to 'them'."

To overcome their oppressive state, the colonized had to take the place of their masters by resorting to a redemptive violence. Fanon suggested that colonial natives, instead of wasting their energy in mutual tribal conflicts and indulging into ecstatic tribal dances, forge a racial solidarity and channel their energy into anti-colonial violence against whites. In fact, violence occupied an essential place in the entire Fanon's liberation philosophy. The writer [42] romanticized violence and attributed to it a pedagogical value: "Violence alone, violence committed by the people, violence organized and educated by its leaders, makes it possible for the masses to understand social truths and gives the key to them." Fanon viewed violence not only as a tool of liberation and education but also as a powerful vehicle of a nation-building and racial consolidation. In the process of their martial mobilization, oppressed natives were expected to nourish the sense of a unified collective: "Individualism is the first to disappear," and "the community triumphs."

Using Marxist class categories and simultaneously filling them with a new content, Fanon argued that in undeveloped colonial countries the only revolutionary class was peasants. These rural masses carried armed struggle from the countryside into cities. This meant that, in contrast to the classical Marxist vision of the industrial working class (proletarians) as the ultimate savior of the world, the class that was to liberate the colonial periphery and the rest of humankind from capitalism was the Third World peasantry.

With such a view, Fanon naturally came to idealize Third World peasant collectives as the cradle of the ideal human commonwealth. Thus, invoking the romantic meme of European primitivism, he contrasted "evil" Western individualism with the "noble" African culture of collectivism represented by village councils, people's committees. According to Fanon, the anti-colonial struggle was to rekindle and strengthen those collectives. Moreover, the racial solidarity nourished during an anti-colonial war was to reeducate a corrupt indigenous bourgeoise – the creature of the West. Through its involvement into the common anti-colonial movement, this bourgeoise would reunite itself with its indigenous soil, merging, along with common people, into a united *Volksgemeinschaft*-type people's community.

Out of anti-colonial sentiments of such Third World intellectuals as Fanon and their colleagues from Western countries, there grew a natural animosity to the West, which was responsible for colonialism. Simultaneously, these spokespeople promoted the idealization of non-Western societies as the holders of a revolutionary potential and better forms of life. The fact that in the 1950s and the 1960s, the West was involved into two bloody colonial wars (France in Algeria, and the United States in Vietnam) amplified those sentiments. As a result, since the 1960s, for the left, the major existential enemy was shifting from capitalism as an economic system to Western civilization that was associated with colonialism, consumerism, and moral decay. In 1967, progressive writer Susan Sontag [45] conveyed well that emerging negative attitude, which later became part of the intellectual mainstream, by saying that the white West was the "cancer of human history."

The 1960s and the 1970s saw much talk on the left about humans being enslaved and alienated by the technology-driven individualistic civilization of the West and less talk about an economic growth, progress, and capitalism robbing workers of a surplus value. In fact, economic progress became a curse phrase. The idealization of the non-Western, tribal, and the primitive became a natural intellectual offshoot of such ideological pursuits. Reflecting on the cultural turn that was launched in the 1960s, Marxist sociologist Harold Bershady [46] stressed that this trend carried obvious reactionary notions: "It was a kind of left-wing conservatism." Gradually ditching the failed argument of the old left, who had insisted that capitalism had been profoundly inefficient and could not provide material affluence, the New Left were switching to the moral and cultural critique of that very affluence that now was declared a major vice.

Ayn Rand [47], a rising countercultural icon on the right who, in contrast to the Marxist ultimate proletarian "savior," invented her own version of a "noble savage" in a form of heroic capitalist entrepreneur-redeemer, responded to those sentiments with a loaded sarcasm: "The old-line Marxists used to claim that a single modern factory could produce enough shoes to provide for the whole population of the world and that nothing but capitalism prevented it. When they discovered the facts of reality involved, they declared that going barefoot is superior to wearing shoes." Among others, unnecessary commodities, which were singled out by the left, included TV sets, comics magazines, soap operas, the variety of household items. Incidentally, in the 1970s and the 1980s, such romantic neo-primitivist attitudes helped the left to find a common ground with environmentalists who began to preach an apocalyptic vision of the global collapse of natural habitat if not arrested by massive government regulations.

7 Conclusions

The incorporation of the non-Western ones and radical environmentalism into the socialist agenda was a natural offshoot of the "going primitive" trend that looked beyond Europe and North America for major revolutionary hubs. The geographical imagination of the left thought collective underwent a profound transformation. Still, exorcising the proletarian messiah class from the popular Marxian socialism and moving toward identity and the idealization of non-Western "others" was not a straight-forward process. In the 1960s-1970s, among the New Left, communist dissidents, and Trotskyite fossils, there was still lingering on a desire to somehow sustain the revolutionary elan of the proletariat. At the same time, among those elements one could detect the growing trend toward romancing the working-class culture and its "organic" anti-bourgeois ways. The Birmingham School of Cultural Studies and dissenting communist historians such as E. P. Thompson, who aspired to cleanse Marxist theory from economic determinism and who elevated the proletarian culture and consciousness, prepared a fertile ground for the later cultural turn in the left thought collective.

Before the current left completely ditched the working class from the pedestal and developed an "acute identity syndrome," the New Left segment acted as an intellectual bridge between classical Marxian socialists and the present-day identitarian left. In the 1960s and the 1970s, the New Left gradually transferred the metaphysical characteristics ascribed to the proletariat to the non-Western "others," domestic people of "color," chronically unemployed, social deviants, women, and gays. Communist bohemian historian Hobsbawm with his bookish "social bandits" and his attempts to probe the revolutionary potential of Latin American peasants is an excellent snapshot of how that process was unfolding.

Just like the proletariat of the old, the new "victim" groups were thought to become the oppressed redeemers – the new "noble savages" who were destined to liberate the world. On the one hand, such revision of traditional Marxism gave an opportunity to the New Left to disentangle themselves from the Old Left that was fixated on economic determinism. Yet, on the other hand, this very same revisionism allowed them to conveniently maintain the familiar Marxist tradition in the new intellectual garb. The new groups designated to the role of the oppressed ones were lumped together in a metaphysical category of the poor and disadvantaged. In the same manner, old Marxism generalized about the proletariat as a homogenous impoverished class, downplaying ethnic, religious, and economic differences within this group.

In the 1960s, the most passionate New Left revisionists who became hooked on the Mills' message of bashing the "Victorian Marxism" cast the newly found surrogates into authentic, uncorrupt and holistic people, the caretakers of the egalitarian ethics and natural goodness. Thus, in a religious-like manner, New Left activist Casey Hayden, the spouse of famous Tom Hayden [48], described her feelings about the new "chosen ones": "We believed that the last should be the first, and not only should be the first, but in fact were first in our value system. They were first because they were redeemed already, purified by their suffering, and they could therefore take the lead in the redemption of us all." Another New Left writer [49] characteristically titled his book about "unspoiled" and "authentic" rural blacks in Mississippi *A Prophetic Minority* (1966).

Those conservative and libertarian authors who are fixated on the Frankfurt School have failed to pinpoint the variety of intellectual fountainheads that contributed to the cultural turn. So have those in the current left mainstream who downplay their own genetic links with Marxian socialism. Besides the Frankfurt School, there were other essential intellectual sources on which the left heavily drew when refurbishing their political religion in the 1960s and the 1970s. This paper has highlighted the role of the British "cultural Marxists" as well as their intellectual predecessors and contemporaries who "racialized" Marxism. Moreover, because of the worldwide hegemony of English language, the writings of C.L.R. James and William Dubois, popular paperback translations of Frantz Fanon, and British Cultural Studies along with *New Left Review* played a far more important role in fomenting the cultural turn than the often-spoken Frankfurt School. In fact, it was *NLR* that should be credited with popularizing "frankfurters'" writings that regular educated readers had a hard time to digest.

The names and schools profiled in this paper do not exhaust other potent sources of the cultural turn on the left. For example, a future researcher cannot bypass the role of secularized Protestantism of northern Europe and North America and its links with the current woke culture. It is impossible to reduce the intrusive moralism of the current woke movement, which is represented by various brands of secular social justice warriors and progressive Protestant denominations in the United States, exclusively to the intellectual evolution of Marxian socialism in the direction of identity issues. Although many elements among the current cultural left, including Black Lives Matter, have been feeding on Dubois, James, Fanon, and New Left neo-Marxist writings, they obviously drew too on the secularized Puritan tradition. In *Multiculturalism and the Politics of Guilt: Towards a Secular Theocracy* Paul Gottfried [12] did briefly outline how via the social gospel tradition, radical Puritanism of some Protestant denominations gradually mutated into virulent cultural moralism of the current mainstream left. Particularly, he stressed how in the present-day political theology of the left, the old Christian notion of sin and salvation became replaced with sensitivity training and social therapy sessions. Still, this whole theme awaits its comprehensive study. In conclusion, I would like to stress that British Cultural Studies, dissenting Marxist intellectuals, who had idealized working class anti-bourgeois culture, Mills along with the New Left, the Frankfurt School, and the latter-day social justice "evangelicals" were cross-fertilizing each other, spearheading ideas that later produced the identitarian woke tradition, which currently represents the left mainstream.

References

1. Fantasia, R.R.: What happened to the US left? Le Monde diplomatique, 7 December (2010). https://mondediplo.com/2010/12/07usleft
2. Davies, I.: British cultural marxism. Int. J. Polit. Cult. Soc. **4**(3), 323–344 (1991)
3. Dworkin, D.: Cultural Marxism in Postwar Britain: History, the New Left, and the Origins of Cultural Studies. Duke University Press, Durham (1997)
4. Kellner, D.: Cultural Marxism and Cultural Studies (2004). https://pages.gseis.ucla.edu/faculty/kellner/essays/culturalmarxism.pdf
5. Garrat, P.: Herbert Marcuse and 'Cultural Marxism'. Verso Blog, 29 March (2019). https://www.versobooks.com/blogs/4285-herbert-marcuse-and-cultural-marxism

6. Mirrlees, T.: The alt-right's discourse on 'cultural marxism': a political instrument of intersectional hate. Atlantis J. 39(1) (2018). http://journals.msvu.ca/index.php/atlantis/article/view/5403
7. Alpers, B.: A far-right anti-semitic conspiracy theory becomes a mainstream irritable gesture. Society for US Intellectual History, 1 December (2018). https://s-usih.org/2018/12/a-far-right-anti-semitic-conspiracy-theory-becomes-a-mainstream-irritable-gesture/
8. Paul, P.: Cultural Marxism': The Mainstreaming of a Nazi Trope. FAIR, 4 June (2019). https://fair.org/home/cultural-marxism-the-mainstreaming-of-a-nazi-trope/
9. McManus, M.: On Marxism, Post-Modernism, and 'Cultural Marxism'. MerionWest, 18 May (2018). https://merionwest.com/2018/05/18/on-marxism-post-modernism-and-cultural-marxism/
10. Hutnyk, J.: Bad Marxism: Capitalism and Cultural Studies. Pluto Press, Ann Arbor (2004)
11. North, D.: The Frankfurt School, Post-Modernism, and the Politics of the Pseudo-Left: A Marxist Critique. Mehring Books, Oak Park (2015)
12. Gottfried, P.: The Strange Death of Marxism: The European Left in the New Millennium. University of Missouri Press, Columbia (2005)
13. Pluckrose, H.: Cultural Marxism" is a Myth. The Threat Comes from Elsewhere. Areo, 30 October (2018). https://areomagazine.com/2018/10/30/cultural-marxism-is-a-myth-the-threat-comes-from-elsewhere/
14. Bolton, K.: Cultural marxism: origins, development and significance. J. Soc. Polit. Econ Stud. 43(3–4), 272–284 (2018). http://jspes.org/samples/JSPES43_3_4_bolton.pdf
15. Breshears, J.D.: A Brief History of Cultural Marxism and Political Correctness (2016). http://www.theareopagus.org/blog/category/articles/
16. Ivry, B.: Deconstructing the Jewishness of the Frankfurt School (2015). https://forward.com/culture/211598/deconstructing-the-jewishness-of-the-frankfurt-sch/
17. Lynd, W.S.: The Roots of Political Correctness. American Conservative, November 19, (2009). https://www.theamericanconservative.com/2009/11/19/the-roots-of-political-correctness/
18. Harvey, D.D.: A Brief History of Neoliberalism. Oxford University Press, Oxford and New York (2005)
19. Monbiot, G.: Neoliberalism–the Ideology at the Root of All Our Problems. Guardian. 25 April (2016). https://www.theguardian.com/books/2016/apr/15/neoliberalism-ideology-problem-george-monbiot
20. Venugopal, R.: Neoliberalism as concept. Econ. Soc. **44**(2), 156–187 (2015)
21. Preston, K.: The Tyranny of the Politically Correct: Totalitarianism in the Postmodern Age. Black House, London (2016)
22. Zubatov, A.: Just Because Anti-Semites Talk About 'Cultural Marxism' Doesn't Mean It Isn't Real. Tablet, 29 November (2018). https://www.tabletmag.com/scroll/276018/just-because-anti-semites-talk-about-cultural-marxism-doesnt-mean-it-isnt-real
23. Mendenhall, A.: Cultural Marxism is Real. Mises Wire, 9 January (2019). https://mises.org/wire/cultural-marxism-real
24. Green, D.: What's Wrong with 'Cultural Marxism'? Spectator USA, March 28 (2019), https://spectator.us/whats-wrong-cultural-marxism/
25. Hall, S.: Life and Times of the First New Left. New Left Review, 61(1) (2010). https://newleftreview.org/issues/II61/articles/stuart-hall-life-and-times-of-the-first-new-left
26. Osborn, J.: Look Back in Anger. Faber and Faber, London (1957)
27. Castoriadis, C.: The Castoriadis Reader. Blackwell, Oxford, 28 (1997)
28. Hall, S.: A Sense of Classlessness. Universities & Left Review 5 (Autumn), 26–31 (1958). https://pdfs.semanticscholar.org/9e87/2b6df78cfef969512af32d2aab8f0d84b453.pdf
29. Williams, R.: Culture and Society, 1780–1950. Doubleday, New York (1960)

30. Hoggart, R.: The Uses of Literacy: Aspects of Working-Class Life. With Special References to Publications and Entertainments. Chatto and Windus, London (1957)
31. Gotzler, S.: Speaking of the Working Class: On Richard Hoggart. Lost Angeles Review of books. 25 April (2018). https://lareviewofbooks.org/article/speaking-of-the-working-class-on-richard-hoggart/
32. Young, D. J.: Neo-Marxism and the British New Left. Survey, January (1967). https://www.marxists.org/history/etol/writers/young/1967/neo-marxism.htm
33. Evans, R.: Eric Hobsbawm: a Life in History. Oxford University Press, New York (2019)
34. Hobsbawm, E.: Primitive Rebels: Studies in Archaic Forms of Social Movement in the 19th and 20th Centuries. Manchester University Press, Manchester, UK (1959)
35. Mills, C.W.: Letter to the New Left. New Left Review, September-October, 18–23 (1960). https://www.marxists.org/subject/humanism/mills-c-wright/letter-new-left.htm
36. Dubois, W.E.B.: The Conservation of Races. Baptist Magazine Print, Washington, DC (1897). https://www.gutenberg.org/files/31254/31254-h/31254-h.htm
37. Dubois, W.E.B.: The Souls of Black Folk. A. C. McClurg and Co, Chicago (1903). https://www.gutenberg.org/files/408/408-h/408-h.htm
38. Dubois, W.E.B.: Dark Princess. Harcourt, Brace and Company, New York (1928)
39. Herman, A.: The Idea of Decline in Western History. Free Press, New York, 217–218 (1997)
40. McLemee, S., Le Blanc, P. (Eds): C.L.R. James and Revolutionary Marxism; Selected Writings of C.L.R. James, 1939-1949. Humanities Press, Atlantic Highlands, NJ (1994). https://doi.org/10.1057/9780230379411_3
41. Avineri, S. (Ed.): Karl Marx on Colonialism and Modernization: His Despatches [sic] and Other Writings on China, India, Mexico, the Middle East, and North Africa. Doubleday, Garden City (1969)
42. Fanon, F.: The Wretched of the Earth. Penguin, New York (1963)
43. Abel, L.: Seven Heroes of the New Left. New York Times Magazine, 5 May, 132 (1968)
44. Kalter, C.: The Discovery of the Third World: Decolonization and the Rise of the New Left in France, 1950–1976. Cambridge University Press Cambridge, UK and New York, p. 218 (2016)
45. Sontag, S.: What's Happening in America? Partisan Review, 34(1), 57–58 (1967). http://johnshaplin.blogspot.com/2012/11/whats-happening-in-america-1966-by_8613.html
46. Harold, J., Bershady, H.J.: When Marx Mattered: An Intellectual Odyssey. Transaction Publishers, New Brunswick, NJ, p.99 (2014)
47. Rand, A.: Return of the Primitive: The Anti-Industrial Revolution. Penguin, New York, Penguin. 168–169 (1969)
48. Ellis, R.J.: The Dark Side of the Left: Illiberal Egalitarianism in America. University Press of Kansas, Lawrence, KS 152–153 (1998)
49. Newfield, J.: A Prophetic Minority. New American Library, New York (1966)

In the Shadow of the Korean Wave: The Political Factors of the Korean Wave in China

Sunyoung Park(✉)

Sejong University, Seoul, Korea
syp@sejong.ac.kr

Abstract. The Korean Wave, or Hallyu, is a term that collectively refers to the international popularity of Korean pop culture. From the mid-1990s, as Korean dramas became popular in mainland China, the Korean Wave spread, and Korean songs and various Korean foods became popular, which contributed to the spread of the Korean Wave worldwide. The popular Korean songs, commonly referred to as K-Pop, accepted without a sense of rejection, because they can connect each other's hearts with music without the interference of politics.

However, China, unlike the general capitalistic countries, can control popular culture according to political conditions, so the political issues in regards to South Korea and Chinese relations have influenced the spread and the vicissitudes of the Korean wave culture in China.

In this paper, the momentum of the formation and development of the Korean Wave in China was examined as well as the concrete aspects through China's political factors that affect the spread of the Korean Wave and the upset of the Korean-Chinese relations. As a result, the cultural aspects regarding the Korean-Chinese international relations and the meaning of the political elements in cultural international relations are examined through this undertaking.

Keywords: Korean Wave · Hallyu · China's political factor · Korean-Chinese relations

1 Preface

Why is the Korean Wave, or Hallyu, in favor of the Korean pop culture popular in China? Can Hallyu continue to affect Chinese culture? One of the ways to grasp the spread of the Korean Wave in China is CNKI, which is a Chinese research site for academic papers. If you look for the theme of "Hallyu" at CNKI, you will find a great deal of data. Except for the simple reports mentioned in newspapers and magazines, the mere presence of hundreds of academic papers and numerous academic papers on the subject of Hallyu could be used as a measure to explain the Korean Wave phenomenon in China.

© The Author(s), under exclusive license to Springer Nature Switzerland AG 2021
R. Bolgov et al. (Eds.): *Proceedings of Topical Issues in International Political Geography*, SPRINGERGEOGR, pp. 355–371, 2021.
https://doi.org/10.1007/978-3-030-78690-8_31

Internet networks are also used to explain why the Korean Wave is so popular in China [1]. From a cultural and sociological perspective, [2] the process of the creation of the Korean Wave is reviewed, and the People's Daily (人民日报), [3] which is the representative organ of the Communist Party of China, discusses why the Korean Wave culture is vital. However, the Korean Wave is not only popular in China, is expanding around the world, so it may be necessary to find out why [4].

Hallyu is a form of cultural exchange, [5] but it is also a genre of cultural diplomacy [6]. The spread of the Korean Wave in China has affected not only the realm of Chinese education, [7] as well as the social and cultural realm, including food, clothing and shelter [8] and the realm of leisure life, [9] but it has also affected the economic realm [10] and the political realm. The popularity of the Korean Wave in China, has created a sense of crisis to the extent that considerations and countermeasures against the safety of the Chinese cultural industry had to be employed [11].

There is a considerable accumulation of research on the Korean Wave and in China in particular. There are various aspects, which include the success factors of the Korean Wave and its impact on each local culture. Hyun-Kyu, P analyzed the success factors of K-pop, which can be called the new Hallyu, and emphasized the importance of using smart media to sustain the Korean Wave [12]. Also Yong Jin, D saw Hallyu as a cultural force from tradition [13]. Many scholars have been interested in the Korean Wave, which has spread throughout the world including Islamic countries, [14] Latin America, [15] Southeast Asia, [16] and Russia [17]. In addition, the Korean Wave is at the level of international cultural exchange, [18] and there has been a great deal of Asian and global responses to the Korean Wave [19]. In other words, the impact of the Korean Wave on the international community has been analyzed [20]. In addition, research was conducted on the strength of Hallyu as a cultural product [21] and the stereotyping of the Hallyu brands as a fixed concept, [22] and in terms of popular culture diplomacy [23].

Specific topics were studied about the export of the Korean Wave and the Korean Wave products in movies, [24] designs, and Korean food [25]. In addition, Chinese academia is interested in the impact and the countermeasures of the Korean Wave on China's various fields [26]. Especially, from an analysis of the articles from 2001, Se-kyung, Y and others showed that the Chinese media mainly used anecdotal Korean Wave phenomena and character information as opposed to in-depth analyses of the Korean Wave phenomenon. It was also determined from the study that there is a frame that emphasizes the policy, economic, and cultural aspects of the Korean Wave, which depended on the policy direction of the newspaper [27].

In this paper, I analyzed changes in the Korean Wave phenomenon in China seldom mentioned in previous research, especially through the political factors of the development of the Korea-China relations and the changes in China's tourism policy.

China's academic research related to the Korean Wave was analyzed methodologically, and statistics from the Korea Tourism Organization were used to examine the relationship between China's tourism policies and the political factors revealed in the media regarding the Korean Wave.

In order to accomplish this, first of all, I would like to look at the formation of the Korean Wave in China, the changing times, and the characteristics of the Korean Wave in China. Then I analyzed the impact of the political factors on the spread of the Korean Wave in the Korea-China relations.

2 The Tourism Policy in China and the Periodic Change of Visitors to Korea

In order to effectively analyze the spread of Hallyu in China, [28] the trends of Chinese tourists visiting Korea were analyzed first in order to see what changes that have occurred during the present times. The Korea Tourism Organization, has recorded statistics about Chinese visits to Korea since 1984. The Chinese have been able to visit Hong Kong, Macau, North Korea and Thailand since the 1980s, visiting overseas relatives and frontier areas. In fact, however, widespread overseas tourism began in the 1990s.

In 1990, Chinese tourists were able to visit Southeast Asia, Malaysia, Singapore, and Thailand for the first time, or make group tours, but with conditions regarding how relatives living abroad could pay travel expenses. However, Chinese tourists' have been free to travel abroad since 1997, when overseas tours were allowed at their own expense. The number of Chinese visitors to Korea changed as China changed its tourism policy, which is shown in Table 1 below.

Table 1. Statistics on the Number of Chinese visitors to Korea according to the Year and the policy changes in China.

Policy changes	Year	Number of visitors to Korea
Visiting relatives and frontier(border)tourism start	1984	108
	1985	308
	1986	801
	1987	362
	1988	7,056
	1989	19,042
Enactment of laws relating to the commencement of de facto overseas tourism and departure	1990	42,516
	1991	78,640
	1992	86,865
	1993	99,957
	1994	140,985
	1995	178,359
	1996	199,604
Enactment of laws concerning permission to travel abroad and departure by oneself	1997	214,244
	1998	210,662
	1999	316,639
	2000	442,794
	2001	482,227
Relaxing the control of overseas tourism and expanding the opening of destinations	2002	539,466

(*continued*)

Table 1. (*continued*)

Policy changes	Year	Number of visitors to Korea
	2003	512,768
	2004	627,264
	2005	710,243
	2006	896,969
	2007	1,068,925
	2008	1,167,891
	2009	1,342,317
Rapid growth of China's overseas travel market and implementation of the leisure law	2010	1,875,157
	2011	2,220,196
	2012	2,836,892
	2013	4,326,869
	2014	6,126,865
	2015	5,984,170
	2016	8,067,722
	2017	4,169,353
	2018	4,789,512
	2019	6,023,021
Restriction by COVID-19 virus	2020.5*	611,421

Sources: Korea tourism organization, Total national statistics sheet as of May 2020 and Korea tourism organization, Analysis of 2019 visit Korea tourism market by country: China, Hong Kong, Taiwan, pp. 189–190. Korea (2020). http://kto.visitkorea.or.kr/kor/notice/data/statis/profit/board/view.kto?id=441795&isNotice=false&instanceId=294&rnum=4, last accessed 2020/07/27.
*: The number of visiting Korea per month in 2020 is limited because of COVID-19 virus: 481,681 by Jan. 2020; 104,086 by Feb.; 16,595 by Mar.; 3,935 by Apr.; 5,124 by May.

The fact that Chinese tourists' became familiar with Korea was a trigger for the establishment of diplomatic ties between South Korea and China in 1992. Previously, China had been hostile to South Korea due to ideological differences, calling it "South Chosen (nanchaoxian, 南朝鲜)," but after Korea-China diplomatic relations, China also changed its official name from "South Chosen (南朝鲜)" to "South Korea (韩国)." As Korean culture flowed into China through various media outlets, the atmosphere of the Korean Wave gradually formed. The Korean Wave can be broadly defined as '*a phenomenon of liking and longing for Korean pop culture abroad.*' [29].

The "Korean Wave" in a narrow sense favors Korean pop music, dramas, entertainment programs and movies. The Korean Wave in a broad sense prefers Korean performances, traditional music, Korean book publishing, Korean animations and characters, Korean food, Korean fashion, Korean beauty, Korean sports, games, medical tourism, martial arts, and Korean traditional clothing [30]. This preference for Hallyu has had a significant influence on the decisions to tour Korea. In other words, those interested in Korea through dramas, music, and mass media are interested in Korean culture as a whole, which leads to Hallyu tourism.

In particular, the Korean Wave, which began in the late 1990s, has expanded from the Korean Wave of consultation to the Korean Wave of broad significance, and the spatial category expanded not only in East Asia, which includes China, but also around the world. This is also collectively called the K-Culture [31]. Along with the spread of the Korean Wave in China, changes in China's tourism policy have led to the expansion of visitors to Korea due to the Korean Wave.

In 1997, the "Temporary Trial of the Management of Overseas Chinese Tourism Management Act" was promulgated, and in May 1998, Chinese in nine regions including Beijing, Shanghai, and Shandong Province ratified Korea as a country where they could freely tour at their own expense. The Korean Wave began to become visible in China in the late 1990s, especially after H.O.T's performance in Beijing in February 2000 was recognized as having a significant cultural impact and not just a temporary syndrome [32].

In 2000, the Chinese government actually opened up all of Korea's tourism programs by ratifying the country as a one in which Chinese can freely tour Korea at their own expense [33]. Before 2003, China was only allowed group tours, but it has introduced the Individual Visit Scheme (IVS) for 49 Chinese cities since 2003, which made China the largest tourist destination in Asia. On October 1, 2013, the first comprehensive law in the field of Chinese tourism was implemented in order to prevent tyrannical practices of travel agencies during tourism, and the Korean government allowed the entry of Chinese visa-free travelers on Jeju Island only.

Due to this change in tourism policy in China, the number of visitors to Korea was 86,865 during the time when diplomatic relations were established between Korea and China in 1992. However, the number of visitors to Korea increased to 442,794 by 2000, which further increased to 8,067,722 by 2016. Even though the overall number of Chinese visitors to Korea has been on the rise since the establishment of diplomatic ties between Korea and China, the number of Chinese visitors to Korea has decreased to a certain extent due to the outbreak of avian influenza in 2003 and the outbreak of MERS in 2015 [34].

In 2012, Korea's exports regarding the cultural industry products surpassed imports for the first time due to the Korean Wave [35]. Also, the number of Chinese tourists that visited Korea nearly doubled from 2012 to 2013.

According to 2014 statistics, the number of overseas Chinese travelers surpassed 100 million. The number of tourists visiting Korea has also increased to 6,126,865, but was almost cut in half to 4,169,353 in 2017, [36] which was because of seven major guidelines for banning travel to Korea were issued on March 16, 2017, and the number of Chinese visitors to Korea was completely suspended through travel agencies. The Terminal High Altitude Area Defense (THAAD) missile crisis in July 2016 played a crucial role in the development of this type of policy in July 2016.

From the Table 1 above, the impact of China's policy changes over time on Chinese visitors to Korea is significant. Of course, they will not visit Korea simply because

China's tourism policy has changed. The popularity of the "Korean Wave" and China's policy changes have greatly affected the volume of Chinese visitors to Korea.

According to a survey, 54.2% of Chinese tourists visit Korea because they are interested in Korean pop culture, 54% are interested in Korean traditional culture/history, 25.8% are interested in Korean tourism promotion/exhibition/festivals, and 25.7% are interested in Korea because it is geographically close. Also, accessibility and the Korean Wave culture were cited as the main reasons [37]. The above statistics confirm that the Korean Wave has influenced Chinese visitors to Korea.

3 Hallyu in China and the Reflections in China

Hallyu's popularity in China has expanded to various fields. Previously, interest was mainly focused on Korean dramas, but it expanded to music, movies, and all areas of Korean culture, even products [38].

The reason for the popularity of Korean dramas [39] and movies [40]was the interest in the influence of moral education on teenagers in dramas [41], especially the influence from Jang Yu Yu Seo(长幼有序), which is idea that the young must obey the elder. This tradition of Confucian culture that is shown in dramas, contributed to the spread of the Korean Wave in China [42]. The Chinese were more interested in this component because Korea's Confucian culture could be explained in connection with China's ancient culture [43] . In other words, even though Korean culture contains a modern aspect of a capitalist society, the ideas included in it are very traditional, valuing filial piety and righteousness [44].

This factor is the reason why the Korean Wave is so popular in China, but there is also the influence of the People's Daily's report. The People's Daily published an article that was interested in the excellence of Korean drama culture, specifically focusing on the drama "*Winter Sonata*" in Japan and the drama "*Dae jang geum*" in China. The People's Daily offered detailed reports of the narrative structure, the production level, and the cultural excellence of Korean dramas, which affected the popularity of the Korean Wave. Of course, the popularity of "*What's Love*," "*Star in My Heart*" and "*Model*," which have been aired on CCTV in China since 1999, has sparked the Korean Wave by raising awareness among various Korean stars who have appeared in the drama.

In 2005, Korean dramas, such as "*Dae jang geum*," "*My Love Patzzi*," "*Be Strong, Geum Soon-ah*" and "*Yeo In cheon ha*" were aired in China, which sparked a craze for Korean dramas. The beautiful scenery and music in the dramas as well as the refined main characters and lines have become factors that have made the Korean Wave a success [45].

A certain drama, which is called "*Descendants of the Sun*," also contributed to creating the Korean Wave in China [46]. However, there is also a movement in Korean poetry to find true "Hallyu," not just Korean dramas [47]. In the past, Chinese was interested in the Korean Wave as a noticeable external factor, which shows that interest in the spirit and the inner culture of Korea is gradually increasing.

Korean popular music is also a major part of the Korean Wave in China, and Korean music has been in the spotlight in various aspects, [48] which the Asian Music International Association explained it as the soft power of the Korean Wave culture in 2017 [49].

It was also used as a pretext to review the current trend of music in China by comparing the Korean Wave and Chinese-style music [50].

When it comes to Korean fashion and designs, [51] it also explains the so-called craftsmanship of Korea [52]. Because of the Korean Wave, Korean fashion has spread not only to major cities in China, such as Beijing [53] and Shanghai, but it has also spread to rural areas. Recently, it was reported in one newspaper that cooking-related broadcasts form the new Hallyu [54].

Even though it is a natural phenomenon in a way, the Korean wave in China has expanded to the craze to learn Korean [55]. The enthusiasm to learn Korean is easily found everywhere in Chinese cities.

Korean institutes have been established at major universities in China to study historical culture related to Korea, but many are interested in modern Korean cultural industries as well. The establishment of the Korean Institute at the Chinese universities was partly due to various forms of support provided by the Korea Foundation for International Exchange, which affected the initial establishment.

The revitalization of the Korean Wave not only led to learning Korean in China and understanding the culture of Korea, but it also led to the craze of studying Korea in person. The Korean Wave trend in China has had a significant impact on the reason why they chose Korea.

In China, the Korean Wave has influenced the manufacturing industry [56] as well as the design space in the beauty industry, [57] the food industry, [58] and the Korean Wave style, which includes children's cartoons, [59] various cultural industries, tourism industries, [60] and plastic surgery [61]. Those who came to Korea due to the Korean wave purchased various items in Korea, and their purchasing behavior confirms their connection to the Korean Wave [62].

The Global Times(环球时报), tried to find out why the Korean Wave was exported abroad [63]. In addition to simply paying attention to what part of the Korean Wave has affected the Chinese society, it has reached the point where Chinese people who are interested in Korean movies and dramas are concerned about the patterns of consuming the Chinese culture [64]. This is because the Korean Wave has affected the consumption behavior of the Chinese college students [65]. This proves that the Korean Wave has established itself as a core cultural content to the extent that it creates a fixed consumer base in the Chinese cultural market.

The Korean Wave has also affected the psychological values of various classes of Chinese society, [66] especially high school [67] and college students [68] in each region. Korean Wave culture has had a significant impact on the formation of values for young Chinese people, so it is argued that China's cultural industry should be developed in the future with the consideration of the Korean Wave impact [69]. Hallyu has also changed the perception of the Chinese college students' uniform culture [70]. Also, the Hallyu fashion, which has become popular in China due to the Korean Wave culture, has also affected the Chinese fashion culture [71]. In particular, the impact on the formation of the values of the female college students is so great that it is necessary to study countermeasures [72].

China understands that the Korean Wave affects the Chinese people and has a profound impact on the Chinese culture and the Chinese society [73]. Therefore, China, which boasts thousands of years of culture, is also moving to reflect on the Chinese culture due to the Korean Wave [74]. They are also reflecting on what Chinese dramas should learn from the Korean Wave and how the Chinese TV dramas should pursue new styles [75]. This is a cross-section where the Korean Wave positively affects the Chinese cultural industry [76]. China is thinking about what the impact of the Korean Wave is on China [77].

The People's Daily reported in an article titled "*The Korean Wave Exclusive Phenomenon Continuing for Seven Years*" on January 14, 2005 that it was disgrace to China that the Korean Wave continued to gain popularity [78]. Although Korea has made good use of and developed Confucian culture, which is a cultural heritage of China, China should reflect on and follow this point, and it was a disgrace that China was not developing its own culture properly. Other articles noted that the production level of the Chinese dramas was not lower than that of the Korean dramas. Furthermore, it was a disgrace to the Korean culture with its vigilance and national sentiment.

The reports of major media outlets, which has been seen before, are bound to affect the Korean Wave such that while they promote national revolt against the Korean Wave, the end result is further deepening and spreading its impact. Even though there are still joint productions with overseas production companies, purchases, remaking of drama copyrights, and the production of imitative programs, there is also a trend in China where the production of contents such as dramas and TV entertainment programs is becoming stronger [79].

4 China's Political 'Restriction (Xianhan, 限韩)' on Hallyu

Although the main role of the Chinese media was to represent the line of the Communist Party of China, some autonomy was created after the reformation and opening era. However, still the Chinese media must basically abide by the socialist principles, and there are no significant changes in the fact that the national policy propaganda is the primary duty of the media [80]. This fact can be seen as an indication that the media coverage may vary at any time depending on China's policy.

With the spread of the Korean Wave in China, the atmosphere of the media, which had praised and praised Korean dramas and products, began to change. As the harmful effects on Chinese culture and society were discussed, the atmosphere of the Korean Wave, which had been taking place in public communications, revealed a hostile atmosphere to the point of reporting things in the newspaper, such as "*the Korean Wave has cooled down a little*" [81] or even "*is it the Korean Wave or a dirty and dirty trend?*" [82] In no time, the atmosphere was changed to "kanghan(抗韩)" against the Korean Wave, [83] the atmosphere of "Xianhan(嫌韩)"hatred of Korea continued to rise [84]. The Chinese who were against the Korean Wave began to consider the patriotism toward China of the Korean Wave fans [85].

There are even voices calling for the education of thought and politics for the college students affected by the Korean Wave culture [86]. The education of ideology, education, and politics, which had long been aimed to achieve political goals in socialist China, has reached the point of having to provide ideological education due to the Korean Wave. In some ways, the capitalist Korean wave is intended to continue the socialist system by tying it up with the ideological verification and the political education. China says it should be careful of the Korean Wave [87] because the Chinese wind (Chinese Wave, 汉风) will soon blow depending on the political atmosphere, but there is also hope that the Chinese wind will intensify soon. They hope that the Korean Wave will disappear [88] due to the clash between the Chinese and the Korean Wave, and the Chinese winds [89].

Anti-Korean sentiment, which reveals the negative views socially, culturally, politically, and diplomatically on Korea's cultural content in China, gradually formed in late 2005. Chinese broadcasting officials have begun to openly criticize or oppose the Korean Wave craze. In 2008, various media outlets in China reported articles that induce anti-Korean sentiment and tried to prevent the Korean Wave from spreading. Based on the logic that Korean TV dramas are highly distorted in history, "*Yeon Gaesomun*," "*Taewangsasingi*," and "*Dae Joyeong*" were banned from being broadcast [90].

On the other hand, in order to prevent the number of Chinese visitors to Korea from continuing to expand, the government tried to put the brakes on the tourist parade to Korea by setting a high price when applying for visas to Korea, but the experts stressed that China's policy to prevent the Korean Wave was not very effective [91]. Despite the atmosphere of "the kanghan (抗韩)" and "the xianhan (嫌韩)" the Korean Wave has still become a cultural trend that Chinese teenagers like and envy.

Even though the Chinese government officially denies that it has ever issued a "xianhan (限韩)," this occurrence be confirmed in a CNKI paper published in China regarding the existence of a "xianhan (限韩)" with respect to this matter. It is worth noting that the discussion about "the xianhan (限韩)" has expanded even more in recent years. It can be said that Korean entertainment has encountered a cold wave in the Chinese market [92]. However, some question whether restricting the Korean Wave in China is possible with certain orders or policies, [93] but other people are trying to ignore it, saying that there is no need to pay attention to whether the law is real [94].

The cause of the so-called anti-Korean wave can be considered by dividing it into the cultural, national, and media aspects [95]. The so-called cultural origin debate on East Asian culture represented by Confucius culture and China's sense of cultural crisis have contributed to opposing the Korean Wave. While nationalism is emphasized in an attempt to restore China's cultural self-esteem, on the other hand, there are trends that oppose the existing social structures, the political culture, and the Chinese national policies. The role of the media is important in any country, but it is also true that the Chinese sentiment will inevitably change if the main media outlets deal a blow to the Korean Wave with a marketing strategy that promotes patriotism.

As a result, Korea, which is the direct target of China's ban on Korean culture, was hit hard [96]. Even though an order banning Korea is being issued, some people criticize Chinese university students for purchasing Korean Wave-related products such as Korean cosmetics from a Chinese national-centered perspective [97]. As it is difficult to import Korean culture directly from China, where an order banning Korea is issued, some in China are calling for self-reflection on Korean entertainment programs [98].

As the Korean Wave spread to the point where measures for the cultural industry had to be taken in China, China applied active implement regulations and restrictions on foreign cultural contents in order to protect and develop their cultural industry. The decisive factor to actively implement this is the THAAD missiles issue that the U.S. deployed on the Korean Peninsula in 2016. As a result, the relationship between Korea and China has changed rapidly [99]. The THAAD missiles issue is a Sino-American one in substance, but the location of the deployment on the Korean Peninsula has created a serious conflict between South Korea and China directly.

First, on September 1, 2016, the headquarters of China Broadcast (中国广电总局) imposed sanctions on imports of Korean goods and services. It banned Korean actors from performing in China, banned investment in the cultural industry, banned large-scale performances involving more than 10,000 spectators, banned joint ventures such as Korean dramas and entertainment, and banned imports of dramas [100]. The so-called "Korean Wave Restrictions" banned the transmission of Korean-made content in China or advertisements that featured Korean celebrities, and wide-ranging measures were implemented to ban all contents that contained Korean elements, such as Korean companies and Korean brands [101].

Specific sanctions in the tourism sector began in October 2016 when China's State Travel Agency issued temporary guidelines that limited the number of people visiting Korea. The main content was to crack down on low-priced products, such as banning the sale of travel to Korea packages under 2,000 yuan and notifying that only 20 percent reduction of the number of people from each travel agency was allowed to visit Korea for six months from November 2016 to April 2017. Relations with China deteriorated further, when South Korea's Defense Ministry signed an agreement with the Lotte Group about the exchange the THAAD missile sites on February 28, 2017.

On March 2, the China National Travel Agency issued seven major guidelines that suspended all work related to Korean group tour packages as of March 16 after convening with local travel agencies and major travel agencies. The seven guidelines that banned travel to Korea in 2017 included [102] ① prohibiting all travel packages to Korea after March 15, ② prohibiting travel group visa applications through travel agencies, ③ marking Korean tour packages on the site were as *sales complete,* and deleting all products if technically difficult to mark them, ④ deleting all Lotte-related products, ⑤ strengthening supervision of low-cost tourism products in Korea, ⑥ disallowing cruise travel to Korea, and ⑦ imposing a fine on those caught purchasing travel packages to Korea after March 15. As a result of these restrictions, the number of travelers from China to Korea drastically reduced. China's Civil Aviation Administration refused to allow new charter flights or increase flights to Korea, the cruise companies in China announced the suspension of the Korean routes in 2017, and sanctions on Korea by surplus and civil aviation authorities continued until 2019.

China's ban on Korean culture has various causes in terms of politics, diplomacy, the economy, and the security, and cultural reasons for the Korean Wave. China's complaints about the unilateral influx of the Korean Wave's political impact on China has affected China's reputation and has become a serious issue. The strained relations between South Korea and China due to the deployment of the THAAD missile also weaken the momentum of cultural exchanges between Seoul and Beijing and undermine the foundation of bilateral relations [103]. In the meantime, however, the number of Chinese games entering the Korean market reached 111 cases, with sales growth reaching 170% [104].

Just as the political element of the deployment of the U.S. THAAD missile on the Korean Peninsula instantly changed the Korean Wave in China and the Korea-China relationship, the opportunity to change the chilly Korea-China relations also played an important role in restoring political relations between the two countries. After the 19th National Congress of the Communist Party of China in October 2017, the leaders of South Korea and China reconciled, which made it possible to advertise Korean goods in China again. The two countries' relations recovered after they agreed to normalize Korea-China relations on October 31 and declared an end to the THAAD missile conflict at the APEC summit on November 1. After political relations between the two countries recovered, Korean model Ji-hyun, Jun could appear as a product model in Taobao, China's largest online commerce, on Nov. 11. In December, Chinese tourists began touring Korea. On December 13, the president of Korea, Jae-in, Moon state visit to China exchanges started to become active [105]. This can be seen as the intervention of political factors plays a greater role than the Korean Wave in China itself, which naturally changes according to Chinese preferences.

The number of Chinese visitors to Korea reached 8,067,722 in 2016, decreased to 4,169,353 in 2017, and rose to 4,789,512 in 2018, which is slightly higher than the previous year, and hit 6,023,021 in 2019. However, the political issues of the deployment of the THAAD missiles in 2016 and the pace of change since the Chinese travel restriction on the Korean Peninsula are very slow compared to the rapid increase in the number of changes since 1997. In addition, the COVID-19 virus, which has been raging since the end of 2019, has reached a type of situational stagnation period where individuals and the nation have to be careful for their own safety.

If there were no political elements that can be big issues between Korea and China, the cultural exchange relationship will recover little by little, but the social and political system of China, which maintains a socialist system, is the most important factor to consider when considering the changes in the Korean Wave in China, because there is room for the national political factors to be involved at any time. This conclusion not only confirmed by statistics about the number of Chinese visitors to Korea, but it is also evidence of the change in the perception of the Chinese tourists using big data [106]. In other words, it is clear that the political and the environmental external factors have a negative impact on relationship.

5 Conclusion

In general, national economic growth gives us leeway to the enjoyment of culture. Since its reform and opening, China has also seen its tourism industry develop a lot with its spectacular economic growth. In particular, China's tourism policy, which is in line with the times, has expanded from limited group travel to free group travel by taking an increasingly open attitude, and the climate of traveling to suit the individual tastes has also been established.

Korea is one of the most visited countries by the Chinese people. However, the Korean Wave, which has been raging since the late 1990s, and it has played a role in making Korean culture more interesting. Therefore, the popularity of the Korean Wave in China and the change in China's tourism policy contributed greatly to Chinese visiting

Korea. The Korean Wave has affected China's living culture, social culture, and even youth college students in various aspects. It has had a wide impact on various fields, which include food, costume designs, architecture, manufacturing, music, dramas, and films.

The Korean Wave has enjoyed considerable popularity in China to the extent that it is said that the Korean Wave has invaded China [107]. China, which had to think about whether it would be able to prevent the Korean Wave from invading China, [108] was also accompanied by a reflection on China's cultural industry. The popularity of the Korean Wave in China has also had a positive impact on China's cultural industry to the extent that it calls for self-reflection. However, it is the political influence of socialist China that can change this atmosphere in an instant.

In particular, it was no exaggeration to say that everything related to Korea was sanctioned due to the deployment of the THAAD missiles in Korea in 2016. China strongly protested the deployment of the THAAD missiles in Korea, which was not between Korea and China, but reflected the U.S. policy toward China. In retaliation, it imposed sanctions on Korean companies in China, banned the Korean Wave, and restricted Korean tourism.

Democratic countries do not impose all-round strong sanctions by utilizing its national power like China, and even if there is a policy to impose sanctions on certain products, not everyone is bound to follow. Such democratic countries are not as unorthodox as China because individuals are given the power to buy according to their free will. Since China, which maintains a socialist system, can cast a shadow over Korea-China relations at any time in this regard, it requires wisdom to respond to the political issues in order to continue developing the Korea-China relations under any circumstances.

References

1. Xi, S., Shengdong, L.: Why is 'Korean wave' popular? Based on Sina Weibo's 'weblog' observation. Brand Research **2016**(4), China (2016)
2. Ling, L.: Cultural Sociology Viewed the Construction Process of Hallyu: Take Korean dramas, TV variety, and K-pop music. Master of Shenyang Normal University, China (2016)
3. Shangwen, C.: Why Hallyu is a Persistent Vitality? People's Daily. 24 Feb 2016
4. Ye, H.: Korean pop music drives the 'Hallyu' culture to the world. Master of the Henan Normal University, China (2015)
5. Aihua, P.: Looking at Sino-Korean Cultural Exchange from 'Hallyu.' Master of Yanbian University, China (2008)
6. Ran, C.: Research on Korean Cultural Diplomacy: focusing on 'Korean wave.' Master of Tianjin Normal University, China (2014)
7. Peiji, J.: Nanchang Junior High School Media Contact and Cultural Identity Study of Hallyu. Master of Nanchang University, China (2010)
8. Jiaxue, L.: Explain the Korean Wave: The Influence of Korean Pop Culture on Women's Fashion in China. Master of Northeast Normal University, China (2009)
9. Liping, T.: A Study on the Leisure Behavior of the Star Chasing Group under the Influence of 'Korean Wave': Take Qingdao City as an Example. Master of Jiangxi Normal University, China (2011)
10. Congren, L.: The Influence of Korean Wave culture on China's Export. Master of Fudan University, China (2011)

11. Yongzhu, W.: The influence of 'Korean wave' on the security of China's cultural industry and Countermeasures. Master of Guangxi Normal University, China (2015)
12. Hyunkyu P.: Success Factor Analysis of New Korean Wave 'K-POP' and A study on the importance of Smart Media to sustain Korean Wave. J. Digit. Convergence **18**(4). Korea (2020)
13. Dalyong, J.: New Korean Wave: Transnational Cultural Power in the Age of Social Media. University of Illinois Press, Urbana (2016)
14. Giyeon, K.: Riding the Korean Wave in Iran. Journal of Middle East Women's Studies **16**(2). Duke University Press (2020)
15. Wonjung, M.; Dalyong, J.; Benjamin, H.: Transcultural fandom of the Korean Wave in Latin America: through the lens of cultural intimacy and affinity space. Media, Cult. Soc. **41**(5), United Kingdom (2019)
16. Taeksoo, S.C.: The Impact of the Mongolian Market's Attitude toward Hallyu on the Image of Korea and the Perception of Korean Products. J. Korean Manage. Assoc. **32**(12) Korea (2019)
17. Kijoo, L.: Russia and Hallyu. Foreign Studies 41, Korea (2017)
18. Haggai Kennedy O., Sungsoo, K. Cultural Exchange and Its Externalities on Korea-Africa Relations: How Does the Korean Wave Affect the Perception and Purchasing Behavior of African Consumers? East Asian Economic Review **23**(4) Korea (2019)
19. Hyunkey, K.: The Korean Wave: An Asian Reaction to Western-Dominated Globalization. Perspect. Glob. Dev. Technol. **12**(1/2) Brill Academic Publishers (2013)
20. Dredge Byungchu, K.: Surfing the Korean wave: wonder gays and the crisis of thai masculinity. Visual Anthropol. **31**(1/2). Routledge (2018)
21. Naeun, K.; Lauren, H.: The power of culture in branding: how the Korean wave can help global brands thrive in Asia. J. Brand Strategy, **6**(3) Henry Stewart Publications LLP, London (Winter 2017/2018)
22. Hyeri, L., Myungsu, C., Yongkyu, C.: The application of categorization and stereotype content theories to country of origin image: Vietnamese perceptions towards Korean wave brands. Asia Pacific Bus. Rev. **26**(3), 336–361 (2020). https://doi.org/10.1057/s41254-017-0076-4
23. Hunshik, K.: When public diplomacy faces trade barriers and diplomatic frictions: the case of the Korean Wave. Place Brand. Public Dipl. **14**(4), 234–244 (2018). https://doi.org/10.1057/s41254-017-0076-4
24. Hyoin, Y.: Coevolution of conventions and Korean new wave: Korean cinema in the 1970s and 80s. Korea J. **59**(4) Korea (2019)
25. Youngjoo, J.: The influence of Korean wave recognition on preference of cheese chicken rib, SNS, image of Korean food, intention to visit Korean restaurant. J. Digit. Convergence **17**(3) Korea (2019)
26. Changuk, H., Jie, W.: Do Hallyu (Korean Wave) Exports Promote Korea's Consumer Goods Exports? Emerging Markets Finance & Trade **53**(6) Taylor & Francis Ltd (2017)
27. Jungju, C., Kai, Y.: Frame study of the Korean Wave Press by Chinese Press: Focused on the People's Daily, the China Youth Daily, and the New People's Daily from 2009 to 2014. J. Sci. Res. **15**(2). Korea (2015)
28. Shaoyu, Y.: The Reason and Analysis of Hallyu Culture Spreading in China. Chizi (Spiritual Leaders) **2017**(1). China (2017)
29. The Korea Tourism Organization (KTO), A Study on the Survey of Hallyu Tourism Market (Theme Survey 2019–01), p. 23. Korea (2019). http://kto.visitkorea.or.kr/kor/notice/data/report/org/board/view.kto?id=441607&isNotice=false&i-stanceId=127&rnum=1. Accessed 28 July 2020

30. The Korea Tourism Organization (KTO), A Study on the Survey of Hallyu Tourism Market (Theme Survey 2019–01), p. 8. Korea (2019). http://kto.visitkorea.or.kr/kor/notice/data/report/org/board/view.kto?id=441607&isNotice=false&instanceId=127&rnum=1. Accessed 28 July 2020
31. The Korea Tourism Organization (KTO), A Study on the Survey of Hallyu Tourism Market (Theme Survey 2019–01), p. 13. Korea (2019). http://kto.visitkorea.or.kr/kor/notice/data/report/org/board/view.kto?id=441607&isNotice=false&instanceId=127&rnum=1. Accessed 28 July 2020
32. Heejin, L.: The impact of the level of use of Hallyu contents on Chinese Anti-Korean sentiment: the proof of secondary sales effect of positive recognition of Korean wave goodwill. J. Korean Content Assoc. **17**(10), 395 (2017)
33. Korea Tourism Organization, An Analysis of the Tourism Market for Korea in 2019: China, Hong Kong, Taiwan. p. 189. Korea (2019). http://kto.visitkorea.or.kr/kor/notice/data/report/org/board/view.kto?id=441207&instanceId=127. Accessed 27 July 2020
34. Kyungyeol, P., Hyerim, H., Seungdam, C.: Comparison of Changes in the perception of Korean tourism by Chinese tourists before and after the deployments of THAAD using Big Data. Tour. Leisure Res. **31**(2), 26 (2019)
35. Heejin, L.: The impact of Hallyu content utilization on Chinese Anti-Korean emotions: The proof of secondary effects of positive recognition of Korean wave goodness. J. Korean Content Assoc. **17**(10), 395 (2017)
36. Korea Tourism Organization, Overall National Statistics, Announcement of Tourism Statistics in May 2020. Korea (2020). http://kto.visitkorea.or.kr/kor/notice/data/statis/profit/board/view.kto?id=441795&isNotice=false&instanceId=294&rnum=4. Accessed 27 July 2020
37. Korea Tourism Organization, An Analysis of the Tourism Market for Korea in 2019: China, Hong Kong, Taiwan, p. 216. Korea (2019). http://kto.visitkorea.or.kr/kor/notice/data/report/org/board/view.kto?id=441207&instanceId=127. Accessed 27 July 2020
38. Jie, C.: Marketing communication mode of Korean wave culture. News communication 21, China (2015)
39. Junnan, P.: Research on the network communication mode and effect of Korean drama in China, master of Jilin University, China (2011)
40. Juan, D.: Korean wave effect: Thoughts on cross cultural communication of Korean images. Movie review 19, China (2018)
41. Yang, Y.: Research on moral education ideas in Korean dramas and their impact on Teenagers. master of Henan University of technology, China (2019)
42. Zhendi, G.: Confucian cultural complex in Korean TV dramas. master of Chinese Academy of Arts, China (2018)
43. Yujing, W.: The origin of Korean culture and Chu(楚) culture, master of Yangtze University, China (2017)
44. Changwan, C.: The Korean wave in China, Its flow and blockage (Korean pop culture in China). Creation Criticism **28**(4), 442 (2000)
45. Sekyung, Y., Seok, L., Jiin, J.: Frame analysis in the Chinese daily's 'Hallyu' report. Korean J. Press Inf. **57**, 218 (2012)
46. Yaohan, Z.: Korean wave phenomenon in China from the success of descendants of the sun, master of Zhejiang University, China (2017)
47. Yu, Q.: Put down Korean dramas and pick up poems, you will find another world: the real 'Korean wave' written by Korean poet Gao Yin. South accent and North tune 10, China (2016)
48. Yuanyuan, Z.: The spread of Korean pop music in Chinese mainland (1998 to 2004). Harbin Normal University, China (2017)

49. Hui, Y.: Korean wave in music and cultural 'soft power': a summary of the 2017 Asian Music Society International Symposium. People's music 7, China (2018)
50. Siqi, W.: 'Korean wave' and 'Chinese style': review of the development of pop music in mainland China in the new century (3). Art of Singing 12, China (2015)
51. Jiaxue, L.: Interpretation of 'Korean wave': the influence of Korean pop culture on Chinese women's fashion', master of Northeast Normal University, China (2009)
52. Zheng, D.: Korea fashion show: interpreting Korean wave's ingenuity spirit. Chinese textile 10, China (2018)
53. Xiao, D.: Korean wave in Beijing, Fashion Beijing 12, China (2016)
54. Dong, J., Huizhen, K., Mumao, T.: Kitchen variety becomes the new Korean wave. Global Times, China. 27 Nov 2019
55. Lin, L.: A brief discussion on the 'Korean wave boom' in Hunan Province and Korean Teaching in Hunan Colleges and universities. Korean teaching and research 3, China (2016)
56. Editor.: Manufacturing industry puts on 'Korean wave coat'. Directors & Boards 5, China (2017)
57. Jia, L.: Korean wave color space design. Fashion Colour 7, China (2016)
58. Yuan, Y.: Riding the Korean wave and speeding up the export of Korean food to China. China trade news, China. 17 May 2016
59. Mengxue, Z.: A study on the influence of idolatry subculture on Tourism Intention: a case study of Korean wave fans in China. master of Nanjing Normal University, China (2017)
60. Marie, F.: Where will the original children's play go under the influence of the Korean wave? Beijing business daily, China. 14 July 2016
61. Cuihua, L.: Taking advantage of the Korean wave during the National Day holiday to promote medical and American travel insurance projects in China and South Korea. Securities Daily, China. 29 Sep 2016
62. Yuzhen, J.: Research on Chinese tourists' shopping behavior in South Korea under the background of Korean wave. Master of Zhejiang University of technology, China (2016)
63. Jiashan, S.: Promoting the Korean wave to go global? Global times, China. 13 Feb 2020
64. Zuqun, Z.: On the influence of 'Korean wave' films and TV works on China's current cultural consumption. Movie review 12, China (2015)
65. Han, L.: A study on the consumption behavior of Korean wave fans in China from the perspective of consumerism. New media research 22, China (2019)
66. Yanni, Z.: Research on the influence of 'Korean wave' on College Students' values and Countermeasures. Research on Transmission Competence 12, China (2019)
67. Lingqian, Z.: Research on the impact of the Korean wave on Chinese college students: a case study of Chongqing University City. Today's Massmedia 12, China (2018)
68. Hongli, Z.: An analysis of the impact of the 'Korean wave' on the psychological development of young people. The Science Education Article Collects 8, China (2017)
69. He, M., Hong, L.: On the impact of Korean wave cultural industry output on the values of Chinese youth: to enlighten the development direction of China's cultural industry. Journalism Communication 20, China (2018)
70. Yingxian, L., Yuanyuan, X.: Research on the change of Korean wave's understanding of Chinese college students' school uniform culture. Modern decoration (Theory) 11, China (2016)
71. Sisi, X.: The influence of Korean wave culture on China's clothing based on women's clothing. Popular literature and art 5, China (2016)
72. Yang, L.: Research on the influence of 'Korean wave' on female college students' values and Countermeasures. Master of Wuhan University of technology, China (2018)
73. Wenyou, J.; The breakthrough of multi culture in the context of Globalization: Taking the Korean wave as an example. Modern communication (Journal of Communication University of China) 6, China (2017)

74. Yue, R.: Selfreflection in the ‘Korean wave’, Western radio and television 3, China (2016)
75. Benqian, L., Sheng, M.: Research on the disruptive innovation mode of Chinese TV variety under the impact of the Korean wave. Science & Technology Vision 36, China (2015)
76. Haihua, Z.: Korean wave phenomenon and its implications for China’s cultural industry. Knowledge base, China (2018)
77. Yuanbao, G.: Revelation of Korean wave. Sichuan drama 10, China (2015)
78. Sekyung, Y., Seok, L., Jiin, J.: Frame analysis in the ‘Hallyu’ report of the Chinese daily. J. Korean Press Inf. **57**, 221 (2012)
79. Jongmin, L.: Finding the way of the Korean wave after THAAD. Chin. Mod. Lit. **88**, 338 (2019)
80. Sekyung, Y., Seok, L., Jiin, J.: Frame analysis in the Chinese daily’s ‘Hallyu’ report. Korean J. Press Inf. **0 57**, p. 209 (2012)
81. Gaochao, L.: Korean wave is a little cold, International Business, China. 24 July 2017
82. Xue, W.: Korean current, dirty flow? Xinmin weekly 12. 07 April 2019
83. Editor.: Will powder related trade be impacted when China sets off a wave of anti-Korea. China powder industry 2, China (2017)
84. Xiaoling, W.: Korean wave and hater of Korean wave. People’s daily, 2013/12/10; Zhijiang, W.: Thinking about the so-called Korean wave. Asia Pacific economic times. 02 March 2008
85. Lingling, Z.: A study on patriotism of Korean wave fans. Beauty & Times 1, China (2017)
86. Huaou, W.: Research on Ideological and political education of College Students under the influence of Korean wave culture. Journal of Daqing Normal University 3, China (2016)
87. Yuqing, D.: Trans.: watch out for Japan’s and South Korean’s cultural wave, China wave coming. Global times 26 March 2019
88. Xuying, L.: “The Chinese wind is hot, why is the Korean wave cool?”. International Business Daily. 26 Nov 2015
89. Zhiya, L.: The collision between the Chinese style and the Korean wave. Economy 21, China (2015)
90. Nakgun, H.: The impact of Anti-Korean wave policy on Hallyu in China: focusing on Korean TV Drama. J. Korean Assoc. Entertainment Ind. **12**(6), 238 (2018)
91. Sihui, L.: Spend 18000 yuan to apply for ‘Korean wave visa’, Experts say that the new policy has little impact on tourism to South Korea. Securities Daily. 23 Dec 2016
92. Yi, T.: Hallyu encounters a cold winter in China’s market. China culture daily. 25 Feb 2017
93. Duanfang, T.: Can the restriction of South Korea depend on order? Law & life 18, China (2016)
94. Longchan, R.: Don’t worry about whether the Korean restriction is true or not. Broadcasting Realm 5, China (2016)
95. Jungmin, J.: The cause and response of Anti-Korean wave on Hallyu contents: focusing on China and Japan. Diaspora Res. **7**(2), 137 (2013)
96. Xiaoxiao, T.: Restrictions on South Korea and its impact on South Korea. Modern Business 29, China (2017)
97. Jinmao, Y.: A study on Chinese college students’ intention to purchase Korean cosmetics under the background of restrictions on South Korea: from the perspective of ethnocentrism. Tax Paying 11, China (2018)
98. Tianjing, Z., Yingping, H.: Reflection on the innovation direction of domestic variety shows in the competition between domestic and imitated Korean variety shows in the trend of restricting South Korea. Cultural and educational materials 9, China (2018)
99. Di, Z.: THAAD issue and Sino ROK relations. master of Yanbian University, China (2018)
100. Restriction on Korea. https://baike.baidu.com/item/%E9%99%90%E9%9F%A9%E4%BB%A4/19875423?fr=aladdin. Accessed 26 July 2020
101. Jungrae, C.: China’s ban on hallyu and challenges of cultural contents industry. Chin. Sch. Assoc. **61**, 335–336 (2018)

102. Korea Tourism Organization, An Analysis of the Tourism Market for Korea in 2019: China, Hong Kong, Taiwan. p. 182. http://kto.visitkorea.or.kr/kor/notice/data/report/org/board/view.kto?id=441207&instanceId=127. Accessed 27 July 2020
103. China Review News: can the restriction on South Korea force South Korea to give up THAAS? China Review News Agency. http://hk.crntt.com/doc/1043/8/2/1/104382116.html?coluid=93&kindid=7950&docid=104382116. Accessed 26 July 2020
104. Jungrae, C.: China's ban on hallyu and challenges of cultural contents industry. Chin. Sch. Assoc. **61**, 344 (2018)
105. Jungrae, C.: China's ban on hallyu and challenges of cultural contents industry. Chin. Sch. Assoc. **61**, 349 (2018)
106. Kyungyeol, P., Hyerim, H., Seungdam, C.: Comparison of changes in Chinese tourists' perception of Korean tourism before and after the deployment of THAAD using big data. Res. Tour. Leisure **31**(2), 40 (2019)
107. Mengxuan, M.: Korean wave assult. China Conference & Exhibition 12, China (2016)
108. Yajie, T.: Is the Korean wave irresistible? China industry and economy news. 26 March 2016

The Issue of Cultural Diversity in the EU Cultural Policy at the Beginning of the 21st Century

Natalia Bogolyubova, Yulia Nikolaeva, and Elena Eltc(✉)

Saint Petersburg State University, Saint Petersburg, Russia

Abstract. The paper reveals the place of the concept of cultural diversity in the cultural policy of the European Union, which gradually replaces the previously dominant concepts of multiculturalism and pan-European cultural heritage. Based on the material of original sources, conventions, and declarations of UNESCO as the most authoritative organization in the field of culture, as well as on the basis of a large body of documentation of the European Union regulating its cultural policy, the authors sought to show the transformation of the fundamental ideas that determine the modern EU policy in the culture sector. This study has examined the documentary foundations of maintaining cultural diversity, it has shown its importance for modern Europe, and also it has characterized the importance of individual programs and projects implemented in modern Europe in this direction.

The paper identifies the reasons that led to a paradigm shift in the cultural policy of the European Union at the beginning of the 21st century. These reasons lie not only in multinational European culture, accordingly there are frequent statements about the unwillingness to lose the national cultural heritage and there have been calls to preserve this heritage and diversity. An equally important reason was the influx into Europe of a large number of migrants who are carriers of their own national cultures, complementing European culture and requiring respect and attention to their own cultural traditions.

Today, there are numerous transformations in the culture of Europe; the very appearance of Europe and its cultural landscape are being changed. In this regard, the issues of cultural diversity are becoming particularly relevant and acute. The material in this paper can be used to develop regulatory frameworks and to prepare projects and programs dedicated to the preservation of cultural diversity not only in Europe but also in other countries of the world.

Keywords: Europe · European Union · Cultural policy · Cultural diversity

1 Introduction

Today, the problem of preserving and maintaining cultural diversity is one of the most urgent both at the global and regional levels. In modern Europe, the preservation of the rich cultural heritage and the maintenance of cultural diversity is the most important

© The Author(s), under exclusive license to Springer Nature Switzerland AG 2021
R. Bolgov et al. (Eds.): *Proceedings of Topical Issues in International Political Geography*, SPRINGERGEOGR, pp. 372–381, 2021.
https://doi.org/10.1007/978-3-030-78690-8_32

area of domestic and foreign policy both at the pan-European level and at the level of individual States.

Europe is a continent with a rich cultural heritage and great cultural diversity. In Europe, people speak approximately 260 languages, over 60 of which are regional and minority ones. The total number of speakers of minority languages is estimated at 40 million. 87 peoples live throughout Europe, 33 of which are the main nation for their States, and 54 are considered to be an ethnic minority in these countries. And all these peoples have their own unique culture, which requires careful preservation and transmission to future generations. In Europe, there are 506 UNESCO World Heritage Sites—this is approximately 50% of the total number of sites in the List [28]. A significant share of the European heritage is also in the UNESCO Intangible Cultural Heritage List: 186 elements of the 508 elements currently making up the List relate to Europe, i.e. about 40% [3].

In this regard, the issue of preserving the diversity of cultures turned out to be extremely significant for Europe. Today in Europe, the ideas of cultural tolerance and coexistence of many different cultures in a single European space are gaining more and more supporters, which is being consistently reflected in European cultural policy.

A wide kaleidoscope of national languages and cultures gathered in Europe, a large concentration of objects of tangible and intangible cultural heritage, a developed network of organizations dealing with cultural matters and cultural heritage, a well-developed documentary base, a solid number of implemented programs—all these circumstances make Europe a unique example of solving the problem of cultural diversity preservation.

The aim of the research is defining the value of discourse of cultural diversity in the cultural policy of the European Union based on the analysis of documentary materials, as well as to determine the factors that influenced the formation of this idea as the main point of European cultural policy.

2 Methodology and Research Methods

The theoretical basis for developing the cultural diversity concept was the work of European scholars O. Calligaro, J. Ferris, D. Frink and C. Geleng, E. Starr and J. Adams, which give a detailed description of the phenomenon of cultural diversity [4, 14, 26].

The research methodology is based on the value approach to cultural diversity arising from its legislative regulation, in particular, from the fundamental provisions of UNESCO's Universal Declaration on Cultural Diversity [27]. When studying the phenomenon of cultural diversity, we have based on the definition formulated by the famous sociologist, Professor of the University of Colorado Caleb Rosado, who defines it as "a system of beliefs and behavior that recognizes and respects the presence of all different groups in an organization or society, recognizes, and appreciates their sociocultural differences" [25]. In exploring the role of the issue of cultural diversity in cultural policy, we have focused on the attitude proposed by Armand Mattelart (professor at the University of Paris-V), who believes that it "should become the common heritage of mankind, since it has unique qualities and can make the world more open, more creative, and more democratic"[19].

When working with the documentary base, a systematic approach and a method of content analysis, which made it possible to determine the place of the cultural heritage phenomenon in international regulatory sources, as well as in the EU cultural policy, have been applied. In order to study the practical activities of the EU in preserving and promoting cultural diversity, methods of institutional and functional analysis, which has enabled to identify the main forms, tools of this activity, and its results, have been used.

The comparative method in this study has been used to identify the general and special in the problem of preserving cultural diversity in the European Union and in the practice of the UNESCO global organization. Also, a similar approach allows us to identify and to note the evolution of the problem of cultural diversity in the documentary and practical aspects of the EU activity.

3 The Culture of Modern Europe: Achievements, Contradictions, and Development Trends

The culture of modern Europe is rich, diverse, it combines the traditions and heritage of the culture of many peoples of the European continent and it is based on the heritage and achievements that determined the progress of human civilization and became the basis for new achievements. European culture includes various democratic rights and freedoms; it is characterized by religious tolerance, freedom of expression. Without prejudice to the contribution of other peoples of the world to the treasury of human civilization, we can responsibly declare that it is Europe that is rightfully considered to be "a citadel of art, a model of democratic and liberal values, an engine of technology, as well as principles of social structure" [12].

Today, European culture is undergoing a transformation that is caused by external and internal factors, modern sociocultural problems, which acquire their own characteristics in Europe. The main trends of modern European culture include:

- Tolerance, which is associated with religious toleration, openness to new trends, a tolerant attitude towards people representing different cultural traditions;
- Glocalization (simultaneous processes of globalization and localization of culture). This trend confirms the peculiarity and the originality of each culture in the multicultural world of European cultures [16];
- Integration that erases borders and harmonizes European culture standards [31].
- Adaptability to streamlined manufacture. IT technologies, together with creativity, are becoming the most important characteristic of the modern European cultural space. It is worth noting that information technology is an integral part of the museum and theater projects, archival and library work, but IT technology is also the basis of independent cultural projects;
- Support and promotion of European cultural heritage, understanding of the value of cultural diversity, which is a significant value of European culture today.

The listed trends are interconnected and they confirm the value of both the common European heritage and the unique achievements of each culture, which are manifested in cultural diversity and they have their own, special significance for world civilization.

4 The Reasons for the Growth of the Idea of Cultural Diversity in the Concept of EU Cultural Policy

A study of the basic concepts of the EU official culture discourse shows significant changes that have occurred in recent years. For a long time, the cultural idea of Europe was based on the idea of cultural heritage. This idea began to gain popularity since the 1970s [17]. With the adoption of the 1972 UNESCO World Heritage Convention, this concept has entered the world stage [6]. Around the same time, in 1974–1975, the foundations of the European cultural policy in the field of heritage were laid, when the European Parliament adopted a number of recommendations in the field of preservation of cultural heritage [20]. Subsequently, the 2007 Treaty of Lisbon, which regulates changes in the EU governance system, also secured a clause on the preservation of cultural heritage [29].

Now, there is a distinct trend when the idea of a world cultural heritage is [21] gradually being replaced by alternative ideas of cultural diversity and intercultural dialogue, which are more and more firmly enshrined in the EU documentation on cultural policy. What reasons led to a change in the dominant ideas of the cultural policy of modern Europe? These changes were influenced by various reasons.

One of the reasons is that after geopolitical changes caused by the collapse of the socialist system and the gradual inclusion of new members in the EU, European culture has become very heterogeneous. As a result, "Old Europe" was replenished with new States that brought with them their national cultural characteristics.

Another reason is the nature of European culture itself, which traditionally included a variety of minority cultures, which, in the context of European integration, found themselves in a disadvantaged position. There is a discussion about this in Europe, the participants of which openly declare their fear of losing their regional and national cultural heritage and demand to preserve and respect it.

But perhaps the most significant was the third reason related to the large number of migrants who have arrived in Europe over the past decade, mainly from countries of the Third World. In 2015 alone, which turned out to be a record, about 1.2 million migrants have entered the territory of the EU countries.

Migrants defend their right to preserve national cultural traditions and characteristics, which should be taken into account in the system of pan-European values. In the early 2000s, tensions between the majority and ethnic and religious minorities, especially Muslims, have intensified in Europe. For example, the conflicts between the indigenous population and Asian migrants in the North of England in 2001, the interethnic and interfaith conflict in Denmark in 2005–2006 after the publication of the images of the Prophet Muhammad (the so-called "cartoon scandal"), disputes over the construction of mosques in Italy, Greece, Germany, France, and the UK in 2005, a referendum on the construction of minarets in Switzerland in 2009, a ban on the wearing of hijabs in France in 2009, an attack by migrants on police officers in French Calais in 2018, etc.

Numerous conflicts on ethnic grounds led to the collapse of the concept of multiculturalism, which was prevalent in Europe in the late 20th – early 21st centuries. During 2010–2012, the leaders of Great Britain D. Cameron, France—N. Sarkozy, and Germany—A. Merkel have announced their rejection of the policy of multiculturalism [15]. The policy of multiculturalism has shown its failure under the influx of a huge number of refugees into Europe. Designed to preserve and to defend cultural pluralism, the policy of multiculturalism was not able to solve new problems related to sociocultural, interethnic, and inter-confessional conflicts between migrants and the indigenous population of European countries. The situation was further complicated by the fact that migrants refuse to integrate into European society, to accept a European culture instead of their national one, and they settle in enclosed enclaves. Some ethnic groups are very radical and aggressive, they riot, attack the local population. In turn, most of the indigenous population of Europe does not seek to recognize as "full-fledged Europeans" migrants who defiantly do not want to integrate into European society. Thus, Europe has become the scene of numerous conflicts, which are based on ethnocultural differences, requiring the leadership of the EU to search for new ideological paradigms.

In turn, this necessitated the search for a different, more applicable to the realities of modern Europe, cultural ideology, which could replace the failed idea of multicultural Europe and reflect the diversity of cultures of modern Europe. Such an idea was the idea of cultural diversity, which gradually began to stand out as the basis for the concept of cultural policy of the EU.

5 Formation of the Idea of Cultural Diversity in the Cultural Policy of the European Union

Preservation, fostering, encouragement of cultural diversity is the most important principle of the external and the internal cultural policy of the European Union. Three dimensions are especially important for European cultural integration: the intra-European cultural diversity of the EU States in a single supranational entity; ethnic and religious minorities; mass external migration from non-European societies. The problems of cultural diversity in the economic sphere and in the field of minority protection are being updated. The EU is developing approaches to promote cultural diversity in the workplace, which are currently being implemented mainly in large companies.

The concept of cultural diversity is closely linked to globalization processes and is actively promoted by organizations such as UNESCO and the Council of Europe. The Universal Declaration on Cultural Diversity (2001) is the first international legal instrument to address this issue internationally. The Declaration recognizes cultural diversity as "the common heritage of mankind" and considers its protection to be the most important task of States and governments [30].

The UNESCO Declaration has formed the basis for the concept of EU cultural policy, reinforcing the notion that cultural cooperation is the "key to the successful implementation and completion of the European integration process" [2]. The European Agenda for the Development of Culture in a Globalizing World [13] (2007) has marked the beginning of the development of EU cultural strategies, has highlighted the interest in protecting human rights, including the promotion and protection of cultural rights, the rights of indigenous peoples, and the rights of persons belonging to minorities. The document emphasizes respect for the principle of subsidiarity and the decisive role of States in the cultural sphere, but already called the European Union a successful social and cultural project [5]. The most important task was the activation of intercultural dialogue at the local, regional, national, and international levels.

The EU has played an important role in the evolution of the discourse of cultural diversity, largely contributing to the emergence in 2005 of a legally binding document—the Convention on the Protection and Promotion of the Diversity of Cultural Expressions [7, 24: 203]. Cultural diversity is defined in it as "the diversity of forms by which the cultures of groups and societies find expression. These forms of self-expression are transmitted within and between groups and societies" [7]. The Convention has secured the sovereign right of States to develop and to implement their cultural policies and to take measures in order to protect and to promote the diversity of cultural expressions, while providing for the development of measures "to protect and to promote the diversity of cultural expressions" and "to create conditions for the flowering and free interaction of different cultures on a mutually beneficial basis". These measures required the creation by the participants of a special environment in their territory, as well as international cooperation.

All EU countries have approved, accepted, ratified the UNESCO Convention on the Protection and Promotion of the Diversity of Cultural Expressions or have acceded to it. This solidarity of the EU members regarding the Convention is associated with the recognition by these countries of cultural diversity and its growing importance in the process of European integration, as evidenced by the adoption by the Council of Europe of the Declaration on Cultural Diversity [22: 223]. The EU assistance in the implementation of the UNESCO Convention was to strengthen the role of culture in the process of European integration.

In the documents of the European Union defining its contemporary cultural policy, cultural diversity, its preservation, and promotion, it occupies an equally significant place as the problem of cultural heritage protection.

6 The EU Programs for the Conservation and Promotion of Cultural Diversity

The EU has developed effective mechanisms to promote the idea of cultural diversity and its practical support, including: the Culture program (2007–2013) to support projects and activities aimed at protecting and promoting cultural diversity and heritage [9]: Creative Europe European Framework Program (2014–2020)—a single mechanism for supporting creative activities in the field of culture and the audiovisual sector (it combined the Culture and Media programs, which had previously been separately funded by the European Commission); the Culture Work Plan 2015–2018 (the third consecutive plan for the implementation of the European Agenda for Cultural Development in a Globalizing World), which emphasized an accessible and inclusive culture [23]; the Europeans program; the 7th EU framework program (7RP) and the Horizon 2020 program.

According to the decision of the European Parliament and the EU Council, which established the Culture program, its goal was to strengthen the European cultural space based on a common cultural heritage by supporting cooperation between cultural participants. The program also had to "contribute to the elimination of all forms of discrimination based on gender, racial, or ethnicity, discrimination based on religion, belief, disability, or sexual orientation" [10]. The protection, development, and promotion of European cultural and linguistic diversity have become the main goal of the Creative Europe program.

7 The "New Cultural Agenda" of the EU and Principles of Cultural Diversity

The EU has traditionally been firmly committed to promoting cultural diversity as an integral part of human rights and freedoms [18: 3–4]. The new European cultural agenda, which began with the adoption in 2018 of two important documents—the New European Cultural Agenda [1] and the Cultural Work Plan for 2019–2022 [11],—includes the issue of preserving cultural diversity among the main priorities both at the intra-European level and in cooperation with partner countries.

The New European Cultural Agenda defines the possibilities of culture in the social, economic, and foreign policy dimensions. The meaning of the New Agenda is to make better use of culture, cultural heritage, and cultural diversity as a way of creating identity and cohesion, a driving force for socioeconomic development and a factor directly contributing to the peaceful relations of Europe with other countries. The Culture Work Plan 2019–2022 identifies five key priorities for European cooperation in cultural policy, including the sustainability of cultural heritage and cultural diversity. Moreover, as noted in the Plan, the implementation of cultural policy should be based on specific actions with clearly defined working methods and target results. The promotion of cultural diversity is also central to the external relations of the EU.

The principle of preserving cultural diversity has become one of the guiding principles of the new EU strategy for international cultural relations, adopted in 2016. Cultural and linguistic diversity is recognized as a key asset of the European Union, and its protection and promotion are central to cultural policy at the European level. Updating this issue requires a holistic and horizontal approach, including taking into account legislation, financing, regular dialogue between EU member States, European institutions, and civil society, as well as thematic cooperation with international organizations and partner countries [18: 4].

Focusing on cultural diversity has become a central aspect of the renewed cultural policy of the EU both at the domestic and foreign policy levels. The protection of cultural diversity is seen as one of the human rights, as a way to restore social cohesion and to resolve crises, especially in protracted conflict situations. Respect for cultural diversity is also important for reconciliation, national dialogue, and recovery [8].

The adoption of the European Agenda for the Development of Culture in a Globalizing World and the entry into force of the UNESCO Convention on the Protection and Promotion of the Diversity of Cultural Expressions, as well as a number of documents characterizing the new EU Cultural Agenda, have opened a new stage in European cultural integration related to the development of the EU strategy to support and to preserve cultural diversity. Measures implemented within the framework of this strategy have become an additional instrument of protection of the rights of ethnic minorities adopted by the EU standards. The problems of the rights of ethnic minorities with an emphasis on the possibilities of digital technologies, the preservation of regional minority languages is reflected in the content of the programs and actions presented in the reports of the European Union on measures to implement the UNESCO Convention. Human rights issues have gained particular importance in the EU foreign cultural policy.

8 Conclusions

The current problems that Europe faces are related not only to the cultural mosaic, which is traditionally inherent in European culture, but also to the growing manifestations of acts of racism and xenophobia against immigrants.

Now, an analysis of the concept of cultural diversity in Europe and related issues occupies an important place on the EU political agenda and it is the most discussed. As the analysis of the main documents of the EU cultural policy showed, a gradual transition to the idea of cultural diversity as the dominant idea of the EU cultural activity took place against the backdrop of a migration crisis and the ensuing crisis of multiculturalism policy. The idea of cultural diversity is in line with basic European values. This allowed us to expand the concept of European heritage, which went beyond the diversity of national cultures and includes a significant cultural segment associated with the culture brought by numerous migrants. The preservation of cultural diversity is perceived as "an opportunity to humanize the ideas of Europe". Since the mid-2000s, cultural diversity has been presented as a value that needs to be promoted and as a factor of social cohesion. The introduction of the cultural diversity concept into the EU cultural policy is a political response to the growing intra-European diversity resulting from international migration and an attempt to use soft means to resolve the conflicts that arise in modern Europe on the basis of ethnic, religious, and cultural differences.

It is obvious that in order to support cultural diversity and differences in Europe itself, both national governments and European institutions have to take certain measures. For instance, education programs of schools, colleges, and higher education institutes should include courses that explain a "new cultural picture of Europe". Various informal approaches, promoting the idea about cultural differences, cultural diversity, and equality of cultures, could be effective on the all-European level. Such approaches can include festivals, exhibitions, performances, and other pan-European events that strengthen the European dialogue of cultures.

Mutually important events in European history, such as memory dates, secular and religious holidays that unite the people of Europe, may become the basis for efficient cooperation. Sport achievements and sport events, based on universal values could also be unitive. Humanitarian aspects of modern life can play the same role, the issues of inclusiveness that are important for all humanity. Europe can become a cutting edge of such ideas and solutions.

On the other hand, for a successful union, the idea of cultural diversity should be supported. This can be practically realized by holding days or weeks of national cultures in the format of multi-genre festivals, the program of which can represent both spiritual and material heritage of various cultures, that became an important part of the modern global civilization. Such events could be held both in traditional and on-line formats, which would drastically increase their audience.

It is important to engage the young audience in these events, as future stability and prosperity in Europe depend on it. In this case, special attention should be paid to innovative approaches and specific formats of cultural events. It is necessary to deploy the potential of creative industries, search for distinctive approaches even to the practice of holding the traditional events. Along with that, it should be understood that all this activity would require both organizational and financial efforts. We presume that the theoretical basis for such practice can be the concept of cultural interpretation, which improves understanding of different cultural forms and materializations. Due to the fact that the cultural space of Europe is the result of the co-existence of societies with different local cultures, the problem of their correct interpretation gains not only theoretical, but also practical significance.

References

1. A New European Agenda for Culture. https://ec.europa.eu/culture/sites/culture/files/commission_communication_-_a_new_european_agenda_for_culture_2018.pdf. Accessed 03 May 2020
2. Belyaeva, E.E.: Cultural Integration as the Main Strategy of the European Union. Moscow City University Publ, Moscow (2012).[in Russian]
3. Browse the Lists of Intangible Cultural Heritage and the Register of good safeguarding practices. https://ich.unesco.org/en/lists?multinational=3&display1=regionIDs#tabs. Accessed 24 Apr 2020
4. Calligaro, O.: From 'European cultural heritage' to 'cultural diversity'? The changing core values of European cultural policy. Politique européenne **45**(3), 60–85 (2014). https://doi.org/10.3917/poeu.045.0060
5. Communication from the Commission to the European Parlament, the Council, the European Economic and Social Committee and the Committee of the Regions on a European agenda for culture in a globalizing world. https://eur-lex.europa.eu/LexUriServ/LexUriServ.do?uri=COM:2007:0242:FIN:EN:PDF. Accessed 31 Mar 2020
6. Convention concerning the Protection of the World Cultural and Natural Heritage, 16 November 1972, https://whc.unesco.org/en/conventiontext/. Accessed 28 Apr 2020
7. Convention on the Protection and Promotion of the Diversity of Cultural Expressions, https://en.unesco.org/creativity/sites/creativity/files/passeport-convention2005-web2.pdf. Accessed 29 Apr 2020
8. Cultural Diversity under Attack High-level meeting and technical conference**,** Brussels, 9–10 June. https://en.unesco.org/cultural-diversity-under-attack-2016. Accessed 10 June 2020
9. Culture programme (2007–2013). https://ec.europa.eu/programmes/creative-europe/previous-programme/culture_en. Accessed 05 Apr 2020
10. Decision No 1855/2006/EC of the European Parlament and of the Council of 12 December 2006 establishing the Culture Programme (2007 to 2013). https://www.eumonitor.eu/9353000/1/j4nvhdfcs8bljza_j9vvik7m1c3gyxp/vi8rm2znu2y7. Accessed 05 Apr 2020
11. Draft Council conclusions on the Work Plan for Culture 2019–2022. http://data.consilium.europa.eu/doc/document/ST-13948-2018-INIT/en/pdf. Accessed 04 May 2020
12. Egoreichenko, A.B.: Modern European culture: features, values, development trends [in Russian]. Soc.: Philos. History Culture **9**, 80–83 (2018). https://doi.org/10.24158/fik.2018.9.14
13. European agenda for culture in a globalising world. https://eur-lex.europa.eu/legal-content/EN/TXT/HTML/?uri=LEGISSUM:l29019. Accessed 18 May 2020
14. Ferris, G., Frink, D., Galang, M.C.: Diversity in the workplace: the human resources management challenge. Hum. Resource Plan. **16**(1), 41–51 (1993). https://www.questia.com/library/journal/1G1-16108086/diversity-in-the-workplace-the-human-resources-management. Accessed 16 May 2020
15. Fokin, V., et al.: Multiculturalism in the modern world. Int. J. Environ. Sci. Educ. **11**(18), 10777–10787. http://www.ijese.net/makale/1464. Accessed 01 June 2020

16. Habibul, H.: Glocalization as globalization: evolution of a sociological concept. Bangladesh e-Journal Sociol. **1**(2), 1–9 (2004). https://www.academia.edu/28585527/Glocalization_as_Globalization_Evolution_of_a_Sociological_Concept. Accessed 28 Apr 2020
17. Hafstein,V.T.: Cultural heritage. In: Bendix, R.F., Hasan-Rokem, G. (eds.) A Companion to Folklore, pp. 501–519. Blackwell Publishing Ltd, Oxford (2012). https://doi.org/10.1002/9781118379936.ch26
18. Joint Communication to the European Parliament and the Council 'Towards an EU strategy for international cultural relations', 8 June 2016. https://eur-lex.europa.eu/legal-content/EN/TXT/PDF/?uri=CELEX:52016JC0029&from=EN. Accessed 23 Apr 2020
19. Mattelart, A.: Bataille a l'UNESCO sur la diversite culturelle. Le Monde diplomatique, 26–27 Octobre (2005). https://www.monde-diplomatique.fr/2005/10/MATTELART/12802
20. Most relevant documents of the European Union concerning cultural heritage. https://www.coe.int/en/web/herein-system/european-union. Accessed 05 May 2020
21. Nikolaeva, J.V., et al.: World cultural heritage in the context of globalization: trends, issues and solutions. Int. J. Sci. Technol. Res. **8**(10), 842–845 (2019). https://www.ijstr.org/paper-references.php?ref=IJSTR-1019-23363. Accessed 18 May 2020
22. Obuljen, N., Smiers, J. (eds.): UNESCO'S Convention on the Protection and Promotion of the Diversity of Cultural Expressions: Making it Work. Institute for International Relations, Zagreb (2006). http://www.culturelink.org/publics/joint/diversity01/Obuljen_Unesco_Diversity.pdf. Accessed 30 Mar 2020
23. Oreshina, M.A.: About some aspects of EU cultural policy [in Russian]. Int. Aff. **12**, 152–170 (2017)
24. Psychogiopoulou, E. (ed.): Cultural governance and the European union. PSEUP, Palgrave Macmillan UK, London (2015). https://doi.org/10.1057/9781137453754
25. Rosado, C.: What makes a school multicultural. Adventist J. Educ. **16**, 41–44 (1997). http://circle.adventist.org/files/jae/en/jae199760014104.pdf. Accessed 16 May 2020
26. Starr, A., Adams, J.: Anti-globalization: the global fight for local autonomy. New Polit. Sci. **25**(1), 19–42 (2003). https://doi.org/10.1080/0739314032000071217
27. The 1972 World Heritage Convention. http://whc.unesco.org/archive/convention-en.pdf. Accessed 01 May 2020
28. The World Heritage List Statistics. http://whc.unesco.org/en/list/stat. Accessed 23 Apr 2020
29. Treaty of Lisbon amending the Treaty on European Union and the Treaty establishing the European Community, signed at Lisbon, 13 December 2007. Official J. Eur. Union **50**, 1–10 (2007). https://eur-lex.europa.eu/legal-content/EN/TXT/PDF/?uri=OJ:C:2007:306:FULL&from=EN. Accessed 06 May 2020
30. Universal Declaration on Cultural Diversity. http://portal.unesco.org/en/ev.php-URL_ID=13179&URL_DO=DO_TOPIC&URL_SECTION=201.html. Accessed 30 Apr 2020
31. Vodopyanova, E.V. (ed.): European Culture: XXI century [in Russian]. Nestor-Istorija, Saint-Peteresburg, Moscow (2013)

The Cultural Expansion of the Persian Gulf States in Europe: Dynamics, Specifics, and Consequences

Regina Biktimirova[1], Konstantin Eidemiller[1](✉), Andrey Anufriev[1], Yury Gladkiy[2], and Valery Suslov[2]

[1] Saint Petersburg State University, Saint Petersburg, Russia
[2] Herzen State Pedagogical University of Russia, Saint Petersburg, Russia

Abstract. Throughout the centuries, irrespective of political changes inside the Muslim and Western worlds, the interaction between these two systems always remained. However, the two civilizations stay in a constant state of antagonism and competition which entail deep misunderstanding and confrontation at all levels in both of these societies. Nowadays, the Gulf states, which dispose of vast hydrocarbon reserves, have impressive financial resources and influence on the international political arena. Such concepts as "The Clash of Civilizations", "Orientalism", etc. inevitably entail new ones, for instance "Arabic alter-globalization", "Arabic version of globalization" and "the cultural expansion (of the Gulf states)". The Gulf states understand the potential weight of the Muslim population of Europe and strive to spread their religious influence and win patronship over the Muslim communities of Europe by various means: sponsoring the creation of mosques and Islamic cultural centers in Europe, educating imams and Muslim scholars who could later promote the principles of the Wahhabi Islam, supporting Muslim infrastructure, Halal industries, the popularization of Muslim culture in European countries, etc. These projects are not only conducted for charitable reasons but practical as well – the Persian Gulf elites, aware of the constant threat of coup d'état and new armed conflicts on religious basis, are forced to search for new potential living environments, and Europe, with its liberal values, open internal borders and large Muslim communities, is an ideal candidate. Therefore, we can conclude that the policy of cultural expansion of the Persian Gulf states is transforming rapidly, adjusting to the ever-changing reality of these states and the world. Moreover, it is obvious that the cultural expansion of the Gulf states has a major impact on the Muslim communities of Western Europe – thereafter, its transformation in the future will inevitably entail changes in the lives of Muslim communities of Europe and the world.

Keywords: Cultural expansion · Multiculturalism · The Persian Gulf states · Europe · Muslim communities · Qatar

© The Author(s), under exclusive license to Springer Nature Switzerland AG 2021
R. Bolgov et al. (Eds.): *Proceedings of Topical Issues in International Political Geography*, SPRINGERGEOGR, pp. 382–394, 2021.
https://doi.org/10.1007/978-3-030-78690-8_33

1 Introduction

The history of interaction between the Muslim world and the Western civilization dates back to the emergence of the first Islamic statehood: during the existence of the Rashidun Caliphate only, the Arab conquests in Europe reached the territory of the modern Southern France, Northern Italy, Greece and the North Caucasus. Throughout the centuries, irrespective of political changes inside the Muslim and Western worlds, the interaction between these two systems always remained [1]. However, the two civilizations stay in a constant state of antagonism and competition which entail deep misunderstanding and confrontation at all levels in both of these societies. This image of "The East" in the eyes of indigenous population of Europe was explained in 1978 by Edward W. Said in his renowned book "Orientalism" [2]. The theory of orientalism depicts the attitude that exists in Western culture towards Muslim culture – it is perceived as strange, poorly developed, practically antagonistic.

The image of "the Stranger" [3], passed throughout the centuries, increases its hold on the minds of members of conservative European societies amid the unceasing armed-conflicts on religious basis in the Muslim world, along with their consequences, such as migration crises and increasing migration flows, which are directed towards Europe [4]. The problem of harmonious coexistence of diverse communities within one state was meant to be solved by the implementation of the multiculturalism policy, but it failed, allowing the emergence of parallel societies in Europe – i.e. Muslim communities – with their own culture, religion, traditions and worldview. The rich states of the Persian Gulf took it upon themselves to provide for the religious and infrastructural needs of these communities.

Nowadays, the Gulf states, which dispose of vast hydrocarbon reserves, have impressive financial resources and influence on the international political arena. Such concepts as "The Clash of Civilizations", "Orientalism", etc. inevitably entail new ones, for instance "Arabic alter-globalization", "Arabic version of globalization" and "the cultural expansion (of the Gulf states)". The Gulf states understand the potential weight of the Muslim population of Europe and strive to spread their religious influence and win patronship over the Muslim communities of Europe by various means: sponsoring the creation of mosques and Islamic cultural centers in Europe, educating imams and Muslim scholars who could later promote the principles of the Wahhabi Islam, supporting Muslim infrastructure, Halal industries, the popularization of Muslim culture in European countries, etc. These projects are not only conducted for charitable reasons but practical as well – the Persian Gulf elites, aware of the constant threat of coup d'état and new armed conflicts on religious basis, are forced to search for new potential living environments, and Europe [5], with its liberal values, open internal borders and large Muslim communities, is an ideal candidate. All of the abovementioned realities determine the relevance of this study.

What distinguishes this paper from most of other works on this subject is that it does not only include the depiction of the phenomenon of the cultural expansion of the Persian Gulf states in Europe along with its premise and consequences, but also provides an unprecedented periodization of this phenomenon with its outcomes to this day and gives a prognosis of its potential development. It is worth mentioning that this subject in particular is yet to be studied in detail, as most of the researchers usually tend to explore either the investment policy of the Gulf states or the position and growth rates of Muslim communities of Europe, but hardly ever are these factors considered altogether and in their interdependence. Therefore, this paper studies the cultural expansion as an instrument of religious and economic influence of the Gulf states in Europe.

2 Literature

The most valuable materials and documents cited in this paper can be found on official governmental websites of Great Britain and the USA (GOV.UK and Law Library of Congress), which contain vast statistical data. Other valuable resources include news articles of such publications as The New York Times, BBC, CNN, etc. Moreover, the paper is based on fundamental works of researchers from all over the world in various fields including international relations, history, philosophy, economy, political and social sciences, i.e. "Jihad: The Trail of Political Islam" (Gilles Keppel) [6], "Multiculturalism that never existed" (A.V. Veretevskaya) [7], as well as research papers created with the assistance of Rand Corporation [8], Network of European Foundations, reports and prognosis of Pew Research Cente [9–12], research conducted by Amnesty International [13].

3 Research Methods

The topic of the cultural expansion of the Persian Gulf countries into Europe is not new in itself, but our main task of this work was to analyze and systematize this process. Our main goal of the research is to determine the role of the multiculturalism policy in the process of cultural expansion of the Persian Gulf states in Europe. In the study, we used the historical method, systems approach, comparative analysis, analytical data, classification, expert assessment method, sociological and psychological method, as well as complex methods of classical theoretical research.

4 The Cultural Expansion: Main Stages and Key Characteristics

After five years since the beginning of the European migration crisis, when the number of Muslim population of Europe is close to reaching 50 million people, there are challenges that affect almost all aspects of life of European society. Not once throughout its long history has Europe demonstrated such high numbers of Muslim population. Today there is no European state that does not have a Muslim community on its territory except for the Vatican. Numerous projects of potential Grand mosques are being negotiated with the authorities of the capitals of Estonia, Latvia, Finland and even Iceland – European states that are significantly distant from the countries of the Muslim world; the Gulf states, Turkey and – at one point – Iran compete for the chance to finance and lead these projects. This half-a-century-long phenomenon is also influenced by almost fifteen hundred years of history of interaction between Europe and the Muslim world. However, it is worth focusing on the events occurring in the Contemporary period and beginning in the 1960s.

The death of the founder of Saudi Arabia King Abdulaziz ibn Abdul Rahman Al Saud, when power was transferred to his sons, coincided with two major factors: social and demographic changes in Europe occurring in the end of the 1960s, and the emersion of a new role of Saudi Arabia in regional and global politics. By the early 1960s, Saudi Arabia, unaffected by the devastation of World War II, began rapid economic growth while practically becoming an oil resource base for the United States and Europe for the next 50 years. The questions of survival and emergence of statehood on controlled territories had been solved by the founder of Saudi Arabia (including most of other states in the region which had gained independence through liberation wars first with the Ottoman Empire, then Great Britain and France), and the end of the 1950s was defined by two key factors: the first concerned final formation of the economy of the Gulf states and the second was the acute need of developed countries in their export raw materials – oil and gas. However, soon enough there appeared to be another export product sent to the United States and Europe – Islam, or so-called "Petro-Islam". From the reign of King Fahd and to this day, almost half a century later, Islam remains the second export product after hydrocarbons. The very first grand projects of mosques in Europe were sponsored mainly by four Persian Gulf states: Saudi Arabia, the UAE, Qatar, Bahrain and to a lesser extent Iran. Turkey also made a significant contribution in this field, however a lot lesser than the monumental cultural expansion of the Gulf states.

The first stage of the cultural expansion of the Gulf states in Europe can be dated back to the dissolution of the Ottoman Empire, when a number of Middle Eastern countries acquired independence, including from the European states. It started with the reinforcement of Saudi Arabia in the mid-1930s and continued until the late-1960s. At this stage, the Gulf states demonstrated their willingness to sponsor and support the creation of mosques in Europe for the first time; however, this stage is mostly characterized by conclusion of contracts and decision-making rather than practical implementation of projects. This period was accompanied by wars in Europe as well as serious unrest in the countries of the Muslim world – these circumstances surely slowing down the delivery of negotiated projects. Moreover, significant numbers of population of Muslim countries (most of them highly qualified) were being forced to leave their homes in dominions and protectorates, which were rapidly turning into independent states, and head to ex-metropolis in search of a better life.

London Central Mosque became the embodiment of this period. The reason of its creation was introduced by the British government as a gift to the Muslim community of Great Britain so that it could have its own proper place of worship. Despite the fact that Winston Churchill's War Cabinet authorized the acquisition of a potential mosque site in London on October 24, 1940, the construction itself only began 30 years later and lasted from 1974 to 1994. The project acquired financial support of a few generations of Muslim leaders, i.e. King Farouk I of Egypt and King Abdulaziz ibn Abdul Rahman Al Saud donated the total of £2 million at different times, most of other costs were covered by Sheikh Zayed bin Sultan Al Nahyan, ruler of Abu Dhabi and President of the United Arab Emirates. Total construction value amounted to £6,5 million. The mosque is comprised of two prayer halls and a minaret. The whole structure can accommodate more than 5,000 people. Al-Azhar University in Cairo delegated the first imam of London Central Mosque and continued to do so for the next decades [14].

The main characteristic of the second stage of the cultural expansion is the beginning of creation of Islamic cultural centers as substantive self-contained constructions in different countries of Europe, predominantly Western Europe. This was a rather forced decision amid the increasing migration flows of low-cost labor from Muslim-majority countries to Europe. The time period of this stage covers the 1970–1980s. One of the very first mosques created during this period was the Great Mosque of Brussels. The second stage of the cultural expansion begins with Saudi Arabian King's 1967 European tour, when European leaders were seeking to conclude profitable treaties with Saudi Arabia concerning oil export [15]. Thus in 1967 the third King of Saudi Arabia Faisal bin Abdulaziz Al Saud, during his official visit to Belgium, received an offer from the Belgian King Baudouin to rebuild one of the buildings (the Oriental Pavilion of the National Exhibition) in Brussels into a mosque that would accommodate the religious needs of the Muslim community of Belgium which was increasing in numbers due to the fact that Belgium was inviting Turkish and Moroccan workers to work on its territory. The site was granted by the local authorities rent-free for the 99-years duration of use. After a long ten-year period of construction, sponsored by Saudi Arabia, the mosque (simultaneously being an Islamic cultural center) was opened in 1978 in the presence of King Baudouin and the forth King of Saudi Arabia Khalid bin Abdulaziz Al Saud. Another important characteristic of this period is that the Persian Gulf states, primarily at the time Saudi Arabia, did not use to build new mosques in large numbers but rather provided financial support for the existing Muslim communities of Europe, helping them pay rent, lease and purchase premises necessary for the creation of prayer rooms (musallas), which would accommodate their religious needs, and also making sure to ensure full equipment and security of these places of worship. Sometimes it would come to a point when almost all the other objects in these facilities (apart from water and electricity), located in the center of Europe, had been brought from elsewhere – from carpets to lighting appliances [16].

However, amid all the cases illustrating the cultural expansion of the Gulf states at this stage, one that really stands out is the Spanish case. In 1975, a few months after the end of the Francoist dictatorship, a decision was made to build the first great mosque in Madrid since its conquest by the forces of Alfonso VI the Brave in 1085. The Madrid Central Mosque was opened in 1988 and named after caliph Abu-Bakr. This mosque also serves as the headquarters of the Union of Islamic Communities of Spain and the Islamic Community of Madrid. In 1976, a year after the beginning of this project, 18 Muslim countries, which had diplomatic missions in Spain, signed an agreement with the Spanish government on the creation of a grand Islamic cultural center in Madrid. However, the implementation of this project was postponed for 11 years until 1987 when the Saudi Arabian King Fahd bin Abdulaziz Al Saud granted the necessary funding for the construction. After 5 years of work, Salman bin Abdulaziz Al Saud then prince and now the seventh King of Saudi Arabia alongside King Juan Carlos I officially opened the Islamic Cultural Center of Madrid. Saudi Arabia immediately began to control the cultural center upon its opening, which caused concern in Madrid even then; in fact, Saudi Arabia was the administrative and spiritual leader of this center until 2001 [17]. The spiritual leadership is currently held by an Egyptian sheikh Monair Mahmud, however this does not change much – religious life in Egypt is almost entirely dependent on grants, donations and direct financial investments from the Persian Gulf states [18].

The third stage of the cultural expansion, the time frame of which covers the years 1990–2000, is characterized by construction of mosques created for more symbolic reasons rather than practical, with potential capacity for the future. This is demonstrated by the creation of the Great Mosque of Gibraltar and the Great Mosque of Rome. The Ibrahim-al-Ibrahim Mosque, also known as the King Fahd bin Abdulaziz al-Saud Mosque or the Great Mosque of Gibraltar was opened on August 8, 1997. The edifice was a gift to the British Overseas Territory from the fifth King of Saudi Arabia Fahd bin Abdulaziz Al Saud. The mosque was built rather quickly in just two years, the whole construction process costing around £5 million. The mosque constitutes a full-fledged Islamic cultural center equipped with all the necessary facilities and infrastructure – i.e. six class rooms, a conference hall, a library, a kitchen, a bathroom, accommodations for maintenance staff, offices for administrative purposes and even a morgue [19]. This is the only mosque in Gibraltar capable of accommodating more than 2,000 worshippers inside and around 7,000 on its entire premises, even though the Muslim community of Gibraltar is comprised of around 1,000 people – which constitutes approximately 4% of the overall population of Gibraltar [20]. The Great Mosque of Gibraltar is one of the largest and the southernmost mosque of continental Europe.

The story of the creation of the Great Mosque of Rome is also noteworthy in this context [21]. The Church ceased to object after the Second Vatican Council in 1965. In 1970s, the Italian government issued a construction permit of an Islamic cultural center, and the municipal government of Rome granted a construction site of 7,5 acres and the cost of $40 million. It was King Faisal bin Abdulaziz Al Saud of Saudi Arabia who managed for the first time to obtain permission to build a mosque in Rome in 1973, when Catholicism was still the official religion of Italy. The construction took 20 years (from 1984 to 1995), was led by a few groups of architects and cost around $50 million, 80% of which was provided by Saudi Arabia [22]. The mosque was opened on June 21, 1995, and at the time of its opening the mosque was considered the biggest one in continental Western Europe. It can accommodate up to 12,000 people. Saudi Arabia continues to cover a significant amount of the mosque's operating costs. At the same time, the implementation of another project was finished – the Edinburgh Mosque and Islamic cultural center [23]. Prior to this construction, there were no mosques in Edinburgh, however, since the Muslim community of Edinburgh kept growing in numbers, a grand mosque became a necessity. The implementation of all construction plans was stalled due to financial difficulties even after the acquisition of the land, but the deadlock was overcome when King Fahd bin Abdulaziz Al Saud of Saudi Arabia was made aware of the problem and donated 90% of the overall project cost. The mosque was opened on July 31, 1998 in the presence of his son, Prince Abdullah bin Abdulaziz Al Saud who was also the patron of the construction project.

The fourth stage of the cultural expansion can be characterized by the policy of "religious outsourcing". Its time frame covers the years 2000–2015. During previous stages, Saudi Arabia itself would handle the construction and maintenance of mosques and Islamic cultural centers, but at this stage it began to introduce other players to assist in these activities, those that it had been preparing for years throughout the countries of the Muslim world. Along with conducting a policy of cultural expansion in Europe, Saudi Arabia was also investing impressive amounts of money in Wahhabi educating centers in Pakistan, Egypt, Jordan, Libya and other Muslim countries in order to prepare religious leaders that would spread the ideology of Wahhabi Islam in their home regions and all over the world. At a rough estimate, more than 7,000 people were educated by the Gulf states in these centers in Middle Eastern and South Asian countries in order to provide for the needs of mosques and musallas in Europe (and other countries outside the Muslim world who had fast-growing Muslim communities on their territories, such as South and North American countries [24], Australia, New Zealand, etc. [25]). However, it is worth mentioning that after the 9/11 terrorist attacks in the USA and the subsequent terrorist attacks in the countries of Western Europe, the Gulf states began to transfer ownership and control over mosques to local European authorities and Muslim communities, formally declining their patronage which in fact was left in the hands of imams that they had prepared and educated in advance.

At this stage, the Gulf states started constructing places of worship in the areas of Europe most remote from the Muslim world – primarily, the countries of Northern Europe [26]. One of such examples is the Stockholm Grand Mosque, re-constructed from a building of an old electric power station. The construction work began in 1996, however, it had to be postponed significantly even at the initial stages of the project's implementation due to numerous appeals and protests on behalf of the Swedish citizens who did not wish to see a mosque in their country. For that reason, extensive construction only began in 1999 and was finished in 2000. The project was implemented at the expense of financial support of the UAE government, and the mosque itself was named after its main sponsor Sheikh Zayed bin Sultan Al Nahyan, the first President of the UAE. The mosque can accommodate up to 2,000 prayers simultaneously, and matches the characteristics of a full-fledged Islamic cultural center. To this day, it is the biggest mosque in Stockholm [27].

The fifth stage of the cultural expansion of the Gulf states in Europe began in 2015, when the richest monarchies of the Persian Gulf decided to assume responsibility to provide for all the religious infrastructure of Muslim communities in those countries of Western and Northern Europe that received the largest numbers of Muslim immigrants during the European migration crisis. Moreover, this is the period of the most ambitious and controversial projects of Muslim infrastructure, during which mosques are being built in areas where Muslim population is not even considered to be a numerous minority. The most debated construction projects of this period are those of the Grand Mosques of Tallinn and Riga (both with the financial support of the UAE, with the first also receiving funding from Turkey), along with the construction of the Grand Mosque of Helsinki (funded by Bahrain) and the construction project of the Grand Mosque of Reykjavik (with financing from Saudi Arabia). All of these projects, like many other Muslim infrastructure-related projects in Europe, caused heated debates in European countries, which undoubtedly affected the timelines of delivery of these projects.

Therefore, we can conclude that to this day, the Gulf states in fact practice the policy of cultural expansion in European countries through the funding of construction of religious objects along with providing for their maintenance and needs – for which they have all the means and capability. However, lately the situation has been changing drastically: since mid-2017, when Mohammed bin Salman bin Abdulaziz al-Saud was appointed the new crown prince of Saudi Arabia, the attitude towards Islam and specifically Wahhabism as state religion started to change. Within the framework of "Saudi Vision 2030" development program, the crown prince has enhanced the rights of women and introduced reforms to the religious police which can work only during office hours now, no longer has the right to detain or make arrests and has to report to civil authorities [28]. Moreover, the Committee for the Promotion of Virtue and the Prevention of Vice, to which the religious police is accountable, can no longer restrict women's right to drive vehicles and cannot prevent them from attending sports events [29]. Mohammed bin Salman says that there shall be no place for xenophobia, gender inequality and religious discrimination in the future Saudi Arabia.

On January 25, 2020 the former Minister of Justice of Saudi Arabia Muhammad bin Abdul Karim Issa announced that Saudi Arabia will seize the funding of mosques in foreign countries [30]. This announcement is not unexpected due to the fact that for many years, ever since the 9/11 terrorist attacks in the USA and a subsequent series of terrorist attacks in Western Europe (in Madrid, London, etc.), Saudi Arabia has started to gradually cede control over the mosques to local authorities. For almost fifty years, Saudi Arabia was preparing Islamic leaders who could profess and promote its ideals; hence the formal gesture of refusal of control and patronage over the mosques is one of no crucial meaning. Saudi Arabia is currently changing its priorities from foreign policy to domestic policy, thus conceding its leadership in the field of cultural expansion to its neighbors and regional competitors – the UAE and Qatar – with whom it used to be in a competition for the chance to gain influence over the Muslim communities of Europe over the last fifty years.

5 Specificity and Consequences

What are the results of all the developments having unfolded in the countries of Western Europe from 1973 to 2020, within the framework of the policy of cultural expansion, or Petro-Islam? Throughout the years, from 1975 to 2020, more than 2,000 mosques, several hundred full-fledged Islamic cultural centers with their proper infrastructure and countless musallas have been built all over the world, financed by Saudi Arabian state funds. The Muslim World League, based in Saudi Arabia, has been playing an innovative role in its support of Muslim associations, mosques and investment plans for Muslim projects for the future "in all the Muslim-populated areas of the world" [31].

Saudi Arabia-funded mosques, wherever they were built, would most of the time be designed against local Muslim architectural traditions, following strict Wahhabi style, decorated with marble and neon lightning. One of the most notorious scandals related to this issue is the restoration of the Grand mosque of Sarajevo, Gazi Husrev-beg Mosque, which was greatly damaged during the Bosnian War (1992–1996). Upon the restoration, which was financed and supervised by Saudi Arabia, the mosque was stripped of its decorative Ottoman tiles and painted wall decorations, to the disapproval of the local Muslim population [32].

Out of those 2,000 mosques, built at the expense of Saudi Arabian government funds, approximately 200 are located in Western Europe. Sometimes the mosques built in certain areas would exceed in scope the actual needs of local Muslim communities – which is demonstrated by the example of Great Britain and the countries of the British Commonwealth. Saudi Arabia financed the construction of almost all the mosques that have been built in the Commonwealth over the past 50 years: from Gibraltar to Edinburgh, from Canada to Australia. Saudi Arabian universities and other religious institutions have prepared countless Muslim scholars, missionaries and imams, who all share the same goal – encouraging Muslims to revive the Wahhabi Islam in their homelands and abroad, along with spreading orthodox values that would sometimes be objected even by Muslim communities and societies themselves. Nowadays, there are Muslim leaders trained by Saudi Arabia all over the world, and they are preaching about the need to rid the religious practices of heretical innovations (bid'ah) and impose strict moral in their societies. According to researcher David Commins, there are approximately 250,000 such people [33].

However, the Wahhabi ideology is currently undergoing transformation inside Saudi Arabia itself. It is generally considered that the discussion on the role of Wahhabism as state religion commenced upon the realization of Saudi Arabia's elite on what they have achieved in their path and actions. This revelation was expressed in the speech of a former imam of the Great Mosque of Mecca (al-Masjid al-Ḥaram), Adel Al Kalbani who said that "the leaders of ISIS drew their ideas from our own books; exploited our own principles, that can be found in our books, among us [Saudi Arabia]" [34]. Therefore, the fact that Saudi Arabia is currently in the process of reconsideration of its own policy of cultural expansion raises a question of who will step in its place and whose concepts could gain momentum.

As it was mentioned above, the main regional competitors of Saudi Arabia during the past 50 years have been the UAE, Qatar and Iran. Due to various factors and circumstances, the UAE has recently become one of Saudi Arabia's ideological partners. Iran was eliminated from the competition for the chance to exert influence over Muslim communities of Europe because of its failure to offer a convenient and acceptable model of progressive Islam. The model that it had been relying on for the last 25 years only resulted in economic turmoil, leading the country into political and epidemiological crisis. Qatar, on the other hand, unlike its regional competitors, has managed to grasp the attitudes, prevailing in the Muslim world and beyond, and started investing in different religious projects. Qatar provided support for the movement of the Muslim Brotherhood throughout the countries and regions of Middle East and North Africa (such as Egypt, Libya, Tunisia, Gaza Strip, etc.), invested in the promotion of the concept of Euro-Islam (in Western Europe) while simultaneously being one of the sponsors of ISIS.

French journalists Christian Chesnot and Georges Malbrunot provide exhaustive evidence on the subject in their book "Qatar Papers: How Doha finances the Muslim Brotherhood in Europe". Their research cites data from bank transfers, checks and official letters obtained from an informant along with documents from the database of the Qatar Foundation led by Sheikha Moza bint Nasser and Qatar Charity led by the relatives of her son, Emir Sheikh Tamim bin Hamad al-Thani. According to these numerous documents, a total of 140 projects in Europe, mostly mosques and Islamic cultural centers, were directly financed by Qatar since 2008 to 2016. Qatar's area of interest includes the countries of Northern Europe, Benelux, Great Britain and France. Qatar financed the construction of an Islamic cultural center on one of the Channel Islands, the Bailiwick of Jersey, even though its Muslim community is very small in numbers. Qatar fully funded the construction of the Islamic cultural center of Lausanne, Museum of Islamic civilizations in La Chaux-de-Fonds in the Swiss canton of Neuchâtel and the Islamic cultural center of the city of Bienne in the Swiss canton of Bern, as well as The Hamad Bin Khalifa Islamic Civilisation Center in Copenhagen [35].

This very research also states that Qatar actively supports the movement of Muslim Brotherhood in Europe and promotes its ideology of Euro-Islam, created specifically for Europe. "One of the documents demonstrates that the Qatar Foundation pays a monthly salary to Tariq Ramadan (who is a grandson of the founder of the Muslim Brotherhood, Hassan Al Banna) in the amount of 35,000 euros". Qatar's influence in France has grown so much over the years that President Emmanuel Macron received a significant amount of criticism over the fact that France has such close ties with the country that is accused of supporting terrorism [36].

Since June 2017, Qatar has been diplomatically isolated from other countries of the Persian Gulf because of its ties to terrorist movements across the Middle East. On June 5, Saudi Arabia, the UAE, Bahrain and Egypt severed diplomatic ties with Qatar upon accusing it of providing assistance to the Muslim Brotherhood and spreading extremist ideology – the diplomatic crisis continues to this day. Apart from supporting the Muslim Brotherhood and their spiritual leader Yusuf al-Qaradawi, as well as Hamas and other movements, Qatar is also reported to have been assisting ISIS at an earlier stage of its formation [37]. Qatar's cultural expansion has also started to cause concern in Europe. The aforementioned journalist Georges Malbrunot stated in an interview that "In some cases, it [Qatar] complied, but in others, it tried to play a double game, like closing the London headquarters of Qatar Charity, then reopened with a rebrand, being Nectar Trust. It just dropped the word 'Qatar' from its name" [38].

Therefore, we can conclude that the policy of cultural expansion of the Persian Gulf states is transforming rapidly, adjusting to the ever-changing reality of these states and the world. Moreover, it is obvious that the cultural expansion of the Gulf states has a major impact on the Muslim communities of Western Europe – thereafter, its transformation in the future will inevitably entail changes in the lives of Muslim communities of Europe and the world.

6 Conclusion

This paper provides analysis of the phenomenon of cultural expansion of the Gulf states in Europe and outlines its main stages. The fact that this policy will continue to develop in one way or another in the future is out of the question, given that the number of Muslim population of Europe only keeps increasing [39; 40], even in those areas of Europe that are geographically the most remote from the Muslim world [41], and its role in the lives of European societies is becoming more and more significant.

The research draws a number of conclusions that fully reflect the essence of the phenomenon of the cultural expansion:

1. The cultural expansion of the Gulf states in Europe only became possible after a major transformation of the Muslim world, when the Persian Gulf states achieved

independence and started to assert themselves on a global scale, each expressing claims to be the leader of the Muslim world, using resources gained from oil revenue to their advantage;

2. The cultural expansion was made possible amid the implementation of the multiculturalism policy, developed by the European states in order to overcome cross-cultural barriers in their own countries [42], which were growing stronger due to increasing migration influx, but this policy failed, unable to solve the problems of segregation and "clash of civilizations";
3. To this day, only three of the Gulf states have enough resources to practice the policy of cultural expansion: Saudi Arabia, the United Arab Emirates and Qatar. Meanwhile, Saudi Arabia has been the ultimate leader in the field until recently: from 1975 to 2020, Saudi Arabian state funds financed the construction of more than 2,000 Grand mosques worldwide (200 of which are located in Europe), along with several hundred Islamic cultural centers. Moreover, Saudi Arabian educational institutions have prepared 250,000 imams and Islamic scholars during the same period of time – who now work in Europe and other countries around the world;
4. The paper outlined five stages of the cultural expansion of the Gulf states in Europe, the most successful of which, judging by the count of mosques and Islamic cultural centers built, could be considered the third and the fifth. Terrorist attacks, occurring in the USA and Europe in the early-mid-2000s, have caused irreparable damage to Saudi Arabia's image, which affected the attitude of European authorities towards Muslim communities of Europe and the amount of aid they received from Saudi Arabia itself;
5. Because of internal social and political changes, Saudi Arabia, which has been the ultimate leader in the field of global cultural expansion over the past years, is currently conceding its leadership to Qatar, who actively supports various religious movements around the world. Over the period of 2008–2016, a total of 140 projects in Europe, mainly mosques and Islamic cultural centers, have been sponsored by Qatar;
6. Due to specific characteristics of Muslim ideology and mentality, paired with the treatment they receive from indigenous population of Europe, the Muslim population of Europe prefers to live in closed communities, with their religious objects playing major roles in their lives – therefore, control over mosques is the key to having influence over public attitudes within the Muslim communities;
7. Investing in European economy is highly important for the Gulf states because of the fact that regional conflicts and potential coups d'état threaten the existence of elites and create the need to search for new living environments.
8. Therefore, the policy of cultural expansion of the Persian Gulf states in Europe is currently undergoing transformation, the results of which will only be known in the future. In 2020, the Gulf states, as well as other oil- and gas-producing countries of the world, are facing a crisis on world energy markets in the light of the coronavirus pandemic, which has entailed manufacture shutdown and termination of air travel, tourism, etc. It is reasonable to assume that the Gulf states, the economy of which is directly dependent on oil revenue, will put their policy of cultural expansion on hold, focusing primarily on economic recovery.

References

1. Bogolubova, N.M., Nikolaeva, J.V., Fokin, V.I., Shirin, S.S., Elts, E.E.: Contemporary problems of cultural cooperation: issues in theory and practice. Middle East J. Sci. Res. **16**(12), 1731–1734 (2013)
2. Edward, W.: Said Orientalism. Western concepts of the East. Saint Petersburg Russkiy Mir (2006)
3. Luchitskaya, S.I.: Mage of the Other: Muslims in the chronicles of the Crusades. St. Petebrug, Aletheia (2001)
4. Skripnuk, D.F., Kikkas, K.N., Bobodzhanova, L.K., Lobatyuk, V.V., Kudryavtseva, R.-E.A.: The Northern sea route: is there any chance to become the international transport corridor? IOP Conf. Ser.: Earth Environ. Sci. **434**(1), 012016 (2020)
5. Fokin, V.I., Shirin, S.S., Nikolaeva, J.V., Bogolubova, N.M., Elts, E.E., Baryshnikov, V.N.: Interaction of cultures and diplomacy of states. Kasetsart. J. Soc. Sci. **38**(1), 45–49 (2013)
6. Kepel, G.: Jihad: The Trail of Political Islam. I. B. Tauris, London (2002)
7. Veretevskaya, A.V.: Multiculturalism, which did not exist: analysis of European practices of political integration of ethnocultural minorities. Moscow MGIMO-University (2018)
8. Benard, C.: Civil Democratic Islam. Partners, Resources, and Strategies. https://www.rand.org/pubs/monograph_reports/MR1716.html
9. Hackett, C.: 5 facts about the Muslim population in Europe. Pew Research Center. https://www.pewresearch.org/fact-tank/2017/11/29/5-facts-about-the-muslim-population-in-europe/
10. Lipka, M., Hackett, C.: Why Muslims are the world's fastest-growing religious group. https://www.pewresearch.org/fact-tank/2017/04/06/why-muslims-are-the-worlds-fastest-growing-religious-group/
11. Rabasa, A., Benard, C., Schwartz, L.H., Sickle, P.: RAND Proposes Blueprint for Building Moderate Muslim Networks. https://www.rand.org/pubs/research_briefs/RB9251.html
12. The Future of World Religions: Population Growth Projections, 2010–2050. https://www.pewforum.org/2015/04/02/religious-projections-2010-2050/
13. Choice and Prejudice. Discrimination against Muslims in Europe. Amnesty International amnesty.org
14. Sherwood, H.: London Central mosque given Grade II listed status. https://www.theguardian.com/world/2018/mar/13/london-central-mosque-given-grade-ii-listed-status-historic-england
15. Cendrowicz, L.: Brussels attacks: How Saudi Arabia's influence and a deal to get oil contracts sowed seeds of radicalism in Belgium. https://www.independent.co.uk/news/world/europe/brussels-attacks-saudi-arabia-influence-oil-contracts-sowed-seeds-radicalism-belgium-great-mosque-a6745996.html
16. Roose, E.: The Architectural Representation of Islam: Muslim-commissioned Mosque Design in the Netherlands. Amsterdam University Press (2009)
17. Nuria, B. : La mezquita de la M-30 se inaugura hoy tras 17 años de espera. https://elpais.com/diario/1992/09/21/madrid/717074656_850215.html
18. Noelia, M.: Un día en la mezquita de la M-30. https://www.elmundo.es/madrid/2016/04/16/571141f8e2704ecb718b463d.html
19. Ibrahim-al-Ibrahim Mosque. https://gibraltar.com/en/travel/see-and-do/europa-point/ibrahim-al-ibrahim-mosque.php
20. Gibraltar. https://www.cia.gov/library/publications/the-world-factbook/geos/gi.html
21. Portoghesi, P., Casamonti, M., Coppa, A., Maggi, M.: La Moschea di Roma (2002)
22. Khalil, S., Giorgio, P., Camille, E.: 111 questions on Islam: Samir Khalil Samir, S.J. on Islam and the West: a series of interviews conducted by Giorgio Paolucci and Camille Eid. Ignatius Press (2008)

23. UK Edinburgh mosque opens. http://news.bbc.co.uk/2/hi/uk_news/143136.stm
24. Kettani, H.: History and prospect of Muslims in South America. Soc. Indic. Res. **115**(2), 837–868 (2014)
25. Kettani, H.: The World Muslim Population: Spatial and Temporal Analyses Pan Stanford Publishing Pte Ltd (2019)
26. Kulik, S.V., Eidemiller, K.Yu., Biktimirova, R.R., Vanicheva, M.N.: Islamic diffusion in the countries of the Arctic region: the processes of Muslim migration to Scandinavian countries in the context of the transforming Islamic world. IOP Conf. Ser.: Earth Environ. Sci. **302**(1), 012070
27. Carlbom, A.: Islamisk aktivism i en mångkulturell kontext – ideologisk kontinuitet eller förändring? https://docplayer.se/148802963-Islamisk-aktivism-i-en-mangkulturell-kontext-ideologisk-kontinuitet-eller-forandring.html
28. Chan, S.: Saudi Arabia Moves to Curb Its Feared Religious Police. https://www.nytimes.com/2016/04/16/world/middleeast/saudi-arabia-moves-to-curb-its-feared-religious-police.html
29. Hubbard, B.: Saudi Arabia Agrees to Let Women Drive. https://www.nytimes.com/2017/09/26/world/middleeast/saudi-arabia-women-drive.html
30. Saudi Arabia to stop funding mosques in foreign countries. https://www.middleeastmonitor.com/20200125-saudi-arabia-to-stop-funding-mosques-in-foreign-countries/
31. Thaler, D.: Middle East: Cradle of the Muslim World. Chapter One. The Muslim World After 9/11. Rand Corporation, 69–146 (2004)
32. Schwartz, S.: The Arab Betrayal of Balkan Islam. Middle East Quarterly, Spring, 43–52 (2002)
33. Commins, D.: The Wahhabi Mission and Saudi Arabia. I.B.Tauris (2006)
34. Former Imam of the Grand Mosque in Mecca, Adel Kalbani: Daesh ISIS have the same beliefs as we do. https://www.youtube.com/watch?v=GWORE6OBfhc
35. Chesnot, C., Malbrunot, G.: Qatar Papers: How Doha finances the Muslim Brotherhood in Europe. Averroes & Cie (2020)
36. Nehme, M.: How Qatar is playing a double game in France with Macron's consent. https://thearabweekly.com/how-qatar-playing-double-game-france-macrons-consent
37. On Wikileaks there was information about the help of Qatar and Saudi Arabia for IS. https://www.interfax.ru/world/565443
38. Moubayed, S.: How Qatar funds Muslim Brotherhood expansion in Europe. https://gulfnews.com/world/gulf/qatar/how-qatar-funds-muslim-brotherhood-expansion-in-europe-1.63386835
39. Hackett, C.: 5 facts about the Muslim population in Europe. https://www.pewresearch.org/fact-tank/2017/11/29/5-facts-about-the-muslim-population-in-europe/
40. Kettani, H.: History and prospect of muslims in Western Europe. J. Relig. Health **56**(5), 1740–1775 (2016). https://doi.org/10.1007/s10943-016-0253-4
41. Eidemiller, K.Yu., Samylovskaya, E.A., A Kudryavtseva, R.-E., Alakshin, A.E.: Social and Islamic diffusion in the Nordic countries with the example of Sweden by year 2050. IOP Conf. Ser.: Earth Environ. Sci. **302**(1), 012071 (2019)
42. Fokin, V., et al.: Multiculturalism in the modern world . Int. J. Environ. Sci. Educ. **11**(18), 10777–10787 (2016)

Russia and Post-Soviet Central Asia in the CIS Framework of Cultural Cooperation

Ksenia Muratshina(✉)

Ural Federal University, Yekaterinburg, Russia
ksenia.muratshina@urfu.ru

Abstract. This paper is intended to analyze the realization of the cultural ties between the Russian Federation and its post-Soviet Central Asian neighbors – the Republic of Kazakhstan, the Republic of Tajikistan, the Republic of Uzbekistan, the Kyrgyz Republic, and Turkmenistan, within the Commonwealth of Independent States (CIS). The research involves the study of the institutions, which are used by the CIS member states in order to develop the cultural cooperation, and of the representation of Russia and five Central Asian post-Soviet republics in the mechanisms of multilateral activities.

Keywords: Russia · Central Asia · Kazakhstan · Tajikistan · Uzbekistan · Kyrgyzstan · Turkmenistan · Cultural cooperation

1 Introduction

Cultural cooperation has already become an essential component of interstate relations at current stage. Both at bilateral level and in multilateral integration frameworks, as a rule, cooperating states carry out cultural exchanges, based on mutual obligations, which are fixed in diplomatic documents – agreements, joint plans and programs. Cultural cooperation (involving exchanges in the areas of arts, music, theatre, cinema, direct contacts between art communities, organization of tours, joint or mutual exhibitions and other art-related projects, assistance to the education in the field of art) acts as a sector of humanitarian cooperation between countries, complementing other humanitarian spheres – education, science, sport, etc. The paper refers to the term "cultural cooperation" in accordance with the documents of the Commonwealth of Independent States (CIS) [1, 2]. Analyzing the level of cultural ties between the states parties can help to evaluate the overall level of their cooperation and the perspectives of their partnership. And, as well as the interstate cooperation, the similar ties within international organizations can also be studied and evaluated.

This paper is aimed at the study of cultural ties between the Russian Federation and its post-Soviet Central Asian neighbors – the Republic of Kazakhstan, the Republic of Tajikistan, the Republic of Uzbekistan, the Kyrgyz Republic, and Turkmenistan, within the CIS. The choice of the states parties subject to analysis is conditioned by the research area of the project carried out by the author, as the project is dedicated to the cooperation

© The Author(s), under exclusive license to Springer Nature Switzerland AG 2021
R. Bolgov et al. (Eds.): *Proceedings of Topical Issues in International Political Geography*, SPRINGERGEOGR, pp. 395–403, 2021.
https://doi.org/10.1007/978-3-030-78690-8_34

between Russia and the post-Soviet Central Asia in different formats. CIS is one of the organizations, uniting the countries in question in its framework, and the purpose of this analysis is to examine, which multilateral formats Russia and post-Soviet Central Asian countries are involved in within the CIS cultural ties.

2 Theoretical Framework

As long as the object analyzed is the work of an international organization, the theoretical and methodological basis of this research is neo-institutionalism. This paradigm is well-known in international and political studies, and widely used for the analysis of organizational processes [3–7]. If we look at the work of the CIS in cultural dimension and the participation of the states subject to analysis in this research, neo-institutionalist approach will be able to demonstrate, which institutions are used by the CIS member states in order to develop their cultural cooperation, how extensive the coverage of cooperation mechanisms is, and how the parties of this cooperation (the states in question – Russia and five Central Asian post-Soviet republics) are presented in multilateral activities.

In terms of methodology, the author combines two main methods. The first is the institutional analysis, which is used due to the neo-institutionalist approach. The second one is the discourse analysis, which has become a common tool for the social studies [8–12]. In the paper, while the institutional analysis addresses the organizational structure of the cooperation within the CIS, the discourse analysis is applied in order to study the cooperation-related multilateral documents, which form the legal basis of cooperation.

3 Results of the Study

3.1 Legal Basis of Cooperation

There are two major documents, in accordance with which the cultural cooperation within the Commonwealth of Independent States is carried out: the CIS Agreement on cultural cooperation and the CIS Agreement on humanitarian cooperation. The first one was signed in 1992 by all member states. It includes the obligations of the states parties to develop ties and exchanges in arts, cinema, theatre, museums' and libraries' work, folk art, and conservation of cultural heritage, providing to the people of each other the access to cultural values, information exchange, organization of tours and exhibitions, direct communications between art communities, as well as cooperation in research and education in the field of culture [1].

The second Agreement was signed in 2006 by 11 member states. The only CIS country, that did not sign it, was Turkmenistan [13]. Among other dimensions of humanitarian cooperation, this document regulates cultural interaction within the organization, too. It qualifies cultural cooperation as the set of the following practices: conducting joint festivals, exhibitions, expeditions and forums, ties between libraries and museums, and exchanging the information about the member states' cultural heritage [2].

Further, three multilateral agreements regulate the work of the administrative bodies: the Agreement on the Council on Cultural Cooperation, the Agreement on the Council on Humanitarian Cooperation and the Treaty on the establishment of the Foundation for Humanitarian Cooperation. The first was signed in 1995 by all CIS members [14]. The second document was signed in 2006 by Russia, Azerbaijan, Armenia, Belarus, Kazakhstan, Kyrgyzstan, Tajikistan and Uzbekistan [13]. The third one was signed in 2006 by Russia, Armenia, Belarus, Kazakhstan, Kyrgyzstan, Tajikistan and Uzbekistan [15].

There are also specialized documents, the adoption of which is a long process, that has already started. E. g., in 2019, CIS Agreement on the cooperation between museums was adopted. This document regulates joint exhibitions and culture-related research, the necessity for cooperation in the education of museum employees, the exchanges of electronic copies of exhibits, and sharing experience in exhibitions, conservation and restoration of cultural heritage [16].

In addition, CIS regularly adopts the so-called Plans of priority measures in the area of humanitarian cooperation. Among other data, they usually contain a section, dedicated to cultural events planning [13].

3.2 Responsible Administrative Bodies

The major questions and aims of humanitarian cooperation, including cultural issues, are discussed at the sessions of the main regular CIS bodies, such as the Council of Heads of States and the Council of Heads of Governments [17]. For detailed review of cultural cooperation, CIS possesses a number of boards, or units, which are responsible for carrying out such communications. The major is the Council on Humanitarian Cooperation. It was established in 2006, in order to coordinate the multilateral cooperation in the areas of culture, education, science, information, media, sport, tourism and youth policy. In the Council, 8 CIS countries are represented: Russia, Azerbaijan, Armenia, Belarus, Kazakhstan, Kyrgyzstan, Tajikistan and Uzbekistan. However, Uzbekistan has not actually taken part in the work of the Council [13]. Furthermore, as we can see, Turkmenistan is not presented in the Council at all.

Cultural ties in detail are regulated by the Council on Cultural Cooperation, established in 1995 [14]. The Council discusses such issues, as the progress in the realization of the program "Cultural capitals of the Commonwealth", the organization of various events, the perspectives of cooperation, and the reports of administrative employees, responsible for the corresponding area of issues. All CIS countries take part in its work [18, 19]. The Council is also responsible for discussing new draft documents, related to culture. For example, in 2018 the Council discussed the draft of the CIS Agreement on museum cooperation. It was later adopted by the Council of Heads of States in 2019 [20].

Another administrative body is the Department of Humanitarian Cooperation of the CIS Executive Committee, which is working at a permanent basis on the realization of the main aims of interaction, selected by the Council on Humanitarian Cooperation, and introduces the corresponding questions in the agenda of other CIS units [13].

Finally, one more administrative body, related to culture, is working in the framework of the CIS Parliamentary Assembly. It is the Commission on Culture, Information, Sport and Tourism. This CIS body is responsible for elaborating model laws and other regulations, which are then adopted by the Assembly and recommended to all CIS member states, in order to unify their legal systems. Some of its initiatives are connected with culture. For instance, in 1995 the Assembly adopted the model law "On the unified policy of an obligatory example of documents", which regulated the distribution of sets of documents for libraries. In 2016 the Assembly adopted the "Model library code" for the CIS. This document regulates the financing of state libraries, and will help the member states to reform their library sphere in a unified mode. In 2020 the Commission started the elaboration of "Recommendations for unified approaches of obligatory presentation of an electronic copy of a print version of a publication in the CIS [20]. Moreover, in 2020 the Commission adopted a draft of Convention on the Conservation of CIS countries' cultural heritage. The document provides a set of unified rights and obligations for member states in the area of cultural heritage conservation, maintenance, and mutual assistance in case of emergency [21]. The Commission unites all CIS member states, except for Turkmenistan. Uzbekistan has also been outside this framework for a long time, however, in 2019 it started participating in the work of the Commission [22].

The multilateral financing for the cultural events in the CIS is aggregated and distributed by the Foundation for Humanitarian Cooperation. This unit was established in 2006 and unites Russia, Armenia, Belarus, Kazakhstan, Kyrgyzstan, Tajikistan and Uzbekistan [15].

3.3 Cooperation Frameworks

The actual cooperation frameworks of the CIS cultural activities can be divided into two groups: permanent and situational, or special.

The first permanent framework of cultural cooperation in CIS is the annual Forum of Art and Academic Community. The forums are held since 2006 and help to organize exchanges between the art and academic communities of the member states, allowing their direct communications. The capitals of different countries celebrate these Forums. Taking the cases of Russia and Central Asia, one can find out that, for example, in 2006, 2010, 2014 and 2017 the Forum was held in Moscow, in 2007 and 2015 – in Astana (Kazakhstan), in 2008 – in Dushanbe (Tajikistan), in 2012 and 2019 – in Ashgabat (Turkmenistan), in 2016 – in Bishkek (Kyrgyzstan) [13]. The forum usually includes plenary session and section discussions on the current state and the perspectives of cooperation. The delegates consider a broad range of topics, including cultural ties and beyond, such as international, environmental and healthcare issues [23, 24].

Another permanent framework is the program "Cultural capitals of the Commonwealth". Taking the case of Russia and Central Asia, we can see that in 2011 the status of the CIS cultural capital was assigned to Ulyanovsk (Russia), in 2012 – to Astana (Kazakhstan) and Mary (Turkmenistan), in 2014 – to Almaty (Kazakhstan) and Osh (Kyrgyzstan), in 2015 – to Voronezh (Russia) and Kulyab (Tajikistan), in 2016 – to Dashoguz (Turkmenistan), in 2020 – to Shymkent (Kazakhstan) [25]. This program is aimed at stimulating the cultural infrastructure development in the cities, and the promotion of national cultures. Each country tries to organize the best presentation of its culture and traditions, as well as to attract more tourists by means of this program. The rank of a cultural capital is an honorary one, so the election of every new city for this program is a series of contests (firstly the national one, then the international one) [26]. Remarkably, in this format we can witness the participation of Turkmenistan. Even this closed country tries to use the status of a cultural capital, in order to promote its cultural heritage. For example, in 2012 Mary held an international conference on history and archaeology "Ancient Merv – a centre of world civilization". And the CIS Foundation of Humanitarian Cooperation supported the publication of an album "Mary – the cultural capital of the Commonwealth – 2012. History and modernity" [27].

One more permanent framework is the annual theatre festival "Meetings in Russia". It welcomes the performances of theatres from the CIS countries, staged in Russian language. Usually, the festival has been organized by means of inviting the theatrical companies on tour, however, in 2020, because of the pandemic, it is held online. The usual participants come from Kazakhstan, Kyrgyzstan and Belarus – the countries, where Russian language has the status either of the second state one, or of an official one [28].

Film festivals are also held in CIS annually. They are featuring fiction films and documentaries, and organized with the support of the Foundation for Humanitarian Cooperation in different member countries [29–33]. The CIS regulation of the festival organization was signed in 1997 by all member states, except for Turkmenistan, which therefore stays outside this framework of cooperation [34].

CIS art awards for the cultural and academic achievements can be classified as a permanent cooperation framework, too. They are given every year, namely "The Stars of the Commonwealth" and "The Commonwealth of debuts" [13]. The organisers of the awards are the Foundation for Humanitarian Cooperation and the Council on Humanitarian Cooperation. The prizes are awarded to actors, researchers, working in the field of culture studies, administrative workers in culture, etc. [35]. One more annual award, named after Chingiz Aitmatov, is given to CIS writers and provided by the CIS Parliamentary Assembly [20].

The first special framework is the organization of Thematic Years. They are announced for CIS countries with a special focus on one of the areas of cooperation. In culture, there were several Thematic Years throughout the last decade. 2008 was marked as the Year of Literature and Reading, 2011 – as the Year of Historic and Cultural Heritage, 2018 – as the Year of Culture [13]. The announcement of the Years usually has the purpose to attract more attention to the development of culture and cultural ties within the CIS, and there are thematic events launched. For example, in 2018 the first Youth Cultural Forum was conducted in the Russian city of Tula.

Another special framework is the celebration of major multilateral events and milestones. For example, the 2011 Intergovernmental Exhibition "20 years of CIS; towards new horizons of partnership" and the 2017 celebrations of the date of the CIS establishment were marked with the concerts of artists from the CIS countries [36, 37]. In 2019, the Council on Cultural Cooperation supported the organization of a book exhibition, featuring the most popular books from all CIS countries, published during the last 25 years. The exhibition was moved to all major libraries of the member states [38].

A new framework was launched in 2018 – the above-mentioned Youth Cultural Forum of CIS. The Forum included discussion sessions on the work of museums, volunteers' cultural activities, theatre, modern cinema, music, the role of Russian language in cultural exchanges, and digital technologies in culture. It was organized by the Russian Ministry of Culture, Russian Federal Agency for Youth Affairs, and Rossotrudnichestvo

(Russian Federal Agency for Cultural Ties with Foreign Partners and Russian Expatriates). The number of delegates for the Forum exceeded 50, featuring the participants from Russia, Kazakhstan, Kyrgyzstan, Tajikistan, Uzbekistan, Turkmenistan and other CIS countries [39].

Finally, there can be classified a completely distinctive framework of cooperation – the work of non-governmental organizations (NGOs), which are acting in accordance with the CIS purposes, are welcome at the Commonwealth's events, but still represent independent units from the legal point of view, as they are the gatherings of companies as legal bodies. One of them is the Association of Book Promoters of the Independent Countries. This professional NGO unites corporate members from Russia (dozens of bookshops and publishing houses), Kazakhstan (one publishing house and one trade company), Armenia and Belarus [40]. As we can see, Central Asia is not represented enough in this framework, although it would probably be beneficial for multilateral exchanges and coordination. In particular, in addition to books distribution on national markets, the Association is an active participant and co-organizer of book exhibitions and fairs in CIS countries [41].

The second NGO is Eurasian Assembly of Libraries. Its corporate members are libraries from the CIS countries. Their contacts are aimed at book exchanges and sharing the experience [42]. In addition, the Assembly coordinates a number of events during Thematic Years of culture. The members of this NGO are the leading libraries of Russia, Kazakhstan, Kyrgyzstan, Tajikistan, Uzbekistan and other CIS countries [43]. Turkmenistan stays outside this format.

The third NGO is the International Assembly of Capitals and Large Cities. It was established in 2009 [44] and unites 87 member cities from CIS countries, including 55 from Russia, 11 from Kazakhstan, 1 from Tajikistan and 2 from Kyrgyzstan. The cooperation in this framework involves the coordination of memorial events, the exchange of urban practices experience (at symposiums and congresses), and the unification of urban law [45, 46].

4 Conclusions

Commonwealth of Independent States, as an international organization, appears to suggest a broad range of cooperation frameworks for its member states to develop their cultural ties. There are as much as five administrative bodies, working in the area of humanitarian and, in particular, cultural cooperation, and both permanent and special frameworks of interaction. The permanent frameworks include the annual Forum of Art and Academic Community, the program "Cultural capitals of the Commonwealth", the annual theatre and film festivals, and the CIS art awards. Special frameworks include the Thematic Years, the Youth Cultural Forum, and celebration of major multilateral events and milestones. In addition, cultural ties are developed by three major non-governmental organizations, affiliated with the CIS.

While Russia takes part in all formats and activities and is the initiator of the majority of them, its Central Asian partners are selective in their participation in cooperation frameworks. The study of the sources demonstrates that Kazakhstan and Kyrgyzstan take part in all frameworks and administrative bodies, while Tajikistan and Uzbekistan are less represented, and Turkmenistan prefers not to be involved in the majority of formats and units. This can be done due to different potential reasons, starting from lack of state financing and finishing with the firm state position to maintain the policy of closeness.

The amount of organizational work, carried out by the CIS bodies, the national governments and the non-governmental organizations, is large indeed. The potential of cooperation can also be characterized as significant and allowing to maintain the long-term cultural ties between the post-Soviet countries. However, at the same time, the frameworks should obviously be widened, involving new participants from the local libraries, museums, theatres, film producers of the member states, and more corporate members, e. g. cities, in the NGOs' framework. This can be done on the basis of constant exchange of opinions, sharing the experience and coordinating the strategy of multilateral cooperation. In addition, the increase in the amount of information in media, related to CIS cultural ties, would evidently contribute into the development of cooperation.

Acknowledgements. This research was supported by the Russian Science Foundation (grant № 19-78-10060).

References

1. CIS Agreement on cultural cooperation (1992). http://docs.cntd.ru/document/1900122
2. CIS Agreement on humanitarian cooperation (2005). http://mfgs-sng.org/sgs/gum_sotr/
3. Spindler, M.: Neoinstitutionalist theory. International Relations: A Self-Study Guide to Theory. Verlag Barbara Budrich, pp. 141–157 (2013)
4. Kraatz, M.S., Zajac, E.J.: Exploring the limits of the new institutionalism: the causes and consequences of illegitimate organizational change. Am. Sociol. Rev. **5**(61), 812–836 (1996)
5. Bloodgood, E.A., Tremblay-Boire, J.: International NGOs and national regulation in an age of terrorism. voluntas: Int. J. Voluntary Nonprofit Organ. **1**(22), 142–173 (2011)
6. Candido, S.E.A., Cortes, M.R., Truzzi, O.M.S., Neto, M.S.: Fields in organization studies: relational approaches? Gestao e Producao **1**(25), 68–80 (2018)
7. De Haven-Smith, L., Kouzmin, A., Thorne, K., Witt, M.T.: The limits of permissible change in U.S. politics and policy: learning from the obama presidency. Administrative Theory Praxis **1**(32), 134–140 (2010)
8. Gee, J.P.: Discourse analysis matters: bridging frameworks. J. Multicult. Discourses **4**(11), 343–359 (2016)
9. Van Leeuwen, T.: Moral evaluation in critical discourse analysis. Crit. Discourse Stud. **2**(15), 140–153 (2018)
10. Tracy, K.: Discourse analysis: bridging frameworks or cultivating practices? J. Multicul. Discourses **4**(11), 360–366 (2016)
11. Roderick, I.: Multimodal critical discourse analysis as ethical praxis. Crit. Discourse Stud. **2**(15), 154–168 (2018)
12. Newman, J.: Critical realism, critical discourse analysis, and the morphogenetic approach. J. Critical Realism **19**(5), 433–455 (2020)
13. Information on the activity of the CIS Council on Humanitarian Cooperation (2020). https://e-cis.info/cooperation/3211/78388/
14. CIS Agreement on the establishment of the Council on Cultural Cooperation (1995). http://docs.cntd.ru/document/1900629

15. Treaty on the establishment of the CIS Foundation for Humanitarian Cooperation (2006). http://mfgs-sng.org/mfgs/dogovor/
16. CIS Parliamentary Assembly news (2020a). https://iacis.ru/novosti/partneri/18_maya_otmechaetsya_mezhdunarodnij_den_muzeev
17. Information on the meeting of the CIS Council of Heads of Governments (2017). http://government.ru/news/27821/
18. Information on the meeting of the CIS Council on Cultural Cooperation (2017). https://www.mkrf.ru/press/news/v_kyrgyzstane_sostoyalos_zasedanie_soveta_po_kulturnomu_sotrudnichestvu_gosudarstv_uchastnikov_sng/?sphrase_id=2714029
19. Information on the meeting of the CIS Council on Cultural Cooperation (2018). https://www.mkrf.ru/press/news/v_sankt_peterburge_obsudili_kulturnoe_sotrudnichestvo_stran_sng/?sphrase_id=2721081
20. CIS Parliamentary Assembly news (2020b). https://iacis.ru/novosti/postoyannye_komissii/predstavitel_mpa_sng_rasskazal_o_modelnom_zakonotvorchestve_dlya_razvitiya_bibliotechnogo_dela_v_stranah_sodruzhestva_
21. CIS Parliamentary Assembly news (2020c). https://iacis.ru/novosti/postoyannye_komissii/proekt_konventsii_o_sokhranenii_obektov_kulturnogo_naslediya_gosudarstv_uchastnikov_sng_odobren_postoyannoy_komissiey_mezhparlamentskoy_assamblei
22. CIS Parliamentary Assembly news (2019). https://iacis.ru/meropriyatiya/meropriyatiya_mpa_sng/zasedanie__postoyannoy_komissii_mpa_sng__po_kulture__informatsii_turizmu_i_sportu
23. Deryabin, A.: CIS in search of new ideas in Ashgabat. Nezavisimaya Gazeta (2012). http://www.ng.ru/community/2012-10-30/9_ideas.html
24. Yuferova, Y.: For whom the Silk Road. Rossiyskaya Gazeta (2019). https://rg.ru/amp/2019/05/21/ashhabad-prinial-umnyh-i-talantlivyh.html
25. CIS Parliamentary Assembly news (2020d). https://iacis.ru/novosti/partneri/prodolzhaetsya_rabota_po_realizatsii_mezhgosudarstvennoy_programmy_kulturnye_stolitsy_sodruzhestva
26. Bogolyubova, N.M., Nikolaeva, Yu.V.: Intergovernmental programme "Cultural Capitals of the Commonwealth" - a model of effective cultural cooperation in the CIS. Dialog: politika, pravo, ekonomika, vol. 1, no. 8, pp. 91–98 (2018)
27. Information on the realization of the programme "Cultural Capitals of the Commonwealth" (2012). http://turkmenistan.ru/ru/articles/37922.html
28. CIS Parliamentary Assembly news (2020e). https://iacis.ru/novosti/partneri/luchshie_spektakli_russkoyazychnykh_teatrov_stran_sng_i_baltii_pokazhut_v_onlayn_formate
29. The CIS and Baltic states film festival "New cinema. 21st century" (2008). https://tvkultura.ru/article/show/article_id/33678/
30. International film festival of the CIS countries in Bishkek (2011). https://regnum.ru/news/cultura/1472683.html
31. Kazurova, N.V.: "Samruk-Ethno-Fest – 2012": the first international ethnographic film festival of the CIS (cultural studies and anthropological discourse). Anthropol. Forum **17**, 361–365 (2012)
32. Information on the film festival "Kyrgyzstan - the land of short movies" (2019). https://barometr.kg/na-kinofestival-kyrgyzstan-strana-korotkometrazhnyh-filmov-podano-svyshe-700-zayavok
33. Information on the film festival "Eurasia.DOC" (2020). https://rg.ru/amp/2020/04/23/reg-cfo/kinofestival-dokumentalnogo-kino-evraziiadoc-nachal-priem-zaiavok.html
34. Decision of the CIS Council of Heads of Governments "On the regulation of the organisation of the CIS international film festival" (1997). http://levonevski.net/pravo/norm2013/num62/d62113.html
35. The award "The Stars of the Commonwealth" (2012). https://www.msu.ru/news/premiya_zvyezdy_sodruzhestva.html?sphrase_id=2758426

36. Information on the international exhibition "20 years of the CIS" (2011). http://turkmenistan.ru/ru/articles/36230.html
37. News of the Ministry of Culture of the Russian Federation (2017). https://www.mkrf.ru/press/news/mastera-iskusstv-stran-sng-soberutsya-v-teatre-estrady20171006173501/?sphrase_id=2714029
38. News of the Ministry of Culture of the Russian Federation (2019). https://www.mkrf.ru/press/culture_life/vystavka_chitaem_vmeste_poznaem_drug_druga_20191120200544_5dd572688a02d/?sphrase_id=2714029
39. News of the Ministry of Culture of the Russian Federation (2018a). https://www.mkrf.ru/press/news/alla_manilova_otkryla_pervyy_molodezhnyy_kulturnyy_forum_sng/?sphrase_id=2716913
40. Association of Book Promoters of the Independent Countries (2020). http://www.askr.ru/index.php/home/members.html
41. Information on the international book exhibition in Almaty (2015). http://evrazia-ural.ru/novosti/na-mezhdunarodnoy-knizhnoy-vystavke-v-almaty-prezentovali-orenburgskuyu-oblast
42. Eurasian Assembly of Libraries (2020). http://bae.rsl.ru/
43. News of the Ministry of Culture of the Russian Federation (2018b). https://www.mkrf.ru/press/culture_life/xxi_obshchee_sobranie_bibliotechnoy_assamblei_evrazii_20181128190516_5bfeaeac37125/?sphrase_id=2714029
44. The Charter of the International Assembly of Capitals and Large Cities (2020). http://www.e-gorod.ru/documents/mag/USTAV_new.htm
45. The International Assembly of Capitals and Large Cities (2020). http://www.e-gorod.ru/documents/main_inf.htm
46. News of the Moscow House of Nationalities (2020). https://mdn.ru/digest/dajdzhest-ot-21-marta-2020-goda

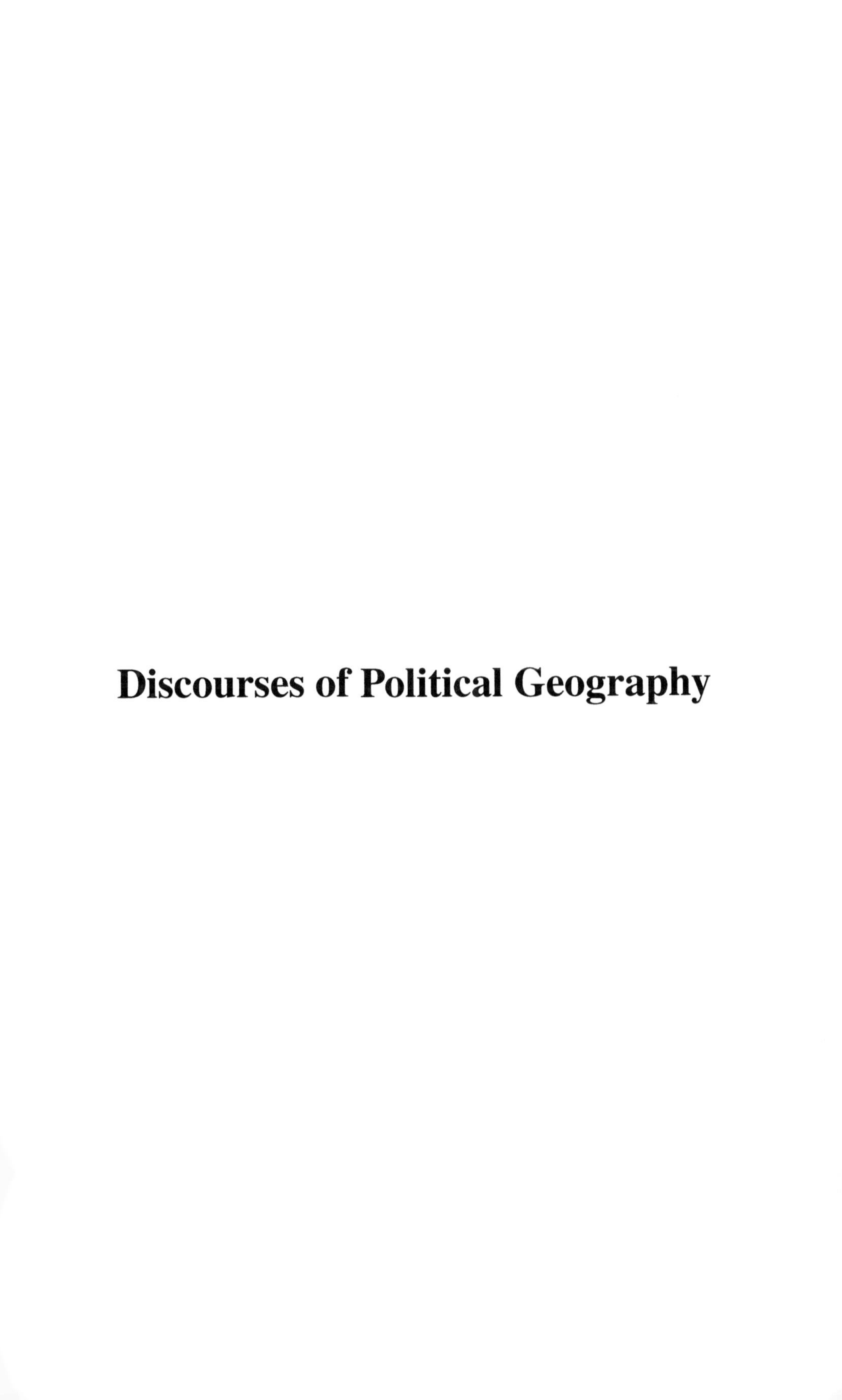

Discourses of Political Geography

From Geopolitics to Linguopolitics: Studying the Language Factor in International Relations (A Pre-history)

Igor Chernov, Igor Ivannikov, and Nikolay Gudalov(✉)

Saint Petersburg State University, Saint Petersburg, Russia
n.gudalov@spbu.ru

Abstract. This paper is aimed at identifying key stages in the development of the so-called 'linguistic-political realism' in the theory of international relations (IR). It is trough language that the paradox of the impossibility of comprehending the objective social world without the aid of its subjective description is lifted. From the viewpoint of linguistics, human language as the basic sign system has numerous functions, the most important of which are the cognitive and the communicative functions. The cognitive function consists in that linguistic signs enable humans to think, i.e., the nature of our consciousness is tied to signs. It is therefore due to this function that the whole real world (both physical and social) around us is accessible for us only through language. The communicative function of language enables any interaction between individuals, and thus creates all social forms (society, state, nation etc.). 'Linguistic-political realism' is understood as a strand of social philosophical enquiry (irrespective of authors' ontological position) which specifically distinguishes an 'objective' role of the language factor in both the development of an individual society and the interaction of societies speaking different languages. Consideration is given to elements of linguistic-political realism in various social philosophical approaches from antiquity to Marxism, which was the last integral philosophical system and predated the emergence of sociology and International Relations theory as separate academic disciplines.

Keywords: International relations · Philosophy of language · Marxism · International relations theory · Political realism · Postmodernism · Linguistic-political realism

1 Introduction. Methodology

Language, as an intersubjective sign system, has two crucial social functions – the cognitive and the communicative. The cognitive function has always been central to studying epistemological problems, be it in the medieval disputes of the realists and the nominalists or in John Locke or Gottfried Leibniz. Moreover, detailed studies of language in epistemology were fundamental for the so-called 'linguistic turn' in philosophy that took place in the 20th century (Ernst Cassirer, Ludwig Wittgenstein, analytical philosophy, poststructuralism etc.). Yet, despite the importance of all these enquiries, they

© The Author(s), under exclusive license to Springer Nature Switzerland AG 2021
R. Bolgov et al. (Eds.): *Proceedings of Topical Issues in International Political Geography*, SPRINGERGEOGR, pp. 407–415, 2021.
https://doi.org/10.1007/978-3-030-78690-8_35

have always provoked doubts as to their linguistic universality applied to humanity as a whole. Therefore, the second, and somewhat collateral, direction in the philosophical study of the cognitive function of language has consisted in attempts to establish a link between a specific language and the thought or ideas of the people which speaks it. Thus, for example, the Greek philosophy has often been presented as a natural product of the classical Greek language, with its grammatical and semantic constructions. At the same time, the Latin language, not being especially favourable for the development of an abstract philosophy, proved to be auspicious for jurisprudence (the Roman law) [12]. Unlike the formal logic tied to the classical Greek, the Chinese language evolved a different methodology, and the so-called numerology took the place of logic in China [15]. Johann Herder pointed to the links between the Chinese language and some features of the Chinese tradition of thought [12]. Likewise, Friedrich Nietzsche connected the similarities found in the Indian, Greek and German philosophies with the shared grammatical features of the respective languages which guide thinking [25]. In the 20th century, the study of 'national forms' of thought was mostly transferred from philosophy to philology and cultural linguistics (Edward Sapir, Benjamin Whorf, Leo Weisgerber, Anna Wierzbicka, Ngugi Wa Tiong'o and many others [24, 34]). Yet, this study kept its main principle, according to which, on the one hand, the whole of language is based on the beliefs of a people (Cicero) and, on the other, a shared language makes 'uniform' the worldview of the members of a linguistic community [33]. Certainly, such an approach has been subject to serious criticisms as well, which hold that confusing the language and culture of an ethnos largely leads to myth-making (see, e.g., [8]). So, the multi-faceted cognitive function of language has been actively discussed in both philosophy (see, e.g., [28, 30]) and other disciplines, and therefore is mainly beyond the focus of our paper. The situation is different with regard to the communicative function of language. The contemporary social philosophy has witnessed precisely the 'confusion of language and culture' noted above, with language gradually 'dissolving' and 'disappearing' in a vague idea of a socio-cultural space. This seems hardly justified, as language is a foundational, systemic element in the formation and functioning of a society. But this obvious idea, sometimes considered as a commonplace, is often sidelined in the contemporary social philosophical and sociological models of social systems, and the phenomenon of language itself is 'outsourced' to sociolinguistics. Although numerous prominent philosophers and sociologists of the past touched, in various ways, upon the role of the linguistic factor in social development, this aspect of their work has not, paradoxically, received sufficient attention. Given all this, our study aims to identify some main milestones in the philosophical study of the role of language in social processes, especially in the interactions of societies speaking different languages. Alluding to Thomas Hobbes, we call a society a 'Linguiathan', a conglomerate of linguistic-communicative networks, i.e. a system of the functioning of these networks which has a specific structure. One direction of the study of the communicative function of language in social philosophy has always been concerned with identifying the role of language (language as a communication tool, language in general) in the formation, structuring and functioning of a society. This study is nowadays continued by linguists (sociolinguistics). However, they focus on the influence of society on language, not on the influence of language on society (J. Gumperz, E. Haugen, G. Fishman and many others). The second direction of study has sought to identify the role of linguistic differences in the emergence and formation of the political divisions between human societies (the different 'Linguiathans') and in their subsequent interactions.

2 From Antiquity to the Early 19th Century

The beginnings of the approach which assigns to language the leading role in the social history of humankind can be discerned already in the antiquity. It may possibly have been the first theory historically which tried to explain the functioning of society and international relations (IR). This is hardly surprising, given that this approach seems quite natural and evident. The main theses of linguistic-political realism were formulated already in the writings of many ancient and medieval authors who reflected on the nature of international relations. According to their viewpoint, it is a shared language that was the basic factor for successful social intercourse, and linguistic difference provoked political divisions and conflicts. Besides, this perspective had been outlined in the biblical legend of the Tower of Babel (written down no later than the 6th century BCE). The tradition of linguistic-political realism was not interrupted in modernity either. For many researchers, the whole human history in all its complexity was mainly a story of struggles between different linguistic communities. Language acts as a factor of integration and disintegration at all the levels of a social system; it marks the boundaries of a society and differentiates it from other linguistic communities [4]. Language also distinguishes human societies from other 'social' organizations in nature. Aristotle considered the state as the supreme, political, kind of communication [1]. Cicero, likewise, thought that, among the many levels of human society, the most natural and broadest was that which was based on language [7]. There were thus no doubts in the ancient world that it was a common language which, by enabling communication, both created and ensured the maintenance of the unity of any society and 'civilization'. Conversely, a disruption of this unity was fraught with political confrontation and war.The separation between 'Us' and 'Them' appeared naturally defined by language. As is well known, the Greek term 'barbaros' alluded to the unintelligibility of alien speech [29]. The Greek language was part of the ancient Greeks' 'identity' and 'solidarity'; 'misunderstanding' led to strife; and differences between languages led to 'problems of communication', which were not at all unfamiliar to Herodotus [27]. Moreover, Herodotus believed that the Athenians, for example, were Greek regardless of their non-Greek origin. The Pelasgi (and the Attic people, being Pelasgian in origin), according to the historian, probably spoke a 'barbaric language', but then those who joined the Hellenes must have changed their language [13]). Thus, even though Herodotus considered all peoples who did not speak Greek as barbarians, the biological origins did not prevent the change of ethnos; the main and necessary ethnic marker was language [13]). Similar views of 'nationality' were probably widespread among other ancient peoples: according to Herodotus, the Egyptians considered as barbarians all those who did not speak their language [13]). From Europe to China, the ancient communities perceived the boundaries of their linguistic communities as the boundaries of civilization. Language was considered a sufficient ground for political cooperation and, potentially, for settling disputes by peaceful means [13]). This led Herodotus to advocate not the interests of any single polis, but the unity of all the Greek-speaking world. Such historical evidence represents a certain challenge or correction to those currently trending views which confuse language and culture and hold, for example, that if the elites and the ruled came to share a common culture only in the last several centuries, they did not belong to a single ethnos before [18]. Rather, the relatively stable features of an ethnos were defined by its language. Similar ideas are found in the Biblical legend of the Tower of Babel, which emerged in the Judaic tradition not later than the 6th century BCE. In the late ancient Latin culture it was addressed, for instance, by the Christian poet Claudius Marius Victor. For him, linguistic unity led to success; disunity, to failure and civil war. He also wrote that even though blood ties may disappear in a tribe, it is not they, but language which shapes a people [31]. Special note must be taken of the so-called Catholicon, the famous encyclopedia of the medieval world (13th century), which was rewritten and reprinted many times and which was not only a Latin dictionary but, to a large degree, the standard of scientific knowledge for many generations of Europeans. The entry for 'language' ('lingua') offers a view which foreshadows elements of both the 'clash of civilizations' and 'end of history' theories. According to it, before the pride of the Tower separated the nations, they all spoke the same language, Hebrew. But probably, only one language would also be spoken in the future. The plurality of languages meant diversity and, hence, discord. In the Church of the saints, there would be no discord, but just one language, joy, love, wisdom, peace, justice, understanding etc. [3] Thus, the ancient linguistic-political tradition continued in the middle ages as well. In the late 18th and early 19th centuries, this conception was given a scientific form in the works of J. Herder, Wilhelm Humboldt, Jacob Grimm etc. But what was problematic was the perception of this theory as just a linguistic one. At the same time, the new social philosophy seemed, at first sight, to focus on the economy as the main driver of social development. This connects with our discussion in the next section.

3 Marxism

Communism promised to destroy the two biblical curses – that of labour in the sweat of one's brow (which results from seemingly 'eternal' economic laws) and that of the linguistic plurality of Babel which caused incessant international strife. After a world revolution, humankind was to escape from 'necessity' into 'freedom'. The economic aspect of Marxism (which is indeed crucial) has always been subject to detailed study. Yet, although Vladimir Lenin wrote that socialism is not reducible to the economy [16], the linguistic-political basis of Marxism has not drawn the attention it deserves. When the Marxist theory was being formed, academic linguistics was thriving. The German school of linguists and historians laid the foundations of comparative-historical linguistics as a rigorous science, which claimed serious attention on the part of anybody concerned with social problematic (sociology, politics, or history). Meanwhile, the foundation of the 'genuinely scientific' Marxist theory was to be a synthesis of the supreme achievements of all the 'genuine sciences'. It was therefore no accident that Karl Marx and Friedrich Engels, who had a good linguistic training and a practical multilingualism, were earnestly concerned with issues in linguistics. Thus, Engels produced, among other things, a serious linguistic study on the Frankish dialect, which was a detailed philological analysis of ancient Germanic dialects and of their influence on the contemporary languages and dialects [6]; and Marx constantly dwelled on etymology even in his economic works (for example, he analyzed the meaning of the word 'value' in Indo-European languages). In this respect, Marx and Engels were no exception among the socialists. The German communist Wilhelm Weitling (whom Marx considered as his teacher in the labour movement) worked on a study dealing with universal logic, grammar and universal language of the humankind. Paul Lafargue (one of the leading French Marxists and Marx's son-in-law) published, somewhat later, the work on the French language before and after the revolution. Issues of language were addressed by Karl Kautsky, Heinrich Cunow and many other German Marxists. Even Pierre-Joseph Proudhon (the founder of anarchism) began his literary career with an essay on universal grammar. This common interest to language was provoked by both theoretical concerns and practical needs of the International's political struggle. It was partly an accident that the linguistic-political theme was less developed in Marxism than the economic one. Thus, for instance, Marx, during the last two years of his life (1881–1883), seriously pursued linguistic-political enquiries concerning human history, but eventually, 'by Marx's will', Engels had to complete this work. Georg Hegel's linear evolutionism of the historical development was adopted by Marxism – according to Lenin, nations pursue the same historical road, yet through very different paths [16]. This national diversity was also necessarily reflected in the national languages, giving a 'national-linguistic' coloring to any universal ideology, including the communist one. Marx, for instance, wrote about the German socialist school, whose main activity was the translation of 'Fourierist, Saint-Simonist and other theories from the French language into the obscure language of German philosophy' [21]. Thus, any idea taken by one nation from another had to be adapted to and changed by the recipient nation [14]. Hegel postulated an intimate connection between language and thinking [10]. According to Marx, this 'spirit' is from the outset 'burdened' by matter (i.e., moving air, sounds); yet, '[l]anguage is as old as consciousness'. Indeed, language is 'practical', 'genuine consciousness', which connects the subject and others and which emerges as a response to the need for communication [22]. A linguistic community was considered in Marxism as a necessary basis for production, as that substance ('matter') where the economic factor acts, for both an individual without society and language without individuals sharing it are 'nonsense' [19]. Language can thus be considered 'a condition and a tool, a cause and consequence of sociality' [26]. Marx therefore viewed any society as a 'form of communication' (Verkehrsform). Moreover, linguistic community is the 'human' form of biological descent which goes beyond its limits: the fact that someone speaks one language rather than another is due to 'circumstances', not to descent [20]. This linguistic community is quite stable and does not disappear in the process of economic development with the emergence of the state. It merely changes its frameworks and form by appearing as a tribe, or a nationality, or a nation. The national and the linguistic questions essentially represented one question. According to Marxism, the linguistic and national diversity of the 'productive societies' hinders economic progress, as it contradicts the international character of the economy. Marx and Engels assumed that, precisely due to the universality of the world's economic development which broke all national boundaries, the linguistic-national question would in the end be duly resolved by a world communist revolution, and Hegel's thesis of the durability of national and racial differences [11] would be refuted. According to Engels, a state emerges on the basis of a conglomerate of linguistically related tribes. Within a single state, large linguistic communities ('nationalities') gradually assimilate small ones. A common language functions as both the ground for the emergence of a state and the necessary condition for its existence as a workable social mechanism. In its most general form, Marxism's linguistic-political theory boils down to the thesis that linguistic communities naturally create a political superstructure (as a form of a community's existence) and are the necessary basis for a nationality and, hence, for a state [6]. Language defines the natural boundaries of a political community. At a certain historical point, only the nation-state is the ideal form for the development of capitalism, as it creates a single national market without internal political and linguistic borders. But the advance of capitalism gradually makes superfluous even the boundaries between nation-states themselves. Mutual ties, interdependence and cosmopolitism come to apply to both the material and the spiritual production [23]. Communism was called on just to hasten this objective process, which, in Marx's and Engels's opinion (indeed reminding of the medieval Catholicon), would abolish the linguistic and national diversity of humankind. The Marxist approach to language as a social factor obviously claimed universality and objectivity, as did positivism. Although human ideas about the forms and relations of objects might have been just 'hieroglyphs' which differed from those forms and relations in themselves, 'these hieroglyphs exactly signify these forms and relations, and this is sufficient for us to be able to study the influences of the things in themselves on us and, in its turn, to affect them' [26]. However, the 20th century saw the emergence of major doubts as to whether researchers can produce knowledge beyond the limits of their historical or cultural context [9] – indeed, beyond their language. This will be addressed in the next section.

4 'Linguistic-Political Idealism'

The emergence of a science of international relations properly, in the early 20th century, was largely influenced by the theory of political idealism (or liberalism), which had a normative character and wished to change the world pursuant to the spirit of the liberal political ideas of the Anglo-Saxon world, about whose opposition to the Kaiser's Germany Alfred Zimmern, the first professor of the first department of international politics, wrote so much. He likened liberalism to a 'political religion' and the main 'credo' of the English-speaking peoples [35]. Yet, as is well known, such theories were unsuccessful in confronting the political realities. Idealism made way for realism, which considered international relations as a system of power struggles between states independent of subjective wishes.

Political realism was close to the geopolitical approach in the study of international relations, which connected the geographical position of a country with its foreign policy. From the viewpoint of geopoliticians, it was geography that had become the main structural factor which determined the behavior and interaction of states on the international arena. 'Sea powers' were doomed to confrontation with 'land powers' in a struggle for leadership in world politics. Mastery over the 'heartland' guaranteed mastery over the world. This approach, however, had certain mythological and schematic features. From our point of view, the main role in any play is exercised by the actors who read their texts, not by the decorations. The factor of language (as discourse) was drawing, in the late 20th century, the ever increasing attention of researchers. The realist methodology often held reality as given independently of the subject. But the paradox is that, even if society functions according to some objective laws, the description of society (its 'self-description') cannot be but subjective, which calls into question the very 'objectivity' of the identified laws. It was therefore no accident that the study of international relations followed philosophy and sociology in the late 20th – early 21st centuries in the 'linguistic turn' to the subject, which has taken the various forms of poststructuralism, postmodernism or constructivism but has invariably foregrounded the role of language (speech, discourse) in the formation of human consciousness and all social reality. Such various approaches might be considered under the term 'linguistic-political idealism'. According to these approaches, the world is governed by ideas (or ideologies), which appear, function and construct reality through speech acts. Any academic enquiry likewise partakes of ideology. Objective facts are replaced by ideas about facts. And the 'reality' of international relations becomes just our idea and endless interpretations, which are inherently tied with our language and ideology. The title of John L. Austin's book, How to Do Things with Words, is telling. Nicholas Onuf, who was the first in IR theory to call himself a constructivist, thinks that social reality does not exist by itself, but is permanently constructed and reconstructed through speech acts [5]. This process can be compared to a constant 'renegotiation' of the social contract, which people strive to discuss and adjust. This applies, of course, to international relations as well. Language and speech acts are the medium through which countries endlessly construct their visions of the world, and the world thus gets constructed; structures, in their turn, also construct agents [2]. According to Onuf, the very notions of an objective world and scientific objectivity disappear. There arises, however, the risk that by following the constructivists (at least 'thick' ones) in studying the subject, we may lose sight of the object of knowledge. Meanwhile, many realists, being wedded to the 'objective', often ignore the linguistic problematic by viewing it as a monopoly of linguistic-political idealism.

5 Conclusions. Towards Linguistic-Political Realism?

Despite the obvious merits of linguistic-political idealism in drawing attention to the problematic of language, which is crucial for any social enquiry, this approach risks throwing the baby out with the bathwater, i.e., repudiating the real world itself. The object has been implicitly dissolved in the 'subject' and 'subjectively' disappeared. But, using Max Weber's terms, 'social science' still needs to be a 'science of reality' [32]. Physical reality is transformed only when people grasp (and do not construct) some objective laws of the world and learn to use them in their interests. And if our subjective ideas depart from those objective features, the external world is bound to somehow correct us. At the same time, focusing on the external world and its regularities must by no means blind the IR scholars to language, for language is also, in a certain way, an objective and crucial element of social reality. Eventually, it is through language that we can overcome the paradox of the impossibility of comprehending the objective social world without a subjective description. From the viewpoint of linguistics, human language as the basic sign system has numerous functions, the most important of which are the cognitive and the communicative functions. The cognitive function consists in that linguistic signs enable humans to think, i.e., the nature of our consciousness is tied to signs. It is therefore due to this function that the whole real world (both physical and social) around us is accessible for us only through language. The communicative function of language enables any interaction between individuals, and thus creates all social forms (society, state, nation etc.). As argued by Niklas Luhmann, language is the structural connection between systems of consciousness and systems of communication [17]. Linguistic-political realism emerges on the basis of political realism, but at the same time uses many achievements of linguistic-political idealism, with its openness and interest towards the problematic of language. The theory of linguistic-political realism, or linguopolitics (analogous to geopolitics), considers language as the basic factor which structures the contemporary world in all its complexity and significantly influences (along with the economy) the processes of globalization and regionalization. But, unlike constructivists, we think that language (and linguistic differentiation) is a part of reality which we are unable to change and which largely programs our behavior. Geopolitics looks into the past, linguopolitics attempts to look into the future.

References

1. Aristotle: Politics. Book 1
2. Azizov, U.B.: Constructivism in international relations. St. Petersburg, Aletheia (2015). (in Russian)
3. Balbi, G.: The Summa grammaticalis quae vocatur Catholicon
4. Bloomfield, L.: Language, Moscow (1968). (in Russian)
5. Chernov, I.V.: Postconstructivism, or the theory of linguo-political realism in the international relations. Vestnik of Saint Petersburg University. Polit. Sci. Int. Relations **11**(1), 86–104 (2018). https://doi.org/10.21638/11701/spbu06.2018.107
6. Chernov, I.V.: Red linguopolicy and the World revolution. Orthodox Marxism on the linguistic factor in historical development. Vestnik of Saint Petersburg University. Polit. Sci. Int. Relations **10**(2), 166–185 (2017). https://doi.org/10.21638/11701/spbu06.2017.206
7. Cicero, M.T.: Philosophical Treatises. About the Laws. Book 1. (in Russian)
8. From linguistics to myth: linguistic culturology in search of "ethnic mentality". Sat articles. Compiled by A.V. Pavlova. St. Petersburg (2013). (in Russian)
9. Gergen, J.K.: Social Construction in Context. Publishing House Humanitarian Center, Kharkiv (2016). (in Russian)
10. Hegel, G.V.F.: Encyclopedia of Philosophy, vol. 1. Science of Logic, Moscow (1974). (in Russian)
11. Hegel, G.V.F.: Encyclopedia of Philosophy, vol. 3. The Philosophy of the Spirit, Moscow (1977). (in Russian)
12. Herder, J.G.: Ideas for the Philosophy of Mankind. Publishing house "Science", Moscow (1977). (in Russian)
13. Herodotus. Story. Translation and notes by G.A. Stratanovsky. Moscow, Ast, the Guardian (2006). (in Russian)
14. Kautsky, K.: Liberation of nationalities. In: Kautsky, K. (ed.) National Problems, Moscow (2010). (in Russian)
15. Kobzev, A.I.: Philosophy for and Before the Chinese wall. https://runivers.ru/philosophy/logosphere/57413/ (in Russian)

16. Lenin on the National and National-Colonial Question, Moscow (1956). (in Russian)
17. Luhmann, N.: Society of Society. Logos, Moscow (2011). (in Russian)
18. Mann, M.: The Dark Side of Democracy: an Explanation of Ethnic Cleansing. Publishing House "Fifth Rome", Moscow (2016). (in Russian)
19. Marx, K., Engels F.: Compositions. Second edition. v .46, Moscow (1968). (in Russian)
20. Marx, K., Engels F.: Compositions. Second edition. v. 3, Moscow (1957). (in Russian)
21. Marx, K., Engels F.: Compositions. Second edition. v. 8, Moscow (1957). (in Russian)
22. Marx, K., Engels F.: German Ideology, Moscow (1933). (in Russian)
23. Marx, K., Engels, F.: Manifesto of the Communist Party, Moscow (1986). (in Russian)
24. Ngugi, W.T.: Decolonising the Mind: the Politics of Language in African Literature. Nairobi, London (1986)
25. Nietzsche, F.: Works in Two Volumes. Publisher Thought. V. 2, Moscow (1990) (in Russian)
26. Plekhanov, G.V.: Selected Philosophical Works. Volume II, Moscow (1956). (in Russian)
27. Rochette, B.: La langue comme facteur d'integration ou d'exclusion. L'Athenes de Pericles et la Rome de Ciceron (2005)
28. Schipkov, A.V.: The concept of "code" in the framework of the modern civilizational approach. Questions Philos. **7** (2018)
29. Shukurov, R.M.: Denomination, ethnicity and Byzantine identity. Religious and ethnic traditions in the formation of national identities in Europe, Moscow (2008). (in Russian)
30. Smirnova, E.D.: Possible worlds and the concept of "pictures of the world". Voprosy Filosofii, vol. 1 (2017)
31. Victor, C.M.: Commentariorum in Genesim
32. Weber, M.: "Objectivity" of Socio-Scientific and Socio-Political Knowledge. Selected Works. Moscow (1990). (in Russian)
33. Weisgerber, J.L.: Native Language and the Formation of the Spirit. Book house "LIBRO-COM", Moscow (2009). (in Russian)
34. Yanzina, E.V., Korneev, O.V.: On the question of the role of grammar in the creation of the language of ancient Greek philosophy. In: Indo-European Linguistics and Classical Philology (2015). (in Russian)
35. Zimmern, A.: Nationality and Government. https://books.google.ru/books/Nationality_Government.html

National Minorities at Saint Petersburg Imperial University in 1905 in the Focus of the Press

Evgeny Rostovtsev[1,2(✉)], Victoria Andreeva[3], and Ilya Sidorchuk[3]

[1] St. Petersburg State University, St. Petersburg, Russia
e.rostovtsev@spbu.ru
[2] Tomsk State University, Tomsk, Russia
[3] Peter the Great St. Petersburg Polytechnic University, St. Petersburg, Russia

Abstract. The objective of this paper is to consider the mapping of the national question in focus of periodical press, referring to the fact that the press not only served as a mirror of the events unfolding at the University in 1905, but it acted as an important lever of influence on the public opinion and the University world, and in this sense acted as a factor of the processes unfolding within the University. The source base of this research was made up of materials published in St. Petersburg newspapers that represented various political trends: "Birzhevye vedomosti" ("Stock exchange news"), "Novoe vremja" ("New time"), "Russkie vedomosti" ("Russian statements"), "Peterburgskij listok" ("St Petersburg sheet"), "Svet" ("Light"). This allowed us to reveal how various political forces related to the national movement of students and what importance they attached to it. The authors turned to the method of content analysis in order to collect data on the events connected with St. Petersburg University, its professors, and students. The research suggests that the beginning of the First Russian revolution in 1905 was a catalyst for the growth of activity and radicalization of the national movement among the students of St. Petersburg University. In the conditions of the general revolutionary rise in the whole Empire, various national and political preferences and a system of self-identification in relation to the Empire and the Russian liberation movement were formed among the national student communities. The right-wing press, seeing national minorities as a potential threat to the monarchy, used news stories about the activity of national student associations as an additional opportunity to discredit them. The liberal press, also not missing such events, thus promoted national movements, considering them as an important element of liberalization. Thus, it promoted their institutionalization and self-identification of national elites.

Keywords: History of education · National movement · National minorities · Russian revolution of 1905 · St.-Petersburg University

1 Introduction

The role of Saint Petersburg University in the events of the First Russian revolution of 1905–1907, as well as higher education in general, regularly attracts the attention of

© The Author(s), under exclusive license to Springer Nature Switzerland AG 2021
R. Bolgov et al. (Eds.): *Proceedings of Topical Issues in International Political Geography*, SPRINGERGEOGR, pp. 416–426, 2021.
https://doi.org/10.1007/978-3-030-78690-8_36

historians [2, 7–10, 21, 24, 27, 29, 36, 42, 43, 46, 58, etc.]. We should also mention that Richard Pipes had seen the roots of the First Russian Revolution in student turmoil [30, p. 31]. Daniel Brower proves that universities were cradles of revolution [6]. A number of studies have also been devoted to the role of student organizations in the events of the First Russian revolution [5, 22, 28, 60, etc.]. At the same time, the topic of St. Petersburg University as the center of the national movement in Russia at the beginning of the twentieth century in the literature devoted to this educational institution has not yet been raised much. The point is probably that for both the Ministry of national education (MNE) and the University administration, as well as for the police Department (documentation related to the activities of these institutions is the basis of research on the history of the University), the national movement was still less important than the activities of the socialist parties [38]. While paying attention to the topic of the student movement, one more important source – the periodical press – is almost ignored. During the period under review, the press was the main means of promoting certain political views and forming public opinion and was one of the main sources of forming a picture of the world. The most prominent political figures of the era, in particular, Vladimir Lenin and Pavel Milyukov, recognized the crucial role in the success of propaganda of their parties' ideas through their party's press. Only the press could broadcast information about the events taking place in the country to a sufficiently large audience; it was the most powerful source of forming public opinion among various segments of the population [37, 40, 41]. The objective of this paper is to consider the mapping of the national question in focus of periodical press, referring to the fact that the press not only served as a mirror of the events unfolding at the University in 1905, but it also acted as an important lever of influence on the public opinion and the University world, and in this sense acted as a factor of the processes unfolding within the University. It can be assumed that popular narratives on the national issue among students contributed to the processes of national self-identification and the formation of national organizations.

Chronological frames of the research are September – October of 1905. Due to the Provisional Rules for the Management of Institutions of Higher Education adopted on August 27, 1905, which restored the University autonomy, the Capital University became the center of country's social and political life, revolutionary movement until the all-Russian October political strike. Public debate on most of the important issues took place inside the walls of the university. Shmuel Galai and Theodor Shanin considered that the returning of autonomy transformed universities into "islands of liberty" [8, p. 261; 42, p. 83]. The national question was not the most significant, neither from the point of view of political forces, nor from the point of view of intra-university life, but it attracted constant attention of the press. The university was the center of St. Petersburg students, and student organizations were of a city-wide character. It was also the center of the revolution and demonstrations, and it was one of the centers of coordination of protest and public activity. Students, including national minorities, took part in revolutionary demonstrations, and revolution served nationalism.

The source base of this research was made up of materials published in the most popular newspapers that represented various political trends: *Birzhevye vedomosti* ("Stock exchange news") reflected mildly liberal views, *Novoe vremja* ("New time") together with *Svet* ("Light") promoted right-wing ideas, *Peterburgskij listok* ("St Petersburg sheet") and *Peterburgskaya gazeta* ("Petersburg newspaper") can be considered as centrist press focused on local news, *Russkie vedomosti* ("Russian statements") was a liberal best-selling Moscow newspaper which covered news in all parts of Russia, and the latter was taken in order to compare its discourse with the Petersburg ones. This allowed us to reveal how various political forces related to the national movement of St. Petersburg students and what importance they attached to it. Official governmental newspapers are not analyzed here as they didn't participate in public debate and didn't cover the student movement and national question issues. The press in that informational situation is one of the main sources for the formation of a picture of the world. It can be assumed that popular narratives on the national issue among the student community contributed to the processes of national self-identification and the formation of national organizations. Liberalization of legislation related to student organizations started in 1903. During the described period, unlike during the period after 1907, there was no practice of official organization of student national circles, and legal registration was not required. That is why the press, but not the documentation of national organizations, is of particular importance.

2 Methods of Research

The authors first turned to the method of content analysis in order to collect data on the events connected with St. Petersburg University, its professors, and students. An article (news, feature or editorial) was taken as a unit. There were 466 units found in the newspapers of September – October 1905. The materials were organized in several categories, including "national question", "student movement", "academic issues", "university charter debates", etc. [36] This article is devoted to the category "national question", which deals with 51 units, concerning Jews (32), Poles (13), Ukrainians (4), Latvian (1) and Chinese (1) as national minorities. There were no leading articles mentioning national minorities, these were mostly chronicle and a few editorials. When implementing the study, the authors turned to a number of primary research and historical methods:

- To reconstruct the key moments in the history of the national communities of St. Petersburg University, the authors used the comparative historical method.
- The historical-genetic method made it possible to identify and trace changes in the consideration and assessment of the national student movements by the press;
- The comparative method allowed us to juxtapose the features of the coverage in the press of the activities of Jewish, Polish, and Ukrainian student organizations of St. Petersburg University.

These methods helped to identify a significant amount of materials that touched on events related to the activities of national organizations of the St. Petersburg University, and to see the dynamics of the presentation of their activity in the press. This approach is particularly relevant, because despite all the differences in the coverage of these episodes of University life in the periodical press, they are characterized by the active independent role of the newspaper word in the formation of the main attitudes of both public and corporate consciousness related to the problems of University life. These approaches helped to analyze the role that print played in the construction of social reality and the image of the University, which was formed in society and among the teachers and students of the University themselves.

3 Results of Research

According to current studies, the number of visiting students (not from the St. Petersburg educational district) at the Capital University at the turn of the century fluctuated in the region of 45–50%, and their base was made up of people from the national outskirts of the Russian Empire. Thus, St. Petersburg University later became one of the centers of formation of the elites of the future independent states formed after the First World war [39]. Unfortunately, the available sources do not allow us to fully reconstruct the national composition of students. According to a study by Dmitry Barinov, at the beginning of the 20th century, numerous student national associations, corporations and clubs were already active here; in 1903–1917, there were 17 nationally and regionally oriented research circles, in 1907–1917 there were 3 national student funds, in 1901–1917, 15 national corporations and circles [3, pp. 243–245]. The most active of them at the time of 1905 were Polish, Jewish and Ukrainian, which is explained by high number of students and the current political situation, when Jewish and Polish issues were widely discussed not only at the University.

A feature of the professors of St. Petersburg University was the commitment of the vast majority of them to liberal and cosmopolitan ideas. They were not members of student associations, but they could sympathize with their activities. In particular, cadet party professors sympathized with the struggle of national minorities for self - determination, but were opposed to radical revolutionary ideas.

The press and the public opinion it created influenced both the university administration and the state authorities. For example, a campaign against the Jewish percentage rate, widely reported in the press, led to its actual abolition in 1905 till 1908. The University Council adopted the first motion to abolish the percentage rate at a session on September 13, 1905 [34, p. 87.]. Subsequently, this issue was continuously discussed, and the university invariably acted in a favorable way for Jewish applicants [35, p. 189]. In 1905, the Jewish percentage norm was de facto abolished, the number of Jews rose significantly from 172 in 1904 to 1200 in 1907, when they became the second largest group after Orthodox students [3, p. 57]. A separate problem was the issue of assigning governmental scholarships to Jewish students – In this regard, the university persistently sought to equalize the rights of "persons of the Jewish faith" with other students [35, pp. 5–6.]. During the ministry Ivan Tolstoy and Petr von Kaufman, the government in general met the needs of universities, allowing admission in excess of the percentage rate, and Tolstoy, following the requests of universities, unsuccessfully tried to achieve the complete abolition of the percentage rate [52]. Tolstoy's petition, supported by the Council of Ministers, was rejected by The Emperor in January 1906 [26; 1, pp. 236–237].

The events of the First revolution contributed to the radicalization of Polish University students. It represented a separate "cultural and political unit" in the capital's University. Among the Polish University organizations, some almost did not participate in University life, among them: the nationalist Union of people's youth (*Sojuz narodnoj molodezhi*) and the Union of youth named after Adam Mickiewicz (*Sojuz molodezhi imeni Mickevicha*). The unions leaders refused to participate in the elections of the Council of elders, stressed their desire not to interfere in Russian affairs [4, p. 63–64].

According to the reports of the "Stock exchange news", a split between internationalists and "chauvinists" was immediately apparent within the Polish students from the beginning of the revolution. Characteristic in this regard is a newspaper article published on September 21: "Yesterday, in the Assembly hall of the University, a meeting of Polish students took place, which brought together more than 600 students in higher education institutions and a lot of invited audience. The debate was very heated and lasted well into the night. By twelve o'clock, the debate on the first of the five issues was still going on. The normal flow of the meeting was hindered mainly by Polish "nationalists" and "chauvinists" [51]. The next day the newspaper published an extensive report from this meeting, which described in detail the "struggle of the parties" and the tactics of "chauvinists" and "nationalists" to delay the meeting, which still managed to pass a resolution on "solidarity with the Polish workers" and an indication of the desirability of introducing autonomy in the Polish region" [48]. Right-wing newspaper "New time" characterized the same meeting somewhat differently, presenting to readers not the differences of the Poles, but the result of discussions: "First of all, upon the choice of the chairman, the question was raised about the attitude of the students of St. Petersburg to political life in the Kingdom of Poland. The essence of the speakers' demand was as follows: full autonomy of the Kingdom of Poland; an elected Sejm in Warsaw, equalization of Poles in rights with the native Russian population, education of the Polish population in the Polish language, etc." [14, 54].

The next meeting of Polish students took place on September 24 and aroused mainly the interest of the right-wing press, which, among other things, focused the readers' attention on the fact that "many outsiders and ladies were present at the meeting", and "the debates were conducted in Polish" [15]. Covering the meeting on September 26, "New time" noted that "most speakers recognized the merger of Russian and Polish parties as quite desirable, provided that the views of the Russian parties on the autonomy of the Kingdom of Poland as a national unit were clarified" [16]. A completely different view of the same meeting is found on the pages of the "Stock exchange news". The newspaper described the confrontation between "true Polish people" ("Polish chauvinists") and the Polish radical party. The latter, following the newspaper's account, won thanks to the passionate revolutionary speech of the Russian worker. However, the fact that the Polish students failed to reach a consensus is evidenced by the final resolution of the meeting, which has a very distant relation to the events in St. Petersburg: "In conclusion, the Assembly decided to send a welcome telegram to Ignatius Dashinsky, a member of the Austrian reichsrat: "Vienna. To Deputy Ignatius Dashinsky. It was with a feeling of irrepressible indignation that we read the telegram from Vienna that the Polish club had voted against universal suffrage. We are happy that there are Poles in the Austrian Parliament who did not participate in this act. Radical Polish youth. At the meeting (*veche*) on September 26" [56]. The polemic between the poles-internationalists and "chauvinists" continued on September 27, which is enthusiastically described as the "most interesting" meeting of the day by the "Stock exchange news". According to the newspaper, the victory was won by the Polish socialists, about whose position the journalist writes very pompously: "In the foreground they have universal interests, class struggle, begging, dark, downtrodden people – the proletariat. Before this world-wide question, purely Polish questions pale. Let it be good for humanity – let it be good for Poland. Our Russian brothers are fighting for the eternal ideals of freedom and equality... Let's give them a hand, let's go together and together, perhaps, create a better new life..." [50].

The Jewish question, or rather the question of the abolition of the percentage limit for Jewish students in universities, has been actively discussed in the press since the first decade of September [11, 12, 19, 20, 32, 33]. The liberal press emphasizes the efforts of the new rector of the University Ivan Borgman to cancel the percentage limit [18, 31, 33]. The norm that restricts the admission of persons of the Jewish faith to the University seems to journalists not only immoral, but also inherently illegal [59]. The need to cancel it was constantly discussed at student meetings, where there were even discussions between students who considered the "Jewish question" so fundamental that it could be the basis for a new strike and the majority who considered it subordinate to the more important question of the revolutionary reconstruction of society, for which the University should be opened [45]. The debate was sometimes very heated. "New time" described the events sarcastically: "Passions flared up to such an extent that the chairman had to stop the meeting due to the "excessive excitement of the audience". The students' outrage was caused by indignation at the scandalous behavior of Jewish students. They were not more than 100 in number, and they insisted that the whole assembly should submit to their wishes in the matter of the open doors of the University. The assembly wanted to pass a general resolution on this issue, while the Jews demanded a separate resolution for themselves on their free admission to universities. When the meeting objected to this, the Jews noisily began to protest and caused a 10-min. break. They used it to develop a further plan of their actions. After the break, they trooped from the IX auditorium, where they had been conferring, to the meeting and continued to behave as before" [13].

The episode with the mentioned meeting on September 15 became an occasion for further speculations of "New time". A week after the meeting, the publication derided the Jewish newspaper "Sunrise" (*Voskhod*), which tries to ""explain" and justify the behavior of those Jewish students who were outraged at the University-wide meeting" and concluded: "The more freedom is given to the Zionists to express their hidden thoughts and carry out their methods of political struggle, the sooner it will become clear what an "unfortunate appendage" of all academic life are such wildly national unions" [23].

Of course, the liberal press described this incident differently, but it also wrote about the accusations of "obstruction by fellow Jews" that were made by the leaders of the meeting [44]. However, as the same newspaper noted, a few days later, returning to this issue, the meeting already adopted without any significant debate a resolution in which "decided to open the doors of the University to all persons, without distinction of nationality or religion" [57]. There is no doubt that the MNE's decision (at least temporarily) to admit Jews to higher educational institutions in excess of the percentage limit helped to calm the situation.

The next time the "Jewish theme" surfaced was in mid-October in connection with an unsuccessful "party" organized by Jewish socialists in the student cafeteria. It is interesting that if "New time" simply stated that "the party failed" [11], "Stock exchange news" gave a brief and very ironic report about it: "The party in the dining room was organized by a Jewish socialist organization and attracted a huge number of guests, so there were not enough tickets or seats. Everyone complained about the closeness, that they could not hear anything. It is clear that the organizers cared more about the income than about the guests. At the party, Prof. Tarle, Prof. K.N. Eristov, attorney-at-law Sokolov, sculptor Ginzburg… Due to the fact that many people who promised their presence at the party did not arrive – the party soon ended" [47].

At the end of September 1905 the University also became the center of the Ukrainian national movement. Newspapers reported in detail about "the first Ukrainian rally in Russia for 250 years". It was originally planned as a meeting of Ukrainian students in the building of the old physical Institute. However, since there gathered about 2000 people, with the permission of the rector, the meeting was moved to the Assembly hall. In addition to University students, students from other higher educational institutions, female students and workers attended the event. However, quite quickly the number of participants thinned out: the organizers of the meeting decided to conduct the discussion "exclusively in the Little-Russian [Ukrainian] language, which was not clear to all those present" and many went to the nearby IX audience to discuss General political issues. As for the main ideas put forward, they largely repeated the slogans of the Polish meetings: the autonomy of Ukraine, protection of the Ukrainian language, accusations against Russian student societies, which "in relation to the liberation movement of the Ukrainian people acted in concert with the Russian government" [17, 49].

The Ukrainians were followed by Latvians who organized their own meeting. According to the "Light" newspaper, the meeting was quite national-revolutionary in nature. The following topics were discussed: "The situation in the Baltic region, the liberation movement in Russia, and the organization of the state Duma". The newspaper succinctly conveyed the mood of the participants: "The State Duma, which under the terms of the election will be dominated by Germans, cannot be expected to benefit the Baltic States. The only useful party for the present is the extreme party" [53].

The authors met only one mention of another national minority at the university – the Chinese. The centrist *Peterburgskaya gazeta* reported on the approval of the charter of the Student Research Society "The Far East" formed by the initiative of Chinese students, which aimed to study the life of China, Manchuria and Korea [55].

In addition to those mentioned above, other national student groups that were holding meetings at the University at that time were mentioned in the press: Lithuanians, Georgians, and others, but these topics did not receive significant attention from press [25].

4 Discussions of Research

The national question was one of the key issues for the Russian Empire, where more than half of the populations were not representatives of the titular nation. It largely determined the relevance of the political agenda and was one of the key events in the subsequent revolution. The enthusiasm with which many minorities joined the revolutionary movement is traditionally seen as one of the most obvious proofs of the failure of the tsarist government's national policy, which also affected the situation of higher education in the country. The diverse ethno-confessional composition of the radical St. Petersburg students, combined with the struggle for national interests, led to an increase in the political activity of students-representatives of national minorities. At the same time, it is not possible to talk about their unity. First, they preferred to defend their national interests, sometimes reluctantly adjusting to the agenda proclaimed by the Russian students. Second, there was disagreement among the communities themselves. In particular, this is clearly seen in the example of a community of Polish students, divided by their political views into two camps.

The press materials allow us to draw a number of original conclusions regarding the national movement at St. Petersburg University in 1905. First of all, we should mention their relatively low attention to the topic, both in the liberal and right-wing press. Apparently, national gatherings were not such an important and interesting news event for readers as the confrontation between the Ministry and the professor's corporation or the confrontation between students and the police. There is also some skepticism about events organized by national student organizations. And if it could be expected from the right-wing press, which used all sorts of ways to criticize Jewish and Polish people, then in the liberal press, which is very sensitive, for example, to the Jewish question, it looks quite unusual.

Some blame for ignoring the student national movement, according to the press materials, lay with the organizers themselves. Their desire to defend exclusively their local national interests, especially in combination with holding meetings in their native languages, turned away from them both the rest of the students and the press. Moreover, the events discussed at their meetings might seem insignificant to the Russian students, who perceived the struggle for national self-determination or the reduction of the percentage limit only as part of the desired revolutionary changes.

5 Conclusions

The research suggests that the beginning of the First Russian revolution in 1905 was a catalyst for the growth of activity and radicalization of the national movement among the students of St. Petersburg University. In the conditions of the general revolutionary rise in whole Empire, various national and political preferences and a system of self-identification in relation to the Empire and the Russian liberation movement were formed among the national student communities. The most active national student associations were Polish, Jewish, and Ukrainian. At the same time, their positions and goals differed significantly from each other. Thus, among Polish students there was a split between those who categorically refused to become part of the general revolutionary movement, putting the priority of defending exclusively national interests, and those who supported the existing socialist movements called "chauvinists". For Jewish students, the most urgent issue was the abolition of the percentage limit, which they planned to achieve be using the revolutionary rise among Russian students. Ukrainians also did not separate themselves from the general student's movement, but used it to promote national culture.

The right-wing press, seeing national minorities as a potential threat to the monarchy, especially when it came to Polish and Jewish, used news stories about the activity of national student associations as an additional opportunity to discredit them. The liberal press, also not missing such events, thus promoted national movements, considering them as an important element of liberalization. Thus, it promoted their institutionalization and self-identification of national elites.

Acknowledgments. The research was carried out with financial support from the Russian Science Foundation, project № 18-18-00121.

References

1. Ananyich, B.V., Ganelin, R.S.: Sergey Yulyevich Witte and his Time, Saint Petersburg (2000). (in Russian)
2. Ascher, A.: The Revolution of 1905: A Short History. Stanford (2004)
3. Barinov, D.A.: Ethno-Confessional and regional organizations of students of Saint Petersburg University (1884–1917). Ph.D. thesis (in history), Saint-Petersburg (2015). (in Russian)
4. Barinov, D.A.: National corporations of students of the Petersburg university at the beginning of the 20th century. Klio **1**(109), 61–67 (2016). (in Russian)
5. Barinov, D.A.: Student communities of Saint Petersburg university during the underground period (1884–1906). Klio **10**(106), 93–100 (2015). (in Russian)
6. Brower, D.: Training the Nihilists: Education and Radicalism in Tsarist Russia. Ithaca, New York (1975)
7. Figes, O.: A People's Tragedy: The Russian Revolution: 1891–1924, London (1996)
8. Galai, S.: The Liberation Movement in Russia 1900–1905, New York (1973)
9. Gusyatnikov, P.S.: Revolutionary student movement in Russia. 1899–1907, Moscow (1971). (in Russian)
10. Harcave, S.: The Russian Revolution of 1905, London (1970)
11. Chronicle. Novoe vremja [New time]. 1905. October 12 (25). No. 10635. P. 4. (in Russian)
12. Chronicle. Novoe vremja [New time]. 1905. September 11 (24). No. 10606. P. 5. (in Russian)
13. Chronicle. Novoe vremja [New time]. 1905. September 16 (29). No. 10611. P. 4. (in Russian)
14. Chronicle. Novoe vremja [New time]. 1905. September 22 (October 5). No. 10617. P. 3. (in Russian)
15. Chronicle. Novoe vremja [New time]. 1905. September 25 (October 8). No. 10620. P. 4. (in Russian)
16. Chronicle. Novoe vremja [New time]. 1905. September 27 (October 10). No. 10622. P. 3. (in Russian)
17. Chronicle. Novoe vremja [New time]. 1905. September 29 (October 12). No.10624. P. 4. (in Russian)
18. Chronicle. The response of the rector to the petition of Jews about their interest-free admission to the University. Birzhevye vedomosti [Stock exchange news]. 1905. September 13 (26). No. 9027. P. 3. (in Russian)
19. Chronicle. Among Jewish students. Birzhevye vedomosti [Stock exchange news]. 1905. September 8 (21). No. 9019. P. 4. (in Russian).
20. Chronicle. At the University. Birzhevye vedomosti [Stock exchange news]. 1905. September 10 (23). No. 9023. P. 3. (in Russian)
21. Ivanov, A.E.: Universities in Russia in 1905. Istoricheskie zapiski, vol. 88, pp. 114–149, Moscow (1971). (in Russian)
22. Ivanov, A.E.: Student corporations of the late 19th – early 20th centuries, Moscow (2004). (in Russian)

23. From Newspapers and magazines. Novoe vremja [New time]. 1905. September 23 (October 6). No. 10618. P. 3. (in Russian)
24. Kassow, S.D.: Students, Professors and the State in Tsarist Russia. Berkeley (1989)
25. Laverov, I.P.: The Emergence of the Caucasian friendly society at St. Petersburg university during the first revolutionary situation. In: Peterburgskij universitet i revoljucionnoe dvizhenie v Rossii [Saint Petersburg University and the revolutionary movement in Russia], Leningrad, pp. 92–101 (1979). (in Russian)
26. Memoria on the question of the admission of Jews to institutions of higher education establishments. In: Sovet ministrov Rossijskoj imperii. 1905–1906 gg. Dokumenty i materialy [Council of Ministers of the Russian Empire. 1905–1906. Documents and materials], Leningrad, pp. 196–199 (1990)
27. Morrissey, S.K.: Heralds of Revolution: Russian Students and the Mythologies of Radicalism, New York (1991)
28. Musaev, V.I.: Estonian Diaspora in the North-West of Russia in the second half of the 19th – first half of the 20th century, Saint Petersburg (2009). (in Russian)
29. Pavlitskaya, N.I.: Petersburg university in the revolution of 1905–1907. Bull. Leningrad State Univ. **11**, 138–156 (1948). (in Russian)
30. Pipes, R.: A Concise History of the Russian Revolution, Knopf (1995)
31. Latest news. Russkie vedomosti [Russian statements]. 1905. September 15. No. 251. P. 2. (in Russian)
32. Newscast. Petition of Jewish students. Birzhevye vedomosti [Stock exchange news]. 1905. September 12 (25). No. 9026. P. 2. (in Russian)
33. Newscast. Meeting of the University Council. Birzhevye vedomosti [Stock exchange news]. 1905. September 14 (27). No. 9030. P. 2. (in Russian)
34. Protocols of meetings of the Council of the Imperial St. Petersburg University for 1905. No. 61, Saint Petersburg (1906). (in Russian)
35. Protocols of meetings of the Council of the Imperial St. Petersburg University for 1907. No. 63, Saint Petersburg (1906). (in Russian)
36. Rostovtsev, E.A., Andreeva, V.V.: Crooked mirror: university life in September-October 1905. Klio **7**(103), 71–88 (2015). (in Russian)
37. Rostovtsev, E.A., Andreeva, V.V.: St. Petersburg university 1905 in the mirror of the periodical press. Kli **10**(94), 3–14 (2014). (in Russian)
38. Rostovtsev, E.A., Andreeva, V.V.: The warring university in the focus of the periodical press (1914–1915). Klio **10**(130), 118–130 (2017). (in Russian)
39. Rostovtsev, E.A., Sidorchuk, I.V.: Exiles of the "Soviet" university: the experience of a collective portrait of the teaching emigration of Petrograd. Bull. Saint Petersburg Univ. History **1**, 64–75 (2016). (in Russian)
40. Rostovtsev, E.A., Sidorchuk, I.V.: The myth of L.A. Casso. Bull. Tomsk State Univ. History **57**, 14–22 (2019). (in Russian)
41. Rostovtsev, E.A., Sidorchuk, I.V.: Saint Petersburg university in 1905. Nauchno-tehnicheskie vedomosti Sankt-Peterburgskogo gosudarstvennogo politehnicheskogo universiteta. Gumanitarnye i obshhestvennye nauki [Scientific and technical Bulletin of Saint Petersburg state Polytechnic University. Humanities and social Sciences], **3**(227), 98–107 (2015). (in Russian)
42. Shanin, T.: Revolution as a Moment of Truth: 1905–1907–1917–1922. Transl. from English, Moscow (1997)
43. Shevyrev, A.P.: Holidays of disobedience: student riots in the urban space of St. Petersburg in 1861–1905. In: Sankt-Peterb. un-t v XVIII–XX vv.: evropejskie tradicii i rossijskij kontekst. Materialy mezhdunarodnoj nauchnoj konferencii. 23–25 ijunja 2009 g. Saint Petersburg, pp. 148–162 (2010). (in Russian)
44. Gatherings of students. A meeting at the University. Peterburgskij listok [St Petersburg sheet]. 1905. September 16 (29). No. 245. P. 3. (in Russian)

45. Gatherings in higher educational institutions. Meeting at the University. Birzhevye vedomosti [Stock exchange news]. 1905, September 16 (29). No. 9033. P. 3. (in Russian)
46. Sladkevich, N.G.: Saint Petersburg university in 1905. Vestnik Leningradskogo gosudarstvennogo universiteta **3**, 23–35 (1955). (in Russian)
47. Student party. Birzhevye vedomosti [Stock exchange news]. 1905. October 12 (25). No. 9071. P. 4. (in Russian)
48. Student Affairs. Meeting of Polish students. Birzhevye vedomosti [Stock exchange news]. 1905. September 22 (October 5). No. 9043. P. 2. (in Russian)
49. Student Affairs. Ukrainian rally. Stock exchange news. 1905. September 29 (October 12). No. 9055. P. 3. (in Russian)
50. Student Affairs. In Saint-Petersburg University. Birzhevye vedomosti [Stock exchange news]. 1905. September 28 (October 11). No. 9053. P. 3. (in Russian)
51. Student Affairs. In Saint-Petersburg University. Birzhevye vedomosti [Stock exchange news]. 1905, September 21 (October 4). No. 9041. P. 2. (in Russian)
52. Tolstoy, I.I.: Note on the regular events of the Ministry of public education and the draft memorial of the Council of Ministers on the abolition of restrictions on Jews entering educational institutions. Department of manuscripts of the Russian national library. F. 781. D. 115. L. 7–8ob. (in Russian)
53. University news. Svet [Light]. 1905. October 10. No. 264. P. 1. (in Russian)
54. University news. Svet [Light]. September 23. 1905. No. 248. P. 1. (in Russian)
55. In educational institutions. Circle for studying the Far East. Peterburgskaya gazeta [Petersburg newspaper], October 8. No. 265. P. 3 (1905)
56. Yesterday's meetings. Meeting of Polish students. Peterburgskij listok [St Petersburg sheet]. 1905. September 27 (October 10). No. 256. P. 3. (in Russian)
57. Yesterday's meetings. At the University. Peterburgskij listok [St Petersburg sheet]. 1905. September 20 (October 3). No. 249. P. 3. (in Russian)
58. Yakovlev, V.P.: Petersburg University in the revolution of 1905–1907. Vestnik Leningradskogo gosudarstvennogo universiteta. History, linguistics, literature, vol. 1, pp. 19–25 (1980). (in Russian)
59. Alexandrovsky, Y.: Against interest rates. Birzhevye vedomosti [Stock exchange news]. 1905. September 16 (29). No. 9033. P. 2. (in Russian)
60. Zavyalov, D.A.: Student scientific societies of Saint Petersburg University in the late 19th – early 20th centuries. PhD thesis (in history). Saint-Petersburg (2006). (in Russian)

Modern Socio-political Crisis in the USA Based on Materials of the Russian Portal InoSMI

Andrei V. Grinëv(✉)

Peter the Great St. Petersburg Polytechnic University, St. Petersburg, Russia

Abstract. This article studies the Russian portal InoSMI, which publishes translations and video materials of foreign media, primarily European and American. Despite the fact that this well-known news resource positions itself as the most neutral and objective, the selection of media and translated materials indicates a clear tendency toward conservatism and rejection of left-wing radicalism and political correctness. This position of the portal is explained by the policy of the state to which it belongs, and the specifics of the Russian audience. In recent months, the portal has focused on the acute socio-political crisis in the United States, which arose as a result of the death of George Floyd and the rise of the Black Lives Matter movement. The article examines some aspects of this crisis and the reaction of Russian visitors to the portal. It is concluded that although the portal is a hidden mechanism of political and ideological manipulation, it nevertheless performs a very useful function, introducing Russian-speaking readers to the point of view of foreign journalists, politicians, and experts on various topical issues.

Keywords: Modern crisis in the USA · International media · Black lives matter

1 Introduction

Now in Russia, with the coronavirus epidemic and the inability to move freely between countries, almost the only way to get information about the situation in the United States after the death of George Floyd and the activation of the Black Lives Matter (BLM) movement is the Internet and television. One of the main sources in this case that can serve is the Internet portal ИноСМИ.ru [InoSMI.ru] (foreign media), whose employees track and translate Russian articles from foreign newspapers and magazines (including online publications and materials from Internet channels), such as the *Washington Post*, *Le Figaro*, *Newsweek*, *Svenska Dagbladet*, and so on. Much less common are materials taken from the Arab press and from the media of Turkey, Japan, China, and other countries. The range of topics published on the portal is very broad, but most of the articles are the analytical materials of Western journalists often related to Russia. Translations are posted daily in real time (online) in five main categories: Politics, Economy, Science, War and the Military-Industrial Complex, and Society. The topics we are interested in are mainly presented in the first and last categories. As a rule, articles are published in

© The Author(s), under exclusive license to Springer Nature Switzerland AG 2021
R. Bolgov et al. (Eds.): *Proceedings of Topical Issues in International Political Geography*, SPRINGERGEOGR, pp. 427–445, 2021.
https://doi.org/10.1007/978-3-030-78690-8_37

full, and only sometimes with abbreviations, and only rarely is information from the foreign press given in the form of digests. However, the portal occasionally publishes videos from various foreign news channels or its own short reviews and commentary.

Being a fairly successful product of media convergence, the InoSMI portal appeared about 20 years ago (2001), when the domain was registered inosmi.ru. Since then the portal has gained a certain popularity among Russian intellectuals who are interested in foreign politics and life in foreign countries in general. Currently, InoSMI is associated with the Russian international information Agency RIA Novosti with the support of the Federal Agency on Press and Mass Communications of the Russian Federation (Rospechat). The peculiarity of the portal is that, unlike the main Russian mass media, political censorship is carried out in a soft, unobtrusive form usually in the form of short summaries of the content of published materials. Articles containing critical statements about President Putin, Russia, Moscow's politics, the way of life of Russians, and so on are allowed. Despite this, the information resource is occasionally criticized for silencing, distorting, or changing the source text of publications, titles, and images. Sometimes this is simply the result of inattention and haste on the part of the translator, which is usually immediately reacted to by readers in their mocking comments on the translation. Various distortions are especially common in translations from Eastern languages, especially Chinese and Japanese. There are many fewer problems with English, and the English-language press, primarily the American press, has become the main provider of information about the situation in the United States and other Western countries for readers of InoSMI.

2 Methods and Materials

In preparing this article, such standard theoretical methods of scholarship were used as induction and deduction, analysis and synthesis, a systematic approach, the method of social modeling, the comparative-typological and comparative-analytical method. In addition to theoretical methods, practical methods such as analysis of printed and electronic sources of information, and the method of content analysis were widely used in writing this article. The use of the latter helped, for example, in identifying the block of articles directly or indirectly related to the socio-political crisis in the USA, posted on the InoSMI portal – see, please, the Table 1.

The table contains materials from various foreign media outlets, as well as data provided by the InoSMI portal itself for the period from late May to early September 2020, which were posted in the "I can't breathe" section. As we can see from the table, InoSMI paid primary attention to the American mass media (24 titles), twice less were represented by the British mass media (12) and even less – the mass media of other countries. This was no coincidence, since it was in the United States that an acute socio-political crisis arose, which then affected the countries of Western Europe. Judging by the statistics given in the table, the peak of the crisis occurred in June, when the events in the United States and Western Europe were widely covered by the media around the world. In July and August, media interest in them began to subside noticeably and their attention began to shift to the beginning of the presidential race in the United States and other world news. What was the cause of the acute socio-political crisis in the United States?

Table 1. Set of articles related to the socio-political crisis in the USA, posted on InoSMI

N	Media name	Country	MAY	JUN	JUL	AUG	∑	N	Media name	Country	MAY	J UN	JUL	AUG	∑
1	ABC	AU		1			1	40	NY Times	US		3	3		6
2	Actulando	BG		1			1	41	Newsweek	US		2			2
3	Al Araby Al Jadid	UK		1			1	42	NoonPost	EG		1			1
4	The American Conservative	US	2	4	1	1	8	43	Nyheter Udag	SE		1			1
5	The American Spectator	US			1		1	44	Obozrevatel’	UA		1			1
6	American Thinker	US	1	3		1	5	45	Open Democracy	UK		1			1
7	Apostrof	UA		1			1	46	Politiken	DK			1		1
8	The Atlantic	US		1			1	47	Polityka	PL		1			1
9	BBC	UK		1			1	48	Project Syndicate	US		2	1		3
10	Berlingske	DE		2			2	49	Rai Al Youm	UK		1			1
11	Breitbart	US	1	3		1	5	50	Rénmín Rìbào	CN		2			2
12	Business Insider	DE		1			1	51	Resett	NO		1			1
13	CNBC	US		1			1	52	Reuters	UK		1			1

(*continued*)

Table 1. *(continued)*

N	Media name	Country	MAY	JUN	JUL	AUG	$\sum$	N	Media name	Country	MAY	J UN	JUL	AUG	$\sum$
14	CNN	US		7			7	53	Russkaya Germaniya	DE		2	2		4
15	Dagens Nyheter	SE		2			2	54	Rzeczpospolita	PL		1			1
16	Daily Mail	UK		3	2	1	6	55	Sabah	TK		2			2
17	Delovaya stolotsa	UA		1			1	56	Sasapost	EG		2			2
18	Duma	BG		1			1	57	Sega	BG			1		1
19	ESPN	US			1		1	58	The Spectator	UK		2	1		3
20	Expressen	SE		1	1		2	59	Der Spiegel	DE			2		2
21	Le Figaro	FR		2		1	3	60	Steigan.no	NO			1		1
22	Focus	DE		1			1	61	Strana	UA		1	1		2
23	Foreign Policy	US		2	2		4	62	Süddeutsche Zeitung	DE		2			2
24	Fox News	US	1	5	2		8	63	The Sun	UK		1			1
26	Glavred	UA		7	1		8	64	Svenska Dagbladet	SE		4	1		5
27	Guancha	CN		1			1	65	The Telegraph	UK		3	1		4
28	The Guardian	UK	1	2	2		5	66	The Times	UK		2			2
29	The Hill	US		2			2	67	TMZ	US		1			1

(continued)

Table 1. (*continued*)

N	Media name	Country	MAY	JUN	JUL	AUG	∑	N	Media name	Country	MAY	J UN	JUL	AUG	∑
30	InoSMI	RU	1	10	3	2	16	68	Trud	BG		1			1
31	The Jerusalem Post	IS		1			1	69	Ukrainska Pravda	UA		1			1
32	La Journada	MX		1			1	70	USA Today	US		1			1
33	Jyllands-Posten	DK		1		1	2	71	Valeurs Actuelles	FR				1	1
34	Los Angeles Times	US		1			1	72	La Vanguardia	SP		2			2
35	Marianne	FR		1			1	73	Washington Examiner	US		1			1
36	Medium	US		1			1	74	The Washington Post	US		3	2		5
37	Milliyet	TK		2			2	75	The Week	UK		1			1
38	NBS News	US	1	1			2	76	Die Welt	DE		2			2
39	New York Post	US				1	1	77	Xīnhuáshè	CN			1		1
									Total		8	128	34	10	180

3 The Death of George Floyd and the Heyday of the Black Lives Matter Movement

The mechanism that triggered the current unprecedented socio-political crisis in the United States was the tragic death of an African-American man, George Floyd, in Minneapolis on May 25, 2020, after being detained by a police patrol. On all the world's leading TV channels and Internet sites, footage was broadcast of a burly police officer, Derek Chauvin, kneeling on the neck of Floyd, who was lying on the pavement for more than eight minutes, repeating "Please" and "I can't breathe" while simultaneously making moans and sobs. These terrible images immediately led to an explosion of indignation and mass demonstrations, not only in the United States, but also in a number of European countries, where a substantial number of people from Africa or their descendants live. In addition, on May 28, the UN High Commissioner for Human Rights, Michelle Bachelet, condemned the murder of George Floyd and called for the law enforcement authorities to be punished. At the same time, the High Commissioner also called on all protesters in Minneapolis to peacefully express their outrage, as reported by InoSMI the next day [29].

The death of George Floyd and subsequent events in the United States and Europe, which received wide coverage in the world press and online publications, led to the appearance of a special section on the InoSMI portal with the catchy title "I can't breathe. Protests in the USA – 2020." In the three months since Floyd's death, about 200 articles and videos have been published in this category, identified using the method of content analysis. The first, May 27 video was posted by Fox News under the headline "In Minneapolis, the Death of an African-American after Police Custody Caused Riots" [16]. InoSMI readers could find out the next day what the riots were about from Rod Dreher, editor of the daily electronic magazine, whose name speaks for itself – the *American Conservative*. His sometimes ironic essay was accompanied by several videos that showed how an excited crowd looted stores and clashed with the police. Dreher wrote on May 28: "Today in Minneapolis, civil rights activists expressed their outrage over the death of George Floyd, who was killed by police. In protest, they relieved the Target store next to the Third precinct of the burden of televisions and other goods". And added: "Tonight, on a local TV station in Minneapolis, a local Black activist pleaded with the protesters to stop the violence and focus on justice. He's right. Riots, robberies, and attacks on police cars will not bring the murderers of George Floyd to justice. What you will achieve is the re-election of Donald Trump" [8].

The next day, InoSMI published a NBC News story with several videos about the protests in the city of Louisville in the United States, during which several people were injured [26]. The baton was passed to the publication of Ethel Fenig in the conservative electronic magazine *American Thinker*. She wrote that the protests in the United States caused by the death of an African-American man after his detention by white police officers revealed an unusual trend. If a Black person is killed by White people, it is a "manifestation of racism." But deadly showdowns between "colored" communities are just material for police reports [14]. On the same day InoSMI published another batch of Rod Dreher tweets, though now the topic was not the robbery but rather the burning of the ill-fated police station in Minneapolis, where the cops responsible for Floyd's death were stationed. According to Dreher, the craziest thing is the reaction of the progressive public in the person of director Michael Moore: "Good citizens burned down a villainous police station" [9].

But the burning of the police station was not limited to protesters in the United States. A few days later their victim was now a historic Episcopal Church in Washington, not far from the White house, where all U.S. presidents since James Madison prayed, as Rod Dreher reported with horror and indignation, accompanying his words with tweets and a report by the Fox News channel: "How long can this be permitted? They are destroying our history and our heritage. And the authorities are showing complete impotence" [10]. Dreher's comments and tweets have led to such thunder and lightning from liberal critics who have accused him of racism that he was forced to make a special minute-by-minute investigation into the death of George Floyd, which conclusively proved that he died not from police violence, but as a result of a drug overdose. The police officer used the neck compression technique not because Floyd was Black, but because he lied and resisted police orders for at least eight minutes [12].

Meanwhile, the world press has become increasingly involved in coverage of events in the United States. On May 31, InoSMI published a reprint of a report from the English *The Guardian*, which said that on Saturday evening, May 30, riots broke out in many U.S. cities due to the murder of George Floyd and the death of Black Americans at the hands of police, and the mayors of major cities imposed a curfew. Donald Trump added fuel to the fire by harshly criticizing "anarchists," as he described the protesters [34].

In fact, the protests were led not by anarchists, but by activists of the far-left anti-racism movement Back Lives Matter. BLM was born in 2013, but its "finest hour" began after Floyd's death. The movement took on a special scale thanks to the support of left-wing radicals of various stripes, who often appear under the collective name Antifa (anti-Fascists) [43]. It was them, not the BLM, that Trump blamed for all the deadly sins, since he did not want to spoil relations with Black voters and get labeled a "racist," and therefore immediately made a proposal to put "Antifa" on the list of terrorist organizations. A similar opinion was expressed by journalist Andy Ngo of the British political weekly *The Spectator*. Ngo wrote that left-wing Marxists in the United States had been preparing for a revolt for decades and they only needed an excuse [40]. By contrast, Marina Meseguer, a journalist for the Spanish left-wing newspaper *La Vanguardia*, suggested that the main internal threat to the United States was not Antifa activists but rather American right-wing radicals [36].

There is nothing surprising in the cooperation between Antifa and the BLM movement. Three years before of the tragic events in Minneapolis political scientist Andrew Cornell wrote about this:

> *As in other parts of the world, anarchists, socialists and Marxists based in the USA have frequently influenced and borrowed from one another over the past century and a half of struggles. More research into these lines of influence is certainly called for. However, any thorough investigation of the cross-pollination of radical traditions in the USA must also consider the many ways in which the autonomous freedom struggles of people of colour have co-mingled with European-origin traditions such as Marxism and anarchism. In fact, I would suggest that it has frequently been on the terrain of campaigns opposed to white supremacy and colonialism*

that anarchists, socialists and Marxists have found common ground to collaborate and to develop synthetic theoretical and tactical paradigms [6].

Meanwhile, in June, the Black Lives Matter movement gained particularly broad scope with the support of left-wing groups and mayors of cities where representatives of the Democratic party were in power (who were actively playing the BLM card in their anti-Trump political war). The confrontation between left-wing radicals and the police reached its greatest intensity in Chicago, Portland, and Seattle, where there was even an independent "Capitol Hill Autonomous Zone", material about which was taken by InoSMI from the magazine *USA Today* [37]. In some cases, police officers and soldiers of the National guard actually sided with the rioters and knelt before them in solidarity and respect for the BLM movement, as well as the victim of police brutality – George Floyd. Democratic members of Congress did the same, and some White Americans, including police officers, went even further and publicly washed the feet of BLM activists and kissed their shoes as a sign of atonement for "the collective guilt of all White people". The InoSMI portal clearly demonstrated the relevant footage to Russian readers by including in the published material a video taken from the *Thirty Mile Zone*, a tabloid news website [40].

In parallel, leftists from Antifa and representatives of the BLM called for the police to be deprived of funding, which can have fatal consequences for the U.S., as noted by Fox News columnist Curtis Hill [27]. However, InoSMI provided a good illustration of what would happen in such a case, using the material of journalist and editor Beth Baumann, published in the online magazine *American Thinker*. The tweet-laden story refers to a completely absurd case in which a Black woman posted a live broadcast on Facebook of Arab merchants armed with AK-47s defending their store in Chicago, which she intended to rob. The woman was convinced that the crowd should be allowed to rob stores since they are insured. For help in the upcoming robbery, she turned to the police (!), but they replied that this was not their station and refused to be complicit in the robbery, which led the woman to extreme rage because of the "sabotage" of the cops [1]. Naturally, such episodes could not add to the popularity of the BLM movement in the eyes of the Russian audience InoSMI.

Much more serious was a demonstration March of about 2,500 heavily armed Blacks from the NFAC (Not F*****g Around Coalition) held on July 25 in Louisville (Kentucky). The NFAC coalition exists separately from the BLM movement and does not set out to transform American society in a Marxist manner. The leader of the coalition nicknamed Grandmaster Jay (his name is John Fitzgerald Johnson) made it clear that his organization is ready to use violence and its goal is to create an independent Black state, for example, in the territory of Texas – writes the British newspaper *The Telegraph* [38], the message of which is translated by InoSMI.

Thus, replacing the poorly organized BLM, which puts economic and social transformations under the Bolshevik slogan "Expropriation of expropriators!" there may come a force that will be much more dangerous for the internal order and stability of the United States. What threatens America with the emergence of such organizations will be discussed below. The political scientist Dr. Ariel Koch warned about future crisis in 2018:

> *The second decade of the Twentieth-First century is characterized by the intensification and increase of violent anarchists, who once again regard violence as a legitimate tool for the purpose of "self-defense" and "struggle" against "fascist" opponents. There might be several reasons for that: a) the availability of the Internet, which enables left-wing extremists, especially among the young generation, to form transnational radical communities; b) the surge in right-wing extremist political parties and street movements; And, c) the existence of battlefronts near Western Europe, such as in Syria and Ukraine, where different extremists from hostile ideologies, including anarchists and anti-fascists, can travel to easily, join a fighting force, and get trained* [33].

4 Historical Monuments as Sacred Victims of BLM

One of the most striking symbols of the current socio-political crisis in the United States is the campaign to demolish historical monuments, which is practiced by the Black Lives Matter movement. The number one goal is considered to be monuments dedicated to political and military figures and pioneers and entrepreneurs who are associated with the history of slavery and colonialism. But sometimes the victims of BLM activists are monuments that have nothing to do with slavery and colonialism, such as the statue of the Wapiti deer in Portland (OR) and even the figure of the abolitionist Frederick Douglass in Rochester (NY), which indicates a poor knowledge of the history of the movement's participants and their propensity for vandalism.

From America, an epidemic of destruction and desecration of monuments to real or imaginary slaveholders, racists, and colonizers spread across Western Europe. In the UK, for example, the statue of Queen Victoria in Leeds was covered in paint, the monument to Winston Churchill in Central London was labeled "racist," and an old statue of the slaver and philanthropist Edward Colston in Bristol was thrown off its pedestal by BLM supporters and drowned in the Bay. Noting these barbaric acts with appropriate tweets, the editor of the *American Conservative* magazine Rod Dreher wrote:

> A college in Liverpool will rename the dorm, striking out the name of William Gladstone, the great liberal premier of the nineteenth century, because his father – his father! – had slaves. All this is nothing more than an attempt to defile, destroy, and erase British history in the name of fighting racism. Where are the men and women who will give battle to this hysterical iconoclasm? There was one in Oxford. Who's with him? If you, dear compatriots, believe that Washington, Jefferson and the other founding fathers will not raise their hand, then you are mistaken. London berserkers from Black Lives Matter even desecrated the statue of Abraham Lincoln, the liberator of slaves [11].

The correspondent of the left-wing Spanish newspaper *El Mundo*, Carlos Fresneda, took a completely different approach to the fight against monuments in the UK. He fully supported the demolition of the monument to Edward Colston in Bristol: "This statue is a reminder, among other things, that it was Edward Colston who led the Royal African Company, which transported more than 80,000 slaves (men, women, and children) at the end of the 17th century" [17]. However, a few days later, Swedish journalist Ronie

Berggren on the pages of the online newspaper *Nyheter Idag* condemned the demolition of the statue of Colston by the hands of supporters of Black Lives Matter, emphasizing: "If some conservative movement doesn't appear soon to put an end to this, the twilight of the West is coming.... Vociferous activists threaten to break our foundation. If they succeed, other countries will rise at our expense. People, who do not destroy their monuments, do not self-reflect and do not engage in complex self-criticism. Then totalitarianism and slavery will return – albeit in a different form" [2].

But this view is alien to many modern American left-wing intellectuals. So, professor of law George Shepherd in an interview with CNN called for the destruction of giant bas-reliefs on stone mountain near Atlanta, dedicated to the military leaders of the southern Confederates. Justifying his claim, Shepherd stated: "One of the means of oppressing African-Americans are monuments to white supremacists" [42]. They eventually included the monument to George Washington in Portland: on June 19, BLM activists toppled his statue from its pedestal and burned the American flag on it, as reported by *Newsweek* correspondent Soo Kim [31].

But the left is not limited to politicians of the past. Three days after the Washington monument was toppled, one of the leaders of the Black Lives Matter movement, named Shaun King, said that all murals and stained-glass windows depicting a white Jesus, his European mother, and their white friends should also be removed, as this is a crude form of white supremacy created as racist propaganda. King expressed his opinion in a series of tweets against the backdrop of large-scale vandalism against historical monuments during the BLM protests [3]. Partly King's idea will be taken to implement in England, where in the Cathedral of St. Alban of the 9th century, as a sign of solidarity with the BLM movement, Christ will be depicted as a Negro [13].

It seems that there are no places left in the United States where the influence of the BLM does not operate. The journalist of the Swiss newspaper *Neue Zürcher Zeitung*, Robert Kindler, informs readers:

> *The fight against monuments has reached Alaska. In the city of Sitka in the south of this American state, a conflict broke out over a monument that was once erected in honor of Alexander Baranov. Indigenous activists are demanding that the monument be removed from the city center—allegedly it has nothing to do with their history and is dedicated to a person who does not deserve their respect. Baranov, who at the beginning of the 19th century was the main representative of Russian colonial power on the American continent, according to them, represents the destruction of their culture as well as the death of a great number of Alaskan residents. The debate about this monument has been going on for several years, but with the rise of the Black Lives Matter movement and the heated debate over historical monuments, the conflict has again escalated* [32].

I must say that although Baranov was a harsh ruler, he was not a slave owner, a racist and did not oppress the local Tlingit Indians [23].

At the same time, American authorities at various levels often follow the lead of the BLM and voluntarily dismantle and remove historical statues and monuments. So, the management of the American Museum of Natural History in New York decided to dismantle the bronze monument to Theodore Roosevelt at the entrance to the museum.

As its management emphasized, this decision is not related to the personality of the 26th U.S. president, but to the compositional decision of the monument: Theodore Roosevelt is depicted on horseback, accompanied by an Indian on foot and a native of Africa. According to the museum's management, the monument depicts the Black and Native American population as subordinate and racially inferior. President Trump and well-known political commentator of the Fox News channel Tucker Carlson, whose opinion is cited by InoSMI, immediately opposed the relocation of the monument. In Carlson's opinion, Teddy Roosevelt was a wonderful, gifted, and very decent man, the best president of America. Further, Tucker Carlson listed the most famous monuments that were victims of the crowd under the slogan of fighting racism. He warns: "This is not a spontaneous public disturbance. This is a serious and highly organized political movement. It is not superficial. This movement is broad and thorough. They are driven by huge ambitions. It is dangerous, and it will grow. Its goal is to end liberal democracy and challenge Western civilization itself" [5].

He was echoed by the famous French philosopher and publicist Alain Finkielkraut, who wrote on the pages of the right-wing news magazine *Valeurs actuelles*:

> *Behind the wave of monument demolitions is racism directed against Whites and hatred of Western civilization, which is blamed for all the sins. . . . At the moment, France is going through a terrible crisis, during which some are attacking visible symbols of our history (especially statues) and advocate their forcible removal from public space. The situation is not new, but such an import of American political correctness to France seems to be a complete disaster. We risk getting involved in a process that will never end. Today it is Colbert, and tomorrow it may be Jules Ferry, who said that it is the duty of the higher races to "civilize the lower". History should be viewed in context, not transferred to the past by modern obsessions* [15].

Thus, on the InoSMI portal Russian readers were able to get acquainted with various points of view regarding the fate and significance of some historical monuments in the United States and Western Europe in light of the aggression against them by BLM activists. Needless to say, the portal's sympathies were clearly not on the side of the latter. And this is no accident, as will be discussed below. The overthrow of monuments has become a visible symbol of the current socio-political crisis in the United States and the West as a whole.

5 The Socio-political Crisis in the United States as the Beginning of the Decline of Western Countries

The mass unrest, pogroms, and looting that swept the United States after the death of George Floyd caused alarm in other countries, especially in the ranks of American satellites such as Poland and Ukraine (as judged by the materials on InoSMI). In early June, the Polish reporter Jerzy Haszczyński published an article in the popular newspaper *Rzeczpospolita*, in which he told about the deep crisis the United States was in as a result of the coronavirus epidemic, the economic downturn, and unrest among the population. He asked his readers anxiously: "The image of the United States, the center of Western civilization, is collapsing before our eyes. Will it be possible to restore it? Isn't it time for us to ask ourselves whether we can pin all our hopes for our future and security on the United States?" [25]. It is obvious that the journalist voiced the concerns of the ruling circles of Poland and the significant part of Polish society that rely on the support of the United States in the confrontation with Russia and Germany.

Following a note from a Polish newspaper, InoSMI published a series of apocalyptic forecasts by Ukrainian economic expert Andrei Golovachev, reprinting his articles from the online magazine *Glavred*. According to the expert, now begins the process of extinction of the American Empire:

> *Any empire begins to die at the moment when it has reached its maximum power. This is quite obvious: when a social system has no opponent, it loses its incentives to develop and decomposes. The same effect can be observed in business. The United States began to die immediately after the collapse of the Communist system. That is, since the end of 1991, the USSR was the last enemy of the United States and with its collapse, America no longer has any enemies and the whole world belongs to it. To understand what happened next to America, it is better to use the analogy of Rome after its victory over Carthage in the Second Punic war. . . . Making America great again is no longer possible. The difference in wealth between the plutocracy and the plebs in the United States is simply prohibitive, and this potential difference is already constantly flaring up, as it has flared up now in Minneapolis. But everything will finally collapse when the U.S. will no longer be able to make the rest of the world feed its plebs by force of arms* [19].

In another publication, Golovachev predicts the inevitability of a color revolution in the United States: "As always, the real stability of America was overestimated. The colossus was on feet of clay. Everything inside has long been rotten and democracy has degenerated into an oligarchy. Well, the oligarchy, as you know, is the most unstable regime. Especially in a country where there are huge interracial problems, as in the United States" [20]. I can agree that in the United States, there is a very significant social stratification, however, the oligarchy is not a political regime but rather a social group of super-rich people. Their political dominance usually leads to the establishment of an authoritarian regime, which, in fact, is rarely stable. But America is still far from authoritarianism: the roots of democracy are still too strong. Only incessant unrest and the imposition of an inherently totalitarian ideology of political correctness [4, 35] can eventually provoke a transition to authoritarianism.

In a new article by Golovachev, reproduced by InoSMI, he again repeats the thesis of the inevitability of a civil war in the United States, which, after the death of Floyd and mass unrest, has lost the moral right to point out to other countries that they violated principles of democracy: "Promotion of democracy around the world, protection of human rights in any corner of the planet have always been the main content of U.S. foreign policy. After the current protests, America lost this right because protesters in European countries surrounded U.S. embassies and accused them of racism and police brutality" [21].

Golovachev's last speech, which was published on the InoSMI portal, smacks of extreme pessimism:

> *If we call a spade a spade, we have to admit that a socialist revolution is beginning in America. On May 25, after the death of drug addict and repeat offender George Floyd, there was an uprising of African-Americans, Antifa, and plebs. The revolution itself will begin later, perhaps this fall. It is noteworthy that the protesting plebs are supported, both morally and financially, by large corporations and banks. The reason for this is that all large corporations and banks have long been united with the state and the prospect of a transition to socialism does not frighten them, but on the contrary, they are quite happy since it will finally consolidate their monopoly status and free them from all competition and threats from private business. Small and medium-sized businesses, which are actually the mainstay of capitalism, are rapidly disappearing. After the epidemic, more than 50% of small and medium-sized businesses will not rise. These people will turn into plebs. Gradually, the social structure of socialism will be finally formed in America. The transition to socialism under this structure will be almost inevitable* [22].

Serious concerns for the future of the United States have been expressed by former and current American diplomats. In a review prepared by CNN commentators, one of them stated:

> *Trump's remarks and his inappropriate and excessive reaction make it difficult for American diplomats around the world to work. In the past, America has been seen as a champion of human rights, a guiding star that calls for restraint and reasonable compromises. And now, instead, her actions are at best causing great concern and alarm, and at worst contempt and ridicule.... When America is torn apart by internal contradictions it cannot be a leader, it cannot be an effective player on the world stage* [24].

The same question was raised three days later in a subsequent CNN story, which was again quoted by InoSMI. The story was a sharp criticism of President Trump, who threatened to deploy armed forces across the country to achieve "total domination", and called the protesters "thugs" and even "terrorists". In addition, CNN reported, China, Russia, and Iran took advantage of the difficult situation in the United States, whose representatives pointed out that the brutal suppression of internal unrest by the American authorities deprives them of the moral right to call themselves a "beacon of democracy". The U.S. allies, including Australia and the Philippines, were also very dissatisfied with Floyd's death and Washington's repressive actions [7].

In a report, Swedish military journalist Magda Gad wrote that what she saw in Chicago after the demonstrations, pogroms, and looting was not much different from the scenes that she repeatedly observed in Iraq and Afghanistan [18]. It is echoed by Stewart Weiss in the pages of the Israeli newspaper *The Jerusalem Post*, who was shocked by the scenes of mass violence, destruction, and looting after the tragic death of George Floyd [46]. In light of this, it is not surprising that the left-wing Swedish journalist Göran Rosenberg expressed doubts about the peaceful transfer of power in the U.S. in the fall of 2020 [41].

As a result, the picture is quite disturbing. A number of publications on the InoSMI portal indicate the deep crisis that American society is going through and contain unambiguous prophecies about the impending civil war in the United States, which will undoubtedly have colossal consequences for the whole world. Although such prophecies do not look convincing enough yet, their appearance in itself is very symptomatic.

6 The Reaction of Russian Readers to the Materials of the InoSMI

It is interesting to follow the reaction of the Russian-speaking audience regarding the materials published on InoSMI. They express their attitude in two ways: by the review of specific articles and by commenting on them. Usually more of the readers' attention is attracted by materials that deal with Russia, which is quite natural and understandable. One of the largest number of views (53,944) was awarded to an interview with BBC reporter Amalia Zatari about problems with racism in Russia when she interviewed several Black people living in the country. From the responses of respondents, it follows that racism exists, but it is different than in the U.S. (not systemic, but domestic) and without extreme manifestations in the form of police violence [47].

Comments of the readers are always hundreds of times fewer than views. Judging by the constantly appearing nicknames of a number of users of the news resource, it seems that the InoSMI portal operates a group of paid commentators – either organized by the portal itself or by state secret services. The statements of these InoSMI regulars are often characterized by intolerance to democratic views, jingoism, and sometimes outright chauvinism with direct insults to representatives of other nationalities and political opposition. However, the portal moderators are not in a hurry to delete such comments, despite the fact that the official "rules of commenting" of the InoSMI portal states:

> *The user undertakes to speak respectfully to other participants of the discussion, readers, and persons appearing in the materials. . . . A user's comment may be edited or blocked during the posting process if it promotes hatred, discrimination on racial, ethnic, sexual, religious, or social grounds, contains insults or threats against other users, specific individuals, or organizations, infringes on the rights of minorities, violates the rights of minors, or causes harm to them in any form.*

In the analysis of responses from readers, the historical past of Russian society, the current political situation, and cultural traditions should be kept in mind. It is impossible for Russians who have never been involved in the Black slave trade and exploitation to understand why they should bear any responsibility for the sins of Western Europeans and Americans, repent before Blacks, and get down on their knees before them. This outraged a Black history teacher from Texas, Kimberly St. Julian-Varnon, who wrote on the pages of *Foreign Policy*:

> *The Black Lives Matter protests have spread around the world, but Russia remains a notable exception. Solidarity demonstrations against racism and police violence were held in Helsinki, Almaty, Kazakhstan, and Vilnius, Lithuania, but not in Moscow. Instead, Russia used the civil unrest in the United States to continue its tradition of mirroring the most unworthy aspects of American life on itself.*

The Russian government and its liberal opposition have used their platforms to discredit the relatively peaceful spirit of the demonstrations and instill the idea of U.S. weakness. . . . The development of the Russian Lives Matter movement is perhaps the clearest example of how the liberal opposition unwittingly helps its government undermine global efforts to combat racism [44].

There is nothing strange about this. In Russian society, there is still a memory of "Soviet political correctness," when the Communists – "fighters for universal equality" – provided benefits to people of proletarian origin and the Communists themselves, as well as representatives of many national minorities, under the pretext of overcoming their age-old backwardness. Naturally, people who were not members of the selected social groups were subjected to direct or indirect discrimination. Therefore, it is not surprising that the vast majority of the Russian population views Western political correctness [28, 39] with poorly concealed skepticism, if not more sharply. With rare exceptions, the death of George Floyd, as well as subsequent demonstrations and unrest, does not bring sympathy from the Russian public. Well aware that the standard of living in the United States is much higher than in Russia, the Russians generally believe that BLM performances in America are due to idleness and democratic permissiveness. Having determined the basic features of Floyd's rich criminal biography using the Internet, Russians watched with extreme surprise on television at his solemn funeral in a gilded coffin and the mayor of Minneapolis sobbing next to him. This was accompanied by images of pogroms, looting, and arson of stores on the streets of American cities, vandalism of historical monuments, and scenes of kneeling before Blacks, which made a very negative impression on the Russian public. Here we should add the obsessive moralizing of the United States, the sanctions imposed by them, and shameless interference in the affairs of other states and violation of their interests, multiplied by the geopolitical rivalry with Russia.

All these factors, of course, affected the content of the comments of the InoSMI portal. They are dominated by irony, ridicule, sometimes gloating, aggression, and quite rarely sound words of reason, sympathy, or self-criticism. Below are examples of comments on a Reuters report that the death of George Floyd sparked protests around the world:

LedOrub

This self-flagellation of White Anglo-Saxons and hangers-on to Black people amuses me. It's like collective insanity. This is especially interesting to see against the background of robbery, looting, beatings, and murders committed by the poor oppressed. This is the extent to which they were brainwashed....

sw34

Amazing logic adds up. It plays out the death of a repeat offender who resisted the cops. At the same time, hundreds of similar deaths of more worthy members of society are ignored. It turns out that all this rabble, who took to the streets, values the life of a criminal above the lives of law-abiding citizens. Conclusion: all the protesters are criminals.

Ivasi2018

I'll say this. The souls of 150 million North American Indians killed by Whites are crying out for revenge, and this revenge at the hands of Negroes will soon take place. I don't feel sorry for anyone in the U.S. In general. Not Negroes, not Whites, not Chinese. They are all enemies. And confident in their superiority. Soon a reservation for Whites will be created, where they with their propaganda-smudged brains, will themselves go to atone for their sins, but first a tax will be imposed on Whites as the race that oppressed Blacks, then they will begin to loot, that is, Negroes are by nature such, the White American liberoids themselves talk about it, like # they are children, well, or # they are Negroes, the inner nature is that, nothing can be done.... We must understand and forgive them!

Correspondents of the Western mass media also caught this mood. The InoSMI portal contains an article by French journalist Sylvie Kauffmann in the newspaper *Le Monde*, where she writes:

For two weeks now, the Russian and Chinese propaganda media have been celebrating. What could be better than riots in big American cities? What could be more effective than a picture of a building burning at night, around which activists of all stripes are dancing joyfully? And is it possible to resist how the protests unexpectedly spread to European cities? . . . Curiously, the Russian and Chinese champions of order do not realize that what they see as the weakness of liberal democracies is actually their strength. That many young people in Russia and China would dream of going out to protest and look at this movement not with horror, but with envy. Black Lives Matter has crossed the Atlantic and is now uniting the West, which was split by Donald Trump with his "America first" doctrine [30].

Here we should disappoint the French journalist: in Russia, there are almost no supporters of the BLM, and given the local culture, traditions, education, and the impact of state propaganda, they are unlikely to appear in the foreseeable future.

7 Conclusion

Can InoSMI be considered an impartial mirror of the foreign press? Obviously not, since the mirror turns out to be a bit askew. Despite the fact that this well-known news resource positions itself as the most neutral and objective, the selection of media and translated materials indicates a clear trend toward conservatism and rejection of left-wing radicalism, which is so popular among a large part of Western media. It is enough to recall the materials of the well-known American conservative Rod Dreher, which appear with enviable regularity in various sections of the InoSMI portal.

This conservative orientation of the portal can be explained in at least two ways: 1) the policy of the state that ultimately owns the news resource and 2) the characteristics of the Russian audience. The ideological policy of the state at the moment relies on the full development of patriotism, since the country after the annexation of Crimea in 2014 came under fire from Western sanctions and political semi-isolation, and therefore the promotion of self-reliance, history, and traditions becomes more relevant than ever. This is in good agreement with another direction of state policy related to the protection of traditional (Christian) values, which means an automatic rejection of Western political correctness with its tolerant attitude toward representatives of the LGBT community. Finally, the authoritarian regime in Russia does not accept full-fledged democracy. In addition, InoSMI, when selecting materials, has to take into account the requests of Russian society and caring about the popularity of the portal, and thus rarely puts translations of articles by Western liberals about the benefits of political correctness and praise of BLM, since a significant part of potential readers treat them negatively, perceiving such articles as outrageous stupidity and the fruit of the decomposition of the Western intellectual elite. This does not mean that such materials do not appear on the portal at all, but frankly they look exotic there.

If we now turn to the specific topic of reflection in the InoSMI portal of materials about the socio-political crisis in the United States, the BLM movement, and political correctness, then a selection of specially selected translations is intended to ultimately demonstrate the crisis of Western society and democracy to the Russian readership. The idea that stability is beneficial for modern Russia is implicitly suggested, the guarantor of which is President Putin, who will not allow mob rule and the wave of pogroms and looting that seized American cities after the death of George Floyd. Thus, InoSMI is actually one of the hidden mechanisms of manipulation of the public consciousness of Russian society.

Despite the above, the portal at the same time performs a very useful function, introducing Russian-speaking readers to the point of view of foreign journalists, politicians, and experts, even in a somewhat prepared form. The materials reproduced on the portal give food for serious reflection on the current moment and the future of the world under the influence of recent events in the United States. It is obvious that the problems that America is currently facing are the result of a whole complex of qualitative shifts within Western society under the influence of the scientific and technological revolution, changes in property relations, and changes in the demographic, social, political, and spiritual spheres. InoSMI, in turn, tries to reflect these problems and trends to the best of its abilities and capabilities within the framework of state policy and the demands of Russian society.

Acknowledgments. I should like to express my gratitude to my good friend and translator, Dr. Richard L. Bland, for the translation of Russian version of the manuscript.

References

1. Baumann, B.: American Thinker (USA): Epic video tirade of a black woman who failed to rob a cell phone store. https://inosmi.ru/social/20200610/247589659.html. Accessed 10 June 2020
2. Berggren, R.: Nyheter Idag (Sweden): the demolition of the western world must be stopped. https://inosmi.ru/social/20200615/247603816.html. Accessed 15 June 2020
3. Betz, B.: Fox News (USA): Sean King: Jesus Christ statues must be torn down as a display of "white chauvinism". https://inosmi.ru/social/20200623/247655022.html. Accessed 23 June 2020

4. Bawer, B.: The Victims' Revolution: The Rise of Identity Studies and the Closing of the Liberal Mind. Harper Collins Publishers, New York (2012)
5. Carlson, T.: Fox News (USA): The Real Reason for Angry Crowds Across the Country Demolishing American Monuments, 24 June 2020. https://inosmi.ru/politic/20200624/247658292.html. Accessed 24 June 2020
6. Cornell, A.: 'White skin, black masks': marxist and anti-racist roots of contemporary US anarchism. In: Prichard, A., Kinna, R., Pinta, S., Berry, D. (eds.) Libertarian Socialism Politics in Black and Red, pp. 167–186. PM Press, Oakland (2017)
7. Dewan, A., Hansler, J.: CNN (USA): Is the USA still the moral leader of the world? Not after what Trump did this week. https://inosmi.ru/politic/20200608/247573604.html. Accessed 08 June 2020
8. Dreher, R.: The American Conservative (USA): (Inadvertently) Pro-Trump riot. https://inosmi.ru/social/20200528/247513216.html. Accessed 28 May 2020
9. Dreher Rod: The American Conservative (USA): the crowd kicked out the police, 29 May 2020. https://inosmi.ru/politic/20200529/247521141.html. Accessed 29 May 2020
10. Dreher, R.: The American Conservative (USA): rioters burn the church of St. John. https://inosmi.ru/social/20200601/247531951.html. Accessed 01 June 2020
11. Dreher, R.: The American Conservative (USA): British courage, 10 June 2020. https://inosmi.ru/politic/20200610/247587696.html. Accessed 10 June 2020
12. Dreher, R.: The American Conservative (USA): why George Floyd died. https://inosmi.ru/politic/20200807/247883827.html. Accessed 07 July 2020
13. Elsom, J.: The Daily Mail (UK): "Last Supper" with Black Jesus to appear in British Cathedral. https://inosmi.ru/social/20200702/247691001.html. Accessed 02 July 2020
14. Fenig, E.C.: American Thinker (USA): Obviously, for some blacks, only some black lives matter. https://inosmi.ru/social/20200529/247523222.html. Accessed 29 May 2020
15. Finkielkraut, A.: Valeurs actuelles (France): "Don't touch my story!" Sign Alain Finkelkraut's appeal. https://inosmi.ru/politic/20200730/247837408.html. Accessed 30 July 2020
16. Fox News (USA): In Minneapolis, death of African American after police arrest sparks riots. https://inosmi.ru/social/20200527/247506724.html. Accessed 27 May 2020
17. Fresneda, C.: El Mundo (Spain): Demolition of monument to slave trader Edward Colston sparks political storm in the United Kingdom. https://inosmi.ru/politic/20200610/247585552.html. Accessed 10 June 2020
18. Gad, M.: Expressen (Sweden): I'm in the US, but I see the Middle East. https://inosmi.ru/social/20200617/247616291.html. Accessed 17 June 2020
19. Golovachev, A.: Glavred (Ukraina): America is living out its last days as a superpower. https://inosmi.ru/politic/20200603/247543707.html. Accessed 03 June 2020
20. Golovachev, A.: Glavred (Ukraina): The riot in the States showed that a big American revolution is inevitable. https://inosmi.ru/politic/20200605/247560702.html. Accessed 05 June 2020
21. Golovachev, A.: Glavred (Ukraina): What is the main result of the protests in the USA. https://inosmi.ru/politic/20200610/247586951.html. Accessed 10 June 2020
22. Golovachev, A.: Glavred (Ukraina): the USA is already finished, Europe is next in line. https://inosmi.ru/politic/20200620/247640426.html. Accessed 20 June 2020
23. Grinëv, A.: Russian Colonization of the Alaska: Baranov's Era, 1799–1818. University of Nebraska Press, Lincoln and London (2020)
24. Hansler, J., Nicole, G., Atwood, K.: (CNN): "We've been betrayed": american diplomats Fear US repression will undermine their overseas efforts. https://inosmi.ru/politic/20200605/247565909.html. Accessed 05 June 2020
25. Haszczyński, J.: Rzeczpospolita (Poland): America's image is crumbling. https://inosmi.ru/politic/20200602/247537178.html. Accessed 02 June 2020

26. Helsel, P., Romero, D.: NBC News (USA): Seven Wounded During Louisville Protests. https://inosmi.ru/politic/20200529/247522192.html. Accessed 29 May 2020
27. Hill, C.: Fox News (USA): Dropping police funding in response to the murder of George Floyd? It's a crazy idea that will make crime worse. https://inosmi.ru/social/20200609/247579267.html. Accessed 09 June 2020
28. Hughes, G.: Political Correctness: a History of Semantics and Culture. Wiley-Blackwell, Oxford (2010)
29. InoSMI, Rossiya: OHCHR: the UN reacted to the death of black in the United States, 29 May 2020. https://inosmi.ru/politic/20200529/247517819.html. Accessed 29 May 2020
30. Kauffmann, S.: Le Monde (France): Beijing and Moscow see Black Lives Matter as the weakness of liberal democracies, but it is their strength, https://inosmi.ru/politic/20200610/247587416.html. Accessed 10 June 2020
31. Kim, S.: Newsweek (USA): in Portland, a statue of George Washington was thrown from a pedestal and covered with a burning American flag. https://inosmi.ru/politic/20200620/247639958.html. Accessed 20 June 2020
32. Kindler, R.: Neue Zürcher Zeitung (Switzerland): the man who conquered Alaska. https://inosmi.ru/social/20200721/247787992.html. Accessed 21 July 2020
33. Koch, A.: Trends in Anti-Fascist and Anarchist Recruitment and Mobilization. J. Deradical. **14**, 1–51 (2018)
34. Luscombe, R., McGreal, C., Levin, S., Wong, J.C., Smith, D., Floyd, G.: Mass Protests and Riots Gripped the United States, and the Authorities of Major Cities Imposed a Curfew (The Guardian, UK). https://inosmi.ru/social/20200531/247527419.html. Accessed 31 May 2020
35. Heather, M.: The Diversity Delusion: How Race and Gender Pandering Corrupt the University and Undermine Our Culture. St. Martin's Press, New York (2018)
36. Meseguer, M.: La Vanguardia (Spain): who are Antifa? https://inosmi.ru/politic/20200603/247544990.html. Accessed 03 June 2020
37. Miller, R.W.: CHAZ, Community Without Cops: What Seattle's Capitol Hill Autonomous Zone looks like (USA Today, USA). https://inosmi.ru/social/20200614/247601222.html. Accessed 14 June 2020
38. Millward, D.: The Telegraph (UK): Armed black militia threatens to create a "nation of its own". https://inosmi.ru/politic/20200727/247826171.htmlAccessed 29 July 2020
39. Moller, D.: Dilemmas of Political Correctness. J. Pract. Ethics **4**(1), 5–25 (2016)
40. Ngo, A.: The Spectator (Great Britain): Mass performances of the American "Antifa". https://inosmi.ru/politic/20200602/247539601.html. Accessed 02 June 2020
41. Rosenberg, G.: Expressen (Sweden): will the change of power in the United States happen peacefully?, https://inosmi.ru/politic/20200707/247710367.html. Accessed 07 Junly 2020
42. Shepherd, G.: CNN (USA): Stone Mountain and other Confederate monuments need to be removed. https://inosmi.ru/politic/20200611/247593266.html. Accessed 11 June 2020
43. Speckhard, A., Ellenberg, M.: PERSPECTIVE: Why branding Antifa a terror group is a diversion. Homeland Security Today, 2 June 2020 https://www.hstoday.us/subject-matter-areas/counterterrorism/perspective-why-branding-antifa-a-terror-group-is-a-diversion/. Accessed 04 June 2020
44. St. Julian-Varnon Kimberly: Foreign Policy (USA): The Strange Story of the "Russian Lives Are Important" Movement (2020). https://inosmi.ru/social/20200714/247751007.html. Accessed 14 July 2020
45. TMZ (USA): In North Carolina, white police officers wash the feet of black priests and demonstrators and ask for forgiveness. https://inosmi.ru/politic/20200609/247584190.html. Accessed 09 June 2020
46. Weiss, S.: The Jerusalem Post (Israel): America is being robbed and silenced. https://inosmi.ru/social/20200619/247633037.html. Accessed 19 June 2020
47. Zatari, A.: BBC (UK): racism in Russia - stories of prejudice. https://inosmi.ru/politic/20200620/247640282.html. Accessed 20 June 2020

Speech Image of a Political Leader: Cases of President of Azerbaijan Ilham Aliyev and Prime Minister of Armenia Nikol Pashinyan

Galina Niyazova and Niyazi Niyazov(✉)

Saint Petersburg State University, St. Petersburg, Russian Federation
{g.niyazova,n.niyazov}@spb.ru

Abstract. Diplomacy and military power are considered to be tools of promoting interests of a state on the international arena. Today interaction in economic and cultural spheres enters the picture due to extension of economic cooperation and information technologies use. Thus, theorists start to speak about the role of strategic communication in the contemporary international relations and global politics. Speech image of a politician plays the pivotal role in the structure of the understanding of strategic communication. The problem posed by language and power interaction acquires a new meaning, attracting great number of scientists. Noteworthy, interdependence of language and power is so great, these notions are so close, that some scholars consider them to be synonyms. Obviously, politicians use their political rhetoric as a tool of domestic and foreign policy conducting. Nevertheless, it is due to new technologies and the Internet that they are able to appeal to the general public not only to their colleagues. Speaking about a foreign language as a tool of strategic communication, we mean interdependence of a language and our ability to extend our influence. Thus, our ability to use a global language implies extending our influence globally. Taking into consideration the current state of affairs concerning the Karabakh settlement, we consider it relevant to resort to English ability of President of Azerbaijan Ilham Aliyev and Prime Minister of Armenia Nikol Pashinyan, because addressing a large audience gives each of them an opportunity to reveal their viewpoints, attracting potential allies and promoting interests of the state each of them presents. Nonetheless, the politicians cope with this task with different results.

Keywords: Speech image · Language policy · Strategic communication · Azerbaijan · Armenia

1 Introduction

The contemporary system of international relations is uncertain. It puts under the doubt the existing world order, destroying agreement, which lasted for decades, and the very principles of interaction of actors of international relations. Under these circumstances, the role of national states as a key element of the system of international relations

© The Author(s), under exclusive license to Springer Nature Switzerland AG 2021
R. Bolgov et al. (Eds.): *Proceedings of Topical Issues in International Political Geography*, SPRINGERGEOGR, pp. 446–452, 2021.
https://doi.org/10.1007/978-3-030-78690-8_38

comes to the fore. Nevertheless, to act successfully on the international arena in the contemporary circumstances a national state needs more than powerful military force and effective diplomacy. Although it is obvious, that against the background of constantly transforming system of international relations the role of military force and diplomacy is growing, it is essential to have developed economy and high technology, including those in information sphere, because the status of strategic communication is getting higher every day in modern international relations and global politics.

When we mention strategic communication, we understand that the center stage in the notion is played by speech image of a politician. It gives us an opportunity to state that the problem of interaction and interdependence of language and power, which has become an object of study for the great number of theorists, is viewed from a different perspective.

It is clear that state political leaders always tried to use the potential of their addresses to promote the interests of the countries they present both in domestic and foreign policy. Previously, published mass media, radio and television were used for this purpose. Today due to the Internet they can appeal not only to politicians and experts but to the general public. "Digital media—including social media—have grown to become integral components of the political communication" [1].

However, in this case it is necessary for the address to be made in the language of international communication, with the English language having this status.

Under contemporary circumstances, using English as a tool of strategic communication gives us an opportunity to speak not only about the interaction of language and power, but also about an ability to extend one's political influence. It leads to the idea of a new qualitative rivalry in information sphere, which can be viewed as a display of interdependence of language and power.

A remarkable example of the mentioned information confrontation in information sphere, picturizing interaction of language and power, was the discussion on Karabakh settlement between Prime Minister of Armenia Nikol Pashinyan and President of Azerbaijan Ilham Aliyev on February, 14 2020 in Munich. It is their recent meeting when both parties were to speak English and there is a striking difference in language proficiency demonstrated by the leaders of these two states. Taking into consideration that the Azerbaijani President showed a good command of English and that the form of communication implied spontaneous speech, we consider it necessary to refer to other Pashinyan's broadcasts, including reading aloud a prepared report in English at the UN General Assembly and answering questions in a face-to-face interview with Euronews in January 2019 at Davos.

Object of the research is presented by speech image of a political leader.

Objective of the research is to point out special aspects of shaping of speech image of a politician, because using a foreign language, accepted as a language of international communication, helps to extend influence of a person, speaking this language, and of a state, this person represents, consequently.

Hypotheses: The given case gives us an opportunity to state that speech image of a politician directly shapes his or her favorable image on the international arena, it leads to shaping of a preferable image of the state a politician represents. Thus, it can be viewed as an effective tool of foreign policy. In case of an international conflict the ability of a state leader to articulate his thoughts in a foreign language speaking directly to the global community, without an interpreter, and laying stress on the facts which should be emphasized, obviously do the groundwork for searching powerful allies.

2 Methods and Materials of Research

To understand the essence of the processes being under study the authors of the research used a number of methods of the humanities. Thus, historical and genetic approach (or a retrospective method) is being used as it resorts to certain historical events. Moreover, the elements of discourse analysis were used. Taking into account that the speakers address various problems in their speeches the authors of the research applied the compositional analysis method.

The empirical base of the research is formed by sources and references that are on open access.

3 Theoretical Framework

In the framework of modern international relations speech image of a political leader is considered to be the essential part of the image of the state the politician presents [2]. Thus, the ability of a political leader to articulate his viewpoint speaking before a large audience and to be clearly understood by the global community contributes greatly to the success of the state on the international arena.

Consequently, the interaction of two aspects – language and politics – is viewed. It is necessary to mention that interdependence of these notions is so great that in the opinion of some theorists they turn to be synonyms. For example, American cognitive linguist R.Lakoff thinks that language is politics, politics implies power, the power one possesses depends on his ability to speak and the way other people understand him [3].

Language is used not only for communication but it's a tool for promoting political, economic and cultural interests of a state and international organizations. That's why ill-judged use of language may hamper the progress on the global political arena. Thus, V. Yag'ya wrote about the problem of Russia's economic and political policy being threatened by inconsiderate publications in the state official press [4].

The role of the language in reflecting reality, formed, among other issues, by political sphere, is indisputable and it can't be overestimated. American anthropologist-linguist E.Sapir wrote about the significance of the language, marking the fact that the environment is not confined to the world of things and public activity and despite wide-spread false opinion people are influenced by the language which is used for communication in the society. He stated that we realize reality through language and it is impossible to understand it without a language. According to Sapir, language can't be considered a tool for just solving social communication problems [5].

Language not only reflects the reality but also forms it through information influence. Language is a system of signs transferring information [6], this definition is unquestionable. In the modern world information about almost everything is essential, because it provides us with the state of security. It is obvious that those, who managed to lose the chance of receiving information in time are treated as underdog in society opposing to those who possess information [7].

This opinion is shared by well-known Russian theorist I.Panarin who states that from the very beginning, from prehistoric times an ability of an individual to survive depended primarily on his capability to gain information about possible danger and to react to it [8].

Information process build-up expanded across continents. As a result, information war in political sphere is waged. Stability of a state political system depends both on the speed with which one manages to gain information and its full disclosure and on perception reaction time. "Master of information has power; the ability to distinguish between more and less important information turns a person into more powerful; the possibility to disseminate information in one's own directing or dissemble it provides one with double power" [9].

The use of language as a power tool was widely discussed in the second half of the XX century. The researchers focused on the reasons for inequality between the speakers, their origin and foundation. Discourse analysis, and its area of focus called critical discourse analysis, emerged [10].

The development of critical analysis into a scientific discipline aimed at uncovering the mechanisms through which the minority, wielding power, with the help of linguistic means manipulate consciousness of the majority which is less powerful. The theory of impact studies all tools and mechanisms which provide means of influencing the process of decision-making [11].

In the framework of interaction and interdependence language and politics it is necessary to emphasize the significance of language identity. In the view of Austrian-British philosopher K.R.Popper, it's more reasonable to use language not as a tool of self-determination, but as a means of rational communication, because language is one of the main institutes of public life and its transparency is a factor of its functioning as a means of rational communication. The use of language is less important for emotion transmission due to the fact that one can express emotions without pronouncing a word [12].

The way language predominance of a state is closely connected with its political power on the international arena is reflected in the opportunity of speech image of a political leader to be linked to success of a state across the globe. Extension of the national language as a language of international communication gives evidence of the high status of the state on the global political arena. A number of theorists, who addressed this issue, state that the success of the state is directly connected with the success of this language speakers. In case they lose in political, economic, social or military competition, the language loses too. A language becomes international for only one reason: due to the political power of those who speak it [13].

The same is for the ability of the leader of any other state which national language is not accepted as a global one, using the international language, attracts large audience and extends influence.

Interaction of language and power, presented in the operating of language policy, demonstrate the power of information influence. If one is able to use language for transmitting information, he turns to be the one who is the most power due to the access to this information only. As a result, we see that information confrontation in political sphere is a typical presentation of language and power interaction.

4 Results of Research

The discussion on Karabakh settlement between President of Azerbaijan Ilham Aliyev and Prime Minister of Armenia N.Pashinyan in Munich on February, 14 2020, serves as one of the examples.

Heads of Azerbaijan and Armenia spoke English at the meeting. Video of the discussion is available in English, Russian, Azerbaijani and Armenian in the internet. We will refer to the original text, because translation tends to miss information or not present it properly. Furthermore, Prime Minister of Armenia N.Pashinyan articulated ideas, which can't be properly translated into other languages from English because Pashinyan's English is not fluent enough to give him an opportunity to speak freely and be understood.

In the beginning of the debate Aliyev claimed that the Caucasian Bureau of the Central Committee of the Communist Party of the Soviet Union made a decision that Nagorno-Karabakh territory was considered to be a part of Azerbaijan. A hundred of years passed but Armenia was not satisfied with that decision of the Central Committee. Taking advantage of the unstable situation in the beginning of the 1990s, Armenia occupied Azerbaijani territories. There were no legal ways to make Nagorno-Karabakh a part of Armenia. And still there are no legal ways to do it.

The main message of Aliyev's speech in Munich was clear. It is necessary to mention that he spoke fluent English. His pronunciation together with perfect grammar and professional terminology make a good impression.

In this context, the speech of the Armenian Prime Minister was spoilt by a great number of mistakes, including elementary level errors, and by unfinished statements. It is obvious that his presentation was less confident and authoritative. Mentioning the same resolution of the Caucasian Bureau of the Central Committee of the Communist Party of the Soviet Union, Pashinyan failed to articulate his idea because of poor English that didn't give him an opportunity to express himself. Moreover, the Prime Minister of Armenia violated regulations, interrupted himself and behaved hesitatingly. This was clearly seen comparing to President Aliyev, who was called "an experienced person in discussing this issue" by the moderator.

I.Aliyev noticed that N.Pashinyan is the third Armenian representative he is discussing the conflict with. Armenia has no intention to change its point of view and it still tries to preserve status-quo. It is perfectly seen that irremovability of people in power from the office in Azerbaijan [14] and "democratic" elections in Armenia provide the same result and has no influence on the conflict settlement.

Another chance to make his state's interests public was given to N.Pashinyan on August, 14 2020, when his interview with BBC Hardtalk was shot and made public. The presenter of the program made an interview with N.Pashinyan, asking him delicate and obviously unexpected questions. As a result, N.Pashinyan switched back to repeating words and word combinations, paraphrasing whole sentences from the very beginning and starting using unfinished sentences, interrupting himself. Finally, he managed to go on just pronouncing separate collocations which made it possible for the anchor man to stop him, not letting him going far from the center topic of their talk.

It is obvious that the panel talk during the Munich conference on security demanded a spontaneous reaction of both speakers. Obviously, it shows that not good command of English offputs N.Pashinyan and prevents him from being a confident political leader and to present his country in favorable manner.

Nevertheless, if we switch to other ways of presenting information like addressing the general public by reading the written text, as N.Pashinyan did at the UN General Assembly in September 2018, or by articulating the prepared answer for the questions the prime minister of Armenia preliminarily knew, the situation doesn't change much. Word pronunciation preventing the general public from recognizing them contributes little to understanding and attractiveness of the politician. Thus, it makes little for making the state he presents advantageous.

It is necessary to note that N.Pashinyan likes to state it multiple times that he came to power after the revolution. And this very word "revolution" is used often in a different meaning. Any slight change in Armenia, any idea that came to the Prime Minister of Armenia is called a revolution. Moreover, this "revolutions" are treated as something extraordinary and extremely new in a global scale.

Another word, so much beloved, is "micro". In the context like "microrevolution" it means that N.Pashinyan acknowledges that these changes are not so great, although he continues to focus on their novelty and on his authorship. But when N.Pashinyan speaks about "microbusiness" it may confuse and don't contribute to understanding the national leader at all. The term "microbusiness" was articulated by N.Pashinyan in his interview with Euronews in January 2019 at Davos.

5 Conclusion

To sum up the research we should state that Nikol Pashinyan tends to call himself the Armenian Prime Minister who is, according to his own words, is the first to state that it is essential to make certain successive steps to resolve the Nagorno-Karabakh conflict. Furthermore, he is the first to understand that the conflict settlement is impossible without "taking into account demands of Azerbaijani people, Armenian people and residents of Nagorno-Karabakh". As a result, such statements gave Aliyev an opportunity to show his ability to react adequately to the claim made by his opponent and say that "the only desire of Azerbaijani people is to come back home, and there is no such entity as "Nagorno-Karabakh Republic".

It is evident that "micro-revolutions", mentioned by the Armenian Prime Minister so often, are new only for him as a person who came to power a short time ago and wants to attract attention by pronouncing pictorial phrases. Such statements can create an illusion that N.Pashinyan has a vague idea about the history of the conflict.

Generally speaking, the way the Armenian Prime Minister speaks a foreign language addressing the global community contributes little to the image of Armenia on the international arena.

Moreover, we should say that perfect English helped President Ilham Aliyev articulate the viewpoint of the Republic of Azerbaijan on Nagorno Karabakh issue as he uses the language understandable by the reference public. Thus, speech image of the Azerbaijani President without any doubt serves the interests of his state. This is the very way to shape the reality via information influence.

References

1. Lalancette, M., Raynauld, V.: The power of political image: Justin Trudeau, Instagram and celebrity politics (2017). https://www.researchgate.net/publication/321396251_The_Power_of_Political_Image_Justin_Trudeau_Instagram_and_Celebrity_Politics
2. Koshkin, P.: The creation of Russia's new image in the Western press between 2014 and 2019. Vestnik Saint Petersburg Univ. Int. Relat. **12**(4), 477–499 (2019). https://doi.org/10.21638/11701/spbu06.2019.406
3. Lakoff, R.T.: Talking Power: The Politics of Language. Basic Books, New York (1990)
4. Yag'ya, V.S.: Instead of the Conclusion. Aktual'nye problemy mirovoj politiki v XX veke, vol. 2. SPbGU Publication, Saint Petersburg (2007). (in Russian)
5. Uorf, B.: Relation of Norms of Behaviour and Thought to Language, Novoe v lingvistike, 1. Inostrannaya literatura Publication, Moscow (1960). (in Russian). http://philology.ru/linguistics1/whorf-60.htm
6. Gerd, A.S.: Introduction to Ethnolinguistics: The Course of Lectures. SPbGU Publication, Saint-Petersburg (2005). (in Russian)
7. Kalandarov, K.H.: Managing Public Opinion/The Role of Communication Processes, Gumanitar. Centr "Monolit" Publication, Moscow (1998). (in Russian)
8. Panarin, I.N.: Information War and Geopolitics. Pokolenie Publication, Moscow (2006).(in Russian)
9. Grachev, G., Mel'nik, I.: Manipulating Personality: Complex of Ways and Technologies of Information and Psychological Influence. IFRAN Publication, Moscow (1999). (in Russian). http://www.gebesh.ru/Knigi/grachev.pdf
10. Bolgov, R., Filatova, O., Tarnavsky, A.: Analysis of public discourse about Donbas conflict in Russian social media. In: Proceedings of the 11th International Conference on Cyber Warfare and Security, ICCWS 2016, pp. 37–46 (2016)
11. Baranov, A.N.: Introduction into Applied Linguistics. URSS Editorial Publication, Moscow (2001).(in Russian)
12. Popper, K.: The Open Society and its Enemies, vol. 2. Feniks Publication, Moscow (1992).(in Russian)
13. Crystal, D.: English as a Global Language, 2nd edn. Cambridge University Press Publication, Cambridge (2004)
14. Letnyakov, D.: Conceptualization of post-Soviet regime changes: Some intermediate results. Vestnik Saint Petersburg Univ. Int. Relat. **13**(3), 374–393 (2020). https://doi.org/10.21638/spbu06.2020.306

Indirect Evidentiality and Its Manifestation in Chinese Language Political Discourse

Aleksandra Nechai and Uliana Reshetneva(✉)

Peter the Great St. Petersburg Polytechnic University, St. Petersburg, Russia

Abstract. The article concentrates on the analysis of indirect evidentiality and its manifestations in the political discourse of the Chinese language in the context of key policy documents related to implementing China's national strategy One Belt-One Road, which became one of the main components of the Chinese foreign policy and foreign economic strategy. Using The International Initiative of China One Belt - One Road put forward by the President of the People's Republic of China Xi Jinping and presented at Nazarbayev University, the program document "Vision and actions on Jointly Building Silk Road Economic Belt and 21st-Century Maritime Silk Road" and office book "Building the Belt and Road: Concept, Practice, and China's Contribution" the authors turn to indirect evidentiality in political discourse, more specifically to its evidential markers. Comprehensive analysis of this material, including the analysis from the perspective of the evidentiality strategies realization, taking into account linguistic and extralinguistic factors, is one of the priority tasks of political geography and political linguistics at the present stage of Russian-Chinese relations.

In the research process, the means of expressing indirect evidentiality of the Chinese political discourse were identified and considered for the first time. The article concludes that indirect evidentiality in Chinese political discourse is characterized by lexical markers: verbs of speech and mental activity, modal verbs, evaluation adverbs, and prepositional constructions. The study showed that the functions of evidentiality markers, in the content of national strategies, are to convey facts and information, express a viewpoint, and hidden belief. The evidentiality category markers indicate the degree of credibility/non-credibility of the information, the reliability/unreliability of the information itself, which is also important in the world of politics.

Keywords: Category of evidentiality · Evidentiality markers · Political discourse · Source of information · Chinese language

1 Introduction

Evidentiality as a language category attracts the attention of not only linguists but also researchers of related humanities, legal studies, social advertising, political science, and other disciplines. In the scientific world, evidentiality is considered from different approaches, restricted and broadside ones. From the viewpoint of a restricted approach,

© The Author(s), under exclusive license to Springer Nature Switzerland AG 2021
R. Bolgov et al. (Eds.): *Proceedings of Topical Issues in International Political Geography*, SPRINGERGEOGR, pp. 453–461, 2021.
https://doi.org/10.1007/978-3-030-78690-8_39

this category is understood as an indication of the speaker's source of information; from the perspective of a broadside approach – not only as an indication of the source of information but also the speaker's attitude to it.

Within the framework of these two approaches, the study of language means that indicate the source of the reported information, and the attitude to it gains special significance in political discourse. The important factor in the relevance of this article is the lack of Russian research studies that consider markers of indirect evidentiality in Chinese political discourse.

In the process of research, we turn to indirect evidentiality in political discourse, more specifically to its evidential markers, as they are the most typical and pragmatically conditioned for this type of discourse.

2 Literature Review

The history of studying the category of evidentiality in world science has been going on since the beginning of the 20th century with the works of the American anthropologist and linguist Franz Boas, who drew attention to the peculiarity of the American Indian language to necessarily include an indicator of the source of information in the sentence structure [1]. A little later, in the middle of the 20th century, the ideas of the American linguist and literary critic Roman Jakobson about evidentiality as an obligatory grammatical category became a new stage in its study by the scientific community [2]. The next significant event in the study of evidentiality was the Berkeley conference in 1981, and the release of a number of collections of articles in the 1980s and 1990s [3, 4].

The achievements of scientists at the turn of the century served as a starting point for further scientific research already in the 21st century in the field of evidentiality based on the material of different languages. A numerical diagram showing a significant increase in the number of foreign publications over the past decade is presented in the work of S. Verhees, and indicates a surge of interest in the study of this category in scientific circles [5].

Currently, the category of evidentiality is fixed in many languages of the world, scientists consider it as a universal linguistic category, in some languages it has a grammatical form of expression, in others it does not, that is, it is expressed by lexical means. Languages in which evidentiality is represented by grammatical forms make up about a quarter of the world's languages [6].

The largest contemporary scholarly works that have inspired current researchers in the category of evidentiality are collections of articles edited by A.Y Aikhenvald and R.M.W. Dixon (eds.) [7, 8], edited by Zlatka Guentchéva [9], and edited by Ad Foolen, Helen de Hoop and Gijs Mulder [10].

In Russia, N. A. Kozintseva was one of the first to study the evidential category. In her works, the scientist sets the task of "developing a holistic typological description of this category" [11: 92], and considers its definition, semantic features, and means of expression, the relationship of evidentiality and other linguistic categories.

Among the domestic works introducing and investigating evidentiality problems, it is worth noting chapters of the monograph and articles by V.A. Plungian [12, 13], a collection of articles based on materials of different-structured languages, edited by V.S. Khrakovskiy [14], and a number of dissertations and scientific articles.

In more detail, the degree of study and approaches to the study of the evidentiality category by Western and Russian scientists on the material of different languages are described in the work of L.B. Kadyrova, published in 2016 [15]. At the same time, it should be noted that we have not identified any scientific works providing information about the evidentiality in the Chinese language. This category is presented only fragmentarily in the works of Russian scientists when considering the perfect in the Chinese language [16]. In recent years, scientists have expanded the range of issues covering the category of evidentiality. Its comparative, corpus and discursive studies became priorities.

In Russian science at the present stage, the results of a comparative study of the evidentiality category are reflected in the work of A.D. Kaksin and L.V. Azarakov [17], and corpus study – in the article by E.V. Bodnaruk and T.N. Astakhova [18].

The category of evidentiality, its types, markers, and correlation with other categories attract the attention of foreign and domestic scientists, and are considered by them in different discourses. In our work, we will focus in more detail on political discourse.

The specificity of actualization of the evidentiality category in the English-language political discourse is presented in the collection of studies edited by J. Marín-Arrese [19], in the works of A.I. Chepurnaya [20], and R.R. Khazieva [21]. Although the number of discursive studies is increasing, this kind of research in Russian science is clearly insufficient. We have found no works devoted to the study of the functioning of markers of indirect evidentiality in Chinese political discourse, which indicates the scientific novelty of our research.

In Chinese linguistics, the study of the evidentiality category begins with the work of Hu Zhuanglin, who published three articles in 1994 and 1995. One of them is dedicated to the analysis of the approaches of the European scientists, Chafe, Anderson, and Willett, to the study of evidentiality 可证性 kězhèngxìng [22]; the second article examines the evidentiality category on the material of news and argumentative discourses [23]; in his third work, the scientist dwells in detail on the study of the evidentiality in the Chinese language, and its discourse analysis [24]. It is fair to say that it was Hu Zhuanglin's scientific works that set the vector for the study of the evidentiality category and its discursive analysis in Chinese science.

At the turn of the century, the attention of Chinese scientists is focused on the terminological apparatus formation, and the definition of ways to express the evidentiality category, first on the material of the English language, then the perspective of research shifts to the study of this category in Chinese.

In 1997, Zhang Bojiang, analyzes the relationship between epistemological modality and evidentiality, terming it 传信范畴 chuánxìn fànchóu, the category of communication [25]. Yan Chensong examines the evidentiality category 传信范畴 chuánxìn fànchóu and its evidentials 传信语 chuánxìnyǔ on the English language material and analyzes the viewpoint of Chafe, and his five basic elements of this category [26]. The study by Niu Baoyi presents an excursus to the foreign study of the evidentiality theory 实据性理论 shíjùxìng lǐlùn and evidentiality markers 实据性标记 shíjùxìng biāojì [27]. Fang Hongmei and Ma Yulei consider the evidentiality category 言据性 yánjùxìng in the English language, highlighting the evidentiality markers 据素 jùsù [28]. Researcher Le Yao defines the essence of the evidentiality category 传信范畴 chuánxìn fànchóu and analyzes its functioning in modern Chinese [29].

Although over time the number of works of Chinese scientists studying the category of evidentiality has increased exponentially, there is still no unified term recognized by all scientists; different synonymous variants are used in their works. Of these, it is the term 言据性 yánjùxìng, according to Zhou Yahong and Mao Hui, that became more widespread in the works of Chinese researchers [30].

At the present stage, Chinese researchers are greatly attracted to strategies of the evidentiality category in the discursive context. Their specificity in scientific discourse is reflected in the works of Sun Zihui [31]; in the study of news discourse - by Chen Yi [32]; in a comparative study of academic discourse in the Chinese and English languages - by the scholars Shui Wudi and Ji Yongjuan [33], in the article on argumentative discourse by Chen Zheng and Yu Dongming [34], and the works of other scholars.

Next, we will briefly highlight the approach of the Chinese scientists Zhou Yahong and Xia Chen, who analyze the views of Chinese and foreign scientists and offers her vision of the elements of the evidentiality category in documents of the national strategy. The author distinguishes the following elements indicating their frequency of use: fact 事实 shìshí (73.5%), report 报告 bàogào (2.4%), induction 归纳 guīnà (7.4%), deduction 演绎 yǎnyì (6.6%), and persuasion 信念 xìnniàn (10%) [35: 94]. Let us dwell in more detail on the facts as more frequent in program documents.

By the facts 事实 shìshí, the scientist understands the elements of zero evidentiality and considers them the most important evidentiality characteristic of the national strategy. The author points out that it is "a large amount of factual data that are used to prove the efforts and achievements of China in the implementation of the One Belt - One Road Initiative" [35: 95].

In the work of Zhou Yahong and Xia Chen, it is indicated that the use of evidentiality markers helps express the foreign policy views of political authorities, eliminate the rejection and doubts of the international community regarding the rapid economic development of the state, recognize and support its development strategy [35: 95].

3 Materials and Methods

When working on the article, we relied on the ideas and methodology of R. Jakobson, A. Aikhenvald, N. A. Kozintseva, and Hu Zhuanglin. We adhere to a broad understanding of the evidentiality category as an indication of the source of information (more broadly – knowledge), and the attitude to the conveyed information (knowledge). In the Russian-language literature, the following synonyms for this concept are used: retelling, obviousness, eyewitness, attestation. Taking into account the views of N.A. Kozintseva and A. Aikhenvald we single out direct and indirect evidentiality, depending on the source of information.

By direct evidentiality we understand that the speakers receive information through their own sensory experience, through the organs of senses. Accordingly, we classify the direct evidentiality into visual (through the organs of vision), and sensor evidentiality (through other sense organs or their combination).

Indirect evidentiality is subdivided into inference and retelling. The inference is obtaining information by the speakers through their own mental activity. Retelling is obtaining information through second- and third-hand evidence. In its turn, it is subdivided into quotative and hearsay.

The study uses a group of methods: comparative-descriptive and discursive methods, and componential analysis. The attention is focused on program documents as genre varieties of political discourse associated with implementing the Chinese national strategy One Belt - One Road, which became one of the main components of the Chinese foreign policy and foreign economic strategy, and entered the Chinese political lexicon as 一带一路 yídài yílù "one belt, one road" [36: 61].

The main sources were: 1) The International Initiative of China One Belt - One Road put forward by the President of the People's Republic of China Xi Jinping and presented at Nazarbayev University in 2013 [37]; 2) the program document "Vision and actions on Jointly Building Silk Road Economic Belt and 21st-Century Maritime Silk Road" developed in 2015, which describes the basic principles and concepts, main mechanisms and priorities of cooperation, as well as China's actions to implement the project [38]; 3) office book "Building the Belt and Road: Concept, Practice, and China's Contribution", 2017, which examined state events aimed at promoting the national strategy and its understanding by the international community [39].

We believe that a comprehensive analysis of this material, including the analysis from the perspective of the evidentiality strategies realization, taking into account linguistic and extralinguistic factors, is one of the priority tasks of political geography and political linguistics at the present stage of Russian-Chinese relations.

4 Results

Analyzing the functioning of the evidentiality category, we take into account the viewpoints of Chinese scientists on the issue of the lexical means of its expression. For instance, the researcher Liu Yun distinguishes epistemic and evidentiality verbs, modal verbs, adverbs of assessment, and modal particles as markers of direct and indirect evidentiality [40: 328]; the researchers Chen Zheng and Yu Dongming distinguish sensory and modal verbs, adverbs and conjunctions [34: 28].

In the course of the research, we identify the following lexical means of expressing indirect evidentiality in political discourse: speech act verbs, mental action verbs, modal verbs, evaluation adverbs, and prepositional constructions. We have not identified modal particles, and we believe that their functioning is not typical of political discourse.

Below are some examples.

1. Speech Act Verbs (阐述, 说, 写, 引自, 曰, 转引自等)
   说– say, explain, tell…off
   "哈萨克斯坦伟大诗人, 思想家阿拜·库南巴耶夫说过: "世界有如海洋, 时代有如劲风, 前浪如兄长, 后浪是兄弟, 风拥后浪推前浪, 亘古及今皆如此" [37]. *The great Kazakh poet and thinker Abai Kunanbayev said:" The world is like the ocean, time is like a strong wind, the first waves are like elder brothers, the last waves are like younger brothers, the wind drives waves of generations to replace each other."*

2. Mental Action Verbs (表示, 参考, 记载, 强调, 认为, 提出, 以为, 知道, 了解等)
   强调 – emphasize, stress, underline
   "中国国务院总理李克强参加2013年中国-东盟博览会时强调, 铺就面向东盟的海上丝绸之路……" [38]. *At the China-ASEAN exhibition, Premier of the State Council of the People's Republic of China Li Keqiang emphasized that it is necessary to form the Maritime Silk Road with the participation of ASEAN countries…*
3. Modal Verbs (能, 能够, 可能, 应, 应该, 该, 会, 愿意等)
   The category of obligation 应该 – should, ought to
   "我们共同的未来应该是更加光明的未来……" [39]. Our common tomorrow should be even better …
   The category of capability 能够 – can
   "只要沿线各国和衷共济, 相向而行, 就一定能够谱写建设丝绸之路经济带和21世纪海上丝绸之路的新篇章" [38]. *If only all states located along this road unanimously cooperate and meet halfway, then they will be able to write a qualitatively new chapter in the history of the economic belt of the Silk Road and the maritime Silk Road of the 21st century.*
   The category of volition 愿意 – willing, ready
   "……中国愿意在力所能及的范围内承担更多责任义务, 为人类和平发展作出更大的贡献" [38]. *China is willing, within its means and capabilities, to take on more commitments and make a greater contribution to the development of humankind.*
4. Adverbs (一定, 必定, 必然, 肯定, 绝对, 估计, 或许, 也许, 大概, 大约, 其实等)
   一定 – for sure, certain
   "我相信, 包括在座各位同学在内的中哈两国青年, 一定会成为中哈友谊的使者, 为中哈全面战略伙伴关系发展贡献青春和力量" [37]. *I believe that the young people of China and Kazakhstan, including all the students present here, will certainly become messengers of Chinese-Kazakh friendship, direct all their efforts and devote their young years to the development of a comprehensive strategic partnership between China and Kazakhstan.*
5. Preposition 根据/据/照/依照/按照 + Person Indication Structures
   根据jù according to + person indication
   "根据中国国家主席习近平的倡议和新形势下推进国际合作的需要, 结合古代陆海丝绸之路的走向, 共建"一带一路"确定了五大方向" [39]. *According to the initiative of the President of the People's Republic of China Xi Jinping, and the need to develop international cooperation in a new environment, the joint construction of the Belt and Road identified five main directions based on the historical example of the Silk Road.*

5 Conclusion

In the research process, the means of expressing indirect evidentiality of the Chinese political discourse were identified and considered for the first time. It is characterized by the following lexical markers: verbs of speech and mental activity, modal verbs, evaluation adverbs, and prepositional constructions. The peculiarities of their functioning include the most frequent use of verbs of speech and mental activity when quoting eminent political leaders, thereby directly indicating the source of information. Appealing to the opinions of reputable persons increases the reliability and credibility of information. The use of precedent phenomena in the speech of politicians enhances confidence and acceptance of information. The analysis shows that there is a combination of evidentiality markers, epistemic modality, and argumentativeness. The use of evidentiality strategies in political discourse is associated with the pragmatics of the message, with the impact on the audience. The evidentiality category markers indicate the degree of credibility/non-credibility of the information, the reliability/unreliability of the information itself, which is also important in the world of politics.

In the future, it is planned to analyze the means of expressing indirect evidentiality, and the features of their functioning in other genres of political discourse.

References

1. Boas, F.: Handbook of American Indian Languages. Bureau of American Ethnology, Washington (1911)
2. Jakobson, R.: Shifters, Verbal Categories, and the Russian Verb. Harvard University, Cambridge (1957)
3. Chafe, W.L., Nichols, J. (eds.): Evidentiality: The Linguistic Coding of Epistemology. Ablex Publishing Corporation, Norwood (1986)
4. Guentcheva, Z. (ed.): L'énonciation médiatisée. Peeters, Louvain- Paris (1996)
5. Verhees, S.: Defining evidentiality. In: Voprosy Yazykoznaniya, vol. 6, pp. 113–133 (2019). (in Russian). https://doi.org/10.31857/S0373658X0007549-2
6. Aikhenvald, A.Y.: Evidentiality. Oxford University Press, Oxford (2004)
7. Aikhenvald, A.Y. (ed.): The Oxford Handbook of Evidentiality. Oxford University Press, Oxford (2018)
8. Aikhenvald, A.Y., Dixon, R.M.W. (eds.): The Grammar of Knowledge. Oxford University Press, Oxford (2014). https://doi.org/10.1093/acprof:oso/9780198701316.003.0001
9. Guentcheva, Z. (ed.): Epistemic Modalities and Evidentiality in Cross-Linguistic Perspective. Walter de Gruyter GmbH, Berlin/Munich/Boston (2018). https://doi.org/10.1515/9783110572261
10. Foolen, A., De Hoop, H., Mulder, G. (eds.): Evidence for Evidentiality. John Benjamins, Amsterdam (2018). https://doi.org/10.1075/hcp.61.01
11. Kozintseva, N.A.: The category of evidentiality (problems of typological analysis). In: Voprosy Yazykoznaniya, vol. 3, pp. 92–104 (1994). (in Russian)
12. Plungian, V.A.: Introducing Grammatical Semantics: Grammatical Values and Grammatical Systems in the World's Languages. RGGU, Moscow (2011).(in Russian)
13. Plungian, V.A.: Types of verbal evidentiality marking: an overview. In: Diewald, G., Smirnova, E. (eds.) Linguistic Realization of Evidentiality in European Languages, pp. 15–58. De Gruyter Mouton, Berlin-New York (2010)
14. Khrakovskiy, V.S. (ed.): Evidentiality in Europian and Asian Languages. Sankt-Peterburg (2007). (in Russian)
15. Kadyrova, L.B.: The study of the category of evidentiality in foreign and domestic linguistics: traditions and modern state. In: Gumanitarnye nauki v XXI veke: nauchnyy internet-zhurnal, vol. 7, pp. 60–71 (2016). (in Russian)
16. Schwartz, A.S.: Indicators of perfect in modern Chinese. In: Acta Linguistica Petropolitana. Trudi instituta lingvisticheskikh issledovaniy, vol. 12, no. 2, pp. 587–609 (2016). (in Russian)
17. Kaksin, A.D., Azarakov, L.V.: Subjective-modal and Evidential Words as a Part of Discursive Vocabulary. In: Vestnik HGU im. N. F. Katanova, vol. 23, pp. 99–102 (2018). (in Russian)

18. Bodnaruk, E.V., Astakhova, T.N.: Corpus analysis of evidential verbs SAGEN and BEHAUPTEN in modern German-language. In: Media Discourse. Nauchnyi dialog, vol. 4, pp. 9–26 (2020). (in Russian). https://doi.org/10.24224/2227-1295-2020-4-9-26
19. Marín-Arrese, J.I.: Perspectives on Evidentiality and Modality. -Editorial Complutense, Madrid (2004)
20. Chepurnaya, A.I.: Epistemic responsibility in political discourse (on Materials of D. Trump's Press Conference). In: Nauchnyy dialog, vol. 7, pp. 35–44 (2017). (in Russian). https://doi.org/10.24224/2227-1295-2017-7-35-44
21. Khazieva, R.R.: The use of evidental markers as information technologies in political media discourse. In: Odin poyas - odin put. Lingvistika vzaimodeystviya: materialy Mezhdunarodnoi nauchnoi konferencii., pp. 188–190 (2017). (in Russian)
22. Hu, Zh.: Evidentiality in language. In: Waiyu jiaoxue yu yanjiu, vol. 1, pp. 9–15 (1994). (in Chinese)
23. Hu, Zh.: Evidentiality, news and argumentative discourses. In: Waiyu yanjiu, vol. 2, pp. 22–28 (1994). (in Chinese)
24. Hu, Zh.: Evidentiality and discourse analysis in Chinese. In: Hubei daxue xuebao (Zhexue shehui kexue ban), vol. 2, pp. 13–23 (1995). (in Chinese)
25. Zhang, B.: Linguistic coding of epistemology. In: Guowai yuyanxue, vol. 2, pp. 5–19 (1997). (in Chinese)
26. Yan, Ch.: Evidentials and evidentiality in English. In: Jiefangjun waiguoyu xueyuan xuebao, vol. 1, pp. 4–7 (2000). (in Chinese)
27. Niu, B.: A survey of some studies on evidentiality abroad. In: Dangdai yuyan xue, vol. 7, no. 1, pp. 53–61 (2005). (in Chinese)
28. Fang H., Ma, Y.: Evidence, subjectivity, subjectivity. In: Waiyu xuegan, vol. 4, no. 141, pp. 96–99 (2008). (in Chinese)
29. Le, Y.: General characteristic and nature of the evidentiality category of modern Chinese. In: Yuwen yanjiu, vol. 2, pp. 27–34 (2014). (in Chinese)
30. Zhou, Y., Mao H.: Analysis of domestic research on evidentiality. In: Haerbin xueyuan xuebao, vol. 5, pp. 110–113 (2018). (in Chinese)
31. Sun, Z.: Evidentiality in scientific discourse. In: Xiuci xuexi, vol. 6, pp. 31–35 (2007). (in Chinese)
32. Chen, Y.: Initial research on evidentiality in newspaper discourse. In: Yuwen xuekan, vol. 3, pp. 7–9, 43 (2013). (in Chinese)
33. Shui, W., Ji Y.: Comparative study of evidentiality in Chinese and English linguistic academic discourses. In: Chongqing youdian daxue xuebao (Shehui kexueban), vol. 1, pp. 152–156 (2015). (in Chinese)
34. Chen, Zh., Yu, D.: Evidential strategies in argumentative discourses. In: Shandong waiyu jiaoxue, vol. 5, pp. 24–32 (2016). (in Chinese)
35. Zhou, Y., Xia, C.: A study of interpersonal meaning of national strategy on the basis of evidentiality. In: Shandong nongye gongcheng xueyuan xuebao, vol. 10, pp. 94–95 (2018). (in Chinese)
36. Kireyeva, A.A.: Belt and road initiative: overview objectives and implications. Sravnitelnaya Politika **9**(3), 61–74 (2018). (in Russian)
37. Xi, J.: Speech at Nazarbayev University (in Chinese). https://www.fmprc.gov.cn/web/ziliao_674904/zt_674979/dnzt_674981/qtzt/ydyl_675049/zyxw_675051/t1074151.shtml. Accessed 15 Aug 2020

38. Vision and actions on jointly building silk road economic belt and 21st-century maritime silk road (in Chinese). http://www.china.org.cn/chinese/2015-09/15/content_36591064.htm?f=pad&a=true. Accessed 15 Aug 2020
39. Building the belt and road: concept, practice, and China's contribution (in Chinese). http://www.gov.cn/xinwen/2017-05/11/content_5192752.htm#1. Accessed 15 Aug 2020
40. Liu, Y.: Functions of evidence in Chinese language and the modes of its expression. In: Hua Zhong Xueshu, vol. 2, pp. 323–331 (2014). (in Chinese)

Geographic Images of the Four Cardinal Directions (East, West, North, South) in the Linguistic Consciousness of Russian and Chinese Students

Galina Vasilieva and Zishan Huang(✉)

Herzen State Pedagogical University of Russia, Saint-Petersburg, Russia

Abstract. The subject of the research is the geographic images of space, related to the basic concept of humanitarian geography. Geographical images as a compilation of the pivotal presentations of a geographic object could be modeled by using different materials: literary texts, survey results, and questionnaires. By using the method of free-associative experiment, our research aims at determining the stereotypical component of the geographic images of the four cardinal points in the linguistic consciousness of Russian and Chinese students. To identify the discrepancies contents of understanding the geographic images in Russian and Chinese cultures, we use the method of associative field comparison, which allows us to define the nonverbal national culture and mentality, which influence the content of various mental structures, including geographic images. To compare the associative potential naming of the four cardinal points we departed the core zone of the associative field and the basic direction of the content in it. The conformance leads to a large number of differences in the directions of content in the association, forming the geographic images of cardinal points in the linguistic consciousness of Russian and Chinese students. The results of the analysis noted that the change of the political background leads to the great change of geographic images of the four cardinal points in the linguistic consciousness of Russian students, and the compliance in the content of Russian and Chinese students leads to the universality of common geographic understanding and presentation. This is because of the relatively closed location of the two countries, the similar ideological presentation of the peoples, and the transition of both countries from a closed/traditional culture to an open culture, which allows students of the two countries to have a wider view of the world. However, the reason for the significant differences in geographic images is that students from both countries regard Russia as a northern and eastern country, while China is considered to be a typical Eastern one. Economic and political changes in the two countries, as well as the culture and domestic traditions of the two countries, are also the reasons.

Keywords: Geographic images · Humanitarian geography · Cardinal points · Stereotypical presentation · Associative potential · The direction of the association · Linguistic consciousness

© The Author(s), under exclusive license to Springer Nature Switzerland AG 2021
R. Bolgov et al. (Eds.): *Proceedings of Topical Issues in International Political Geography*, SPRINGERGEOGR, pp. 462–474, 2021.
https://doi.org/10.1007/978-3-030-78690-8_40

1 Introduction

The integration trends in modern science give rise to a lot of interdisciplinary direction related to cultural-geographic analysis. It first appeared in the work of foreign researchers in the 20th century [1, 2]. As we all know, the subject of the classic physical geography is the description of the nature of the earth around people. But there is also the natural which is not related to humans. Within the framework of cultural and geographic direction, geography was interpreted as the science of the humanitarian cycle, the main object of which is a person perceiving the surrounding nature.

The first appearance of the study of cultural-geography happens in the 1990s in Russia. At the moment, the work associated two basic disciplines: "cultural geography" and "humanitarian geography". Researchers pointed out that the second concept is used more often than the first one because the broader of it is wider [3].

One of the founders of the humanitarian geography study in Russia is Zamyatin D.N. He defined it as the most general direction of cultural-geographical research, a part of the interdisciplinary scientific field that studies various ways of representing and interpreting terrestrial spaces in human activities, including thinking (mental) activity [4].

An important concept of humanitarian geography is the geographic image, which is "the collection of distinctly, characteristic, centralized signs, symbols. the key presentation that describes real space (territories, localities, regions, countries, landscapes, etc.). During the process of interpretation shows not only the geographical differences but also para-geographic concepts. This relates to cultural, historical, political, economic, and other concepts and presentations, including the spatially marked components [5]."

Representatives of humanitarian geography accomplishing the goal of reconstruction, modeling geographical images on the material of the various text [6–8]. Meanwhile, when reconstructing the content of geographical image we are not only focusing on the individual author's understanding (interpretation) contained in various texts but also the essential role of stereotypical presentations of its collective nature. Within the framework of humanitarian geography, stereotypical presentation to a particular geographical object could be considered as the base for the formation of a basis geographical image that shows the characteristic of a certain ethnic or social group.

In recent years the researches on modeling geographic images of various cities, regions have appeared. However, the images of the most common geographic concepts such as cardinal directions (east, west, north, south) have not become the subject to be studied yet.

The common methods for modeling geographic presentations are questionnaires, directed surveys, and compilation of mental maps. Since the task of reconstructing geographic images is related to its general stereotypically representation. The practical method is the study of linguistic consciousness, reconstructed through psycholinguistic experiments, and the most important, through a free-associative experiment. The usage of this method for reconstructing images of various countries in the consciousness of Russian students is described in our published article [9]. As we all know, the associative field of a word-stimulant represent not only the fragment of a verbal memory of a person but also the fragment of a worldview of an ethnic group, reflected in the consciousness of the "average" border of a particular culture, its motives and perspectives so as its cultural stereotypes.

The article aims at analyzing the associative potential of the names of the most common geographical concepts – the four-cardinal direction (east, west, north, south), defining the stereotypical component of their geographic images in the linguistic consciousness of the two socially and ethnically marked groups: Russian and Chinese students.

2 Materials and Methods

According to researchers, the methods of free-associative experiments, which is widely used in psycholinguistic helps us uncover non-verbal knowledge of native speaker about objects around the world [10, 11]. This knowledge stored in their everyday consciousness and is reflected in the associative potential of words of their mother language. The source material for the study of naive geographical representations of the cardinal points of the world chosen from the «Russian associative dictionary» (RAD) edited by Yu.N. Karaulov, is based on a wide range of psycholinguistic experiment [12]. In the first volume of the dictionary (from stimulus to reaction) three names plus an adjective are mentioned (east, north, south, and western). In the reverse volume (from reaction to stimulus) all four names are included. To focus on the information contained in the dictionary, we used both "forward" and "reverse" sections.

The analysis of the meaning of the four sides in the explanatory dictionary of Russian appeared before analyzing materials from RAD. Here we have to point out, in the tradition of Russian lexicography, two designations are used: "directions" and "cardinal points". The use of the second name shows the trend of study in recent years.

To clarify those concepts mentioned above, and identify the original verbalized code of presentations of these geographical objects, we grab the material from one of the latest explanatory dictionaries – «Great universal dictionary of the Russian language» [13]:

East: 1. Direction, in which the sun rises, also the side of that direction. 2. The area in this direction also means the eastern part of something. 2.1. (with a capital letter). Countries in this direction, the opposite of West, Europe, and America.
West: 1. Direction, in which the sun goes down. Also, the side of that direction. 2.0. The area in this direction also means the western side of something. 2.1 (with a capital letter) West Europe, countries of West Europe and North America, also the countries opposite to East Europe and other regions, previously referred to the socialist camp.
North: 1. One of the four cardinal points, the opposite to the south. The direction is shown by compass, the side of this direction. 2. The area in this direction also means the northern part of something. 3. The area located in the pole of cold and extreme climate. 3.1 (often with capital letter) Territory located in the zone of cold climate on the outskirts of Eurasia and North America, also means the polar region of the earth.
South: 1. One of the four cardinal points, opposite to the north. Also means the side in this direction. 2. The area in this direction, also the southern part of something. 3. The area located in the zone of the warm and hot climate.

From those interpretations, the names of the side in the Russian language are polysemous. In the first definition, all of them are the cardinal points, directions, and the opposites of other sides and the areas or countries located in the direction. In this case, we can divide a series of particular meanings. Thus, the interpretation of east and west include the indication of sunrise and sunset. The particular meaning of the north relates to the compass while south and north relate to their climatic feature (hot climate, cold climate). A special ideologically and political motivated meaning of the west can be considered to the former countries of the socialist camp. This meaning has been recorded in the most recent dictionaries. These meanings served as the source material for identifying the content features of the corresponding associative fields.

Because the receiving information of the geographical space and its partitional objects (including the cardinal points) might change, in our study, to identify the changes of the naive geographical picture of the world of Russian students, which have occurred since the release of RAD, the free-associative experiment was conducted among modern students. In 2020, by using the method based on RAD, an experiment was conducted among the students of various specialties at RSPU named after A.I. Herzen (from philology, economic and technical), who carries the naïve geographic presentation (native speaker). The experiment involved 100 students aged from 17 to 25. They were asked to report their first reaction to the stimulant (see Table 1).

Table 1. Numbers of reactions to stimulant-names of sides in the RAD and in the modern experiment (ME)

Stimulant	Total reactions to stimulant		Number of different reactions to stimulus		Number of single reactions to stimulus		Number of refuses	
	RAD	ME	RAD	ME	RAD	ME	RAD	ME
East	183	100	69	62	55	31	0	0
West	183	100	60	53	31	29	0	0
North	209	100	50	46	39	24	0	0
South	317	100	108	53	88	49	0	0

3 Results of the Research

The core of associative fields was identified by working the results from quantitative analysis, which represent frequency and repetitive reactions. The core associations of the participants in the experiment are given according to frequency (see Table 2).

Since the geographic images are not even in their content, to reconstruct we need substantial segmentation of the associative field. Following the proposition of Karaulov, who uses «semantic gestalt» to reflect the internal semantic organization of the composition of the associative field [14], correlating its structure with the structure of real-life contains in it. By working with the data we indicate the maximum number of parameters

Table 2. The core associations of the participants

Stimulant	RAD	Modern experiment
East	Skinny, China, Dalni, Japan, part of the world, Blizhniy, direction, other culture, rocket, tartar, Asia, ancient, sage, exotic, tea, Shanghai, Ashkhabad	China, tea, sun, tradition, ancient, philosophy, Japan, Asia, market, crafty, sweets, pilaf, exotic, Muslims, Buddhists, Emirates, desert, Sheikh, Arabic, skinny, tea, raisins, Turkish delight, perfume, kimono, spices, spaceship, hookah, character
West	America, western, skyscraper, England, Europe, New York, capitalism, bourgeois, film, Germany, sunset, cloth	USA, America, Europe, civilization, cowboy, fast food, tolerance, individualism, side of the world, McDonald's, Protestants, progress, European Union, Schengen, business, sanctions, travel, western, restaurant
North	Krayniy, Cold, Siberia, Arctic, pole, cold (adj.), snow, frost, sever, cold (adv.), ice, compass, icebreaker, cafe, iceberg, romance, cigarettes, homeland, deer, taiga, mammoth	Cold, tundra, pole, polar bears, aurora borealis, Scandinavia, Arctic Ocean, deer, Chukchi, North Pole, fish, minority, calmness, white, snowboard, ice, icebergs, yurt, Vologda, White Sea, husky, walruses, skiing
South	Vacation, relax, Sochi, Black Sea, hot, heat, sea, warm, sun, Bulgaria, Sahara, beach, palm. Israel. Melon, travel, Tashkent, thirst, compass, market, Persian, health resort, side of the world, crowded	Sea, vacation, beach, sun, heat, warm sunburn, palm trees, fruits, wine, Italy, Spain, part of the world, Africa, poverty, delicious food, mountains, enthusiastic, sand, antiquity, siesta, peaches, Black Sea, swimwear, lazy people

that reflect the nature of students' knowledge and para-geographic (cultural-geographic) characteristics of the sides that cause the appearance of the corresponding direction of the association. Thus, we have a narrow geographic presentation and cultural-geographic presentation.

In associative fields recorded in the RAD, the following narrow geographical and cultural-geographical directions are distinguished:

Geographical names and objects: East (China, Dalni, Japan, Blizhniy, Asia, Russia, Shanghai, Ashkhabad); West (America, USA, England, Germany, Berlin, Europe, New York); North (Utmost, Siberia, Arctic, Handy-Mansiysk, Omsk); South (Sochi, the Black Sea, Antarctica, Bulgaria, Sahara, Tashkent, Crimea, Asia, Persian).
Climate: East (no reaction); West (no reaction); North (coldness, cold, severe, wind, iceberg, ice, snow, snowstorm); South (hot, heat, sea, sun, warmth, warm, very hot, hot and moist).
Botanic and zoic world: East (no reaction); West (no reaction); North (deer, taiga, birch, penguin, clover); South (palm, pineapple, blooming, melon, peach, grapes, flying crane).

In associative fields recorded in RAD, the following cultural-geographical directions are distinguished:

Culture and traditions: East (ancient, sage, concubine, other culture); West (film, western); north (no reaction); south (no reaction).
Daily culture: East (concubine, tea, silk); west (cloth); north (cafe); south (market, shorts).
Politic and ideology: east (no reaction); west (bourgeois, decaying, curtain, capitalism); north (no reaction); south (no reaction).
Psychological perception, evaluation: East (skinny, exotic, remote); west (decadent); north (romance); south (beautiful place, blooming, very beautiful, want it).

The associative fields, which were recorded by modern experiment could be divided into the following narrow geographic directions:

Geographical names and objects: east (China, Japan, Asia, Arabic, Far East, Vladivostok, eastern Siberia, Emirates); West (USA, America, Europe, European Union, Switzerland, Mediterranean Sea); North (Arctic, Arctic Ocean, White Sea, Arkhangelsk, Scandinavia, Bergen, North Pole); south (Italy, Spain, Crimea, Africa, Antarctica, South Pole).
Climate: east (sun, heat, ocean); west (no reactions); north (frost, cold, snow drifts, snow, glaciers, fur coat, iceberg, snowstorm); south (sea, sun, heat, warm).
Botanic and zoic world: East (sakura, fruits, tigers, camels); west (no reactions); north (fish, deer, bear, husky, tundra, polar bears, walruses, penguins, dwarf birches, moss); south (fruits, flowers, peaches, figs, palm trees, plantations).
Culture and traditions: East (traditions, antiquity, ancient, design pattern, abundant culture, papyrus, sheik, character, Muslims, Buddhists, kimonos, patterns, masjid, burkas, fairy tales, art, poverty, «The Thousand and One Nights»); West (Western, popular culture, Protestants, the Catholic Church, film, art, higher education, civilization, architecture, literature, philosophy, education, the Bologna system); North (chanting); south (antiquity).
Daily culture: East (sushi, farm, sweets, market, geisha, tea, hookah, belly dance, curry, fruit, fan, kimono, silk); west (jeans, maple syrup, sneakers, McDonald's, fast food, car, restaurants); north (warm clothes, fur coat, yurt); south (delicious food, coffee, fruit, wine, pizza, slippers, churchkhela, baked corn).
Politics and ideology: east (war, race); West (Schengen, sanctions, freedom, democracy, pseudo-democracy); North (no reactions); south (no reactions).
Economy: east (no reactions); West (progress, business, money, dollar, euro, pseudo-democracy, Audi, comfortable life, high technology, unemployment); North (no reactions); south (poverty, unemployment).
Psychological perception, evaluation: east (crafty, skinny, beauty, exotic); West (freedom, progress, pragmatism, decay); North (slowness, calm); south (lazy, emotional).

To compare the geographical images of the cardinal points existing in the minds of Russian and Chinese students, we conducted a free-associative experiment among Chinese students (100 people from 16 to 27 years old).

In Xinhua dictionary [15] the cardinal points defined as follows:

东(east): 1. Direction, the side where the sun rises; opposite to the west. 2. Host, owner. A person who treats others. (In ancient China, the room of the owner of the house is located on the eastern side while the guests room in the west).
西(west): 1. Direction, the side where the sun sets, opposite to the east. 2. Europe and the USA. 3. Western cultures, which in content or form belong to the Western-style: west food (European food), Western medicine, Western costume (business suit), graduation from a university in western countries, etc.
北(north): 1. Direction, the left side when facing the sun in the morning, opposite to the south. 2. North of China (In China. Areas located in the north of China were called north in general). 3. (literal translation) Defeat north (Retreat after being defeated).
南(south):1. Direction, the right side when facing the sun in the morning, opposite to the north. 2. South of China (In China, areas located in the south of Yellow River were called south in general).

The Explanatory Dictionary points out that the names of the cardinal points in Chinese are also ambiguous. The first (main) meanings of them in Russian and Chinese coincide. East and north carry cultural meanings inside. Regions related to the four cardinal points relate first to areas in China (except west).

A free-associative experiment taking place among Chinese students from the preparatory department of St. Petersburg Institute for Painting, Sculpture, and Architecture named after I.E. Repin, as well as students and doctoral students from the Herzen State Pedagogical University of Russia of various faculties, took part in. The Table 3 demonstrates the following results obtained.

In the associative fields getting from the result of the experiment, the following narrow geographical vectors\groups of association were identified:

Geographical names and objects: East (China, Asia, Japan, the Pacific Ocean, Shanghai, Hainan, Shandong Province, the East China Sea, East Sea, Middle East, the Great Wall, homeland); West (Europe, USA, Milan, Aegean Sea, European Union, Wall Street, Spain, Roman Empire, Atlantic Ocean); North (North Pole, Beijing, Russia, Finland, Iceland, Siberia, North America, Denmark, Norway, Murmansk, Canada, Helsinki); south (Hainan, Australia, Guangdong, Thailand, Tibet, Guangzhou, Vietnam, Antarctica, south of the Yangtze River, South Pole).
Climate: east (sun, east wind); west (wind, Aegean Sea); north (ice, snow, cold, strong wind, winter, fog, iceberg, cold river, high mountains); south (hot, damp, high temperature, sun, warm, humid, heat, stuffiness, dry, typhoon, rain, summer, vitiated air, south wind, sea, mountains, island).
Botanic and zoic world: East (lotus, peony); west (no reactions); north (huge polar bear, wolf, horse, moose, dog, penguins, forest, steppe, trees without leaves, arctic fox); south (wild goose, durian, elephants, tropical fruits, marine products, seafood, coconut, seagull, crocodile, mosquitoes, southern bear, swallow).

In the associative fields getting from the result of the experiment, the following cultural-geographical vectors\groups of association were identified:

Table 3. Most frequent and repetitive reactions

Stimulant	Most frequent and repetitive reactions
East	China, tzarevich, sun, Oriental Pearl, dragon, sunrise, ancient, yellow race, empire food, prosperous culture, Buddhism, red, Pacific Ocean, wisdom, civilization, East sea, Asia, porcelain, compass, direction, peony, silk, master, Confucius, lotus, prosperous, mysterious, east wind, the East China Sea, eastern sales market, chopsticks, homeland, hidden, Guohua (traditional Chinese painting) complete luck in everything, traditions, the Great Wall, Mao Zedong, crowd, cuisine, Shanghai, Middle East, war, Chairman Mao, renew, Japan, Chinese characters
West	Sunset, Europe, democracy, freedom, USA, Journey to the West (novel), currency, NATO, developed countries, direction, compass, white race, art, fashion, Milan, technology, western countries, the dark side, Eiffel Tower, cowboy, Indians, human rights, opposition, english, Aegean Sea, queen, magnificent, cathedrals, fast food, European union, cheese, economy, costume, fascism, portfolio, European architecture, bars, fat people, gold, Wall Street, stock market, the film "Good, Bad, Evil", painting, betrayal, lies, wine, Christianity, Spain, Jesus, forests, capitalism, tough
North	Ice and snow, cold, aurora, Beijing, polar star, north wind, cardinal points, north pole, Finland, penguin, Russia, Iceland, polar bear, winter, Santa Claus, icebreaker, arctic ocean, Siberia, vast, polar bear, North America, Denmark, Northern Chinese, accent, northern wolf, Norway, Canada, Helsinki, Murmansk, fur coat, vodka, warm hat, death, white, seven-star, beer, climate, location, moose
South	Hot, sun, sea, the Southern Hemisphere, Hainan, antarctica, Australia, south wind, rain, Guangdong, compass, wind direction, wet, Guangzhou, Vietnam, warm, tropical fruits, global warming, south pole, delicious food, tea, water sport, summer, seafood, mountains, heat, beach, summer, island, swallow, green rice, water, beach, heat, forest, Tibet, emperor, Thailand, seagull, embankment, crocodile, stuffiness, fishing, Eastern Jin Dynasty, southern bear, street food, folk music, rich, businessman

Culture and traditions: east (ancient, emperor, mainstream culture, Buddhism, red, wisdom, civilization, Chinese characters, chairman Mao, renew, Mao Zedong, crowd, traditions, master, Guohua (traditional Chinese painting), all go well (Chinese idiom), Confucius, master; West (Christianity, the film "Good, Bad, Evil", painting, European architecture, Jesus, the Queen, English, "Journey to the West" (novel), democracy, art, cowboy, Indians); North (other traditions, Santa Claus, a Chinese dialect, another accent, death, fairy tale, white); south (the Eastern Jin Dynasty, folk music, fishing, emperor).
Daily culture: East (food, porcelain, compass, silk, cuisines, chopsticks); West (wine, cheese, costume, fashion, briefcase, fat people, fast food); North (noodles, warm hat, fur coat, Chinese vodka, sleigh, down jacket, fire, knife); south (market, shorts, beer, tea, delicious food, water sports, fishing, street food, green rice).
Politics and ideology: east (war, Chairman Mao, regain, Mao Zedong), west (fascism, revolt, capitalism, human rights, democracy, freedom); north (aggression); south (no reactions).

Economy: East (no reaction); West (capitalism, Wall Street, stock market, gold, economy, currency); North (no reaction); South (rich, businessman).
Psychological perception, evaluation: East (hidden, mysterious, wisdom, civilization, red, prosperity); West (lie, betrayal, magnificent); North (white, death, powerful, man; South (no reaction).

To find out which part of the world the two countries belong to, in the students' perception, we raised the question: What kind of countries are Russia and China (eastern, western, northern, southern)? We received the following answers:

Russian students (100 people): Russia is an eastern country (52), northern (38), almost northern (5), north-eastern (4), almost western (1); China is an eastern country (97), southern (3).
Chinese students (100 students): Russia - a northern country (80), north-western (12), northeastern (6), Eastern European (2); China is an eastern country (100).

4 Discussion and Conclusions

It should be mentioned that the perception of the cardinal points of Russian students of different ages has undergone some diachronic changes.

We can conclude that the changes in ideas of a different order (narrow geographical and cultural-geographical) were influenced by a political reason. In other words, the appearances of those changes in the life of the people of a country reflected not only in the narrow way (fading the ideologically mark of west, etc.), but also determined the dynamics of other association vectors. Therefore, the change of the Russian culture to an open one [16] lead modern students to see the other countries located in the other half of the earth, to learn their cultural-geographical features and to know their human and living beings. The narrow geographical direction of association of geographical objects, climate, Botanic and zoic world (chosen in RAD) has significantly expanded. This mainly happens to those vectors relate to sphere of relaxation and tourism of Russians. The understanding students of economic of countries and regions in the other hemispheres shows that the political isolation has been moved. The biggest change took place in the sphere of daily culture. This vector is not actually represented in the RAD, and has become a necessary component of modern geographical images of the directions. Its content has been widely expanded mainly due to the increase of students' understanding of the national cuisine of other nationalities.

The results allow us to conclude that political changes along with the development of the modern society have influenced the geographical and cultural-geographical ideas of students about the geographic directions.

The narrow association of geographic directions (recorded in RAD) related to geographic objects, climate, the botanic and zoic world has been expanded. This mainly concerns those vectors that fall into the rage of active leisure and tourism of Russian.

Perceptions about the economies of countries and regions on earth reflect the economic crisis, political changes, and various rates of economic growth. The biggest change of perception appears in the sphere of daily culture. This vector of association, which rarely shown in RAD, has become an essential part of modern geographic images of the cardinal points. Its content has been expended mainly because of the increase in students' understanding of the cuisine in other countries.

Those results allow us to conclude that the political and economic changes representing world development have influenced the cultural-geographic perception of students. The transformation of Russian (traditional) culture into an "open" one has played an important role. It helps modern students to see countries on different sides of the world, to understand their cultural-geographic characteristics and to know the people's lives in them. Therefore, the stereotypically geographic images about the cardinal points have lost their most common "bookish" features and gained lively, daily ones, thanks to the impression of tourists.

Comparing the results with the answers from Chinese students, both coincidences and differences were found.

The first explanation of the cardinal points written in the dictionaries of both languages is very close. The common parts of the reactions of Russian and Chinese students in the narrow-geographic sphere (side, direction, sunrise, sunset, compass, South Pole, North Pole, and others) refer to the association, which reflects the conceptual content of the stimulants.

Among the narrow-geographic perceptions, both similarities and differences were noted.

In the section of geographical names and objects, the rection of "East" in the core of the associative field of both Russian and Chinese students are alike (China, Asia, Japan). The reactions of Russian students pointed out both foreign countries and regions (Arabic, Emirate, Persian) and the eastern part of their own country (far East, Vladivostok, Eastern Siberia), while Chinese students call only to the regions in China (Shanghai, Hainan, Shandong province, East China sea, East sea, Great wall, homeland). In the section of "West" students from both countries relates only the foreign territories: Russian students (USA, America, Europe, the European Union, Switzerland, the Mediterranean sea) and Chinese students (Europe, USA, Milan, the Aegean sea, the European Union, Wall Street, Spain, the Roman Empire, and the Atlantic ocean). This pointed out the fact that students from both countries regarded "west" as an area out of their own. The reaction of Russian students to "North" (Antarctica, Arctic ocean, White sea, Arkhangelsk, Scandinavia, Bergen, North pole) shows that: they connect it to the Northern part of Europe as well as Russia, while Chinese students (North pole, Beijing, Russia, Finland, Iceland, Siberia, North America, Denmark, Norway, Murmansk, Canada, Helsinki) refer to a wider understanding, including north-European countries, Russia and its regions, the United States, Canada and the north part of China. The perception of "South" of Russian students (Italy, Spain, Crimea, Africa, Antarctica, the South pole) relates to the south of Europe, continents in the southern part of the earth and the southern part of their own country, in the associative field of Chinese students (Hainan, Australia, Guangdong province, Thailand, Tibet, Guangzhou, Vietnam, Antarctica, the area in the south of the Yangtze River, the South pole) didn't mention the geographic objects and countries in southern Europe, contains only the south part of China, countries in the south-west of Asia as well as the continents in the southern part of the earth.

We can conclude that Russian students relate the cardinal directions and their corresponding direction to places located both in their own country and abroad, while Chinese students see more in China (except for West).

Students from both countries lack significance in their response to "east" and "west". About "north" and "south" they showed an almost equivalent potential: south—hot and north—cold.

In the section of the botanic and zoic world, both students practically don't respond to "north", and in "south" contains a huge amount of associations, showing the obvious and concrete presentation about the flora and fauna in north and south.

From the result we can see, the perception of the climate, flora, and fauna characteristics of the cardinal directions of students are not symmetrical. Comparing with the obvious, multi-dimensional perception of the nature of south and north, the east and west don't have any associative potential in the consciousness.

The great difference recorded in the contents of cultural-geographic associations.

In the associative section of "cultural and tradition", the common reactions to stimulant "east" are antiquity, traditions, rich culture, wisdom, hieroglyphs. Besides, Russian students carry clear religious characteristics (Muslims, Buddhists, mosques), while Chinese students relate more to the names of prominent figures of their own country (Chairman Mao, Mao Zedong, Confucius, the Emperor). "West" evoked similar reactions from students in both countries, regarding art, architecture, literature and higher education of the West. The "North" and "South" sections barely recalled any associations among Russians, while the Chinese associate the south with the culture and traditions of their own (Eastern Jin Dynasty, Emperor).

A large number of reactions turned out to be in the daily cultural section. The stimulant "East" caused Russian students numerous specific reactions related to cuisine (sushi, sweets, tea, curry, fruits, spices) and daily tradition (hookah, harem, market, belly dance, fan, kimono), and Chinese students more relate to general reference (food, porcelain, cuisine, chopsticks). The daily culture of the West is associated with Russian students with jeans, sneakers, cars, as well as maple syrup, McDonald's, fast food, and restaurants. To Chinese students is wine, cheese, fast food, and fat people, as well as fashion, costume, and suitcases. Most of the associations cased by the stimulant "North" are different types of warm cloth. For Russian students, "South" is mainly about a holiday on the beach (in Russia or abroad), and that arouses other associations such as delicious food, relaxation, coffee, fruit, wine, pizza, swimwear, slippers, cafes, churchkhela, baked corn, and sunglasses, while Chinese students the connection between south and beach holiday haven't been expressed.

Section politic and ideology of the students from both countries don't react to "south" and "north"; the stimulant "east" linked to war, while the stimulant "west" carries the strongest ideological characteristics. Except for the common reaction (democracy, freedom), to Russian students, it calls Schengen, sanctions, and pseudo-democracy, and for Chinese students, it is fascism, counteraction, and capitalism.

The section economic don't reflect stimulants "east" and "north". To Russian students "west" is progress, business, money, dollar, euro, comfortable life, high-tech, and unemployment. As for Chinese students, it is capitalism, Wall Street, stock market, gold, economy, currency. The stimulant "South" caused Russian students to associate with poverty and unemployment, and Chinese students with the rich and the businessman. In other words, the economic section of "South" understood by students in different ways.

Most of the perceptions of the cardinal directions in the psychological section do not coincide. Russian students call both positive and negative evaluations to "East" (crafty, skinny, beauty, exotic), while Chinese are mostly positive (mysterious, wisdom, civilization, red, prosperity). West section receives completely different responses from both Russian (freedom, progress, pragmatism, decadent) and Chinese students (lies, betrayal, magnificent). For Russians, the North is associated with slowness and calmness, and for the Chinese with white color and death. South for Russian students is with laziness and emotionality, or Chinese students it does not cause evaluation associations.

Therefore, the significant coincidence of geographical images in Russian and Chinese students' cognition is because of the common geographical knowledge and perceptions, the relatively close geographical location of the two countries, the similarity of the ideological and politically understanding of the people, and the transition of the two countries from closed tradition culture to open allows both groups of students to see the wider world. However, the discrepancies due to the location of the Russian Federation and China (Russia located in Europe and Asia, China in Asia). The cognition of Russian students is closer to northern and eastern countries. But China stands as a typical eastern country. What's more, the differences are also caused by the cultural and daily traditions of the two countries.

References

1. Gilbert, E.W.: The idea of the region. Geography **XIV**, 208 (1960). Part 3
2. Minchull, R.: Regional geography: Theory and Practice. Leningrad (1967)
3. Uvarov, M.S.: Cultural geography in cultural perspective (analytical review). Int. J. C. Res. **4**(5), 6–18 (2011). (in Russian)
4. Zamyatin, D.N.: Humanitarian geography: space, imagination and interaction of modern Humanities. Sociol. Rev. **9**(3), 26–27 (2010). (in Russian)
5. Zamyatin D. N.: Humanitarian geography: space and language of geographical images. Saint-Petersburg, Alethea, pp. 49–50 (2003). (In Russian)
6. Shareho, I.N., Tupitsyna, N.B.: Pushcha Belarus – an element of cultural and geographical image of the country. Bull. Priamursky State Univ. Sholom-Aleichem **3**(24), 79–90 (2016). (in Russian)
7. Lavreneva, O.A.: Geographical space in Russian poetry of XVIII – early XX century (geocultural aspect). Institute of heritage, Moscow (1998). (In Russian)
8. Loshakova, G.A., Smirnova, L.E.: Germany in the works of Russian writers of the late XVIII – the first half of the XIX century: the transformation of cultural and geographical image. Phil. Sci. Theory Pract. **12**(3), 389–392 (2019). (In Russian). https://doi.org/10.30853/filnauki.2019.3.81
9. Vasileva, G.M., Vinogradova, M.V., Miachina, V.W.: The dynamics of cultural-geographic perceptions in the linguistic consciousness of Russian students (as exemplified by country images). Perspect. Sci. Educ. **5**(41), 262–270 (2019)
10. Kiss, G.: Words, associations, and networks. J. Verbal Learn. Verbal Behav. **7**, 707–713 (1968)
11. Balyasnikova, O.V., Ufimtseva, N.V., Cherkasova, G.A., Chulkina, N.L.: Language and cognition: regional perspective. Russian J. Linguist. **22**(2), 232–250 (2018). (In Russian). https://doi.org/10.22363/2312-9182-2018-22-2-232-250

12. Karaulov, Y.N., Cherkasova, G.A., Ufimtseva, N.V., et al. (eds.): Russian associative dictionary in 2 volumes. vol. 1. From stimulus to reaction. AST. Astrel, Moscow (2002). (In Russian)
13. Morkovkina, V.V.: Large universal dictionary of the Russian language. Dictionaries of the XXI century (2016). (In Russian)
14. Karaulov, Y.N.: Indicators of the national mentality in the associative-verbal chain. Linguistic consciousness and the image of the world. In: Ufimtseva, N.V. (ed.) Digest of articles, Moscow, pp. 191–206 (2000). (In Russian)
15. Explanatory dictionary of the Chinese language "Xinhua". Beijing: Commercial publishing house, p. 1411. (In Russian)
16. Triandis, H.C.: Cross-cultural studies of individualism and collectivism. In: Berman, J. (ed.) Nebraska Symposium on Motivation, Lincoln, University of Nebraska Press, pp. 41–133 (1989)

Author Index

© The Editor(s) (if applicable) and The Author(s), under exclusive license
to Springer Nature Switzerland AG 2021

R. Bolgov et al. (Eds.): *Proceedings of Topical Issues in International Political Geography*, SPRINGERGEOGR, pp. 475–476, 2021.
https://doi.org/10.1007/978-3-030-78690-8

9783030786892